D1566457

CONTROL OF ELECTRIC MOTORS

HARWOOD'S CONTROL OF ELECTRIC MOTORS

Fourth Edition

RALPH A. MILLERMASTER

Vice President, Engineering & Development
Cutler-Hammer, Inc.
Milwaukee, Wisconsin

WILEY INTERSCIENCE,

a Division of John Wiley & Sons

NEW YORK • LONDON • SYDNEY • TORONTO

10 9 8 7 6 5 4 3
Library of Congress Catalogue Card Number: 72-94919

SBN 471 60620 0

Printed in the United States of America

PREFACE

The fourth edition of Harwood's *Control of Electric Motors* endeavors to continue and enhance the broad presentation of authoritative and useful technical material in motor application and control which accounted for the worldwide acceptance of the three previous editions as the foremost reference in the field.

Recognizing the ever-increasing rate of change and improvement in the technologies of systems and motor control, we have included significant new material dealing with power conversion, both from alternating current to direct current and from alternating current to alternating current, and with static logic.

These technologies, in the areas of hardware as well as software, and substantially improved subsystems and components, have made possible new levels of precision in controlling processes, machines, and motors.

The text has been reworked and expanded to reflect all of these improvements; the photographs and tables have been similarly revised.

The character and format of previous editions have been retained. A few of the early chapters provide general information and are applicable through the remainder of the book. Later chapters examine the character of the system and its driven machines, factors governing motor selection, the types of control usually employed, and the performance which will result.

The preparation of the fourth edition represents the effort of several executive electrical engineers employed at Cutler-Hammer, Inc. They are recognized authorities on motor control and specialize in the subject matter treated in the chapters which they have written. These men are:

Robert N. Eck
Manager of Controller Development

Verne H. Simson
Manager of Process Control

Blakeslee G. Wheeler
Manager of Adjustable Speed Drives and Systems

I express my appreciation to these men, and to Jesse E. Jones, recently retired Vice-President and General Manager of the Industrial Systems Division of Cutler-Hammer, Inc., who continued his substantial contributions by coordinating the multiple inputs necessary to this fourth edition.

I am also indebted to Charles H. Prout, Vice-President and Secretary of Cutler-Hammer, Inc., for his counsel, and to those companies and organizations that have so willingly supplied information, photographs, and permission to reprint material.

Finally, and very importantly, to Paisley B. Harwood, who encouraged us to continue this series covering the important field of electric control which he began in 1935, and in fond memory of the pleasant decades of our professional careers which we spent together, I dedicate this fourth edition.

Ralph A. Millermaster
Fellow, Institute of Electrical and Electronic Engineers
Vice-President Engineering and Development
Cutler-Hammer, Inc.

Milwaukee, Wisconsin
July, 1969

CONTENTS

CONTROL OF ELECTRIC MOTORS

1

INTRODUCTION

The success of any installation of electrically driven machinery is dependent on the proper selection and correlation of the machine, the motor, and the controller. Each is important, and the improper application of a motor or a controller will lower the efficiency of the installation and may cause it to be a failure.

In order to make successful installations, it is necessary to understand the factors which enter into the selection of the machine, motor, and controller, and therefore to have a good working knowledge of the characteristics of the three devices. The requirements of the machine with respect to speed, torque, and special functions must be known. An understanding of motor characteristics is necessary to proper selection of a motor which will accomplish the desired results, and a knowledge of control apparatus is required to insure the application of a controller which will cause the motor to perform the functions required of it.

The five factors which enter into the study of a motor and control application fall logically into a definite order. They are

The machine.
The power supply.
The motor.
The operator.
The controller.

The Machine. The machine is a known factor. It has been designed to accomplish certain results, and it has certain speed and torque requirements. These are known to the designer and builder, and information concerning them is therefore available. Specific information should be obtained covering the following functions.

Check List of Machine Characteristics

(For use as an aid in the selection of control apparatus)

Starting

Will the machine be started by an operator?
If so, what type of starting device is preferred:

Pushbutton?
Safety switch?
Drum controller?
Other means?

If not, what type of automatic starting device will be used:

Float switch?
Pressure switch?
Time clock?
Thermostat?
Interlock in another machine?
Other means?

How often will the machine be started?
Is the starting torque light or heavy?
Has the machine a high inertia?
How much time will be required for starting?
Is a particularly smooth start necessary?
Will the machine always start under load, or always unloaded, or will conditions vary?

Stopping

Will the machine be stopped by an operator?
Will it be stopped by a limit switch, or some other automatic device?
Is a quick stop essential?
Is an accurate stop essential?
Is a particularly quick stop in an emergency required?
Is a magnetic brake required for stopping, or for holding the machine stationary?

Reversing

Will the motor be reversed in regular operation? If so, how often will this occur?
If the machine is essentially nonreversing, will emergency reversing be required?
Will the motor be plugged?
Must the machine perform the same functions in the reverse direction as it does in the forward direction?

Running

Will the machine run continuously or intermittently?
Will the load be overhauling at any part of the cycle?
Are any special functions required during the cycle?

Speed Control

Has the machine essentially a single constant speed?
Has the machine two or more constant speeds?
Is adjustable speed required? If so, over what range?
Will the speed be adjusted while running, or will it be preset?
Are different speeds required in different parts of the cycle?
Must the running speed be relatively constant under different load conditions, or may it vary with the load?
Is slowdown required during any part of the cycle?

Safety Features

Are any of the following protective features required:

- Overload protection?
- Open field protection?
- Open phase protection?
- Reversed phase protection?
- Overtravel protection?
- Overspeed protection?
- Reversed current protection?

At first sight, this seems to be a very formidable list, but it will be evident that only a few of the items will apply to any one machine. Furthermore, the machine builder who must supply the information will be entirely familiar with his machine and will be able to supply the data without a great deal of effort. Even though the requirements are apparently simple, it is a good plan to check against a list of this nature, simply to be sure that nothing is overlooked.

Power Supply. The question of power supply is sometimes definitely settled by the motor requirements and is always closely tied in with the motor selection. Direct current is particularly valuable where a wide range of speed control is required. Another advantage of its use is the ease of obtaining dynamic braking. One of the chief advantages of alternating current is that it is available almost everywhere. Furthermore, the voltage of the supply may easily be changed by means of transformers. Alternating-current machines have relatively constant speeds, and it is difficult to obtain satisfactory speed variations.

A crane hoist is an ideal application for direct current. Heavy loads must be hoisted at low speeds, but it is desirable that light loads be hoisted at considerably higher speeds. Similarly, heavy loads must be lowered slowly, with safety, and light loads may be lowered at a higher rate of speed. A dc series motor has exactly the right speed-torque characteristics to accomplish these results and is easily arranged for dynamic lowering at any desired speed.

Most woodworking machines use ac motors since relatively high speed and constant speed are desired. Dynamic braking is not often required.

Where there is only one kind of power supply it will usually be possible to select a suitable motor and controller. If the requirements are such that the available power cannot be utilized, it will be necessary to arrange to get the required type of power by using a motor-generator set or some other means of transformation.

The quantity of power available and the rules governing its use should also be investigated. It may be found necessary to limit the current inrushes during starting, in order to avoid voltage disturbances on the line, even when such a limitation is not required for satisfactory starting of the machine.

The Motor. The power supply may be determined by the necessity of using a particular motor, or the motor may have to be selected to suit the power available. Each type of ac or dc motor has definite speed and torque characteristics. A motor which is exactly suited to one application may be entirely unsatisfactory for another. Since a motor having the proper characteristics must be selected, a knowledge of the characteristics of the available types is essential. Also the methods which may be employed to take advantage of the motor characteristics, to start and stop it properly, and to control its speed must be understood.

The Operator's Function. There is generally a choice in the selection of the method of operation of a machine. Sometimes it is necessary that the operator be called upon to exercise a considerable amount of judgment and to take a great deal of responsibility. He may be required to initiate every motion of the machine himself by pushing buttons or by moving levers. In the other extreme, it may be desirable to make the entire operation automatic, taking the responsibility entirely out of the hands of the operator. Even the starting and stopping may be automatically controlled. Between these two extremes are many arrangements giving an operator a greater or lesser degree of responsibility.

As an example, consider the installation of a large crusher which is driven by a motor and which is lubricated under pressure. The oil is pumped by a separate small motor. Failure of the lubricating system may ruin the crusher. The simplest arrangement will be to have a starting controller for the crusher and a separate starter for the oil pump. Obviously, however, this does not provide any safety feature, since we are dependent upon the operator to start the oil pump before he starts the crusher, and to stop the crusher any time the oil pump stops. A better arrangement would be to interlock the two controllers, so that the oil pump must be running before the crusher can be started, and also so that, if the oil-pump motor stops at any time, the crusher will automatically be shut down. Still better would be an arrangement using only one starting button which, when pressed, would first start the oil pump, then, after the pressure was up, would start the crusher. This arrangement would include interlocking to shut down the crusher if the oil pressure fell below a safe value. A number of other schemes giving fair degrees of safety are possible. For instance, instead of shutting down the crusher when the oil pressure is low, it would be possible to ring a bell, warning the operator, whose responsibility it would be either to restore the oil pressure or to shut down the crusher.

The final solution of this problem, or of any one like it, would be a matter of judgment. Some of the factors to be considered would be the necessity of safety, the amount of damage which might be caused by an operator's mistake or neglect, the mental capacity and skill of the operator, and the relative cost of the various control schemes.

The Controller. When the characteristics of the machine are known, and when the motor has been selected and the operator's part has been determined, it is possible to select the proper control. If the requirements are simple, this may be a standard catalog device. For more complicated machines, the controller will have to be specially designed. When the great variety of machines in use in all the industries is considered, it will be evident that many special controls will be required. Their number is further multiplied by the many different types of motors and by the various sizes required. Still further complication is introduced by the many voltages and frequencies which are standard, and the use of single-, two-, or three-phase power.

A study of the requirements will enable the engineer to decide how the control is to function and to determine approximately the apparatus required. An elementary or line diagram will then be made and the exact scheme worked out. This diagram will determine the number and type of contactors and relays required, and the power requirements of the installation will determine their current rating. The next step is to combine the parts into a unit by arranging to mount them on a common base, which is usually a steel panel. The panel, in turn, will be mounted on a framework, or in an enclosing cabinet, which may also be arranged to mount any resistor material that is required. Pilot devices are usually complete in themselves and separately mounted.

Controllers are built in accordance with the rules of the National Electrical Code and the standards of the Institute of Electrical and Electronic Engineers. The majority of them also meet the standards of the National Electrical Manufacturers' Association (NEMA).

Some controllers are required to meet the standards of the National Fire Protection Association, those of the Association of Iron and Steel Engineers, or those of departments of the federal government or of the Bureau of Mines.

It is the purpose of this book to describe some of the control devices that are in common use, and to show how they are assembled and connected to make a complete controller. The characteristics of various kinds of motors are described, and ways of using these characteristics in the control of the motor are discussed in detail. The design of magnetic structures has not been included, as that information is available in many textbooks.

2

ELECTRICAL DIAGRAMS

Purpose of the Diagram. Wiring diagrams serve a number of specific purposes, all of which must be kept in mind to insure that the diagrams are properly made. The initial purpose is to document the engineering of the circuitry involved. From this engineering standpoint, they will serve as the official record of equipment designed and supplied for a specific purpose. They will also be used as engineering reference in the later design of other equipment that may have similar specifications or functions. In the manufacturing operations the connection diagram or wiring list will be used by the wireman to connect the wires as required. In testing the equipment the test technicians will use the diagrams as engineering instructions as they check the equipment for performance in accordance with the diagrams. The installing electrician will follow the diagrams in installing the external wiring between the various separate units of the equipment (control cabinets, operator stations, pushbuttons, limit switches, motors, and so on). Finally the diagrams will be used for field service work by technicians in locating trouble in the equipment at any time during its life, or in modifying existing equipment to allow its operation to meet new conditions.

Types of Diagrams. The following definitions of diagram types are established in NEMA Standards ICS 1-101.02.

Schematic Diagram or Elementary Diagram. A schematic or elementary diagram is one which shows all circuits and device elements of an equipment and its associated apparatus or any clearly defined functional portion thereof. Such a diagram emphasizes the device elements of a circuit and their functions as distinguished from the physical arrangement of the conductors, devices, or elements of a circuit system.

Circuits which function in a definite sequence should be arranged to indicate that sequence.

A control sequence diagram or table, which is a portrayal of the contact positions for each successive step of the control action, may be included.

Wiring Diagram or Connection Diagram. A wiring or connection diagram is one which locates and identifies electrical devices, terminals, and interconnecting wiring in an assembly. This diagram may be (1) in a form showing interconnecting wiring by lines or indicating interconnecting wiring only by terminal designations (wireless diagram), or (2) a wiring table, or (3) a computer wiring chart, or (4) a machine command tape or cards. The term does not include mechanical drawings, commonly referred to as wiring templates, wiring assemblies, or cable assemblies.

Interconnection Diagram. An interconnection diagram is one which shows only the external connections between controllers and associated machinery and equipment.

Symbols. Graphic symbols are used as a shorthand means of illustrating and defining the elements and functions of an electrical circuit. Such electrical symbols divide into two classes: those used on elementary diagrams and those used on connection and interconnection diagrams. Although these two classes of symbols are distinct, there is a significant relationship between them.

On elementary diagrams the symbols are representative of the basic elements of the circuit, that is, relay contact, coil, switch, pushbutton contact, fuse, and so on. These basic symbols are graphical representations of the electrical functions occurring in the circuit. Figures 2.1, 2.2, and 2.3 show many of the most commonly used elementary diagram symbols.

On connection and interconnection diagrams these basic symbols are grouped together to represent graphically a complete electrical component such as a three-pole contactor. Historically, the symbols for complete devices, such as contactors, relays, or circuit breakers, were line drawings of the device. As such they were very complex, being nearly a picture reproduction of the device. This has now been greatly simplified to reduce the time and expense of making diagrams while still retaining the essentials of a functional symbol. For this purpose the symbol terminals (represented by small circles) must be arranged as the terminals actually appear on the device. This will enable a wireman reading a diagram to follow it easily and correctly, and to make the indicated connections to the proper terminals of the device. For devices having terminals on their sides, top, or bottom, as well as the front, some of the principles of drawing projection are applied in flattening the symbol

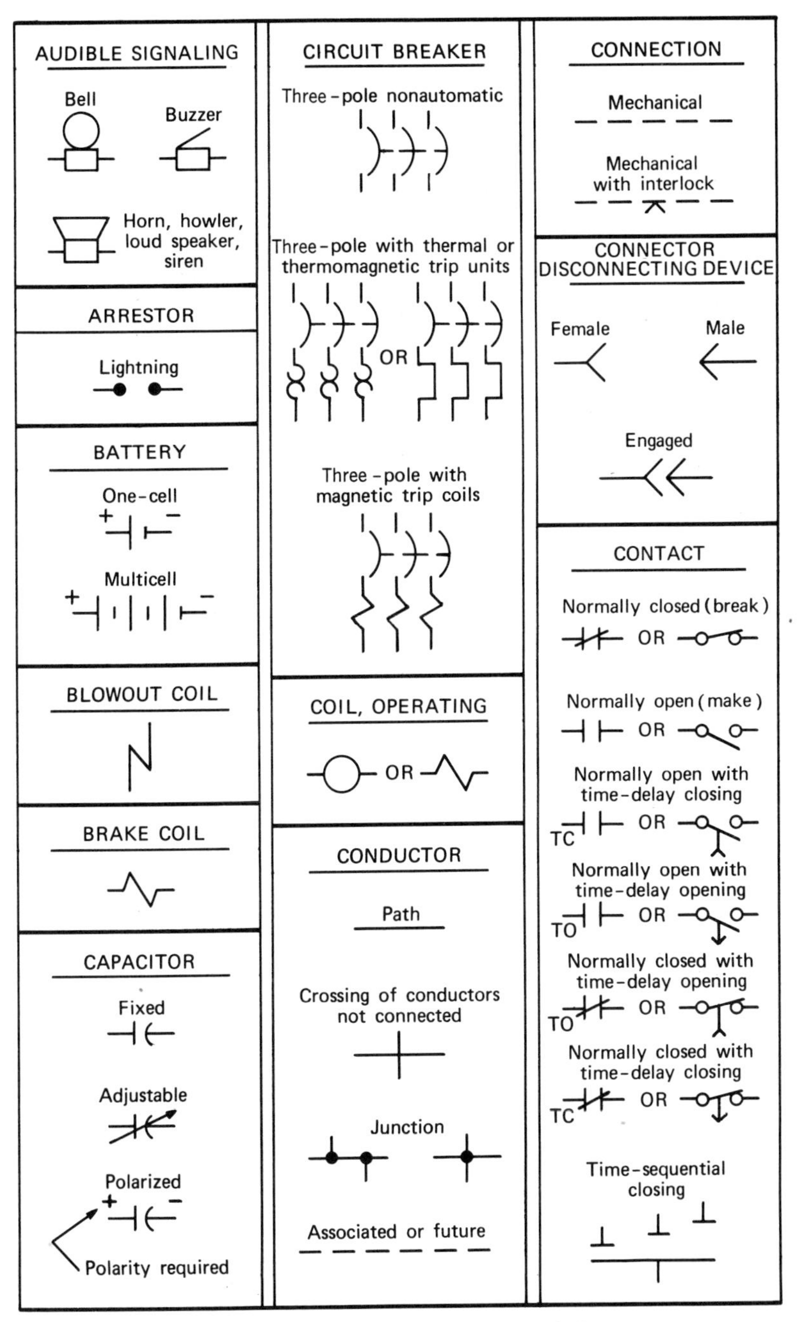

FIG. 2.1 Elementary Diagram Symbols

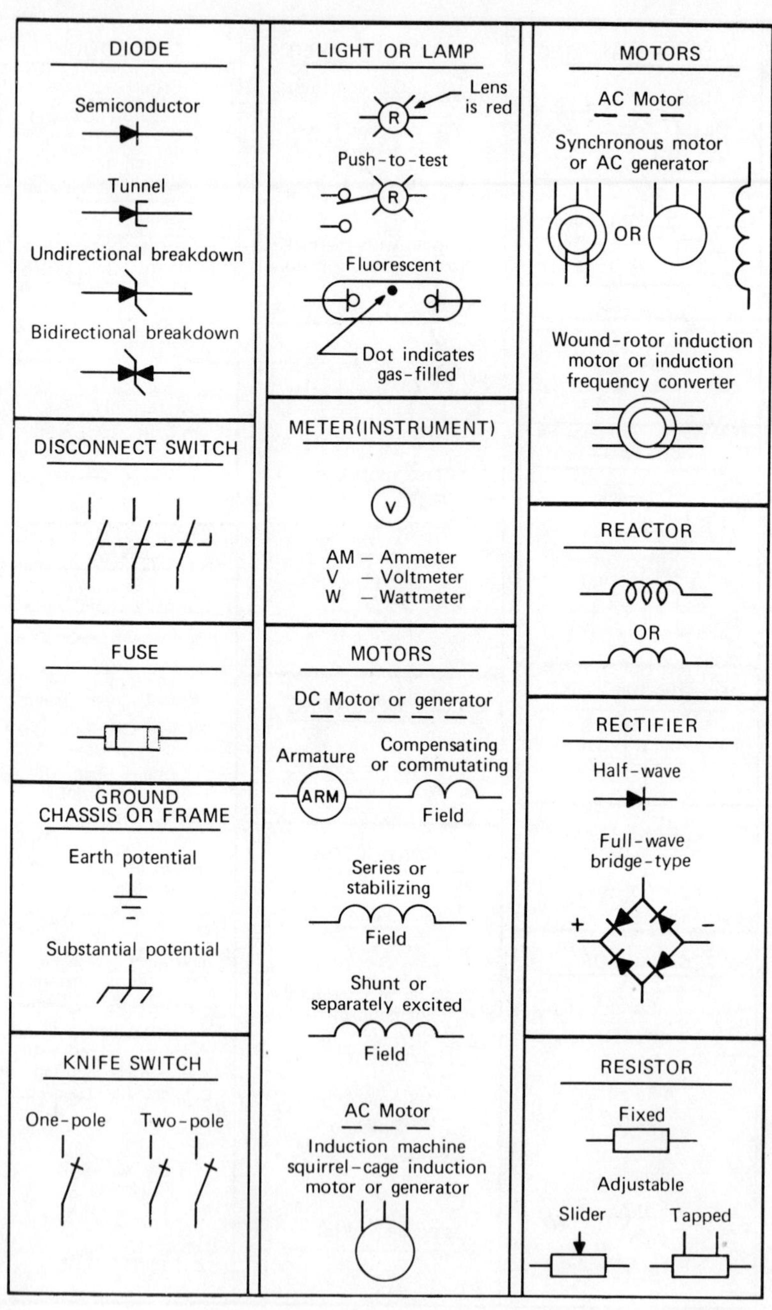

FIG. 2.2 Elementary Diagram Symbols

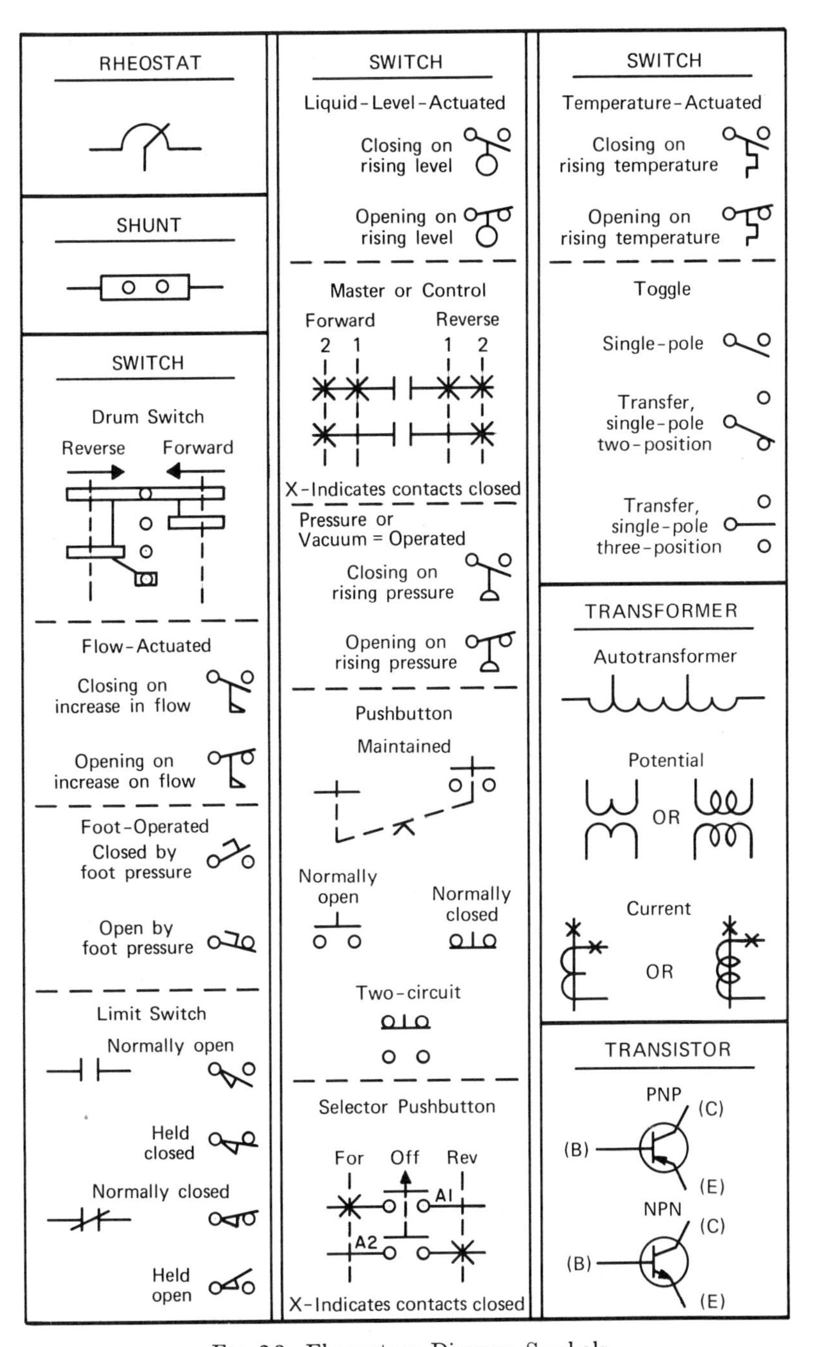

FIG. 2.3 Elementary Diagram Symbols

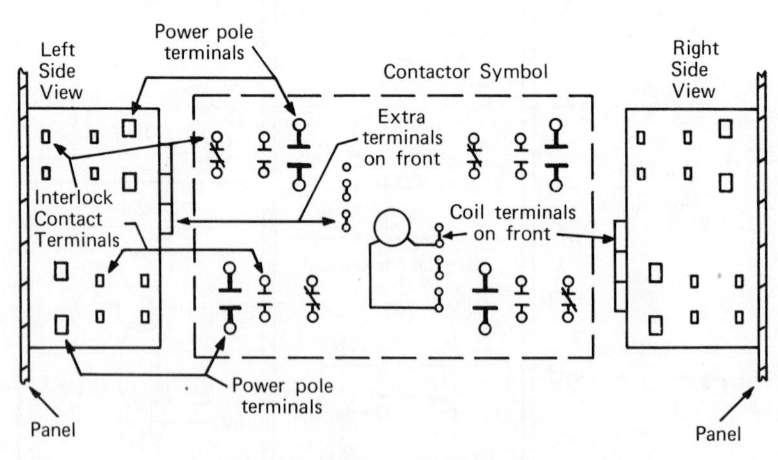

FIG. 2.4 Four-Pole AC Contactor

into one plane on the paper. For an illustration of this, see Fig. 2.4. Sometimes explanatory notes such as "top," "bottom," or "rear" are added for clarity. With the terminals drawn, the functional parts of contacts or coils are now added between the terminals to which they are actually connected. The completed symbol shows the device nearly as it physically appears, except that a function symbol such as a relay contact symbol has been substituted for a picture drawing of each relay pole. Figure 2.5 shows a representative group of connection diagram symbols.

The difference between elementary and connection diagram symbols may now be defined another way. Let us consider as an example a seven-pole relay having five normally open contacts, two normally closed contacts, and the operating coil. In the connection diagram these contacts would all be grouped together in a complete symbol including the coil, with each open or closed contact in its correct position with respect to all others, just as the device is actually constructed. In the elementary diagram, however, these eight functional circuit elements would be completely separated from each other, with each individual contact or coil appearing in the particular line of the circuit in which it functions. The only relation to each other in the elementary diagram would be their markings. If the device designation for this relay were MCR, then each of the eight elements of the relay would be marked MCR. This is illustrated in Fig. 2.6.

The illustration applies similarly to external pilot devices shown in interconnection diagrams such as pushbuttons, limit switches, and cam-type master switches. In the interconnection diagram a twelve-contact cam-type master switch would be drawn as it physically appears, usually including a target or table showing the positions in which each contact opens or closes. On the elementary diagram the symbol might be separated into twelve contacts, with each in its respective functional circuit.

The standard graphical electrical symbols have developed over many years. They are in general use throughout the electrical industry. These symbols appear in United States of America Standards Institute publication USAS Y32.2 and National Electrical Manufacturers Association standards publication, NEMA ICS 1-101. Many electrical symbols have worldwide recognition and accordingly are recommended by the International Electrotechnical Commission (IEC). Both the USAS and the NEMA publications listed above are necessary references for anyone involved in any way with graphic symbols as they are used on electrical and electronic diagrams. Many commercial customers of electrical equipment, as well as the military services, will specify that electrical dia-

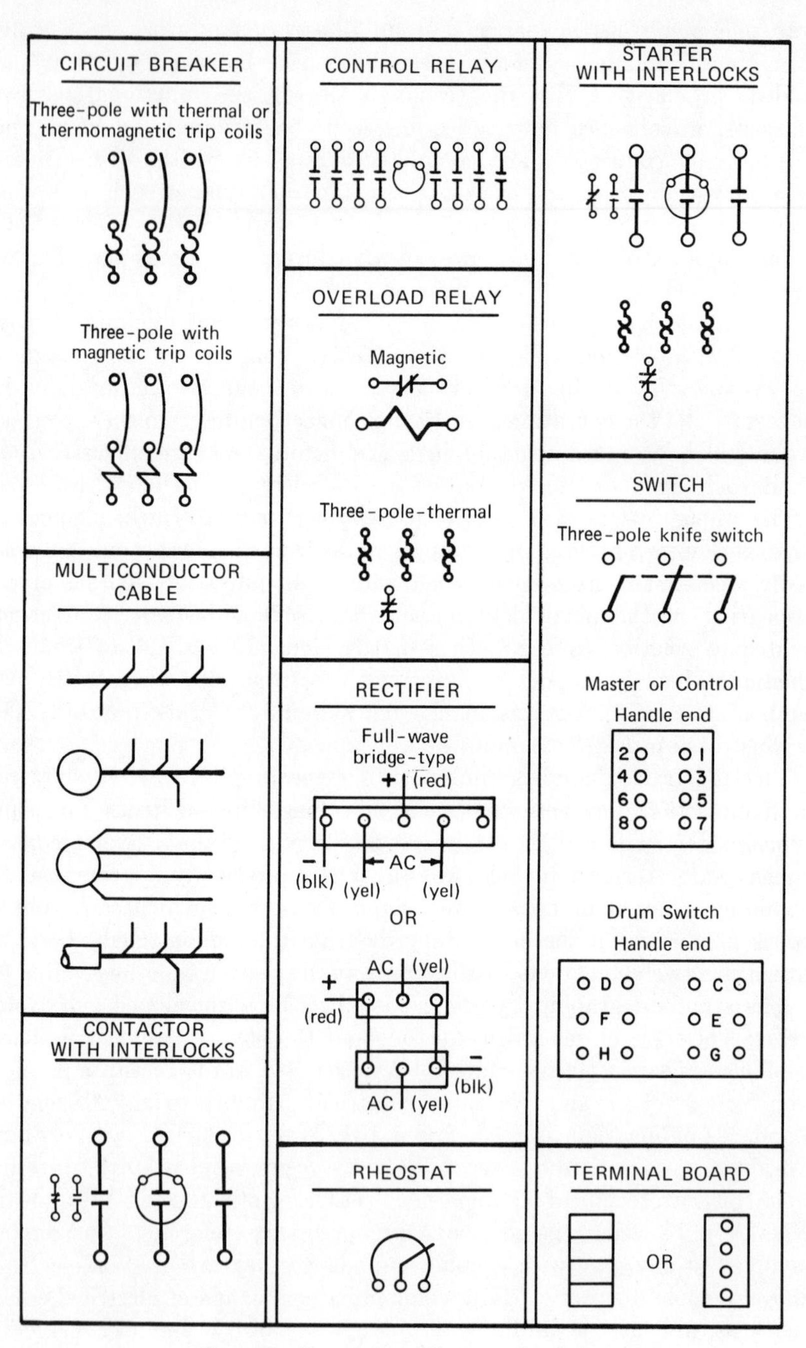

Fig. 2.5 Connection Diagram Symbols

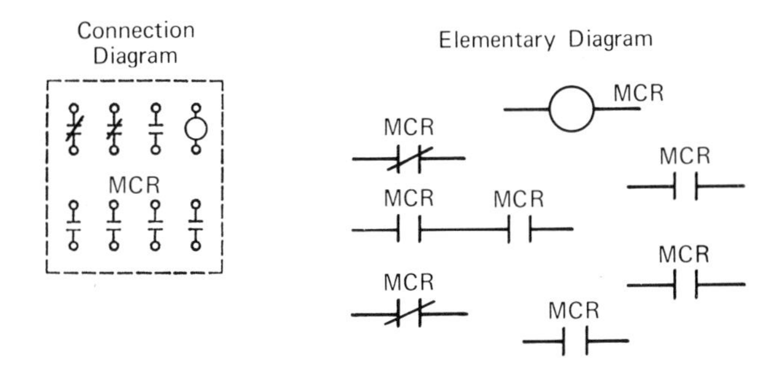

FIG. 2.6 Seven-Pole Relay

grams for any of their equipment must be in conformance with these standards.

The number of varieties of electrical devices for which graphic symbols have been established is very large. There are so many that all the easily recognizable geometric shapes and all the variations and combinations of those shapes have been assigned. As new electrical devices are created, especially those which provide a combination of operations or functions, there is a trend for electrical draftsmen to try to create new symbols or combine portions of present symbols. Such a practice is discouraged, because such symbols do not appear in any approved standard publication and will be unknown or misinterpreted when they appear in diagrams. To cover the symbology for such devices there has come into use the term "uniform-shaped symbol" (see Fig. 2.7). This symbol is simply a rectangle to which incoming and outgoing wires are connected. These wires are numbered at the outside edge of the rectangle. Inside the rectangle the function is designated by a word or statement. The device marking should also be shown inside the rectangle.

This uniform-shaped symbol has NEMA approval in NEMA

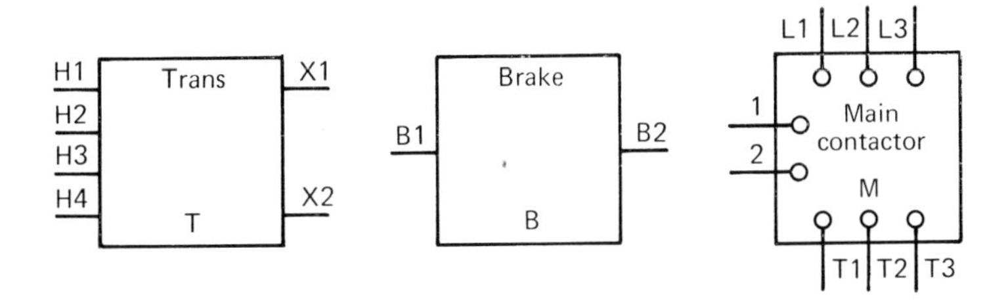

FIG. 2.7 Uniform-Shaped Symbols

ICS 1-101.08.3. Such uniform-shaped symbols may be used as an alternate for any of the distinctive-shaped symbols shown in the standards.

It should be emphasized that the advantage of distinctive-shaped symbols is that they represent an electrical function by their graphic shape. Therefore, when we read a diagram using distinctive symbols, we see functions by recognizing the significant shapes and can really read a whole line of symbols at a time. In contrast, a line of uniform-shaped symbols would be a line of rectangles with words in them. Such symbols cannot be read in groups, because each descriptive word must first be read to understand the function of each symbol.

Device Designations. Device designations are used on diagrams to mark each of the graphic symbols of a circuit. Device designations are formed by the assignment of a letter or letters representing the principal function performed by the device.

The letters usually are the first letters of the words describing the function. For example, field accelerating would be designated FA. Table 2.1, from NEMA ICS 1-101.05, lists device designations that are in

TABLE 2.1

DEVICE DESIGNATIONS IN STANDARD USE

Device or Function	Designation
Accelerating	A
Ammeter	AM
Braking	B
Capacitor	C or CAP
Circuit breaker	CB
Control relay	CR
Current transformer	CT
Demand meter	DM
Diode	D
Disconnect switch	DS or DISC
Dynamic braking	DB
Field accelerating	FA
Field contactor	FC
Field decelerating	FD
Field-loss	FL
Forward	F or FWD

TABLE 2.1 (Continued)

Device or Function	Designation
Frequency meter	FM
Fuse	FU
Ground protective	GP
Hoist	H
Jog	J
Limit switch	LS
Lower	L
Main contactor	M
Master control relay	MCR
Master switch	MS
Overcurrent	OC
Overload	OL
Overvoltage	OV
Plugging	P
Power factor meter	PFM
Pressure switch	PS
Pushbutton	PB
Reactor	X
Rectifier	REC
Resistor	R or RES
Rheostat	RH
Reverse	REV
Selector switch	SS
Silicon-controlled rectifier	SCR
Solenoid valve	SV
Squirrel cage	SC
Starting contactor	S
Tachometer generator	TACH
Terminal block or board	TB
Time-delay relay	TR
Transformer	T
Transistor	Q
Undervoltage	UV
Voltmeter	VM
Watthour meter	WHM
Wattmeter	WM

standard use. It is recommended, when making a diagram having device functions other than these, that a chart be included in the diagram listing those additional functions and their corresponding device designations.

When a panel contains a multiple number of devices performing the same function, the device designations should be prefixed or suffixed with suitable numbers and/or letters. On a panel having three main contactors, the contactors would be marked 1M, 2M, and 3M. Their related overload relays would then be marked 10L, 20L, and 30L. For large systems having multiple panels it is helpful to number the panels and then use the panel number as the prefix of all device designations for the components mounted on that panel.

Coil and Contact Designations. The following designations are commonly used to identify special functions of certain coils or contacts of a device. They are not a part of the device designation, but are added near the symbol as additional information. Table 2.2, from NEMA

TABLE 2.2

STANDARD COIL AND CONTACT DESIGNATIONS

Function	Designation
Closing coil	CC
Holding coil	HC
Latch coil	LC
Time-delay closing contacts	TC or TDC
Time-delay opening contacts	TO or TDO
Trip coil	TC
Unlatch coil	ULC

ICS 1-101.06, lists coil and contact designations that are in standard use. Any other special function should be indicated in full without using abbreviations.

Terminal Markings. The terminals of a control panel that are provided for the connection of external wiring must be marked. This identifying marking enables the installing electrician to properly connect all external wiring coming from other panels or accessories. The individual terminal marking is really a circuit number and as such appears at that point of the circuit on the elementary diagram and the connection diagram as well as on the actual terminal on the panel. The marking is therefore a most important feature in assisting maintenance personnel to identify circuits when servicing existing installations.

Power circuit markings should follow American Standard C 6.1-1956. This publication lists the standard terminal markings for all types of ac and dc electric motors.

Control circuit markings may be any arbitrary system that follows some logical order. It must assist anyone reading the diagram, so that he may easily follow the circuits through their many connections. Each control circuit number appears at its respective place on the elementary diagram and at each point on the connection diagram to which that circuit connects. It has also become a common practice to label each end of each wire on the controller with a marking tag giving its control circuit number. This, again, is very helpful for maintenance personnel in servicing the controller. Such wire marking is frequently specified by the purchaser when the equipment is a complex systems installation.

Preparation of Diagrams—Order of Procedure. The diagrams necessary for a complex control equipment are usually made in the following order: (1) the elementary diagram, (2) the control sequence table, if one is required, (3) the connection diagram, (4) the interconnection diagram. This order is logical because when an engineer starts to design a controller, he must first work out the elementary circuit design in order to fulfill the specifications and provide an equipment that will perform the functions required. If a control sequence table is desired, it is made next, giving a graphic representation of circuit conditions at various points in the time cycle of the controller operations.

The next operation is to analyze the elementary diagram to determine the various electrical components that will be required to build that equipment. The elementary diagram is also used as a guide in determining an optimal arrangement of the electrical components on a panel, to achieve short and simplified wiring between related electrical components. After the physical and mechanical arrangement of these electrical components has been determined, the required mechanical drawings are made for the manufacture of the control panels, switchboards, enclosures, operators consoles, and so on. With these drawings for reference, the electrical draftsman then makes the connection diagram or wiring list for each unit of the equipment. These diagrams show an electrical symbol for each electrical component, and they are arranged in their proper position relative to their actual physical location on the panel or enclosure.

On the connection diagram the draftsman draws lines representing wires between the respective terminal points of the electrical symbols that should be connected together. He does this by following the circuitry as shown on the elementary diagram and applying this to the various elements of the electrical components that are mounted on a panel or in the enclosure. Instead of drawing in lines for all the wires, he may choose to make a wireless connection diagram in which all terminals are identified with terminal markings and all similarly marked terminals are connected together by the wireman. Sometimes these terminal markings are tabulated in a list called a wiring list.

All wiring of circuits leading to other cabinets or external devices are brought to terminal boards for convenience in connecting the external wiring when installing the equipment. If the diagram is of a rather simple controller, the external wiring to pushbuttons, limit switches, motors, and so forth is included in the connection diagram. For complex systems equipment consisting of multiple panels, and consequently many panel connection diagrams, the electrical draftsman now makes an interconnection diagram which shows only the external or field installation wiring between all panels and all other electrical accessory items.

Coding Methods for Complex Diagrams. Most drafting departments limit the drawing size to 34 × 44 in. maximum. The preferred, and most universally used, size is 22 × 34 in., commonly called "D" size. The limiting of sheet size for convenience of paper handling has forced the adoption of multiple sheet diagrams for large complex control systems. The multiple sheets make it even more necessary for a satisfactory numbering system to be worked out.

In a preceding section on terminal markings, it was stated that control circuit markings may be any arbitrary system that follows some logical order. As a diagram becomes larger and more complex, the choice of a numbering system becomes extremely important. It is the only key to assist any user of the diagram in finding his way through the circuits, from the elementary diagram to the connection diagram to the actual panel wiring, or from the panel wiring back to the elementary diagram.

Numbering of devices and circuits by some defined method may be termed coded numbering. Coded numbering of devices and circuits is usually related to either the physical arrangement of panels and components on the panels or to the location of a circuit within the elementary diagram. Many variations of coding can be created, and each may have certain advantages. The following paragraphs describe one general coding system that is applicable, and distinctly advantageous, to large complex control systems. Basically, the system provides two numbering methods, one for the device designations and another for the circuit wire numbers.

The coded numbering of the devices should be related to the physical location of the device within the control. That is, each major part of the control equipment, such as a panel, should be assigned a basic number. Then each device on a specific panel would use the panel number for a prefix to the standard functional device designation. If a subpanel is mounted on the main panel, it may be assigned an additional letter to be included in the prefix to cover all items mounted on that subpanel. If the subpanel consists of further definite subassemblies, they should be assigned additional suffixes. Numbers and letters should be assigned to alternate levels of equipment—numbers to the first and third unit levels, and so on; letters to the second and fourth unit levels, and so on. This is sometimes referred to as the location coding method and as such is completely described in USAS Y32.16, Section 6. The objective of this type of device coding is to enable all users of the elementary diagram to easily locate all devices on the connection diagram and on the actual control assembly.

The coded numbering of circuit wire numbers should be related to the location of the circuit in the elementary diagram. This is accomplished by numbering all of the circuits of the elementary diagram in sequential order as they occur through the entire diagram. If the elementary diagram consists of multiple sheets, then each sheet has the same base drawing number followed by a sheet number suffix, that is, D12479-1, D12479-2, D12479-3. Each horizontal line of circuitry is assigned a sequential number prefixed with its respective sheet number. Each circuit point across the line from left to right is assigned the line number plus an additional sequential suffix digit. This is illustrated in the elementary diagram, Fig. 2.8. The corresponding circuit of the connection diagram is shown in Fig. 2.9.

From these illustrations and the description of the coding, the advan-

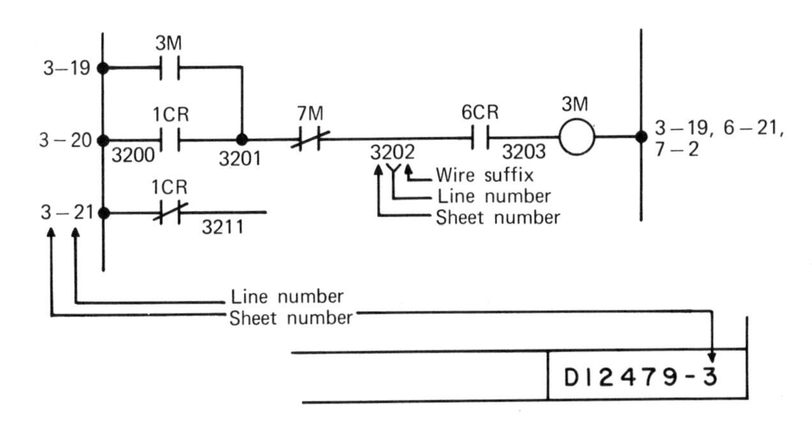

FIG. 2.8 Elementary Diagram

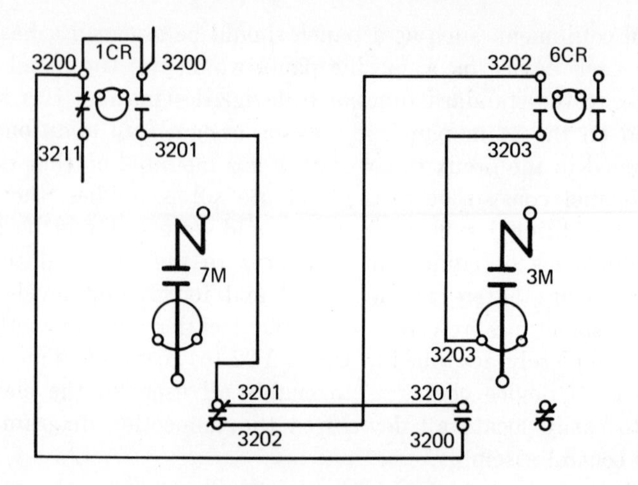

Fig. 2.9 Connection Diagram

tage of this method of coding circuit wire numbers can be demonstrated. If a person servicing the equipment and observing the operation of contactor 7M raised the question of the circuit function of the 7M normally closed interlock, he would simply check a wire number on that interlock, which is 3201. The circuit for that interlock could then be found quickly and easily by interpreting that wire number to sheet 3 of the elementary diagram, line 20. The 1 is the second point from the left on that line. On large complex system diagrams, which may often have one hundred or more sheets of elementary diagrams, this method is of the greatest advantage. This coding technique minimizes laborious searching for a 7M interlock contact somewhere, anywhere, on the one hundred sheets of diagrams.

Another coding which may be used on diagrams gives contact location information. In Fig. 2.8 there are three sets of numbers just to the right of the 3M coil. These figures show where the three contacts of that 3M contactor are located on the elementary diagrams. Here 3-19 indicates that there is a normally open contact on sheet 3, line 19, and 6-21 indicates that there is a normally closed contact on sheet 6, line 21. The underlining of the number indicates a normally closed contact. The other contact is on sheet 7, line 2. In analyzing circuit functions, this coding is most helpful in locating the circuits which will be affected by the energizing of the 3M coil. This contact reference code is in general use throughout the electrical industry on multiple sheet diagrams or on diagrams for controls having enough relays and contactors to make its use beneficial.

Problems

Make schematic diagrams for the following controllers.

1. A nonreversing pushbutton-operated controller for a shunt motor, consisting of

DPST knife switch
2 single-coil overload relays
2 main line contactors
3 accelerating contactors without timing
Resistors
Start-stop pushbuttons

2. A controller duplicating that of problem 1, except with the addition of dynamic braking and a field discharge resistor.

3. A controller duplicating that of problem 1, except with the addition of a field rheostat, a field-loss relay, and some means of short-circuiting the rheostat during acceleration.

4. A controller duplicating that of problem 1, except with the addition of a jogging pushbutton.

5. A reversing pushbutton-operated controller for a series motor, consisting of

DPST knife switch
2 single-coil overload relays
1 main line contactor
4 SP reversing contactors
3 accelerating contactors without timing
Resistors
Forward-reverse-stop pushbuttons

6. A controller duplicating that of problem 5, except operated by a four-speed, drum-type master controller instead of by pushbuttons, and having a no-voltage relay.

7. A controller duplicating that of problem 1, but with the addition of three series relays for current-limit acceleration.

8. A controller for starting and reversing two series motors in parallel, consisting of

DPST knife switch
3 single-coil overload relays
1 main line contactor
8 SP reversing contactors
2 SP plugging contactors
3 SP accelerating contactors without timing
No-voltage relay
Resistors
Four-speed, drum-type master controller

9. A controller duplicating that of problem 5, except equipped with a shunt brake, and with 2 SP limit switches, one for stopping in each direction of travel.

10. A controller duplicating that of problem 1, except with the addition of a normally closed armature shunt contactor and resistor, and with individual dashpot timing relays for controlling the accelerating contactors.

11. A reversing magnetic controller, master-operated, for a 50-hp 230-V series motor driving a crane bridge. Use inductive timing without inductor.

12. A reversing controller for a squirrel-cage motor, consisting of

TPST disconnect device
3 single-coil overload relays
2 three-pole contactors
Shunt brake
Forward-reverse-stop pushbuttons

13. A reversing autotransformer starter for a squirrel-cage motor, consisting of

1 three-pole circuit breaker
3 single-coil overload relays
2 three-pole contactors for reversing
1 five-pole contactor for starting
1 three-pole contactor for running
2 autotransformers
1 dashpot timing relay operated by starting contactor
Forward-reverse-stop pushbuttons

14. A nonreversing, primary-resistance increment starter for squirrel-cage motor, consisting of

1 three-pole line contactor
3 single-coil overload relays
3 three-pole resistor commutating contactors
3 dashpot timing relays operated by contactors
Resistor in three phases
Start-stop pushbutton

15. A selective controller for the motor of Fig. 16.20.

16. A variable-frequency control for three squirrel-cage motors. This consists of a dc motor with a nonreverse three-step time-limit controller, and an alternator driven by the motor. Each squirrel-cage motor is equipped with an across-the-line magnetic starter with overload and disconnecting circuit breaker. A field rheostat controls the speed of the dc motor.

17. A controller for a wound-rotor motor, similar to the controller of Fig. 17.8, except reversing, and operated by a four-step drum-type master controller. Note that undervoltage relay must be added.

18. A controller for a synchronous motor, similar to the controller of Fig. 19.7, except with single-step primary-resistance starting.

19. A controller of the part-winding type as in Fig. 16.13, with the addition of a step of resistance in each leg of the first section of the motor winding, and a timed contactor to short-circuit the resistor before the second section of the winding is energized.

20. A controller for a wound-rotor motor like the controller of Fig. 17.8, except with three-phase series relay acceleration, and with a four-point master controller instead of a pushbutton.

21. Complete the diagram of Fig. 17.9 by drawing the master and control circuits.

22. Make a sequence table for the controller of Fig. 17.9, showing the contactors which are closed on each point of the master controller.

3

CONSTRUCTION OF CONTROL APPARATUS

Material for Panels. The material used for control panels is generally slate, asbestos composition, or steel. The best variety of slate is known as Monson slate; it is free of iron and other impurities and may safely be used for up to approximately 1600 V. Other varieties of slate are satisfactory for many purposes provided that they are reasonably free of impurities. In general, their use is limited to approximately 1000 V. The slate, after being drilled, is either sprayed with an insulating finish or oiled.

The asbestos composition is used for many purposes and especially for voltages above 1600. It has a slightly greater tendency to take up moisture. The asbestos composition excels slate chiefly in its resistance to impact and volume resistance. The impact strength of the composition is two to four times that of slate. On the other hand, slate has about twice the resistance to rupture. The cost of working the materials is about the same.

Fabric-base phenolic material is sometimes used for control panels because of its strength. It has a tendency to char under arcing; also, it is too expensive for general use.

Steel panels, being strong and relatively inexpensive, are being used in increasing quantity. The apparatus mounted on steel must be of dead-back construction having terminals on the front. Steel panels are wired on the front.

Arrangement of Panels. A number of factors enter into the arrangement of apparatus on a panel. The common method of making such layouts is to use paper templates cut to the exact size and shape of the various pieces of equipment which are to be mounted. These templates may then be laid out on a table and shifted around until the best arrangement is found. One of the first considerations is to make the size of the panel as small as possible, both from a cost standpoint and also because space is important in most installations. Appearance is another consideration. The parts should be arranged symmetrically and should be lined up as nearly as possible.

Both the internal wiring of the panel and the external wiring, which the user must install, should be considered in making a layout. It is important to have the wires and bus bars on the panel as short and as straight as possible, and also to have a minimum number of heavy connections. The terminals to which the user will connect his cables should be placed in an accessible position; it is generally advisable to group them either at the top or at the bottom of the panel.

It is preferable to have the heavier pieces of apparatus mounted near the bottom of the panel, so that it will not be top-heavy. It is also advisable to mount oil dashpots, or other devices using oil, near the bottom so that if there is leakage the oil will not run down onto other pieces of apparatus. Dashpots and other devices which require adjustment should be located so that the adjustment is accessible.

Knife switches should be arranged to open in the direction of gravity and should be mounted at a convenient height. For ease of reading, meters and instruments should be mounted at approximately eye level. Current and potential transformers, if mounted on a panel, are usually mounted on the rear, metal straps or supports being used for the purpose.

Thermal overload relays are preferably mounted at the bottom of the panel, especially if the panel is enclosed. Heat generated by other devices will rise to the top of the enclosure and so will not be likely to affect the operation of the thermal devices.

TABLE 3.1

CLEARANCES BETWEEN ARC-RUPTURING PARTS AND ENCLOSURE

Horsepower Rating	Distance from Contacts in Direction of Blowout in Inches DC and AC Circuits		Vertical Distance above Contacts, without Blowout in Inches				Horizontal Distance from Contacts and Distance below Contacts in Inches DC and AC Circuits	
			DC Circuits		AC Circuits			
	300 V	600 V	300 V	600 V	300 V	600 V	300 V	600 V
5	1¾	3	4	Barriers	1¾	3	¾	1½
10	2	4	5	Barriers	2	4	¾	1½
50	3	5	6	Barriers	3	5	1	2
100	4	Barriers	Barriers	Barriers	4	Barriers	2	3
Above 100	Barriers	Barriers	Barriers	Barriers	Barriers	Barriers	Barriers	Barriers

Note. All distances shall be measured from the contact tips or arc horns.

If panels are enclosed, adequate ventilation must be provided, and for dust-tight equipment it may be necessary to make the enclosure relatively large in order to prevent overheating of the apparatus in it.

Line contactors, or other contactors in which considerable arcing occurs, are preferably mounted at the top of the panel, so that the arcing will not affect other apparatus on the panel. It is necessary to provide adequate clearance in the direction of the arc. Minimum clearances are given in Table 3.1.

Definitions. The following definitions, applying to the construction of the enclosures of controllers, are taken from NEMA Industrial Control and Systems Standards:

Acid-resistant: So constructed that it will not be injured readily by exposure to the acid fumes.

Corrosion-resistant: So constructed, protected, or treated that corrosion caused by a specified corrosion test will not exceed specified limits.

Drip-proof: So constructed or protected that vertically falling dirt or drops of liquid will not interfere with the successful operation of the apparatus.

Drip-tight: So constructed or protected as to exclude falling moisture or dirt.

Dust-proof: So constructed or protected that the accumulation of dust will not interfere with its successful operation.

Dust-tight: So constructed as to meet the requirements of a specified dust-tightness test.

Fume-resistant: So constructed that it will not be injured readily by exposure to the specified fumes.

Moisture-resistant: So constructed or treated that it will not be injured readily by exposure to a moist atmosphere.

Nonventilated: So constructed as to provide no intentional circulation of external air through the enclosure.

Oil-resistant: (as applied to gaskets): Made of a material that will not be damaged by exposure to oil or oil fumes.

Oil-tight: Constructed to exclude oils, coolants, and similar liquids, when subjected to specified tests.

Proof (used as a suffix): So constructed, protected, or treated that successful operation of the apparatus is not interfered with, when subjected to the specified material or condition.

Rainproof: So constructed, protected, or treated that subjection to a beating rain test will not interfere with the successful operation of the apparatus.

Rain-tight: So constructed or protected that exposure to a beating rain will not result in the entrance of water.

Resistant (used as a suffix): So constructed, protected, or treated that it will not be damaged when subjected to the specified material or conditions for a specified time.

Rust-resistant: So constructed, protected, or treated that rust caused by a specified rust resistance test will not exceed a specified limit.

Spash-proof: So constructed and protected that external splashing will not interfere with its successful operation.

Sleet-proof: So constructed or protected that the accumulation of sleet under specified conditions will not interfere with its successful operation (including external operating mechanism).

Sleet-resistant: So constructed that the accumulation of sleet under specified conditions will not cause damage.

Submersible: So constructed as to exclude water when submerged in water under specified conditions of pressure and time.

Tight (used as a suffix): So constructed that the enclosure will exclude the specified material under specified conditions.

TABLE 3.2

CHARACTERISTICS OF NONVENTILATED ENCLOSURES FOR INDOOR NONHAZARDOUS LOCATIONS

	Enclosure Capability							
Provides Protection Against	Type 1	Type 2	Type 4	Type 4x	Type 6	Type 11	Type 12	Type 13
Accidental contact with enclosed apparatus	Yes	Yes	Yes	Yes	Yes	Yes	Yes	Yes
Falling dirt	Yes	Yes	Yes	Yes	Yes	Yes	Yes	Yes
Falling liquids		Yes	Yes	Yes	Yes	Yes	Yes	Yes
Dust, lint, or fibers			Yes	Yes	Yes		Yes	Yes
Hose-down or splashing water			Yes	Yes	Yes			
Oil or coolant seepage							Yes	Yes
Oil or coolant spraying or splashing								Yes
Corrosive agents				Yes		Yes		
Occasional submersion					Yes			

Ventilated: So constructed as to provide for the circulation of external air through the enclosure to remove heat, fumes, or vapors.

Watertight: So constructed as to exclude water applied in the form of a hose stream under specified conditions.

Weather-proof: So constructed or protected that exposure to the weather will not interfere with its successful operation.

Types of Enclosures. NEMA defines standard types of nonventilated enclosures and establishes design tests which are used to demonstrate conformance to these standards. Certain types can also be furnished in ventilated designs, although protection against some environments may have to compromise with the need for ventilation. Tables 3.2 and 3.3, taken from the NEMA Industrial Control and Systems Standards, show protection afforded by various NEMA types of nonventilated enclosures for nonhazardous locations. Enclosures for hazardous locations should be given very special consideration, and for these requirements consultation with competent authorities is recommended.

Most of the NEMA types of enclosures are illustrated in the following figures.

TABLE 3.3

CHARACTERISTICS OF NONVENTILATED ENCLOSURES FOR OUTDOOR NONHAZARDOUS LOCATIONS

Provides Protection Against	Enclosure Capability			
	Type 3	Type 3R	Type 3S	Type 6
Accidental contact with enclosed apparatus	Yes	Yes	Yes	Yes
Rain, snow, and sleet[a]	Yes	Yes	Yes	Yes
Sleet[b]			Yes	
Windblown dust	Yes		Yes	Yes
Hose-down				Yes
Occasional submersion				Yes

[a] External operating mechanisms are not required to be operable when enclosure is ice-covered.

[b] External operating mechanisms are operable when enclosure is ice-covered.

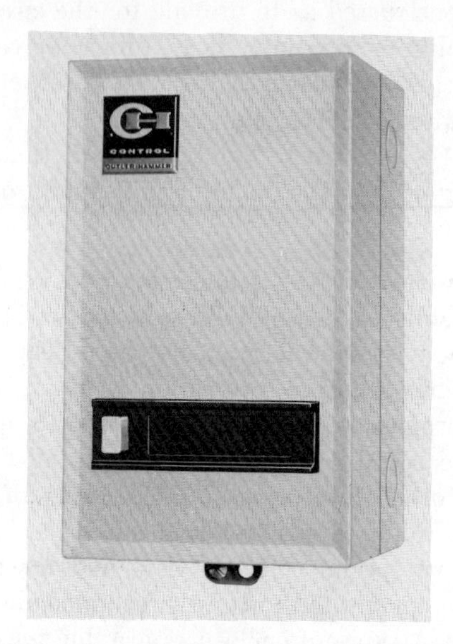

FIG. 3.1 General-Purpose Enclosure NEMA 1

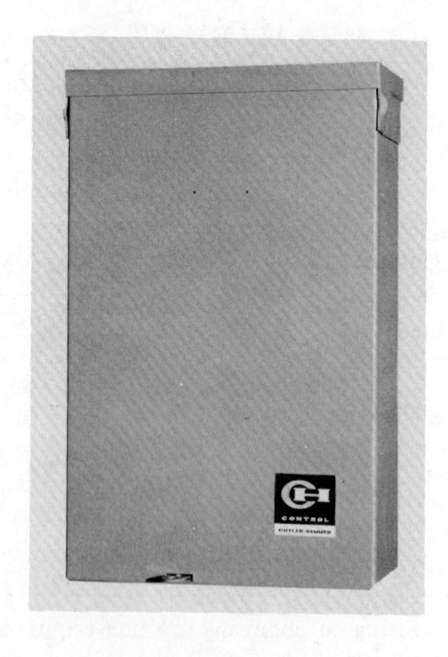

FIG. 3.2 Weather-Proof Enclosure NEMA 3

Type 1. General Purpose (Fig. 3.1). Enclosure is intended for use indoors to prevent accidental contact with the enclosed apparatus.

Type 3. Outdoor, Dust-tight, and Rain-tight (Fig. 3.2). Enclosure is intended for use outdoors to protect the enclosed equipment against wind-blown nonhazardous dust and water.

Type 4. Watertight (Fig. 3.3). Enclosure is intended for use indoors to protect enclosed equipment against splashing or hose-directed water.

Types 7 and 9. Hazardous Locations (Fig. 3.4). Enclosures are intended for use indoors in specified atmospheres as defined in the National Electrical Code. The type of hazardous location must be indicated by a letter A through G as covered in the above code.

Type 12. Industrial Use, Dust-tight, and Drip-tight (Fig. 3.5). Enclosures are intended for use indoors to protect the enclosed equipment against nonhazardous fibers and dust, and light splashing, seepage, dripping, or external condensation of noncorrosive liquids.

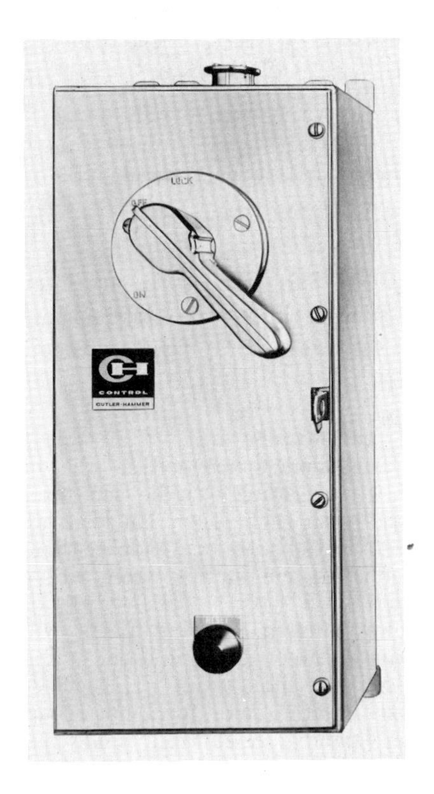

FIG. 3.3 Watertight Enclosure NEMA 4

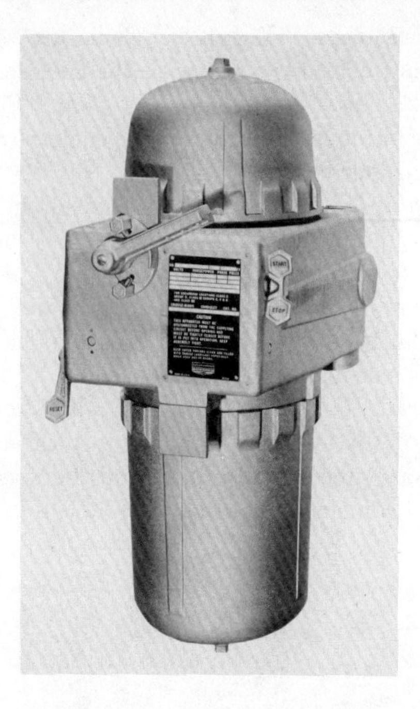

FIG. 3.4 Hazardous-Location Enclosure NEMA 7 and 9

Type 13. Oil-tight (Fig. 3.6). Enclosures are intended to house pilot devices used indoors, such as limit switches, foot switches, pushbuttons, and to protect against seepage or spraying of water, oil, or coolant.

Design of Enclosures

1. There must be sufficient space within the enclosure to permit uninsulated parts of wire terminals to be separated so as to prevent their coming in contact with each other. Enclosures must be such as to permit proper wire connections to be made with adequate spacing of the terminals and ends of conductors from adjacent points of the enclosures.

2. Exposed nonarcing current-carrying parts within the enclosures must have an air space between them and the uninsulated walls of the enclosure, including conduit fittings, of at least ½ in. for 600 V or less. Enclosures of sizes, material, or form not securing adequate rigidity must have greater spacing. A suitable lining of insulating material not less than $\frac{1}{32}$ in. in thickness may be considered acceptable where the spacing referred to above is less than ½ in.

Exception: For fractional-horsepower controllers, and other small devices, of 300 V or less, where the enclosure is rigid, an air space of

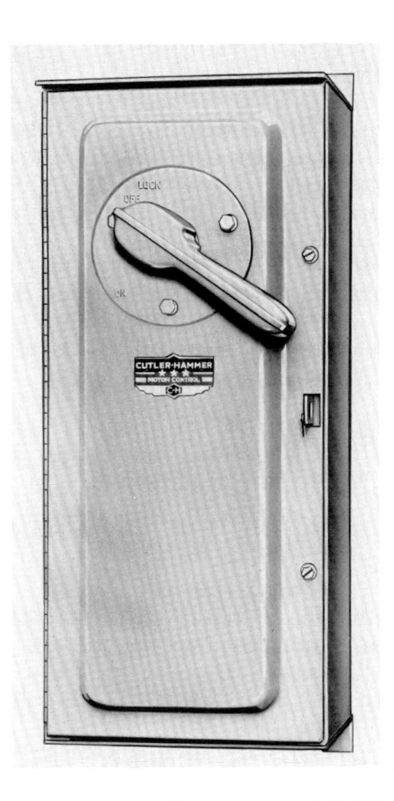

FIG. 3.5 Industrial-Use Enclosure NEMA 12

FIG. 3.6 Oil-tight Enclosure NEMA 13

¼ in. is permitted between nonarcing current-carrying parts and the uninsulated part of the enclosure.

3. All enclosures and parts of enclosures such as doors, covers, and tanks must be provided with means for firmly securing them in place. Among the available means are locks, interlocks, screws, and seals.

4. Where the walls of the enclosure are not protected by barriers or by a lining of noncombustible insulating material, the arc-rupturing parts of the controller should have air spaces, as indicated in Table 3.1, between them and the walls of the enclosure, unless a test on any specific device demonstrates that a smaller space is safe for that particular device.

Material

1. In the following paragraphs it is assumed that steel (or gray iron for castings) will be the metal employed. Copper, bronze, and brass are sometimes used, but the requirements given for steel must still be complied with.

2. Thickness of castings. Cast metal for enclosures, whether of iron or other metal, shall be at least ⅛ in. thick at every point and of greater thickness at reinforcing ribs and at door edges, except that die-cast metal may not be less than $\frac{3}{32}$ in. thick for an area greater than 24 in.^{2} or having any dimension greater than 6 in. and may be not less than $\frac{1}{16}$ in. in thickness for an area of 24 in^2. or less or having no dimension greater than 6 in. Cast metal shall be at least ¼ in. in thickness.

3. Sheet-metal thickness. The minimum thickness required for sheet-metal enclosures varies with the size. For solid enclosures without slot or other opening, and for solid enclosures except for a slot for the operating handles or openings for ventilation or both, the sheet metal must be of gauge not less than given in Table 3.4, except that metal shall not

TABLE 3.4

Minimum Thickness of Sheet-Metal Enclosures

Maximum Area of Any Surface (in.^{2})	Maximum Dimension (in.)	Without Supporting Frame (Minimum USS Gauge)	With Supporting Frame or Equivalent Reinforcement (Minimum USS Gauge)
90	12	20(0.0359)	24(0.0239)
135	18	18(0.0478)	20(0.0359)
360	24	16(0.0598)	18(0.0478)
1200	48	14(0.0747)	16(0.0598)
1500	60	12(0.1046)	16(0.0598)
Over 1500	—	10(0.1345)	16(0.0598)

be less than No. 20 USS gauge in thickness at points where rigid conduit is to be connected.

4. All enclosures composed of wire mesh, perforated screens, or grillwork must be provided with supporting frames.

5. Ventilating openings in an enclosure, including perforated holes, louvers, and openings protected by means of wire screening, expanded metal, or perforated covers, shall be of such size or shape that no opening will permit passage of a rod having a diameter greater than ½ in., except that, when the distance between live parts and the enclosure is greater than 4 in., openings may be larger than those previously mentioned, provided that no opening will permit passage of a rod having a diameter greater than ¾ in. The wires of a screen shall not be less than No. 16 AWG when the screen openings are ½ in.^{2} or less in area, and shall be not less than No. 12 AWG for larger screen openings.

Except as noted in the following paragraph, sheet metal employed for expanded metal mesh, and perforated sheet metal, shall not be less than No. 18 USS gauge in thickness when the mesh openings or perforations are ½ in.^{2} or less in area, and shall not be less than No. 13 USS gauge in thickness for larger openings.

In a small device where the indentation of a guard or enclosure will not affect the clearance between uninsulated, movable, current-carrying parts and grounded metal, No. 24 USS gauge expanded metal may be employed, provided that (*a*) the exposed mesh on any one side or surface of the device so protected has an area of not more than 72 in.^{2} and has no dimension greater than 12 in., or (*b*) the width of an opening so protected is not greater than 3.5 in.

A floor-mounted controller for use on circuits not in excess of 600 V may be built without a covering for the bottom, provided that the surrounding enclosure is within 6 in. of the floor, and exposed live parts are not less than 6 in. above the lower edge.

Spacings

1. The distance between nonarcing, uninsulated live parts of opposite polarity and between nonarcing, uninsulated live parts and parts other than the enclosure which may be grounded when the device is installed, shall be not less than given in Table 3.5.

2. The spacings in snap switches, lamp holders, and similar wiring devices supplied as part of industrial control equipment need not comply with the requirements of these standards, provided that such devices are not employed in the motor circuits.

There are a number of conditions under which other spacings may be used, and anyone interested in these requirements is referred to NEMA Industrial Control and Systems Standards.

Special Service Conditions. A standard line of control equipment is designed to meet the requirements of the average installation, where the ambient temperature does not exceed 40°C, where the altitude does

TABLE 3.5

ELECTRICAL CLEARANCES

Rating in Volts	Distance in Inches			
	Through		Across Clean Dry Surfaces	
	Air	Oil	Air	Oil
0–50	0.125	0.125	0.125	0.125
51–150	0.125	0.125	0.250	0.250
151–300	0.250	0.250	0.375	0.375
301–600	0.375	0.375	0.500	0.500
601–1000	0.550	0.450	0.850	0.625
1001–1500	0.700	0.600	1.200	0.700
1501–2500	1.000	0.750	2.000	1.000
2501–5000	2.000	1.500	3.500	2.000

not exceed 6000 ft, and where only an ordinary amount of moisture and no acid or conducting or abrasive dust is present. If the conditions do not conform with these specifications, special construction or special protection should be provided as outlined below.

Unusual conditions of installation usually incorporate a number of factors which must be carefully considered before a definite recommendation is made. It is therefore advisable to give complete details of the conditions to be met when the equipment is ordered, in preference to designating any particular condition as listed below. With such information each point affecting the design can be considered, and the equipment best suited to the conditions can be supplied.

A. Exposure to Damaging Fumes. Use cast-iron or welded-steel enclosing cases with red-lead paint and acid-resisting paint. Joints between case and cover should be gasketed, the material of the gasket depending somewhat on the nature of the fumes, but rubber and felt are the most common. In certain instances, oil-immersed equipment may be necessary, but it cannot be used on direct current. For hydrogen sulphide gas (sewage plants), use blue-lead instead of red-lead. Cadmium plating should be avoided.

B. Operation in Damp Places. Large iron and steel parts should be red-lead-painted before regular painting. Bearings should be of bronze or brass. Enclosing cases should be of corrosion-resisting material (such as brass or aluminum) or should be red-lead-painted before final painting. Joints between case and cover should be gasketed with rubber in extreme instances. Small iron and steel parts should be well plated with zinc or cadmium.

C. Exposure to Excessive Dust. Enclosing cases should be of cast iron or welded steel with dust-tight joints, gaskets of felt or similar material being used.

D. Exposure to Gritty or Abrasive Dust. Same as Item B above.

F. Exposure to Excessive Oil Vapor. This is similar to Item A above. Special enclosing cases should be used, and in some instances oil-immersed equipment will be required. If gasketed joints are not considered sufficiently gas-tight for the particular installation, a very expensive case with wide flanges, carefully machined and fitted, will be required.

G. Exposure to Salt Air. Same as Item B above.

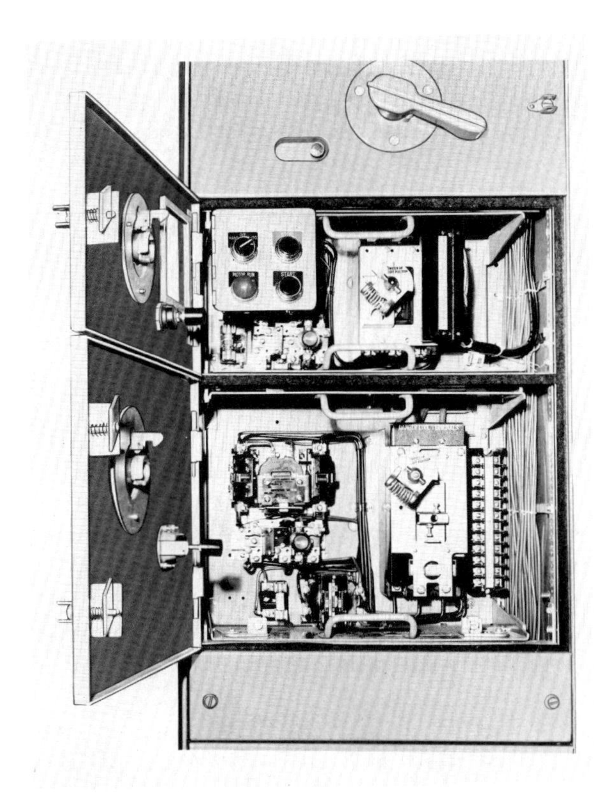

Fig. 3.7 Combination Motor Control Units in a Control Center

FIG. 3.8 Typical Motor Control Center

H. Subject to Vibration, Shock, and Tilting. Special equipment and construction will be required. Although modifications of standard parts may be used, precautions will be necessary in order to prevent breakage of certain parts and loosening of others. This applies to resistors as well as to other parts of controllers. The nature of the equipment required will depend upon the degree of the special conditions to be met.

I. Exposure to Explosive Dust or Gases. This is similar to Item A above. Special enclosing cases should be used, and in some instances oil-immersed equipment will be required. If gasket joints are not considered sufficiently gas-tight or dust-tight for the particular installation, a very expensive case with wide flanges, very carefully machined and fitted, will be required.

J. Exposure to Weather or Dripping Water. Same as Item B above.

K. Ambient Temperatures in Excess of 40°C but not Exceeding 80°C. High air temperature must be taken into consideration in the rating of all current-carrying parts. Unless special parts are used, this will result

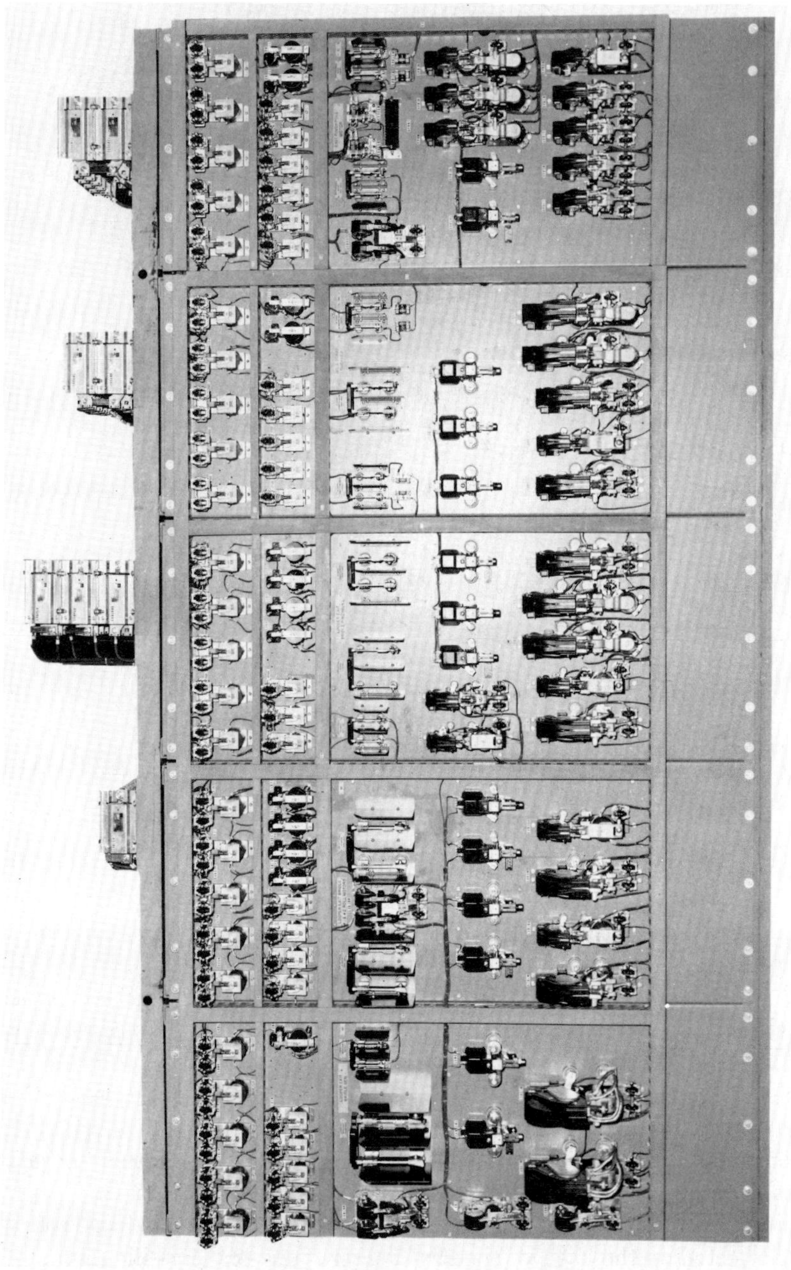

FIG. 3.9 Mill Control Floor Frame Assembly

FIG. 3.10 Mill Control Built into Control House

in the derating of contacts, blowout coils, and so on. Shunt coils must also be protected or rated.

L. High-Altitude Installations. General-purpose and special-service controllers designed in accordance with the usual standards are satisfactory for use at altitudes of 6000 ft or less.

For altitudes greater than 6000 ft, control equipment should be selected as follows.

Continuous-duty resistors should be derated to 75% of their normal wattage rating.

Intermittent and starting duty resistors should be applied on a duty cycle selected on the basis of the next higher "time-on" classification.

A magnetic contactor should be used to open the main line circuit when drum controllers are applied.

Autotransformers and control circuit transformers should be derated to 75% of their normal kilovolt-ampere rating.

Control Centers. A motor control center provides a means of assembling a number of motor starters or other electrical equipment into a single flexible structure. This centralizes all the electrical control apparatus for a given installation in a convenient housing for easy field installation and maintenance. Vertical sections are generally 20 in. wide by 90 in. high, and most designs can accommodate up to six motor-starter units per section. Horizontal bus feeders connect to the vertical bus in each section, and the control units stab onto the vertical bus to provide the line connection. The control units are readily removable and interchangeable. Any number of vertical sections can be furnished for one installation.

Figure 3.7 shows two motor control units installed in a control center. Each unit includes a combination circuit breaker and motor starter.

Fig. 3.11 Operator's Benchboard for Mill Control

The upper unit contains integral operating buttons and a pilot light. Figure 3.8 shows a complete control center installation for the air-conditioning system of a building.

Mill Construction. Many large control systems are built on open frames which include all contactors, resistors, relays, and other control functions. Where shipping considerations limit the size, these frames are broken down into "shipping splits" which can be wired together on location. This construction is often supplied to mills where the control installation is made in an environmentally controlled room. Figure 3.9 illustrates this type of construction.

All the control equipment for a mill installation can be supplied by the control manufacturer complete in a "control house." This is a complete enclosure including doors, interior lighting, and so on. A portion of the inside of such a control house is shown in Fig. 3.10.

The operation of a large control system generally calls for some type of operator's console from which one or more operators can monitor and direct all operations. Such a control desk is shown in Fig. 3.11.

References

National Electrical Code Hazardous Locations, Articles 500, 501, 502, 503.
Standards for Industrial Control UL 508, Underwriters' Laboratories, Inc.
NEMA Industrial Control and Systems Standards. National Electrical Manufacturers Assoc.

4

PILOT DEVICES AND ACCESSORIES

Pilot Device Contact Ratings. The electrical control industry, through NEMA, has suggested a design standard for rating the contacts on pilot and control circuit devices as shown in Table 4.1. The letters "A" and "B" in the designations denote what is generally considered to be "heavy-duty" and "standard-duty" respectively. The numerical suffix in the designation gives the maximum voltage design value. For example, a contact rating designation A600 indicates a permissible contact-carrying capacity of 10 amp, clearance and creepage distances for 600 V, and make and break ratings as shown in this figure. These ratings are for ac and are adequate for practically all industrial control applications, such as direct operation of contactors, relays, and control solenoids.

TABLE 4.1

CONTACT RATINGS FOR PILOT DEVICES

Contact Rating Designation			Maximum Contact Ratings per Pole					
			Maximum A-c Voltage, 50 or 60 Cps	Amperes		Continuous Carrying Current	Voltamperes	
				Make	Break		Make	Break
A600	A300	A150	120	60	6	10	7200	720
A600	A300	—	240	30	3	10	7200	720
A600	—	—	480	15	1.5	10	7200	720
A600	—	—	600	12	1.2	10	7200	720
B600	B300	B150	120	30	3	5	3600	360
B600	B300	—	240	15	1.5	5	3600	360
B600	—	—	480	7.5	0.75	5	3600	360
B600	—	—	600	6	0.6	5	3600	360
C600	C300	C150	120	15.0	1.50	2.5	1800	180
C600	C300	—	240	7.5	0.75	2.5	1800	180
C600	—	—	480	3.75	0.375	2.5	1800	180
C600	—	—	600	3.00	0.30	2.5	1800	180

Ratings in the "C" category are intended to make a lower range available where it can be useful to the design engineer.

Direct-current ratings are not as well standardized as ac ratings although NEMA and the Underwriters' Laboratories have established certain test values when a manufacturer wishes to design a device for dc control. The preponderant usage of pilot devices is in ac control circuitry, and it is recommended that the manufacturer's data be consulted when dc ratings are required.

Conventional contacts used in pilot circuit devices are usually of a bridging type and are made of silver in order to minimize the contact resistance. A unique class of contact-making elements employ magnetic reeds which are sealed in a tube and which respond to a magnetic field around this tube. For several years the electrical ratings of these reed switch contacts was limited to very low-power circuity such as that found in communication or computer applications. Reed switches have only recently become available with electrical ratings in the conventional industrial control area, and thus they can now be employed in control relays or in other pilot circuit devices. Such a device, known as the Cutler-Hammer "Powereed," is shown in Fig. 4.1.

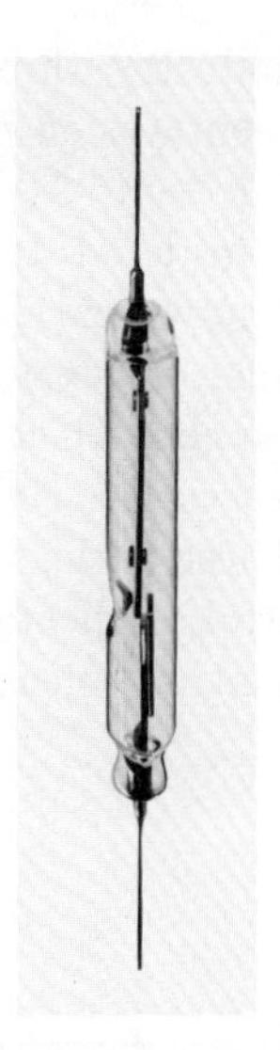

FIG. 4.1 Cutler-Hammer Industrial Control "Powereed" Switch

Pushbuttons. Undoubtedly the most frequently used pilot device is a pushbutton station. Such a station may include one or more buttons of the momentary-contact type only, or the buttons may be of the maintained type, or both types may be included.

Pushbuttons for motor control are made in two general classes. Standard-duty buttons are suitable for most general-duty applications where the current to be handled is not excessive, and where the service is not extremely severe. They are available in one-, two-, and three-button styles. Figure 4.2 shows various two- and three-button standard-duty pushbutton stations. Heavy-duty pushbuttons are, as the name indicates, of heavier construction, and are able to handle higher currents and withstand severe service. They are available in almost any desired combination of contact elements and operating mechanisms. In the usual construction the operator and the contact block

Fig. 4.2 Standard-Duty Pushbutton Stations

are separate units, thus giving great flexibility in assembling the desired contact arrangement with the operating head. Figure 4.3 shows various operators and pilot lights of the one-hole mounted oil-tight type. Contact blocks mount behind the pushbutton or selector switch operators to give the desired number of normally open or normally closed electrical circuits. Figure 4.4 shows a six element operator's station consisting of one-hole mounted heavy-duty elements placed in a sheet-metal enclosure.

For service in wet locations, as on shipboard, watertight pushbuttons are required. These have a heavy case, fitted with gaskets to make it completely watertight. A flexible rubber or leather diaphragm is mounted on the inside of the cover, and the buttons are operated through

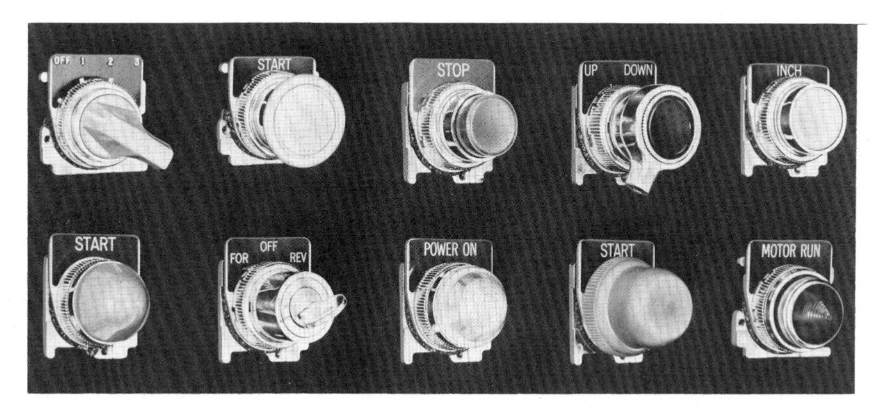

Fig. 4.3 Heavy-Duty Pushbutton Operators and Lights

Fig. 4.4 Heavy-Duty Operator's Station

this diaphragm. Another possible construction is a shaft extending out of the case, through a watertight stuffing box. Cams for operating the buttons are mounted on the shaft, inside the case, and a lever or button is fastened to the shaft on the outside.

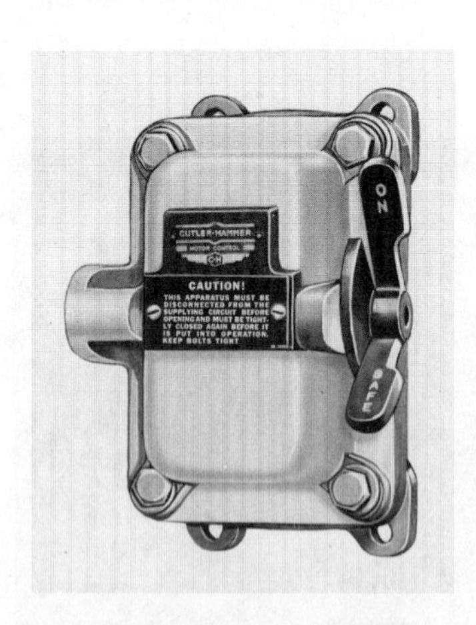

Fig. 4.5 Pushbutton Station for Hazardous Locations

Explosion-resisting buttons are required in hazardous locations, as where gasoline or other explosive fumes are present. To meet this requirement heavy cast cases are made with wide flanges on both case and cover. Figure 4.5 shows such a pushbutton station designed for use in hazardous locations.

In connection with hazardous locations, another system is available which allows the use of General-Purpose NEMA I pushbutton stations on other pilot devices with safety in these areas. This scheme places such low-power levels on the pilot devices that no contact arcing or resulting hazard is possible. These relays, known as intrinsically safe devices, generally operate through transistor amplifiers or reed switches requiring very little power. The output of this low-power device operates a relay which in turn handles the higher control circuit power. The intrinsically safe device, with its output relay, must be either suitably enclosed for the hazardous area or located in a safe area. Several lower-priced general-purpose pilot devices can often be used in hazardous areas through the use of one intrinsically safe relay, with a saving as a result.

Transfer Switches. When it is desired to operate from either of two masters, or from either of two pilot devices of any type, a transfer switch may be provided. By this means, either device may be connected into circuit. The transfer switch is simply a double-throw pilot switch of the required number of poles. If only a few circuits must be transferred, and if an open switch is satisfactory, a small double-throw knife switch may be used. Generally an enclosed device is preferred, and the small drum-type switches, used for starting multispeed motors, are widely used as transfer switches. They may be separately mounted, or mounted on the controller panel. Figure 4.6 shows a transfer switch for flush or cavity mounting.

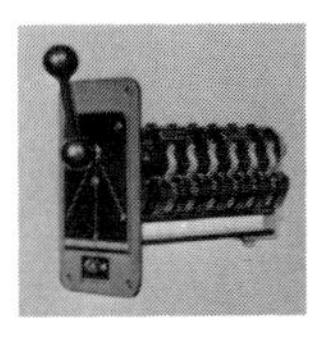

Fig. 4.6 Transfer Switch for Flush or Cavity Mounting

Automatic Starting Devices. It is often desirable that a motor start and stop automatically, without the attention of an operator.

Automatic starting devices accomplish this result, the type of device depending upon the nature of the requirements. Liquid-level switches are used to maintain the level of water, or other liquid, in a tank or basin. Pressure-operated switches maintain a desired pressure on a closed system, as, for instance, automatic sprinklers, or compressed-air tanks in a filling station. There are also switches responsive to temperature and to humidity. They may control motors driving fans or stokers, or furnace draft doors. Time-clock switches serve to start and stop motors at desired intervals. Speed governors can be equipped with electric contacts to open or close at a required speed. Limit switches of many types are employed to open or close circuits after a desired function has been performed. As an example, a piece of material moving on a conveyor may be arranged to strike a limit switch at a certain point in its travel, and thus start a machine which will push the material off the conveyor belt.

Liquid-Level Switches. Various means are available to detect the level of a liquid either in a pressurized tank or in an open tank or sump. A float is the most common method of operation. This can be made to operate electrical contacts through a rod, counterbalanced chain, or lever system. Figure 4.7 shows a switch for flange mounting onto a tank.

Liquid-level switches are often used to connect small motors directly to the supply lines without magnetic contactors. For this purpose they may be two-, three-, or four-pole. Pilot liquid-level switches are single- or double-pole.

For such applications as controlling the level of water in a sewage settling basin, where the inflow may vary widely, a number of pumps are necessary, and some of them may be required to run at several different speeds. The number of pumps in service at any one time will be determined by the level of the water, so that, as the level rises, one after another will be brought into service, and as the level falls, the pumps will be cut off one at a time. Such an application requires

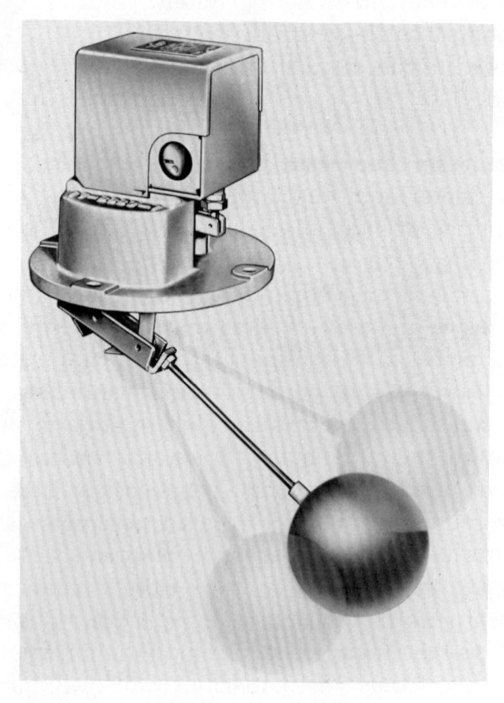

Fig. 4.7 Float Switch

a multipoint switch, which will generally have to be designed to suit the particular application in question.

Another method of liquid-level detection is by means of probes so placed that the liquid comes into contact with the probe at a predetermined level. Liquid in contact with the probe causes a change of state, such as a resistance change between two electrical connections. A sensitive relay detects this change of state in the probe and provides an output which can be used to operate a pump motor starter.

Pressure Regulators. Pressure regulators are used on closed systems to maintain a desired pressure. The regulator has a rubber or metal diaphragm against which the pressure of the fluid is exerted. A heavy spring opposes the liquid pressure, and the device is set for the desired pressure by varying the spring tension. A lever is operated from the diaphragm and so arranged that it multiplies the amount of movement of the diaphragm sufficiently to operate an electric contact. The contact is opened and closed by means of a toggle mechanism, in order to obtain a quick make and break. This arrangement prevents chattering of the contact with each stroke of the pump or compressor.

Pressure regulators are often equipped with an unloading device which relieves the back pressure against the compressor during the starting period and so lessens the torque required of the motor. The unloader is a valve, which is held closed by the air pressure while the regulator switch is closed, and is opened when the switch opens. It is connected to the compressor discharge line by means of a small pipe. A check valve in the line between the unloader and the tank must be used in connection with the unloader valve, to prevent discharging the tank when the unloader valve is open. Figure 4.8 illustrates a pressure regulator of this type.

Some designs of diaphragm pressure regulators are double-pole devices which are suitable for pilot circuits but are also able to handle small motors directly. NEMA standard horsepower ratings when handling motor circuits are given in Table 4.2.

When installing a pressure regulator, care should be taken to connect it in a part of the system which is not subject to fluctuations in pressure; this will increase the accuracy and permit close setting. It is preferable to connect the regulator to the storage tank and not to the discharge pipe from the pump. If the device must be connected to the discharge pipe, an air chamber should be interposed between the pipe and the regulator, and the feed to the air chamber should be throttled by means of a needle valve or stopcock. This arrangement will smooth out the pulsations from the discharge pipe.

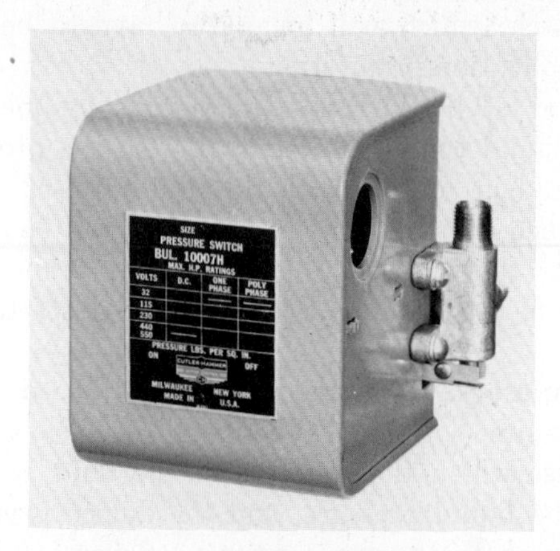

FIG. 4.8 Pressure Regulator

Where visible indication of the pressure is required, a regulator of the gauge type is used. This type of regulator is similar in construction to the ordinary pressure indication gauge except that, in addition to moving an indicating needle, the pressure tube is used to move an arm which will make an electric contact at desired high and low pressures. The device usually includes a relay, because the current capacity of the gauge contacts is small, and the movement is not quick make and break. As the pressure tube expands and contracts, the contact arm moves back and forth, making contact with the stationary contacts at either end of its travel. When the arm makes contact with the stationary contact at the low setting, the relay is energized and closes. This ener-

TABLE 4.2

HORSEPOWER RATINGS FOR PRESSURE-OPERATED SWITCHES

Maximum Pressure (psi)	Single-Phase AC			Two- or Three-Phase AC			DC		
	115 V	230 V	440–550 V	110 V	220 V	440–550 V	32 V	115 V	230 V
60	1	1	—	1	1	—	—	—	—
80	1½	2	3	2	3	3	¼	½	½
200	2	3	5	3	5	5	½	1	1
250	2	3	5	3	5	5	½	1	1
250	2	3	5	3	5	5	½	1	1

gizes the coil of the motor-starting switch, and the motor is started. The relay remains closed until the swinging arm makes contact at the high setting, when the relay coil is short-circuited and the relay drops open, stopping the motor. The high and low settings are adjustable by moving the stationary contacts. With the arrangement described, the equivalent of quick make and break is secured, and the light contacts in the gauge never have to break the circuit.

The pressure tubes in these regulators are made of brass or of steel, depending upon the nature of the fluid with which they are to be used. A brass tube should be used for water, air, steam, sulphur dioxide, acetylene, carbon dioxide, carbon monoxide, helium, hydrogen, and illuminating gas; a steel tube for ammonia, alcohol, benzol, chlorine, creosote, cyanide, gasoline, nitrogen, oxygen, ethylene, and ethylene oxide.

Thermostats. Thermostats are well-known control devices which respond to a change in temperature. Common types of power elements are bimetallic strips or enclosed vapor systems.

Some thermostats use simply a single-pole make-and-break contact, so that the contact is closed to make the circuit and opened to break it. Since the thermostat contacts are of necessity quite light, it is essential that there should be no arcing on them. A positive method of preventing arcing on the contacts is to arrange them so that they do not break the circuit. This may be done by means of the relay scheme described in connection with gauge pressure regulators, which of course requires that the thermostat have three point contacts.

Speed Governors. Centrifugal speed governors are employed to protect machinery against overspeed or underspeed and also to control some functions in the operation of a magnetic controller which must be performed at a given speed of the motor or machine. For instance, a governor can be used with a controller for a two-speed ac motor. When slow speed is desired, the high-speed winding of the motor is disconnected from the line and the slow-speed connected into circuit. If the slow-speed winding were connected immediately upon disconnecting the high-speed, a severe jar would occur and the motor would take a high inrush of current. To avoid this, the connection of the slow-speed winding can be controlled by a governor set to close its contacts at the slow speed of the motor. Then, when the high-speed winding is opened, the motor is allowed to drift down in speed until its slow running speed is reached, and at that exact speed the governor will operate to connect in the slow-speed winding. There will then be no inrush of current and no jar to the machine.

Plugging Switches. If it is desired to obtain a certain control function when a motor is running and to discontinue that function when the motor stops, a plugging switch can be used. The switch can also be used to obtain a function when a motor is running in one direction and to discontinue the function if the motor is reversed.

The switch illustrated in Fig. 4.9 has a double control arm held in a center position by a spring. In this position, both contacts are opened. A belt connected to the arm passes over a pulley driven from the motor. If the motor starts to run in a clockwise direction, the belt pulls the contact arm over until the right-hand contacts are closed. The arm is prevented from moving further, and the belt slips on the pulley, holding the contacts closed by means of friction. When the motor is stopped, the centering spring has sufficient strength to move the arm back to the center position. Counterclockwise rotation of the motor will close the left-hand contacts.

With a reversing controller, particularly in an ac installation, it is often desirable to stop the motor quickly by connecting it to the line in the reverse direction until it stops. The forward and reverse magnetic contactors are mechanically and electrically interlocked, so that only one can close at a time. When the forward contactor is closed, the motor starts to turn, and the friction switch closes its contacts to energize

Fig. 4.9 Mechanism of a Belt-Type Friction Switch

the reverse contactor. This contactor, however, cannot close because of the interlocking; but as soon as the stop button is pressed the forward contactor opens, and then the reverse contactor is free to close. In this way the motor is connected to the line in a reverse direction until it stops, when the friction switch moves to the center position and opens the circuit to the reverse contactor.

Another type of plugging switch has an Alnico rotor, permanently magnetized in four poles, and mounted on a shaft for connection to the motor. The frame of the switch is steel. Between the rotor and the frame is an aluminum cup which is free to turn through a limited arc and to which a contact arm is mounted. The driven Alnico rotor produces a rotating magnetic field that induces eddy currents in the walls of the cup, causing it to turn and bring the contacts into engagement. An adjustable spring determines the force required to turn the cup and so determines the rotor speed at which the switch contacts will close. The switch is provided with a safety latch to prevent closing of the contacts when the switch is turned by hand.

Limit Switches. A limit switch is a device mechanically operated from a motor or machine in its operation, and arranged to perform some electrical function. The term "limit switch" indicates its primary function, namely, to provide a limit to the travel of a machine and to stop it at that point. This is usually done by opening the circuit to magnetic contactor coils, and so allowing them to open and disconnect the motor from the line. If the installation includes a magnetic brake, the circuits will be arranged so that the brake is set in the limit of travel. Dynamic braking may be applied by the limit switch.

In addition to the function of stopping, limit switches may serve many other purposes. They are often used to provide slowdown at a point ahead of the stopping point, so that the stop may be more accurate. They may also act to vary the speed of a machine in different parts of its cycle. They frequently provide an interlocking means between two or more machines. In this capacity they may be made to start one machine when another is at a certain point in its travel, or to prevent operation of one machine unless another machine has reached a desired point.

The electric contacts of a limit switch may be made either normally closed or normally open. To insure safety, normally closed contacts are generally used for stopping and also for any other function which must be performed without fail. The reason for this is obvious, for with such an arrangement, any electric failure like a broken wire or a burned-out

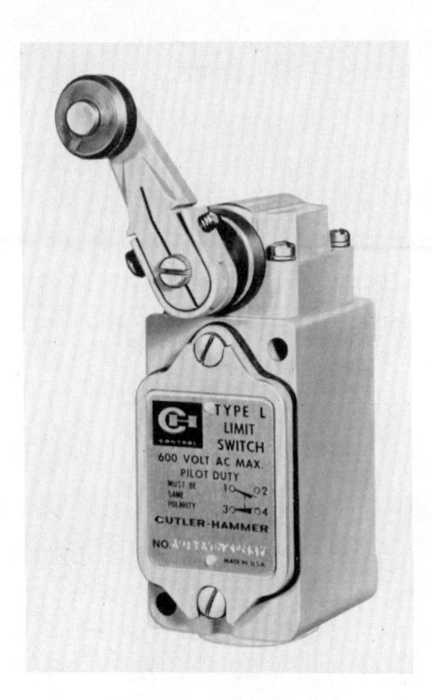

FIG. 4.10 Machine Tool Limit Switch

coil or a failure of the power supply will stop the equipment. With a normally open limit switch contact, any of these failures would prevent a stop. The normally closed limit switch may fail because of a ground around it. To guard against this, stopping limit switches are often made double-pole, one pole being connected in each side of the control circuit. Installations where failure to stop is very hazardous, as, for instance, elevators, make use of two sets of stopping limit switches, one set ahead of the other. In normal operation the first switch will always stop the motor, but in case of failure the car will run through the first limit and be stopped by the second, or overtravel, limit switch. Normally open contacts are generally satisfactory for obtaining slowdown, as failure of this function is not so vital. Also, it is usually necessary to close magnetic contactors to obtain slowdown, so that a normally open limit contact does not add any hazard, as a circuit failure would prevent closure of the contactor in any event. Where slowdown is vital to safety, the controller should be arranged to use normally closed slowdown limit switches.

Figure 4.10 shows a typical limit switch adaptable to many machine applications and commonly referred to as a machine tool limit switch. It is protected against environments where splashing oil or dirt may be present and is mechanically and electrically very rugged. Many different types of operating heads can be applied to the basic switch mechanism, such as roller arms, push rods, or wobble sticks. A roller arm operator is shown in Fig. 4.10.

Traveling-Cam Switches. When a machine travels forward and reverse in a fixed cycle, and within fixed limits, it is possible to use a limit switch having a rotating shaft geared directly to the machine. Elevators and skip hoists are typical examples. These are driven from cables on a winding drum, and a limit switch can be geared to the drum. Geared limit switches cannot be used on such applications as crane trolleys or bridges, because any slip of the wheels would cause the switch to trip at the wrong point of travel. Figure 4.11 illustrates one form of geared switch, known as a traveling-cam switch. The operating shaft is threaded and has a nut, or crosshead, mounted on it. This crosshead moves along the shaft as the shaft turns, and acts as a cam to operate electric contacts. Each electric contact is a double-pole quick-acting switch, which may be either normally open or normally closed or may have one pole open and the other pole closed. Each switch with its operating mechanism is mounted on rods parallel to the main shaft. The tripping point of the switch is adjusted by moving it along on the mounting rods, until the crosshead operates it at the desired point. The device shown has a total of four sets of electric contacts, two on each side of the shaft. Similar devices are made with eight elements for more complicated control requirements.

It is evident that the accuracy of a switch of this type is dependent

FIG. 4.11 Traveling-Cam Limit Switch; Cover Removed

upon the length of the shaft. For a skip-hoist installation the switch is geared directly to the winding drum of the machine. The total travel of the crosshead must, therefore, occur in the same number of turns as is required for the full travel of the hoist. If the shaft is short, a small movement of the crosshead will represent a considerable movement of the hoist, and an accurate setting will be difficult to obtain. A longer shaft will allow the crosshead to move a greater distance for the same movement of the hoist, and so more accurate settings can be secured. For very long hoists, and where great accuracy is required, a compound limit switch has been used. This switch has a mechanism that greatly multiplies the movement of the crosshead when the hoist is approaching the limit of travel. Its use is too infrequent to warrant a complete description.

Rotating-Cam Switches. The rotating-cam limit switch, illustrated in Fig. 4.12, is a form of geared limit switch particularly applicable to eccentric drives making one revolution, or a part of one revolution, for a complete cycle. An example of such a drive is a metal-forming or stamping press. The shaft of the limit switch is geared to the driving shaft of the machine.

The switch consists of a number of electric contacts operated by cams on the switch shaft. The length and location of the cam face determine the portion of the cycle during which the contact is open or closed. Adjustment is made by rotating the cam around the shaft and by changing the length of its face. An easy means of making adjustments around

FIG. 4.12 Open View of Rotating-Cam Limit Switch

the shaft is necessary, as this determines the tripping point of the electric contacts. It is better practice to use two cams and two contacts to get a function in each direction rather than to use one cam and depend upon its length being accurate. In designing switches of this type, the principal points to keep in mind are ease of adjustment and the use of fairly large cams for accuracy of setting and of hardened cams and rollers to prevent excessive wear.

Problems

1. It is desired to ring an alarm bell and light a signal light when the pressure in a tank reaches a critical value. The ringing of the bell may be stopped by pressing a pushbutton, but the light is to remain on until the pressure goes down. Draw a circuit for this arrangement, using a single-pole pressure regulator, two relays, and a pushbutton.

2. The lights in a building are turned on and off by a time clock, which has an electric contact. The control circuit may be transferred to momentary-contact pushbuttons, which will then control the lights. If power fails while the control is with the pushbuttons, control is to be automatically returned to the time clock. Draw a circuit for this arrangement, using such relays as may be necessary.

3. The float of a float switch is a hollow cylinder of copper having a diameter of 12 in. It is rigidly fastened to a vertical rod, which in turn is rigidly fastened to the arm of the float switch. If the float and rod weigh 5 lb together, and it takes 3 lb to operate the switch, how far will the water rise before the switch operates, if it is assumed that the water level at the start is at the bottom of the float?

4. If the water level starts down as soon as the switch operates, how far will it recede before the switch resets?

5. A rotating-cam limit switch having six circuits is equipped with a double sheave wheel on the shaft. A rod is hung from a cable on one sheave wheel and a counterweight from a cable on the other sheave wheel. A float is arranged to slide up and down on the rod as the water level in a tank changes, and two adjustable stops are provided on the float rod. The device is to be used to start and stop three motors, as follows:

At water level	10 ft	start motor 1
	11 ft	start motor 2
	12 ft	start motor 3
	4 ft	stop motor 3
	3 ft	stop motor 2
	2 ft	stop motor 1

Make a sketch showing the arrangement of the device, and the location of the stops on the rod.

6. Draw the electric circuit for problem 5, using circles to represent the motors.

7. Draw a circuit to show how a solenoid may be energized and deenergized from any one of three locations, using momentary-contact pushbuttons, and having a single-pole snap switch at each location which, when it is open, will prevent operation of the solenoid from any of the three locations.

8. A solenoid is operated from a three-point thermostat, using one double-pole normally open relay, and one single-pole normally closed relay. Draw a circuit which will close the solenoid when one contact of the thermostat closes, maintain the circuit as the thermostat moves away from its contact, and open the circuit when the other contact of the thermostat closes.

9. Draw a circuit using a master controller, a relay, a solenoid, and a limit switch, and arranged so that, if the limit switch opens when the solenoid is energized, the master controller must be returned to the OFF position before the solenoid can again be energized.

10. A switch in the track of a model railroad is operated by a device consisting of two solenoids and a single plunger. When one solenoid is energized, the plunger moves to the right, moving the track switch for main line operation. When the other solenoid is energized, the plunger moves to the left, moving the track switch for branch line operation. In each direction of travel the plunger closes an auxiliary contact, like a normally open limit switch. Two momentary-contact pushbuttons are used, and two signal lights to show the track switch position. It is also necessary to connect a circuit across the switch points in each position of the track switch, to insure that there is a circuit through the track to run the train.

Draw a circuit diagram for the arrangement, using such relays as may be necessary.

5

DIRECT-CURRENT CONTACTORS AND RELAYS

The standards of the National Electrical Manufacturers' Association define magnetic contactors and relays as follows.

A contactor is a device for repeatedly establishing and interrupting an electric power circuit.

A magnetic contactor is a contactor actuated by electromagnetic means.

A relay is a device which is operative by a variation in the conditions of one electric circuit to effect the operation of other devices in the same or another electric circuit.

The electromagnetically operated contactor is a useful mechanism devised for closing and opening electric circuits. Since controlling an electric motor resolves itself largely into opening and closing electric circuits, and since the magnetically operated contactor is extremely versatile and flexible in its forms and applications, each manufacturer of control apparatus has available standard contactors and relays in a wide variety of types and sizes.

Advantages of Using Contactors. A number of advantages are gained by the use of magnetic contactors instead of manually operated control equipment. One of the most important is the saving of time and effort. Where large currents or high voltages have to be handled it is difficult to build suitable manual apparatus; furthermore, such apparatus is large and hard to operate. On the other hand, it is a relatively simple matter to build a magnetic contactor which will handle large currents or high voltages, and the manual apparatus controls only the coil of the contactor. Also, where there are many functions to perform, or where one operation must be repeated many times an hour, a distinct saving in effort will result if contactors are used. Controllers may be so arranged that the operator has simply to push a button and the contactors will automatically initiate the proper sequence of events. The operator is then saved the trouble of constantly concentrating on doing the right thing at the right time, so far as the motor is concerned, and can put his entire attention on his work.

Magnetic contactors considerably increase the safety of an installation. High voltage may be handled by the contactor and kept entirely away from the operator. The operator also will not be in the proximity of high-power arcs, which always involve the danger of shocks, burns, or perhaps injury to the eyes.

A third advantage of contactors is the saving of space, which is often valuable in the vicinity of a motor-driven machine. With contactors the power control equipment may be mounted at a remote point, and the space required near the machine will be only that necessary for the pushbutton. This is a particular advantage where the power control equipment is large or where the operation of a number of equipments is under the control of one man.

With contactors it is possible to control a motor from a number of different points. A good example is the control equipment for a newspaper printing press, which must be started and stopped from many different points around the press. It would be difficult to do this by means of manual control, because all the control stations would have to be mechanically interlocked against each other to insure proper operation and foolproof control, and such an interlocking would involve a large amount of mechanical equipment. The problem is relatively simple with contactors, since it is possible to control one contactor from as many different pushbuttons as are desired, with only the necessity of running a few light control wires between the stations.

The control of such equipment as pumps and compressors, which are automatically started and stopped from pilot float switches or pilot pressure regulators, is greatly simplified by the use of contactors. It is evident that pilot devices of this nature, and also many other types, such as thermostats and sensitive gauges, are limited in power and size, and it would be difficult to design them to handle heavy motor current directly.

Automatic acceleration is readily accomplished by means of contactors, and thus acceleration is taken entirely away from the operator's control. The motor is started in successive steps automatically, and in the proper time for safe acceleration. The acceleration may either be directly set for a definite time, or be under the control of the current which is drawn by the motor, and which is a measure of the rate of acceleration. Any danger of damage to the motor or machinery because of improper starting is thereby avoided.

Automatic control by contactors results in a saving in wiring expense if the point of operation is at any distance from the motor and controller.

With manual control it is necessary to run a heavy power circuit wire to the point of operation. With contactors it is necessary to run only the small control wires.

To meet the requirements of the functions outlined above, several types of magnetic contactors and relays have been developed. Some of the more common forms for dc work will now be described.

Shunt Contactors. Shunt contactors are so called because they are operated by a shunt coil, which is supplied with energy from a constant-potential circuit. They are usually made single-pole or double-pole for dc work. The contactor is said to be normally open when it is arranged to open its contacts if the coil is deenergized and to close them if the coil is energized. A normally closed contactor opens its contacts when the coil is energized. Contactors may also be arranged to have one set of contacts normally open and one set normally closed, or in fact with almost any desired arrangement of normally open and normally closed contacts.

A typical shunt contactor of the normally open type is shown in Fig. 5.1. The magnet frame is a steel casting, and the coil is mounted on

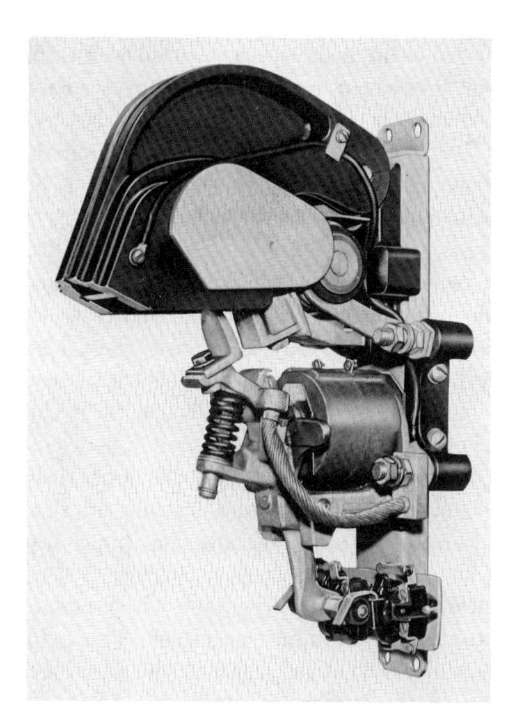

Fig. 5.1 Mill-Type DC Contactor with Electrical Interlocks. Arc Shield Raised

a core of steel. When the coil is deenergized, the contactor is opened by gravity. Energizing the coil attracts the movable armature to the stationary core, and the contactor closes. The current-carrying tips close first, compressing the spring under them and building up a heavy contact pressure as the armature moves in to seal against the core. The current-carrying capacity of the contactor depends on the area of contact surface and also on the pressure of the contacts. The stationary contact is mounted on a brass or copper post through which the current flows to a terminal stud where a wire or bus bar is connected. The movable contact has a heavy flexible cable, or pigtail, which is connected to an independent stud, so that the current does not flow through the steel frame of the contactor. The stationary contact is provided with a metal arcing horn which serves to direct the arc upward away from the contacts. The movable contact has a metal guard to prevent an arc from striking the contact spring. The shaft and moving parts are of hardened steel to prevent wear.

Magnetic Blowout. At the top of the contactor a heavy bar-wound coil may be seen. This coil, wound on a core of steel, is mounted between two pole pieces, also of steel. The inside of the pole pieces is lined with a refractory insulating material. This structure, called a magnetic blowout, is used to extinguish the arc rapidly. In the illustration the blowout has been raised to make the contacts visible, but in actual operation the blowout is lowered down over the contacts, which are thus enclosed in an insulated box. The arcing tip on the stationary contacts guides the arc into the box. The blowout coil is connected in series with the contactor itself, so that motor current is flowing through the coil as long as the contactor is closed, or as long as there is an arc between the contacts. The current sets up a magnetic field through the core and pole pieces of the blowout structure and across the arcing tips of the contactor. When an arc is formed, the arc sets up a magnetic field around itself. The two magnetic fields repel each other, and the arc is forced upward and away from the contacts. It thus becomes longer and longer until it breaks and is extinguished. The extinguishing action is extremely rapid, and it not only speeds up the operation of the contactor but also greatly reduces the wear and burning of the contacts. Proper blowout design, therefore, is an important factor in contact life.

Contacts. Contacts are ordinarily made of copper and may be either left plain or plated with cadmium or silver. The oxides of copper are poor conductors, and an oxidized contact tip is subject to overheating. Of course, plating at the surface where arcing occurs does not last very long, and it is necessary to arrange the contacts so that they will be self-cleaning. The plating is of value on the bottom and back of the contact tip, where it joins the post on which it is mounted. The plating here assures a good electric contact at the joint. Self-cleaning of the contacts is accomplished by arranging them so that they strike first at the tip and then, as the contactor closes, they come together with a rolling motion, finally resting together close to the bottom of the tip. Small pin arcs which may occur when the contacts close, and which pit and oxidize the tips, are thus confined to the top of the contacts, while the current is actually carried at the bottom where the arcing has not occurred.

If a contactor is to be closed for a long period of time with infrequent operation, the contacts will not get a chance to clean properly and may eventually give trouble as a result of heating. In such applications silver contact tips are used, or copper tips with a silver plate welded into the face of the tip.

When a contactor closes it does so with a considerable force, and

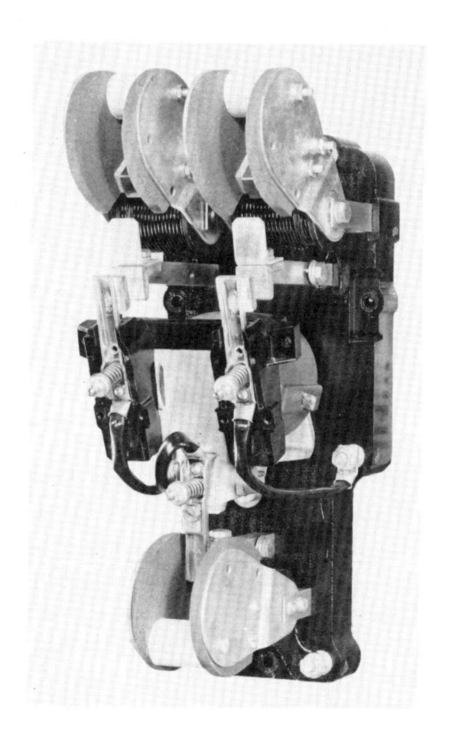

FIG. 5.2 Fifty-Ampere DC Contactor with Two Normally Open, One Normally Closed Contacts. Arc Shields Raised

there is always a certain amount of rebound of the contacts. Small arcs are formed during this rebound, and, if the contactor closed again at exactly the same point, there would be a likelihood of its welding closed. However, since the armature of the contactor has moved in farther toward the core during the time of the rebound, when the contacts come together again they touch at a different point and so do not weld. High spring pressures are used on the contacts to insure that they open. In spite of all precautions, copper contacts occasionally weld. A common cause is low line voltage. If the voltage is just high enough to cause the contactor to close part way, the contacts may come together, but without sufficient pressure on them to seal completely. When this happens the contacts touch only lightly, and an arc is formed, which, owing to the light contact pressure, is likely to weld the contacts together. The same result is sometimes obtained by inching or jogging a controller, as the coil circuit may be interrupted at a time when the contacts are just about to touch. At the instant of touching they may be practically stationary, with very little force either to close or to open them. Welding is almost certain to occur under these conditions.

Another cause of welding contacts is improper relationship among the pull of the closing coil, the initial pressure when the contacts first touch, and the pull required to seal the armature magnet. This is most likely to occur with large contactors, the coils of which have a relatively high inductance and so are slow in building up to full power. If the contactor is relatively easy to start, the armature may move far enough to close the contacts by the time the pull of the closing coil is only partly built up. Then, when the contacts touch there may not be enough pull to make the armature continue its travel, and it will hesitate until the coil pull builds up enough to pull it in. During this hesitating period the contact pressure is relatively light and the contacts may weld. Care must be taken in the design of the contactor and the coil to insure that, once started, the armature will pull all the way to the fully sealed position.

Electrical Interlocks. It is often necessary to make the operation of one contactor dependent upon the operation or nonoperation of another contactor, or to interlock one against the other. For this purpose auxiliary contacts are mounted on the contactor (Fig. 5.1) and arranged to open or close a circuit when the contactor is operated. An auxiliary interlock is said to be normally open if the auxiliary circuit is open when the contactor coil is deenergized, and normally closed if the circuit is closed when the contactor is deenergized.

By using different combinations of operators and contacts, different results may be accomplished. Normally closed contacts may be arranged to open the instant that the contactor starts to close, or they may be made so that they will not open until the contactor has nearly sealed. Similarly, normally open interlocks may be arranged to close immediately upon energization of the contactor, or not to close until the contactor is practically sealed.

Electrical interlocks are designed in several configurations, but have ratings comparable with heavy-duty relays. Frequently the electrical interlocks on contactors utilize the same contact blocks as used on mill relays (Fig. 5.6). Generally the movable contacts are operated by a cam. A bridging contact bar completes the circuit between two stationary contacts. The contacts proper are of silver.

Dynamic Braking Contactors. Normally closed contactors are used in braking circuits to insure fail-safe operation in case of power failure. As shown in Fig. 5.3, a spring with sufficient strength to close and seal the contactor pushes out on the bottom of the armature to close the contactor. Application of power to the coil pulls the armature in, compressing the spring and opening the contacts. Contactors of this type,

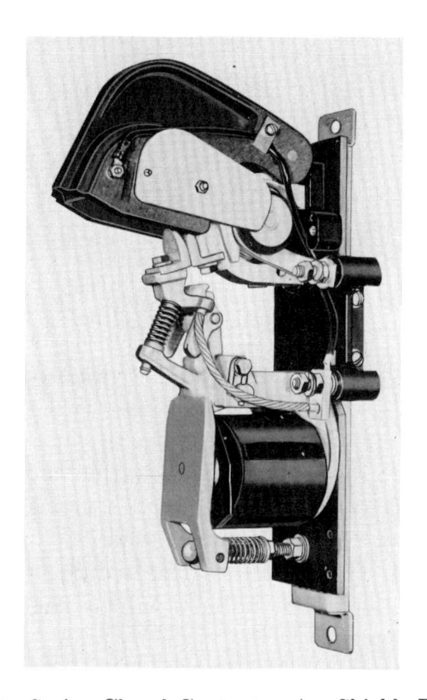

FIG. 5.3 Spring-Closed Contactor. Arc Shields Raised

in series with power resistors, are frequently connected in parallel with the motor armature for dynamic braking. Removal of control power causes the main contactor to open and this contactor to close to allow the motor to regenerate into the resistor.

Mechanical Interlocks. It is often desirable to insure that two or more contactors close together, as for instance two single-pole contactors which are serving in the place of a double-pole contactor. To accomplish this positively, the contactors are mechanically tied together by means of an insulting strip between their armatures. The strip is rigidly fastened to one contactor, and a little play is allowed at the other in order to compensate for inequalities in wear of the two contactors.

When it is necessary to tie together a series of more than two contactors, the amount of play allowed must be carefully considered, so that the total play between the first contactor and the last contactor will not be enough to prevent proper interlocking.

It is also often necessary to insure that a certain contactor does not close when another is closed. For this purpose a mechanical interlock is used. The mechanical interlock may consist of a piece of metal hinged at the center, and arranged so that with both contactors open the center piece is in a neutral position. When one contactor closes it strikes one end of the interlock, moving it in toward the panel and, of course, causing the other end of the interlock to move out into such a position that it will interfere with the closing of the second contactor. Such a device is called a walking-beam type of interlock. There are many other forms of mechanical interlocks, all of them arranged to operate so that, when one contactor closes, it will move the interlock parts into such a position that they interfere mechanically with the closing of the second contactor.

Coils. The coil of a contactor must provide enough ampere turns to operate the contactor properly, not only on normal line voltage but also under low- and high-voltage conditions. NEMA Standards say that dc contactors shall withstand 110% of their rated voltage continuously without injury to the operating coils, and shall close successfully at 80% of their rated voltage. In testing, consideration must also be given to the fact that the resistance of a coil will increase as the coil heats, so that the test voltage must be lower than 80% of rated voltage if the coil is tested cold. If the coil is expected to increase in resistance by 20% when hot, the cold test voltage will be 0.80/1.20, or 0.67 times the normal rated voltage.

Shunt-wound contactor coils have a high self-induction, which is sometimes troublesome, especially on higher voltages, but which may sometimes be put to valuable service. The coils are wound with many turns of fine wire, and they are used on a nearly closed iron magnetic circuit. Consequently, when the circuit to the coil is suddenly broken, high voltages, sometimes reaching 15 times normal values, are induced. These induced voltages tend to puncture the insulation of the coil. It is possible to reduce them by connecting a discharge resistance across the coil terminals or encircling the iron core of the magnet with a copper sleeve. In general, however, these methods are unsatisfactory, because they slow up the speed of operation of the contactor. The most satisfactory solution for this trouble is to provide adequate insulation, with a large factor of safety where such voltages are likely to occur.

The phenomenon of self-induction can be used advantageously to delay the opening of a contactor for a short time after the current is cut off from the coil. With almost any contactor half a second may be secured by connecting a discharge resistance across the coil, or by having a copper sleeve on the magnet core. If a longer time is desired, it may be obtained by special design. Comparatively little delay in the time of closing can be secured by such electromagnetic damping.

In contactor design, residual magnetism must be guarded against or it may cause the contactor to remain closed after the coil is deenergized. The usual method of guarding against residual magnetism is to provide a thin nonmagnetic spacer in the back of the contactor core, and to use a soft grade of steel or malleable iron for the magnet circuit, or for at least part of it. Care must be taken particularly when the pressure on the contacts is light, as there is then less tendency for the contactor to open when the coil is deenergized.

Most contactor coils are wound on a form and thoroughly impregnated with some compound which will hold the windings in place and prevent moisture from entering the coil. Other coils are wound on a phenolic spool or bobbin. It is generally possible to get a winding of greater wattage capacity into a given space by the use of a bobbin type of coil, which, however, is more expensive than the form-wound type.

On occasion it is desirable to use coils which are not wound to stand full line voltage continuously. Such coils are less costly and have the further advantage that they take a larger size of wire than a coil which is wound for continuous duty at full voltage. A heavier wire, of course, has less tendency to break under vibration or rough handling. Furthermore, such intermittent-duty coils, having relatively low inductance, operate contactors more quickly. It is common practice to use coils wound for intermittent service, and then to arrange to insert a protecting resistance in series with the coil after the contactor has closed. This may be done by means of a late-opening interlock on the contactor itself or by interlocks on succeeding contactors of the controller.

Lockout Contactors. Magnetic lockout contactors are primarily designed for use as an acceleration means for dc motors. For such service the contactor is normally open and is required to remain open when an inrush of current occurs and until the current has fallen to a predetermined value. Normally, lockout contactors have separate magnetic circuits for the closing and lockout functions (Fig. 5.4). By various combinations of series and shunt coils, the type of iron in each magnetic path, and the length of air gap, the contactor is adjusted to close at a predetermined point. For example, the closing coil may be excited from a constant voltage bus and the lockout coil connected in series with the armature of the motor. Assuming current is allowed to build up in the series holding coil before energizing the closing coil, the contactor will not close until the armature current has decayed to an acceptable point for the next step of acceleration.

A type of lockout contactor based on the inductance principle is illustrated by Fig. 5.5. The lockout magnet of the inductive contactor is provided with a heavy iron magnet circuit. The relative strength of the closing and holdout coils is also adjusted so that the contactor will remain open with the full line voltage on the closing coil and approximately 1% of the full line voltage on the holdout coil.

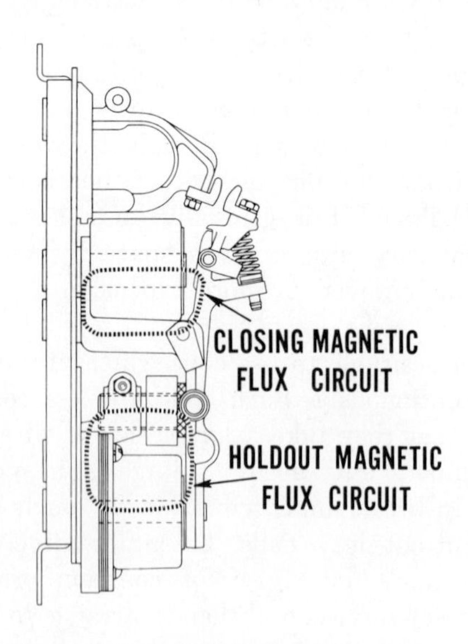

Fig. 5.4 Two-Coil Lockout Contactor

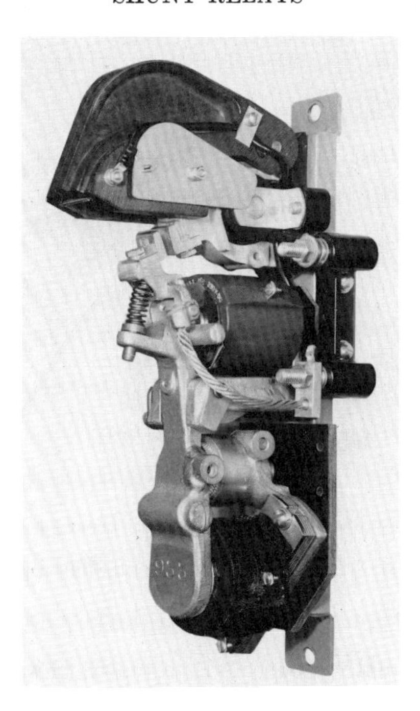

FIG. 5.5 Inductive Time-Delay Contactor. Arc Shield Raised

Another type of accelerating operation is obtained by connecting the shunt-type lockout coil directly across the step of resistance which will be shorted by the contactor. This coil will then hold the contactor open until the current in the resistor decays to a predetermined level.

Shunt Relays. Shunt relays are similar to shunt contactors, except that they are required to handle only small amounts of current, since they commutate the coils of other relays and contactors and do not handle motor current. Shunt relays are of many different sizes and forms. Some are very small and have light contacts, being so arranged that they will require a minimum current in the operating coil. The contacts of pressure regulators, thermostats, and other automatic pilot devices are often very light and incapable of handling much current. Consequently, low-wattage relays are required in connection with those devices. Other shunt relays are made larger and heavier for hard service as shown in Fig. 5.6. Some designs may have a blowout magnet on any or all of the contacts if necessary. They may also be provided with mechanical interlocks. Shunt relays are available in many different

FIG. 5.6 Mill-Type Eight-Circuit Shunt Relay

combinations of normally open and normally closed contacts, to suit the requirements of the control designer.

Overload Relays. The purpose of an overload relay is to protect a motor against excessively heavy loads. The relay usually has a series coil, which is connected in the motor circuit and carries motor current. The coil is arranged to lift an iron plunger when a certain value of current has been reached. The plunger is pulled up into the center of the coil and, when it is lifted, acts to trip a set of contacts which, in turn, disconnect the coil of a motor contactor and open the circuit to the motor. Adjustment for tripping on different values of current is made by varying the initial position of the plunger in respect to the coil. If the plunger is farther up in the coil to start with, it will, of course, move at a lower value of current.

It is often desirable to set an overload relay to trip at a value of current which is below the inrush current obtained when the motor is started. To accomplish this result, the overload is provided with a small oil dashpot, which causes a time delay in tripping. The time delay can be so adjusted that the overload will not trip when the motor is started, but if an overload continues for a short time the relay will trip and disconnect the motor.

For some types of control circuits only a single-coil overload relay is required, and the circuit is so arranged that, once it has been opened, the operator has to push a button to start the motor again. The overload relay will, of course, reset as soon as it has disconnected the motor, since there will no longer be any current in the relay coil. Other arrangements require that the overload relay remain tripped after it has operated. This is sometimes accomplished by making the contact of a latching type, so that it is necessary to reclose it by hand. More frequently it is accomplished by supplying the overload relay with a small shunt coil, in addition to the series coil. This shunt coil has just sufficient strength to keep the plunger lifted after it has once operated. In order to reset, it is then necessary to break the circuit to the shunt coil, which will allow the plunger to drop back to its initial position.

Figure 5.7 shows the Westinghouse Electric Corporation type-TI-2 overload relay, which has both time delay and instantaneous tripping features.

The relay is made up of magnetic and mechanical parts mounted on a molded base to form a unit common to all ratings, on which are assembled coils and heaters that vary with the rating.

The relay operates according to a combination of magnetic and thermal principles. A clapper-type magnet is magnetized by a series coil and carries a horizontal armature, the free end of which may take an upper or lower position depending on the magnetic and thermal conditions. The armature is normally biased to its lower position, where it is held by the magnetic attraction of a strip of nickel-iron alloy called Invar. Under tripping conditions, the lockout effect of the Invar strip is neutralized or overpowered and the armature is drawn to its upper position by the magnetic attraction of an upper pole formed by the bent end of the rear frame. In moving upward, the armature lifts a push rod which opens a normally closed contact at the top of the relay. A spring is arranged to engage a notch in the push rod in its upper position and thus hold the contact open for "hand-reset" operation until the latch is disengaged by depressing the reset pushbutton. If "automatic-reset" operation is wanted, the spring latch is permanently held out of engagement by depressing the reset pushbutton and giving it one-quarter turn clockwise.

The time-delay features of the relay depend on the special physical property of Invar by which it loses its magnetic permeability at a temperature of about 240°C. This property is utilized by connecting the Invar lockout strip or "heater" in series or parallel to the coil and passing the load current or a fraction of it through the heater. On moderate sustained overloads the internally generated heat is sufficient to raise the temperature of the heater to its demagnetization point, and the lockout effect of the heater is neutralized, allowing the relay to trip.

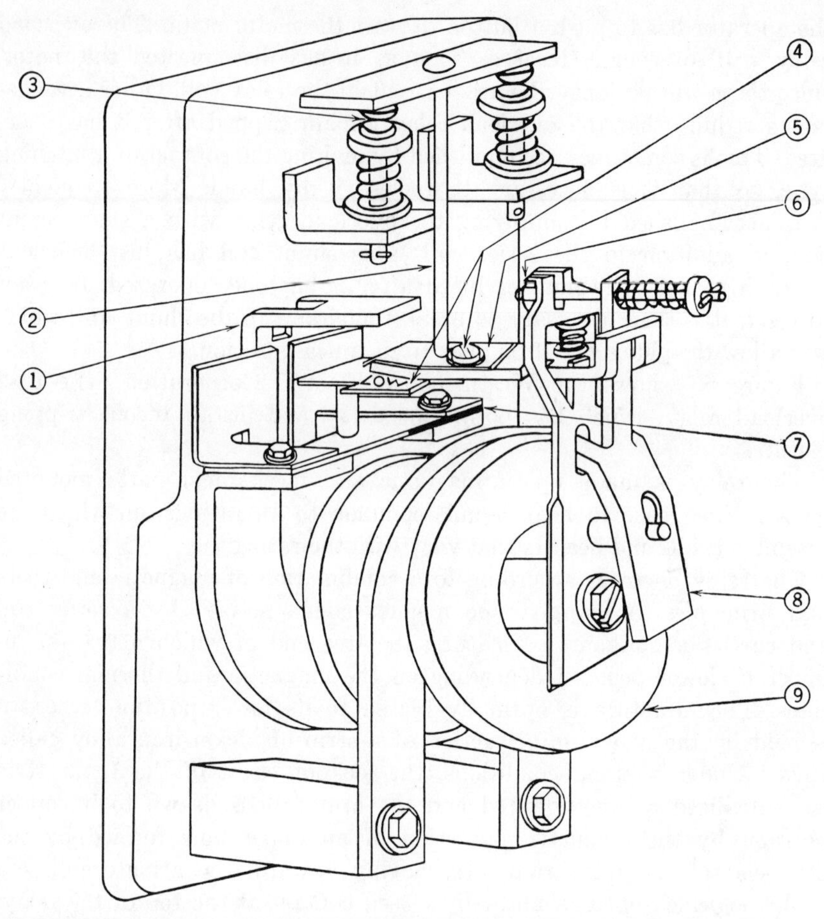

Item	Description or Name of Part
1	Bent end of rear frame
2	Push rod
3	Normally closed contact
4	Rating adjustment—loosen screw, and turn adjusting plate so that "low" or "high" marking is exposed, as desired
5	Horizontal armature thermal trip
6	Instantaneous trip adjustment—increase spring tension for higher tripping current
7	Invar heater
8	Vertical armature instantaneous trip
9	Coil

Fig. 5.7 Westinghouse Type-TI-2 Overload Relay

For overloads which exceed the instantaneous trip setting of the relay, a vertical auxiliary armature, attracted toward the coil, strikes the horizontal armature and raises it from the Invar $\frac{1}{32}$ to $\frac{1}{16}$ in., which is sufficient to break the lockout, allowing it to trip.

The relay has two adjustments: (1) to vary the rating, and (2) to vary the instantaneous tripping current.

The adjustment for rating is made by turning an adjusting plate attached to the horizontal armature, so that the lockout pin registers with a hot spot (low current rating) at the right or with a cool spot (high current rating) at the left of its motion. The change of current rating which may be expected between "low" and "high" adjustments is about 12% for large current heaters and 20% for small ones. This adjustment is changed by loosening a screw.

A nut adjustment is provided for varying the spring tension on the auxiliary armature to vary the percentage overload at which instaneous trip occurs. The travel of the nut is limited, to provide adjustment between about 200 and 600% of rating. The adjusment is set at the factory to give instantaneous trip at 300 to 400%.

Series Relays. Except that the coils must carry power circuit current, series relays are similar to shunt devices. The coil is connected in the motor circuit, and the contacts are arranged to handle the coils of a magnetic contactor. The armature of a relay of this type is light and the magnetic gap is very small, so that the relay is fast in operation. It may be set to operate on any desired current by adjusting the tension spring which holds the armature open.

Latched-In Relays. Sometimes it is necessary to have a relay so arranged that, once it has closed, it will remain closed even though voltage fails. To accomplish this result, relays of the latched-in type are used. Such relays usually have two coils, one for closing the relay and the other for opening it. When the relay closes, a latch operates to lock it in the closed position. The tripping coil is arranged to raise the latch and permit the relay to open.

Timing Relays. Direct-current timing relays frequently are arranged to close a number of circuits in sequence within a definite time after the coil has been energized. The coil of the relay pulls upward on a plunger, which is prevented from instantaneous operation by means of a dashpot to which the plunger is connected. The dashpot may be of either the oil or the air type. It consists of an iron pot in which a piston operates. The piston has a valve which permits the oil or air to flow through only very slowly in one direction but very rapidly in the other direction. Consequently, the plunger of the relay moves upward slowly and downward instantaneously. The timing is adjusted by varying the opening of the valve.

As the plunger goes upward, it moves a finger board on which a number of pilot fingers are mounted. The mountings are staggered, so that the fingers close in rotation from left to right. In other words, the first finger may have to move only $\frac{1}{8}$ in. to close, whereas the last finger may have to move $\frac{1}{2}$ in. to close, and the others somewhere between these values. The fingers may be connected to the coils of contactors, which will then close in a definite sequence.

By the use of a bypass, which will permit the plunger to move upward a short distance before the dashpot has any effect, it is possible to close the first finger immediately and then obtain timing for the closure of the other fingers.

Relays of this type are also available with all contacts normally closed, so that they open in sequence in a definite time. Such relays are used for inserting successive steps of resistance in series with the shunt field of a motor, to increase its speed.

Field Relays. Figure 5.8 presents a relay of a very useful type. As shown, it has a series and shunt coil with a normally open contact. It has a relatively light armature, an adjusting spring which may serve to vary the closing and opening points, and an adjustable air gap. The armature is prevented from sealing against the iron core by means of a brass screw. This screw may be adjusted to vary the gap. If the air gap is wide, the opening point of the relay will be close to the closing point. If the air gap is small, the relay will remain closed on a current much less than that which was required to close it. By varying the air gap, the range between the opening and closing values may be adjusted. The value of both opening and closing points may be adjusted by varying the tension spring.

With series coils, the relay may be used for automatic acceleration and deceleration of motors by means of field weakening. This is described in Chapter 8.

A relay of this same type may serve as a jamming relay. For such a purpose it would have a series coil connected in the motor circuit and would be so arranged that when the current reached a definite value the relay would open. Normally closed contacts would, of course, be required. When the relay opened, it would act upon the coil of a magnetic contactor, which in turn would insert a step of resistance in series with the motor to slow it down somewhat. This arrangement is often used

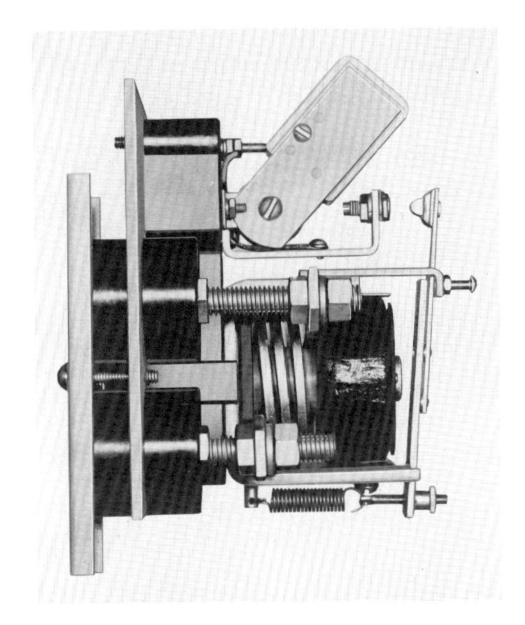

FIG. 5.8 Vibrating Field Relay

on such installations as electric shovels, where, in digging, an obstruction may be met which will tend to overload the motor and perhaps stall it. It is not desired to have an overload relay cut off power under these conditions, but instead a jamming relay is used to insert a certain amount of resistance and slow the motor down, at the same time preventing the current from reaching an excessive value.

The same type of relay could be used for a voltage relay, since its ease of adjustment would permit its being set to open and close on definite voltages.

There are many other forms of control relays, and manufacturers are almost constantly having to design new types to meet special applications. However, the majority of them are modifications in one form or another of the types which have been described above.

Maintenance of Contactors. Contactors should be kept reasonably clean. If dust or scale accumulates on the controller, it should be blown off with compressed air.

The contacts should not be lubricated, because oil or Vaseline will be decomposed by the heat and will increase the destructive effect of the arc.

The contacts should be kept smooth by filing occasionally with a fine file, care being taken to remove no more metal than is necessary to secure a smooth surface. Contact tips should be replaced when they are worn approximately half through. Definite instructions on this point are generally sent out by the manufacturer with each contactor. As the tips wear, the spring pressure behind them is reduced. If they are allowed to go too long, the contactor will not be able to handle the load properly and may be seriously damaged by overheating or by welding of the contacts.

Hinge pins and bearings should be lubricated with a light machine oil.

Interlock contacts should be replaced when the contact surface is worn nearly through. Cams and operating bars or linkages must be replaced when mechanical tolerances may cause improper seating of the contacts.

Arc shields should be replaced before the arc has burned through them. If a shield is burned clear through, the arc will strike the metal pole piece of the blowout and the contactor will be damaged.

The surfaces of the core and the armature which seal together when the contactor is closed should be kept clean and free from rust, oil, or grease. Rubbing these surfaces occasionally with emery cloth will accomplish the desired results.

Standard Ratings. The NEMA standard 8-hour open ratings of dc contactors are 25, 50, 100, 150, 300, 600, 900, 1350, and 2500 amp. Their intermittent rating in amperes is 133⅓ of the 8-hour rating.

For general-purpose magnetic controllers and machine tool controllers, the rating of the line, reversing, and final accelerating contactors is given in Table 5.1. Intermediate accelerating contactors are selected

TABLE 5.1

CONTINUOUS-DUTY RATINGS OF DC CONTACTORS

NEMA Size Number	8-Hour Open Rating (amperes)	115 V		230 V		550 V	
		Horse-power	Number of Accelerating Contactors	Horse-power	Number of Accelerating Contactors	Horse-power	Number of Accelerating Contactors
1	25	3	1	5	1	..	..
2	50	5	2	10	2	20	2
3	100	10	2	25	2	50	2
4	150	20	2	40	2	75	3
5	300	40	3	75	3	150	4
6	600	75	3	150	4	300	5
7	900	110	4	225	5	450	6
8	1350	175	4	350	5	700	6
9	2500	300	5	600	6	1200	7

TABLE 5.2

INTERMITTENT RATINGS OF CONTACTORS

8-Hour Contactor Rating	Contactor Mill Rating	Horsepower Mill Rating	Minimum Number of Accelerating Contactors
100	133	35	2
150	200	55	2
300	400	110	2
600	800	225	2 or 3
900	1200	330	3
1350	1800	500	4
2500	3350	1000	5

so that the open 8-hour ampere rating will not be less than one-fourth of the peak current obtained during acceleration.

For intermittent service in steel mills, the ratings are given in Table 5.2. The smaller contactors are not used for this service. For crane service the ratings are the same, but an additional accelerating contactor is used.

Problems

1. A dc coil circular cross-section has the following characteristics:

Diameter of inside turn	1.12 in.
Diameter of outside turn	2.86 in.
Terminal volts	240
Wire used	No. 35 cotton-covered
Wire resistance	329 ohm/1000 ft, at 25°C
Number of turns in coil	N

Calculate the ampere turns which the coil will produce at the specified voltage, at 25° C.

2. The length of the winding space of the coil of problem 1 is 2.4 in., and 9243 turns of the wire specified can be wound in 1 in.^{2} Calculate the resistance of the coil at 25°C.

3. Calculate the watts in the coil at 25°C.

4. Calculate the resistance of the coil, and the watts, at 100°C.

5. It has been found that this coil will dissipate heat at a rate which will prevent its exceeding a safe temperature of 100°C if its continuous wattage is determined on the basis of 0.7 watt/in.^{2} of surface, figuring on the cylindrical surface plus the area of one end. What is the continuous wattage rating of the coil?

6. A coil of this type is wound with No. 28 cotton-covered wire, which has a resistance of 64.9 ohm/1000 ft, and a winding factor of 3462 turns/in.^{2} Calculate the resistance of the coil.

7. Calculate the continuous voltage rating of the coil of problem 6.

8. A contactor having the coil of problem 6 is used on a 250-V circuit. The coil is protected by a series resistor which is cut into circuit by interlock contacts on the contactor. What is the minimum value of the protecting resistor, allowing for a possible 10% overvoltage?

9. A certain shunt contactor will pick up and close on 2020 amp turns, and drop out at 530 amp turns. If this contactor is equipped with the coil of problem 1, and it is desired that it operate at 80% of rated voltage at 100°C, what is the minimum rated voltage on which it may be used?

10. At what voltage will the contactor open?

11. If the contactor is equipped with the coil and resistor of problem 8, what will be the minimum rated voltage on which it may be used?

12. At what voltage will the contactor open?

13. An overload relay has a single series coil. The calibration range of the relay is 950–3750 amp turns. The desired maximum current density in the coil is 1500 amp/in.^{2} at the motor rating. Calculate the number of full turns, and the cross-sectional area of a coil for use with a 100-hp 230-V 375-amp motor when the relay is to be calibrated at 125–150–175% of the motor rating.

14. Sketch a series coil blowout magnet and its coil, and show the contactor contact tips. Indicate the direction of current in the turns of the coil around the magnet and between the contact tips, and show which direction an arc will be blown from the tips. Show the direction of the magnetic flux lines.

6

AUTOMATIC ACCELERATING METHODS FOR DIRECT-CURRENT MOTORS

Although it is satisfactory to connect some small motors directly to the line when starting, the majority of installations require a more gradual application of power. Resistance is connected in series with the armature to limit the initial current, and this resistance is then short-circuited in one or more steps.

Various devices are employed to short-circuit resistance steps automaticallly as the motor accelerates. A satisfactory device must meet a number of requirements.

1. It must properly protect the motor against high peak currents and against high stresses and strains. Some years ago this was the principal requirement, but today motors are built to withstand hard service and a great deal of abuse, so that this function is not so important as it formerly was.
2. The accelerating device must protect the machinery by applying the starting torque gradually and in limited amounts.
3. The current taken from the line must be limited in order to prevent line disturbances. Most power companies have definite rules regarding the increments of current that may be drawn from the line when starting a motor.
4. Smooth acceleration is a requirement of many installations, as, for instance, elevators, and the trolley or bridge motions of a crane.
5. The rate of acceleration is important to many drives. Rapid acceleration saves time and power. Frequently the operation of a machine is very rapid, and the motor is required to get up to speed in a short time.
6. The accelerating device should be reliable and should operate for long periods of time without maintenance. It should be simple, easy to understand and adjust, and easy to maintain and repair.

Obviously, these requirements vary in importance, depending on the application. A centrifugal casting machine requires an extremely smooth acceleration. Speed is essential in accelerating the motor that drives the manipulator fingers of a rolling mill. The current taken from the line may be of no importance if the power system has a relatively large capacity. In selecting the method of acceleration, therefore, the application must be considered and the vital requirements determined. An accelerating method which will best meet these requirements can then be adopted.

Thoery of Acceleration. Figure 6.1, showing what happens during the acceleration of a shunt motor, will serve to illustrate the theory for any motor. At no load the motor runs at a speed near 100%. A slight drop in speed is caused by the voltage drop in the armature and leads. At any given load the speed will fall at a corresponding point on the curve *ae*. If the motor were connected directly to the line to start, the initial inrush current would be limited only by the resistance of the armature and leads and would reach point *e*. To limit the inrush to a reasonable value, external resistance is connected in series with the armature. The resistance is designed to allow an inrush which will be sufficiently above full load to insure starting. In the diagram the initial inrush is shown at *s*. When the motor starts, it begins to develop a countervoltage opposing the line voltage and in proportion to the motor speed. Consequently, as the motor speed increases, the current in the armature decreases along the line *sa*. If the motor had no load, it would

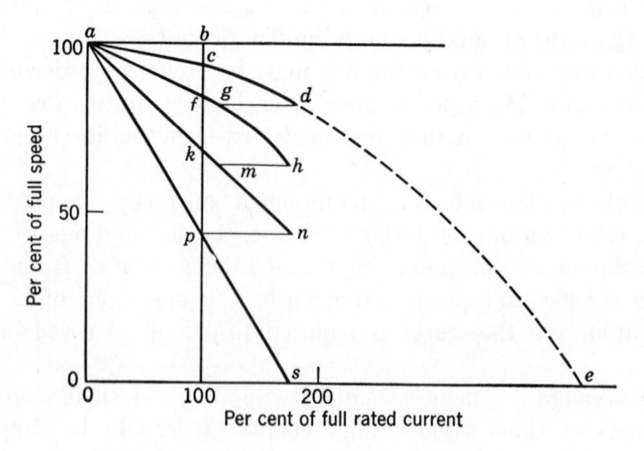

Fig. 6.1 Acceleration of a Shunt Motor

accelerate to full speed with the resistance in circuit; but if the motor is loaded, it will accelerate only until the armature current is just enough to provide the necessary running torque. With a fully loaded motor the maximum speed would be reached at point *p*. When or before this speed is reached, it will be necessary to cut out some of the resistance if further acceleration is to take place. It is customary to cut out just enough to produce a second inrush *n*, equal to the initial inrush. The motor then accelerates along the curve *na* until point *m* is reached, where it is necessary to cut out another step. This will be selected to give a third equal inrush *h* and permit the motor to accelerate along the curve *ha*. At point *g* the last step is cut out and the motor is across the line.

The choice of the initial inrush value *s* determines the location of point p, and obviously, if the inrush is increased, the speed at *p* will be higher and the length of the line *bp* will be shorter. Line *bc* represents the voltage drop in the motor and leads, and *cp* that across the external resistance. It will be evident that *kp* represents the drop across the first step of resistance, *fk* that across the second step, and *cf* that across the third step. Since these voltages are proportional to the resistance values, the current being the same in all steps, the lines *kp, fk,* and *cf* are proportional to the ohmic value of the resistance steps. The curve, therefore, gives a graphical method of determining the value of the resistance and of dividing it properly. The total ohmic value is found by dividing the line voltage by the initial inrush current *s*. The first step is then

$$\frac{kp}{bp} \times \text{Total ohms}$$

The other steps may be found similarly.

Note that the sum of the steps will be less than the total ohms by the amount $bc/bp \times$ Total ohms, which is the armature and lead resistance.

Practically all methods of automatic acceleration fall into one of two general classes known as current-limit acceleration and time-limit acceleration.

Current-Limit Acceleration. This method of acceleration makes use of the fact that, as a motor increases in speed, the current taken from the line decreases. Relays or other devices operating on current are used to control contactors which cut out the resistance steps. The relays are set to permit operation of the contactors successively as the current approaches the full-load value. Obviously, the relays must be set to close above the value of load current to insure that they will always operate. If the load is variable, the relays must be set high enough to insure acceleration under the heaviest load encountered. If the load at any time requires a current higher than the relay setting, the relay will not operate, the resistance will not be cut out, and the motor will not accelerate any further.

With current-limit acceleration the time required to accelerate will depend entirely on the load and will be constant only when the load is constant. When the load is light, the motor will accelerate quickly, the current will drop rapidly, and the relays will permit the resistance commutating contactors to operate quickly. When the load is heavy, the motor will require a longer time to accelerate, the current will fall more slowly, and a longer time will be required for short-circuiting the resistance.

Time-Limit Acceleration. Time-limit acceleration may be defined as a starting method which permits short-circuiting the starting resistance and connecting the motor to the line in the same time for each start, regardless of the load on the motor and of the time which it actually takes for the motor to accelerate that load. The accelerating relays or devices are entirely independent of motor current. The timing should be adjusted so that under normal load conditions each resistance step will be cut out just as the motor has reached its maximum acceleration with that step in circuit.

Comparative Advantages. With time-limit acceleration properly adjusted, the current peaks will be lower than with current-limit acceleration, because the operation of the accelerating device is independent of current, and the motor may be allowed to accelerate on each step as far as it will. With time-limit acceleration, and a variable load, the resistance may be made high enough to limit the current to just enough to start a light load smoothly. Then on a heavy load the motor will not start on the first point, or perhaps not on the second point. The accelerating device will continue to cut out resistance until the motor does start. In either event the motor will be connected to the line in the same time, but power will be saved with a light load.

In many instances a controller is adjusted to work properly when first installed, and when the machine is new and relatively hard to drive. With current-limit acceleration the relays will have to be set high enough to take care of this condition. Later, when the machinery is worn in, much less current may be required to start it, and the relays could be set to operate at a lower value. The chances are, however, that this adjustment will not be made, since there will be nothing to call the operator's attention to the possibility. The controller will then be wasteful of power during starting.

For the reasons given, time-limit acceleration has the advantage of saving power during acceleration.

The time-limit controller generally has the advantage from the standpoint of simplicity, since fewer relays and electric interlocks are required. The series lockout contactor requires no interlocks or auxiliary interlocks, and so provides a simple current-limit control, but it can be used only in the simple form for single-speed applications.

The time-limit controller has an additional advantage which is a strong argument for its use on motors subject to a large number of operations in a manufacturing process. This advantage is that, since it always accelerates the motor in the same time, it is possible to time a number of operations definitely with respect to one another. For instance, an operator in a steel mill may control a number of motors driving various parts of the mill. Since the response of each motor is always the same, he soon discovers just when to operate the starting switch for each drive to obtain the results he desires. The sequence and time of operating the various levers become almost automatic; and the operator's efficiency is improved. If he had current-limit controllers, the time of response would vary with changes in load, and it would be difficult, if not impossible, to arrive at the same efficiency of operation.

It would appear that the time-limit controller has a definite advantage when employed with drives having varying loads. The chief obstacle to its general use was, for many years, the lack of any simple, reliable form of timing device. This difficulty has been overcome, and the use of time-limit acceleration is almost universal.

Forms of Control. The principal forms of current-limit acceleration are those using

Counter-emf acceleration
Series relay acceleration
Lockout contactors
Voltage-drop acceleration

The principal forms of time-limit acceleration are those using

Time of contactors only
Individual timing elements
Multicircuit dashpot relays
Inductive time-limit acceleration
Capacitor discharge

Counter-emf Acceleration. A controller using counter-emf acceleration is shown in Fig. 6.2. The run button is pressed to start, energizing

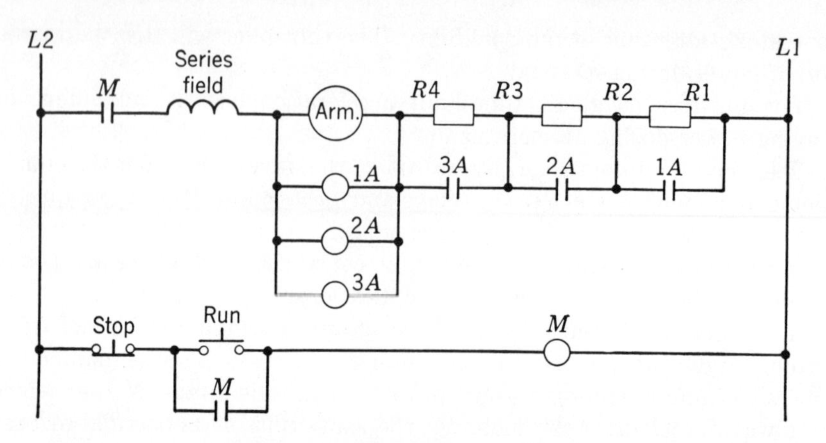

FIG. 6.2 Counter-emf Acceleration

the closing coil of the main contactor *M*. When *M* closes, it connects the motor to the line through the starting resistance and also provides a maintaining circuit for its own coil by closing an auxiliary interlock to bypass the run button. The accelerating contactors have their closing coils connected across the motor armature. The coil of the first contactor, 1*A*, is designed to close the contactor on approximately half of normal voltage. The coil of the next contactor, 2*A*, is designed to close the contactor at, say, 80% voltage, and that of the last contactor, 3*A*, at approximately 90% voltage. Consequently, the first contactor will close when the motor has reached approximately 50% speed, and the third at 90% speed. Contactors for this purpose are provided with an adjustable tension spring or some other means of adjusting the closing point.

This simple method of acceleration is satisfactory for controllers of small capacity. It is not generally used above 3 hp. Larger contactors, when closed on a low rising voltage, will close slowly and with light pressure. Under these conditions, the contacts are likely to weld together so that they cannot be opened.

Series-Relay Acceleration. Figure 6.3 shows a controller using series-relay acceleration. The relays have normally closed contacts connected in series with the coils of the accelerating contactors. The coils of the relays are connected in the main motor circuit as shown. The relays are designed to be very fast in operation. They are provided with an adjustment, so that they can be set to close on a selected value of current.

When the run button is pressed to start, the main contactor *M* closes.

Its main contacts connect the motor to the line, and current flows through the coil of relay *SR*1. An instant later the auxiliary contacts of *M* close to provide a maintaining circuit around the run button and also to energize the closing coil of contactor 1*A*. However, the series relay is fast enough to open its contacts before the auxiliary contacts of *M* close, and so contactor 1*A* is not energized. When the motor has accelerated enough to bring the line current down to that for which the relay is set, the relay will close. A circuit is now provided for 1*A*, which closes, cutting out the first step of resistance and short-circuiting the first series relay. Current now flows through contactor 1*A* and relay *SR*2. This relay opens its contacts in the interval between the closing of the main contacts of 1*A* and the auxiliary contacts. Consequently, no circuit is provided for coil 2*A* until the motor has accelerated enough to reduce the current to normal again. At this point relay *SR*2 closes, providing a circuit for coil 2*A*. Contactor 2*A*, in closing, cuts out the second step of resistance, short-circuits relay coil *SR*2, and allows current to flow through relay coil *SR*3. This relay operates in the manner described above, finally allowing contactor 3*A* to close and connect the motor directly to the line.

At first thought it may seem difficult to insure that the relay will operate in the relatively short time between the closing of the main contacts and the auxiliary fingers of a contactor. When it is considered, however, that the speed of the relay may be as high as 100 times that of the contactor, the matter does not seem so difficult. In fact, this method of control is very reliable and has been widely used for many

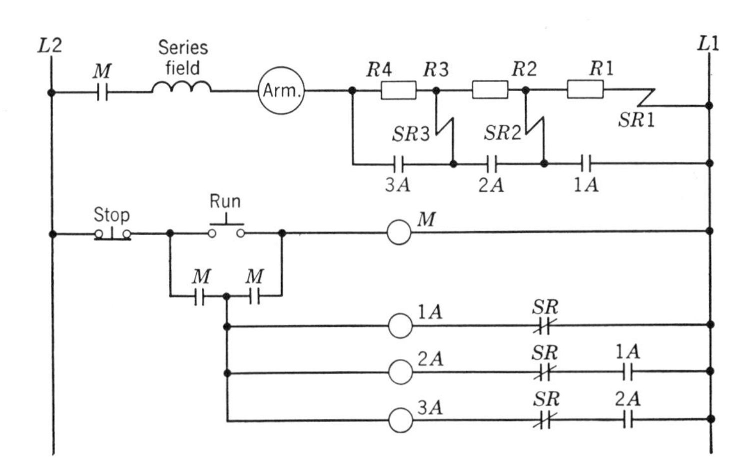

Fig. 6.3 DC Series-Relay Acceleration

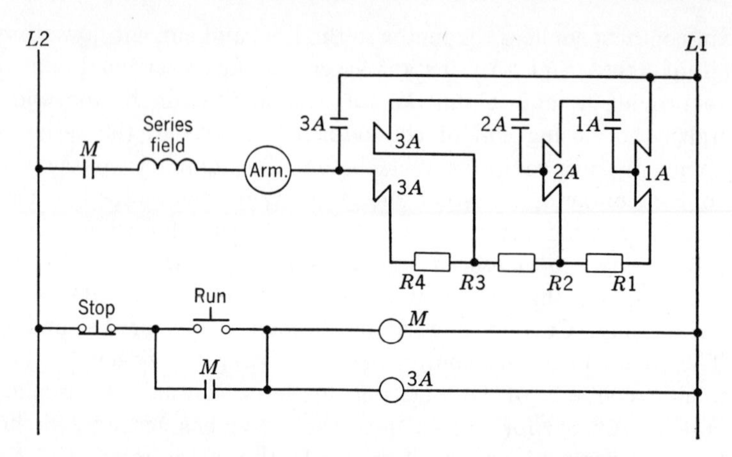

FIG. 6.4 Acceleration with Lockout-Type Contactors

years. Its chief disadvantages are that it requires a relatively large number of electric interlocks, which are undesirable from a maintenance standpoint, and that the relays themselves must be light to insure speed.

Lockout Acceleration. Figure 6.4 shows a controller using one form of series-lockout acceleration. The contactor operation is described in Chapter 5. Contactors 1*A* and 2*A* are each provided with two series coils, one being a closing and the other a lockout coil. Contactor 3*A* has in addition a small shunt coil to hold it closed if the running load is light and the pull of the series coil is low. Referring to the diagram, the initial inrush current flows from line *L*1 through both coils of contactor 1*A*, locking it open, and then through the resistance and the motor to line *L*2. The current also flows through the lockout coil of contactor 3*A*. This arrangement prevents any tendency of 3*A* to close on the pull of the shunt holding coil. When the current falls to normal, 1*A* closes, short-circuiting step *R*1–*R*2 and also its own lockout coil. Current now flows through the closing coil and the contacts of 1*A*, and through both coils of 2*A*. Contactor 2*A* is locked out until the current inrush, resulting from the closure of 1*A*, has fallen to normal, when it closes. The closing of 2*A* short-circuits step *R*2–*R*3 and the lockout coil of 2*A*. Current now flows through the closing coils only of 1*A* and 2*A*, and through both series coils of 3*A*. When this inrush has fallen to normal, 3*A* closes, short-circuiting all the resistance, and also all series coils, on all contactors. Contactor 3*A* is now held closed by its shunt holding coil until the stop button is pressed.

There are other forms of lockout contactors, some having only one coil and obtaining the lockout effect by means of a restricted iron circuit which becomes saturated on inrush current. The chief advantage of lockout acceleration over series relays is its simplicity, as no relays or interlock fingers are required. Its chief disadvantage is that it is not very adaptable to multispeed controllers because of the series coils. Another disadvantage is that, as the coils must carry motor current, they must often be wound of heavy copper strap or bar, and a number of different windings will be required to take care of different motor sizes. It is possible to use shunt closing coils, and also shunt lockout coils, without electric interlocks, and this is sometimes done.

Voltage-Drop Acceleration. Figure 6.5 shows a current-limit controller using the voltage-drop method of acceleration. The accelerating

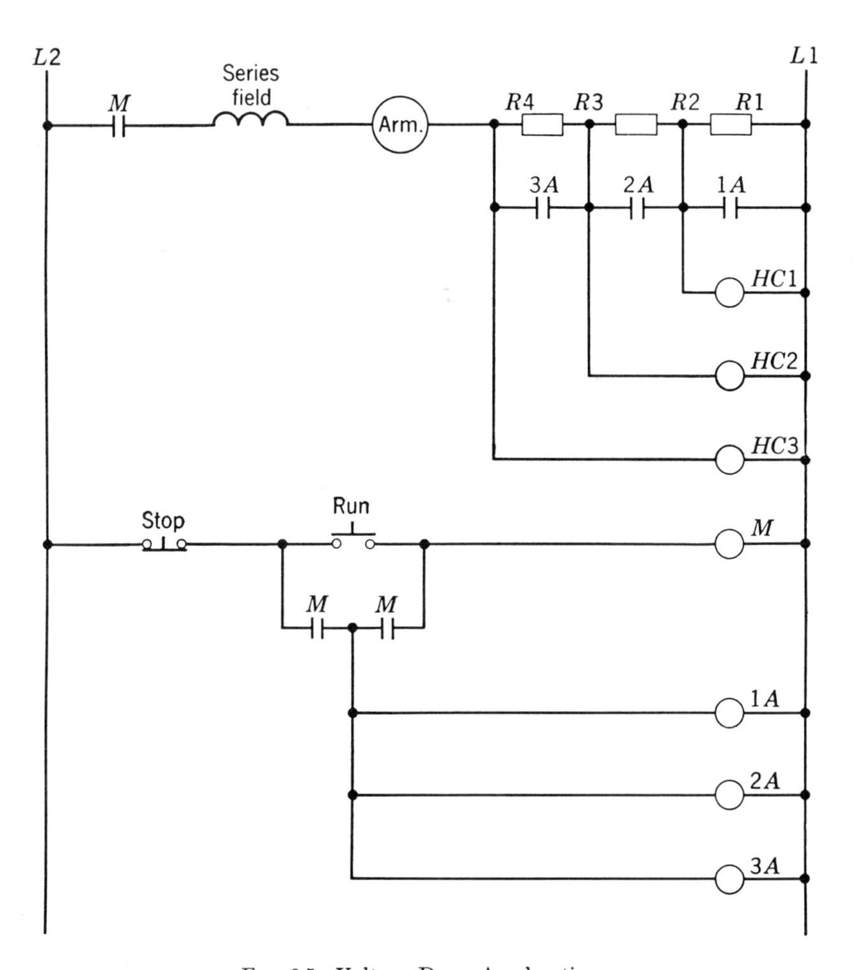

Fig. 6.5 Voltage-Drop Acceleration

contactors are provided with two coils, one for closing and the other for holding the contactor open. The closing coils are connected to line voltage through interlock fingers on the main contactor. The holdout coil of *1A* is connected across the first step of resistance, and it is designed to permit the contactor to close on the voltage obtained when normal full-load current is flowing through the resistance. Similarly, contactor *2A* is arranged to close when normal current flows through step *R2–R3*, and *3A* when normal current is flowing in step *R3–R4*.

The initial starting inrush, being higher than normal current, holds all accelerators open. When the current falls to normal, contactor *1A* closes. Contactor *2A* does not close because its holdout coil is connected across both steps *R1–R2* and *R2–R3*. The closing of *1A* short-circuits step *R1–R2*, and the resulting inrush of current increases the voltage across *R2–R3*. The holdout coil of *2A* is now connected across that step only, and when normal current is reached, *2A* will close.

Step *R2–R3* is now short-circuited, and the holding coil of *3A* is connected across step *R3–R4* only, so that, when the inrush has again fallen to normal, contactor *3A* will close. This method of connecting the contactors avoids the necessity of using any electric interlocks between them. If each holding coil were initially connected across its own step of resistance, it would be necessary to connect each closing coil behind interlocks on the preceding accelerator, for otherwise all accelerators would close as soon as the initial inrush had fallen to normal.

Time-Limit Acceleration with Contactors. The simplest means of time-limit acceleration is the use of contactors only, without any auxiliary devices. Of course, the time required for a contactor to close is short, but the time required for acceleration is sometimes only a fraction of a second. If the accelerating time required were half a second, and the accelerating contactor required one-tenth of a second to close, then the use of five contactors would give the required time and would provide a very simple and dependable control. Although this method is not common, it has its application in high-speed drives, particularly in steel mills.

Individual Timing Elements. Separate timing elements are available either as timing relays or as timed interlocks to be added to contactors. The timing relay comes in many forms, one of the most common being the pneumatic dashpot type. When equipped with a dc coil, this relay can be used as the timer for an accelerating contactor.

A more economical form of the pneumatic timer is available as an adjustable timed interlock which can be attached to any of the starter

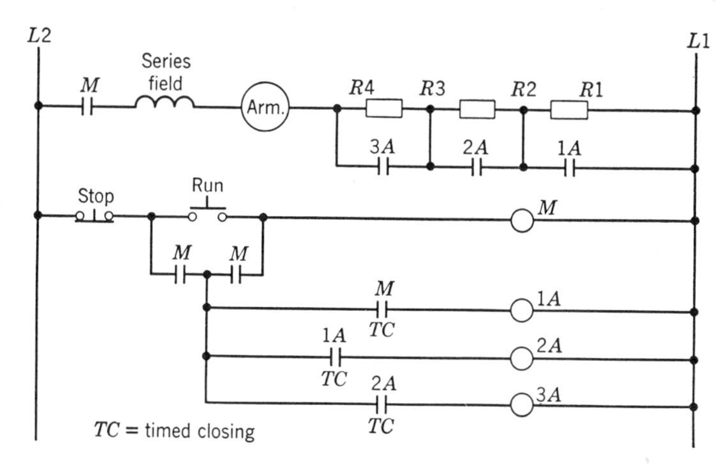

FIG. 6.6 Acceleration Using Individual Timing Interlocks

contactors. A controller with such interlocks is shown diagrammatically in Fig. 6.6. The timed interlocks are attached to contactors *M*, 1*A*, and 2*A*. Pressing the run button closes the main contactor M. This operates timed interlock *M*, which closes after a preset time to pick up accelerating contactor 1*A*. The next two steps occur in a similar way, picking up 2*A* and 3*A*. The control is simple and easily understood for service or maintenance.

Multicontact Timing Relays. Where it is not required to open and close the accelerating contactors individually for speed regulation, a multipoint dashpot timing relay may be employed. Such a device is used in Fig. 6.7. The relay consists of a solenoid, the plunger of which is retarded in its motion by a dashpot. The dashpot may be of the air, oil, or mercury type. The plunger of the solenoid is adapted to close a series of contact fingers, which are staggered so that they will close in a definite order. The contact fingers are individually adjustable, and the timing of the whole cycle may be varied by a valve adjustment on the dashpot. Since this device has its own operating coil, mechanical ties to the contactors are avoided. There is also an advantage in having only one dashpot to maintain and adjust.

The control circuits are simple. Interlocks on the main contactor energize the coil of the timing relay. The relay closes slowly, making circuits for the coils of 1*A*, 2*A*, and 3*A*, in order and at desired time intervals. When the stop button is pressed, the relay resets itself immediately.

A device of this type, if properly designed, is sturdy and reliable. For small motors up to approximately 15 hp, the contacts of the relay

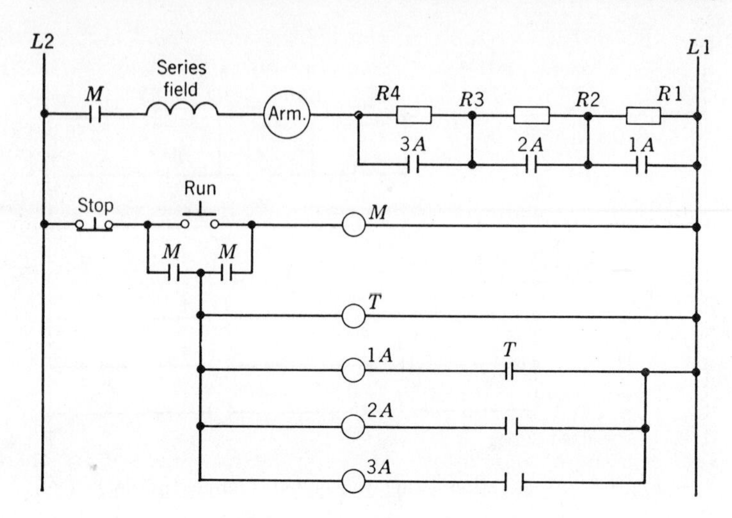

FIG. 6.7 Acceleration Using a Multicircuit Timing Relay

may be built heavy enough to commutate the resistance directly, without the use of contactors.

Inductive Time-Limit Acceleration. An important development in time-limit acceleration was the use of magnetic induction to give the time delay. When a constant direct voltage is impressed on a circuit having resistance, but neither inductance nor capacitance, the current will instantly reach its maximum value, and any change in the voltage will instantly produce a corresponding change in the current. When the voltage is removed, the current will instantly drop to zero. However, if the circuit contains either inductance or capacitance, a period of time will be required for the current to reach its final value when voltage is applied, and also for it to drop to zero after the voltage is removed. This is the basic principle of the inductive time-limit control.

The accelerating contactors used have two coils, one for closing the contactor and the other for holding it open. The closing coil operates on line voltage, but the holding coil is connected across one or more steps of the starting resistance. The iron circuit for the holding coil is highly inductive, and the air gap is small. The strength of the coils is so proportioned that, with full line voltage in the closing coil and only 1% of line voltage on the holding coil, the contactor will be held open.

A simple form of this control is shown in Fig. 6.8. The coils marked *HC*1, *HC*2, and *HC*3 are holding-out coils. Those marked 1*A*, 2*A*, 3*A* are closing coils. In the OFF position the holding coil of $1A$ is energized, since it is connected to the line through a resistance A–B. When the run button is pressed, the contactor M is energized, connecting the motor to the line through the starting resistance. The motor current, flowing through the resistance, produces a voltage drop across the steps $R1$–$R2$ and $R2$–$R3$, so that contactors $2A$ and $3A$ are held open. Interlock fingers on M provide a circuit to energize the closing coils of the accelerating contactors, but they remain open because of the holdout coils. The interlock fingers of M also short-circuit the holding coil of $1A$. Because of the inductance of this coil, the current flowing in it does not drop to zero immediately but requires a period of time to do so. The contactor is held open until the current has dropped to practically zero, after which it closes. When $1A$ closes, the first step of resistance is short-circuited, and also the holding coil of contactor $2A$. Time is again required for the current in this coil to drop low enough to allow $2A$ to close. When $2A$ closes, the holdout coil of $3A$ is short-circuited, and after a time $3A$ closes, completing the accelerating cycle.

When an inductive contactor of this type is open, the magnetic gap between the core and the armature of the closing circuit is relatively large, and that of the holding circuit relatively small. The ratio is about ⅜ in. for the closing gap, as against a few thousandths of an inch

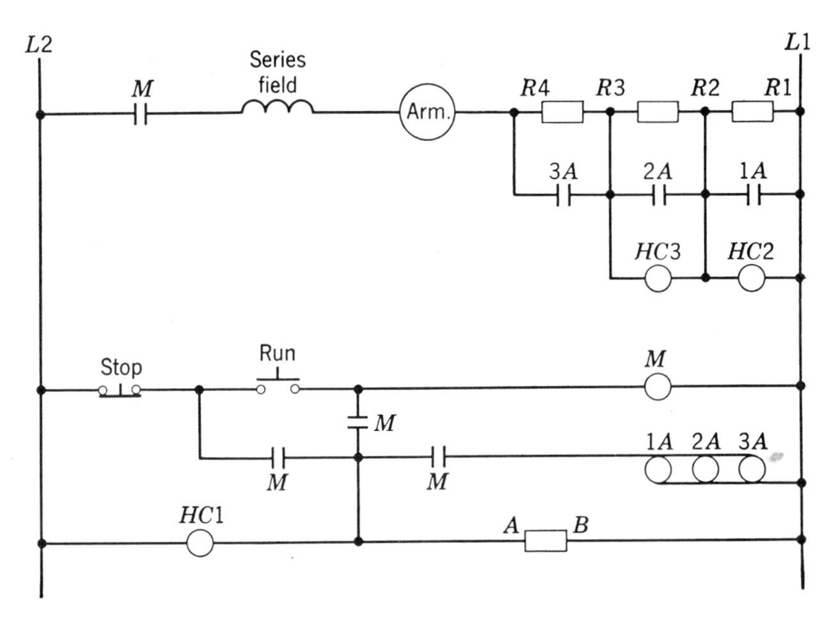

FIG. 6.8 Inductive Time-Limit Acceleration

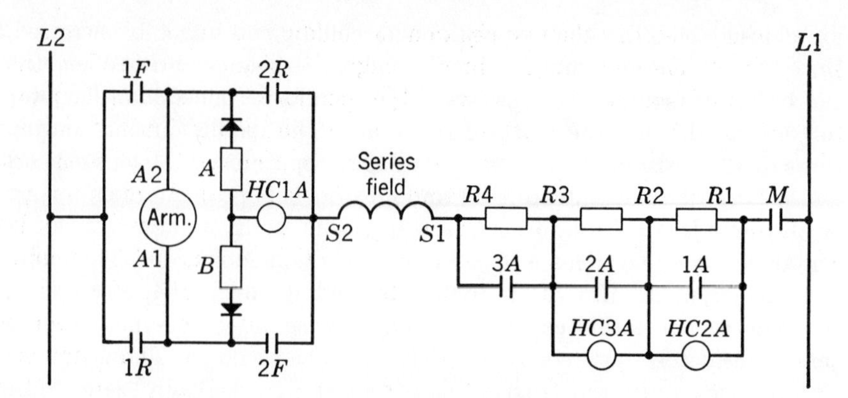

FIG. 6.9 Reversing Plugging Inductive Time-Limit Controller

for the holding gap. Consequently, the holding coil is able to keep the contactor open with a very low current in the coil. When the contactor has closed, the magnetic gap of the closing circuit is zero and that of the holding circuit large, so that the holding coil, if energized, would have very little effect. It follows that, as the contactor closes, the pull of the closing coil increases and that of the holding coil decreases, so that the closing is definite and the closing pull strong.

Adjustment for the time is made by changing the magnetic gap in the holding coil circuit, a small gap giving the greatest inductance and the maximum time.

Reversing Inductive Control. Since the inductive method of acceleration has found its greatest application in connection with the auxiliary drives in steel mills and on cranes, it seems proper to include here a description of a reversing plugging control of the type used for that purpose. An additional contactor and resistance step are used to keep down the inrush when the motor is plugged, that is, connected to the line in the reverse direction while it is running full speed in the forward direction. Under this condition the armature countervoltage, which has been opposed to the line voltage, is added to line voltage, so that the total voltage across the resistance is approximately 180% of line voltage. As the motor decelerates, the voltage drops, until at zero speed only line voltage is left. The motor then accelerates in the reverse direction, just as if it were starting from rest. The plugging contactor should, then, remain open until the motor stops, and close at the instant of reversal.

When the motor is started from rest, the plugging contactor should close at once, without any time delay. Figure 6.9 shows the connections for such an arrangement. The holding coil of the plugging contactor is connected at the midpoint of a circuit which consists of two small rectifiers and two resistors A and B. When the motor starts from rest, there is no countervoltage across the armature, and contactor $1A$ closes at once. When the motor is plugged, the countervoltage is high at the instant of plugging and causes current to flow through one of the rectifiers, holding out the plugging contactor. As the motor decelerates, the voltage decreases, finally becoming low enough to allow the plugging contactor to close at, or near, zero speed. The rectifiers are used to insure that the current in the holding coil will pass through zero.

Timing with Capacitors. Timing can be obtained by allowing a capacitor to discharge through a contactor coil. To obtain sufficient timing for most purposes, the contactor or relay must be designed specifically for this purpose. This can be accomplished by providing a large, high-resistance coil and an adjustable ampere-turn pickup on the armature lever.

The timing of a capacitor circuit is given by the equation

$$T = KRC\,(\log E_1 - \log E_2)$$

where T = time in seconds
K = a constant (0.00000230). If logs are to the base e, $K = 1.0$
R = the circuit resistance in ohms
C = the capacitance in microfarads
E_1 = the initial voltage across the capacitor
E_2 = the final voltage across the capacitor

The contactor shown in Fig. 6.10 has two adjustable armature levers which carry contacts that can be used directly to short resistance in

Fig. 6.10 Accelerating Contactor for Capacitor Timing

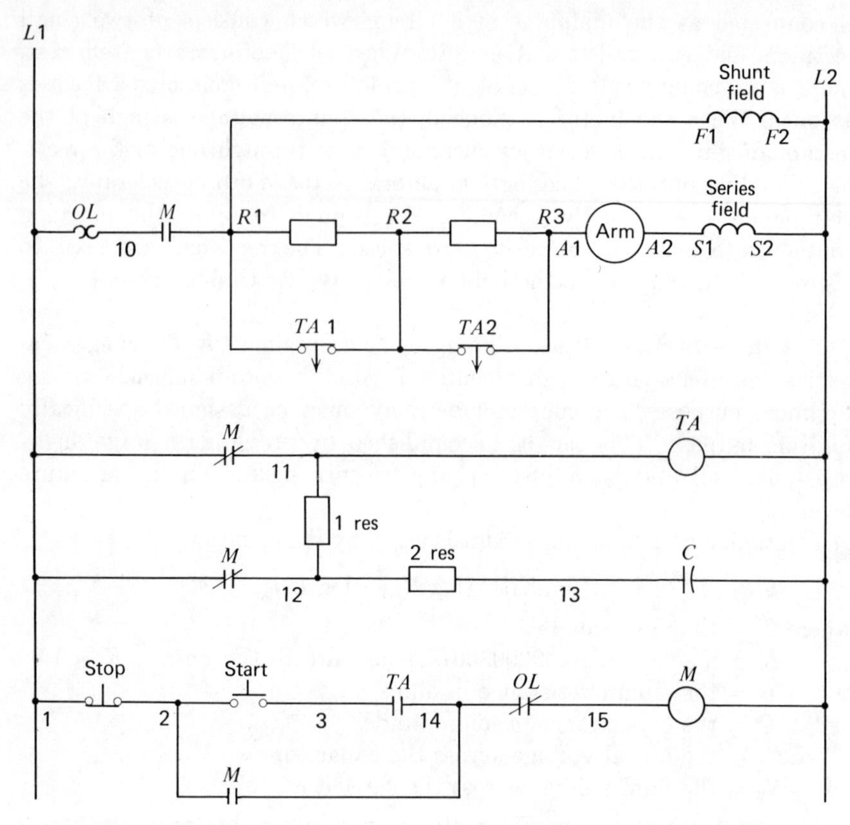

FIG. 6.11 Method of Timing Acceleration by Capacitor Discharge

the motor-starting circuit. A screw adjustment varies the sealed magnetic gap on these armatures. A third armature lever carries a control circuit interlock and operates without appreciable time delay.

The accelerating contactor in Fig. 6.11 is designated *TA*. It has interlocking contact *TA* and timing contacts *TA*1 and *TA*2. Before starting the motor, the contactor *TA* must be picked up so that the resistor is inserted in the armature circuit. This is insured by having the normally open interlock *TA* in series with the start pushbutton. Thus the timed contacts *TA*1 and *TA*2 are open, capacitor *C* is charged, and the timer is ready to function.

When the start pushbutton is operated, contactor *M* picks up, maintaining its coil circuit and connecting the motor to the line through the two steps of starting resistance. At this time the discharge of capacitor *C* through contactor coil *TA* is initiated by the opening of the normally closed interlocks on M. Contacts $TA1$ and $TA2$ will close successively at preset time intervals, first shorting out resistor step $R1$–$R2$ and then step $R2$–$R3$. Timing on contacts $TA1$ and $TA2$ occurs because of the timed decrease of the voltage across coil TA.

The double interlocks and resistors in the capacitor circuit provide a fast-charging circuit so that timing will not be reduced on a rapid restart of the motor. Typical values in this circuit are 1000 ohms for $2RES$, 50,000 ohms for $1RES$, and 25 microfarads for C. The high resistance of $1RES$ is required in the discharge circuit to give timing up to 2 seconds for each step. Longer timing can be obtained by increasing the size of the capacitor.

Double-coil contactors, as used for inductive time-limit control, may also be timed by capacitors, as shown in Fig. 6.12. The circuit is similar to that of Fig. 6.8 except that the holdout coils are of high resistance and are connected across the line to hold the accelerating contactors

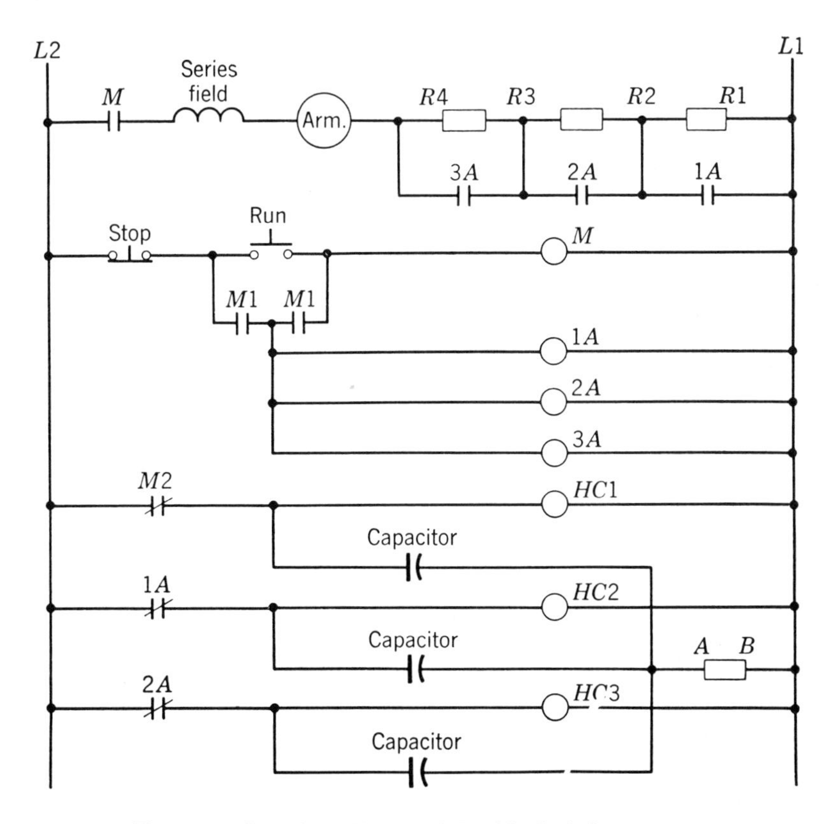

FIG. 6.12 Capacitor Timing of Double-Coil Contactors

open. The capacitors are charged in the OFF position of the controller. In starting, the interlock $M2$ opens the circuit to $HC1$, and the capacitor then discharges through that coil. When the condenser voltage has dropped low enough, the contactor closes, and an interlock on it opens the circuit to the next holdout coil. The timing may be adjusted by varying the magnetic gap of the holdout coil, just as for inductive time-limit acceleration.

Calculation of Accelerating Time. It is often necessary to calculate the time which will be required to accelerate or decelerate rotating machinery, when the available torque is known, or to calculate the torque which will be required to accelerate or decelerate in a desired time.

If the torque available for accelerating or decelerating is constant throughout the speed range being considerd, the time required may be calculated from the equation

$$\text{Time in seconds} = \frac{WR^2(N_2 - N_1)}{308} \times \frac{1}{T} \tag{6.1}$$

where W = weight of the rotating part in pounds
R = radius of gyration of the rotating part in feet
N_1 = the lower speed in revolutions per minute
N_2 = the higher speed in revolutions per minute
T = accelerating or decelerating torque in pound-feet

In calculations involving motor-driven machines, it is often convenient to express the torque in per cent of the motor full-load rated torque. The equation then becomes

$$\text{Time in seconds} = 0.62 \times \frac{WR^2}{H} \times \frac{(N_2 - N_1)}{1000} \times \frac{S}{1000} \times \frac{100}{T} \tag{6.2}$$

In this equation,

W = weight of the rotating part in pounds
R = radius of gyration of the rotating part in feet
H = horsepower of the motor
N_1 = the lower speed in revolutions per minute
N_2 = the higher speed in revolutions per minute
S = speed of the motor at full load in revolutions per minute
T = accelerating torque in per cent of full-load rated torque

It should be noted that, in any of these calculations, the rotating part is assumed to be driven directly from the motor shaft, and to be rotating at the same speed as the motor. If there is gearing between the rotating part and the motor, this must be taken into account, and the figure used for the WR^2 of the rotating part must be adjusted to a WR^2 equivalent to direct connection to the motor shaft.

$$\text{Equivalent } WR^2 = WR^2 \times \left(\frac{N}{N_b}\right)^2 \tag{6.3}$$

where W = weight of the rotating part in pounds
R = radius of gyration of the rotating part in feet
N = speed of the rotating part in revolutions per minute
N_b = speed of the motor shaft in revolutions per minute

By the use of equation 6.3, it is possible to calculate the WR^2 of a system including several rotating parts, which are rotating at different speeds. The WR^2 of each part is adjusted to its equivalent WR^2, and the equivalent WR^2 figures are added together to obtain the WR^2 of the whole system.

Equations 6.1 and 6.2 apply when the accelerating or decelerating torque is constant, as, for example, when a mechanical brake is used to stop a rotating system. When a motor is used to accelerate a machine, or when dynamic braking is used for stopping, the torque in most cases is not constant. The equations will still apply if the term $1/T$ is replaced by the term $(1/T)$ average, where this term is the average of the reciprocals of the torques available during the accelerating or decelerating period. It is relatively easy to calculate the average of the torques available, particularly since, in most cases, the current peaks during the acceleration are equal. Sufficient accuracy is usually obtained if the speed-torque curves are assumed to be straight lines. However, if the average of T_1 and T_2 is calculated, and this value is used in the equations instead of the average of the reciprocals of the torques, a very appreciable error will result. The time calculated could easily be incorrect by 25%. The average of the reciprocals of the torques may be determined from the average of T_1 and T_2 by multiplying the latter by a factor K. The equation for the factor K is

$$K = \frac{2\left(\dfrac{T_1}{T_2} - 1\right)}{\left(\dfrac{T_1}{T_2} + 1\right)\log_e \dfrac{T_1}{T_2}} \tag{6.4}$$

where T_1 = highest value of the accelerating torque
T_2 = lowest value of the accelerating torque

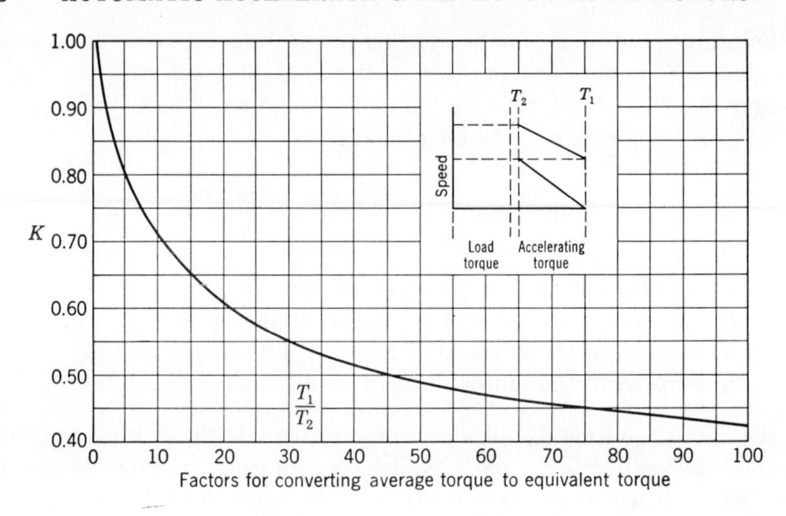

FIG. 6.13 Factors for Converting Average Torque to Equivalent Torque (Reprinted by Permission from *Factory Management & Maintenance,* February 1937)

The use of factor K results in an equivalent torque T_e, which may be used in equations 6.1 and 6.2 with accuracy.

$$T_e = KT \tag{6.5}$$

where T_e = equivalent torque
T = average of T_1 and T_2

Figure 6.13 is a curve showing values for the factor K, plotted against values of T_1/T_2.

If the values of T_1 and T_2 or either of them, vary on the different speed points of a controller, it is necessary to calculate the accelerating time on each step, and then add the times together.

To illustrate the method of using these equations, consider a 100-hp 1750-rpm motor which is to accelerate a constant-torque load having a total WR^2, for motor and load, of 500 lb/ft². The controller limits the torque peaks to 150% of motor full-load torque, and on each step the motor accelerates to a speed where the motor torque is 5% greater than the load torque.

The average of the accelerating torques is then

$$T = \frac{50 + 5}{2} = 27.5\%$$

$$\text{The ratio } \frac{T_1}{T_2} = \frac{50}{5} = 10$$

From the curve,

$$K = 0.71$$

From equation 6.5,

$$T_e = 0.71 \times 27.5 = 19.5\%$$

From equation 6.2,

$$\text{Time in seconds} = 0.62 \times \frac{500}{100} \times \frac{(1750 - 0)}{1000} \times \frac{1750}{1000} \times \frac{100}{19.5} = 48.6$$

If the average torque of 27.5% had been used, the calculated time would have been 34.5 sec, representing an error of 29%.

The radii of gyration of a few frequently encountered shapes are given here for convenience.

Solid cylinder about its own axis,

$$R^2 = \frac{r^2}{2}$$

where r = the radius of the cylinder.

Solid cylinder about an axis through its center,

$$R^2 = \frac{L + 3r^2}{12}$$

where L = length of the cylinder
r = the radius of the end

Solid cylinder about axis at one end,

$$R^2 = \frac{4L^2 + 3r^2}{12}$$

Solid cylinder about an outside axis,

$$R^2 = \frac{4L^2 + 3r^2 + 12dL + 12d^2}{12}$$

where d = distance from center of gyration to bottom of cylinder

Hollow cylinder about its own axis,

$$R^2 = \frac{r_1^2 + r_2^2}{2}$$

where r_1 = the inner radius of the rim
r_2 = the outer radius of the rim

Problems

1. A 50-hp 230-V 180-amp motor has an armature resistance of 5% of E/I. Plot a speed-torque curve similar to Fig. 6.1, for this motor, using four steps of resistance, equal inrushes on each step, and steps cut out at 100% load.

2. What is the total resistance required (motor and controller) in per cent of E/I?

3. What is the resistance required in the controller in ohms?

4. Calculate the ohms required in each step of the resistor.

5. Calculate the resistance of each step in per cent of E/I.

6. Using the values calculated in problem 5, determine the ohms required in each step of a resistor for a 15-hp 115-V motor.

7. Using the curve prepared for problem 1, and the same resistances, determine the current inrushes which will be obtained on each step if current-limit starting is used, and the relays are set to cut out the steps when the current drops to 120% of full load.

8. A capacitor having a capacitance of 100 microfarads is charged at 230 V and then discharged across a resistor of 500 ohms. What is the time required to discharge to 20 V?

9. Plot a speed-torque curve similar to Fig. 6.1, for a shunt motor, the starting peak currents being limited to a maximum of 130% of full-load current and the contactors closing at 100% of full-load current. How many resistance steps are required?

10. Plot a speed-torque curve for a shunt motor driving a fan, assuming that the initial running torque required is 30% of full-rated torque and that the torque increases directly with the speed until it reaches 100% at 95% of full no-load speed. All accelerating peaks are to be 130% of full-rated torque, and the contactors are to close when the torque is equal to the load torque required at the speed reached. How many resistance steps are required?

11. Draw an elementary diagram of a reversing controller for a shunt motor, including the following:

1 line contactor
4 reversing contactors
1 plugging contactor controlled by a series relay
2 accelerating contactors controlled by mechanically operated timing relays
Forward-reverse-stop pushbuttons.

12. Draw an elementary diagram of a nonreversing controller for a shunt motor, using three accelerating contactors, one a countervoltage contactor, one a lockout contactor, and one controlled by a series relay.

13. A controller having three resistor steps uses series-relay acceleration. If the adjustment of the second series relay is changed so that it closes at a higher current, what will be the effect on the inrush peak on each of the four starting points?

14. A controller having three resistor steps uses series-relay acceleration. If the resistor is changed so that there are more ohms in the first step and correspondingly fewer ohms in the second step, and the relay settings are unchanged, what will be the effect on the inrush peak on each of the four starting points?

15. Draw a speed-torque curve for a shunt motor driving a fan, assuming that the initial running torque required is 20% of full-rated torque, and that the torque increases directly with the speed until it reaches 100% at 95% of full no-load speed. Each accelerating peak is to be 50% more than the required running torque at that point, and the accelerating contactors are to close at 100% of the torque required.

16. If the total resistance required for this controller is 2.5 ohms, determine from the curve how many ohms will be required in each resistor step.

17. A shunt motor equipped with a series-relay controller is driving a machine which has a flywheel. It is found that at certain points in the machine cycle the overload on the motor is too great, and it is desired to increase the flywheel effect. Will it help to

(*a*) Increase the ohms of the starting resistance?

(*b*) Set the series relays to close at a lower current?

(*c*) Connect a small amount of resistance permanently in series with the motor?

(*d*) Drop out the last accelerating contactor when the peaks occur?

18. A 250-hp motor has a full-load speed of 420 rpm, and a WR^2 of 900 lb-ft^2. Calculate the rated torque of the motor, and determine how long it will take to accelerate to full speed if it is provided with an electronic controller which will hold the accelerating torque constant at 150% of rated torque.

19. If the motor of problem 18 is geared directly to a flywheel which is a solid cylinder having a diameter of 4 ft, and weighing 2000 lb., how long will it take to accelerate?

20. If the speed of the flywheel is doubled by gearing it through a 2-to-1 gear train, how long will it take to accelerate the motor?

21. If the motor is provided with a contactor-type controller which allows an average accelerating torque of 25%, how long will it take to accelerate under the conditions of problem 19?

22. Under the conditions of problem 19, how long will the accelerating period be if the 250-hp motor is replaced by two 125-hp motors, each having a WR^2 of 300 lb-ft^2?

23. Under the conditions of problem 19, calculate the accelerating time if the speed of the flywheel is increased 30% by the use of gearing, and the 250-hp motor is replaced by two 125-hp motors.

7

THE DIRECT-CURRENT SHUNT MOTOR

General Description. Direct-current motors are made in many forms, and with many types of windings, but they all fall into one of two general classes. When the main field winding is designed for connection in parallel with the armature, the machine is a shunt motor. When the main field winding is designed for connection in series with the armature, the machine is a series motor.

The field of the shunt machine is not affected by changes in the armature current, and the motor speed is relatively constant with different loads.

The armature current of a series motor also passes through the field, and so the field strength, and the speed, will vary widely with the load.

Many motors are built with both shunt and series fields, and their characteristics may fall anywhere between those of the shunt machine and those of the series machine, depending on the relative strength of the two fields. Figure 7.1 shows the general characteristics of shunt, series, and compound types.

Motors having a predominating shunt field and a relatively light series field are called compound motors. The light series field is supplied to increase the starting torque and to cause the speed to decrease a little under heavy load. Compound motors are used for the same general purposes as shunt motors and are controlled in the same manner. They are treated in this chapter.

Motors having a predominating series field and a relatively light shunt field are called series-shunt motors. The light shunt field is supplied to prevent excessive speed under light load. Series-shunt motors are used for the same general purposes as series motors and are controlled in the same manner. They are treated in the chapter on series motors.

Shunt Motor Ratings. Two organizations have adopted standards for shunt-wound motors. Mill motor standards are adopted by the Asso-

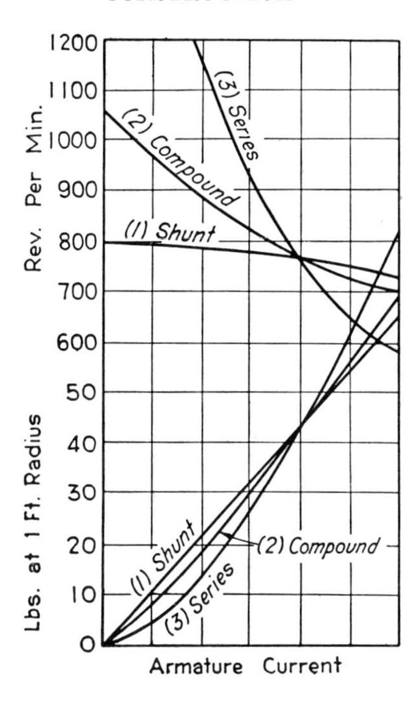

FIG. 7.1 Speed-Torque Characteristics of DC Motors

ciation of Iron and Steel Engineers and are shown in Chapter 11. NEMA has adopted standard ratings for industrial dc motors. The ratings in Table 7.1 are taken directly from NEMA standard MG 1-10.

Construction. A dc shunt motor has a stationary field member and a rotating armature member. The frame of the motor may be cast or fabricated steel, and to it are bolted steel pole members, around which the field coils are placed. The pole pieces are usually laminated to reduce the inductive effect and to make the field flux build up quickly when the field is energized. The main field coils are wound of insulated wire, either in a form or on a bobbin, and are mounted on the pole members. Figure 7.2 shows the general construction. It also shows a stabilizing winding, which is a light series field wound on the ends of the main pole pieces, next to the armature. The stabilizing winding is used on adjustable-speed motors, to insure stable operation when the main field is weakened. If it were not used, armature reaction at heavy loads might nullify the main field flux. Between the main field poles are mounted commutating field poles, which are wound with a bar or strap winding

TABLE 7.1

NEMA Standard Ratings of Industrial DC Motors

The horsepower, voltage, and base speeds for industrial dc motors and the speeds which may be obtained by field control of straight shunt-wound or stabilized-shunt-wound industrial direct-current motors shall be as follows:

Horsepower at Base Speed	Horsepower at 300% of Base Speed and Higher Speeds (Drip-Proof Motors Only)	Base Speed (rpm) 3500	2500	1750	1150	850	650	500	400	300	
		Speed by Field Control (rpm)									
½	0.65	—	—	—	—	3000	2600	2000	1600	—	
¾	1	—	—	—	3200	3000	2600	2000	1600	—	120 and
1	1.3	—	—	3500	3200	2800	2600	2000	1600	—	240 V
1½	2	4000	4000	3500	3000	2800	2600	2000	1600	—	
2	2.6	4000	4000	3300	3000	2600	2600	2000	1600	1200	
3	4	4000	3700	3300	2800	2600	2600	2000	1600	1200	
5	6.5	3700	3700	3000	2800	2600	2400	2000	1600	1200	
7½	10	3500	3500	3000	2800	2600	2400	2000	1600	1200	
10	13	3500	3500	3000	2800	2500	2200	2000	1600	1200	

15	20	3500	3300	3000	2600	2500	2200	2000	1600	1200	
20	26	3500	3300	3000	2600	2400	2200	1800	1600	1200	
25	33	—	3100	3000	2600	2400	2000	1800	1600	1200	240 V
30	40	—	3100	3000	2600	2400	2000	1800	1600	1200	
40	52	—	3100	2700	2400	2200	2000	1800	1600	1200	
50	65	—	—	2700	2400	2200	1800	1800	1600	1200	
60	80	—	—	2400	2200	2000	1800	1600	1600	1200	
75	100	—	—	2400	2200	2000	1800	1600	1600	1200	
100	130	—	—	2200	2000	1800	1600	1600	1600	1200	
125	165	—	—	2000	2000	1800	1600	1600	1600	1200	
150	200	—	—	2000	2000	1800	1600	1600	1600	1200	
200	260	—	—	1900	1800	1700	1600	1600	1200	1200	
250	—	—	—	—	1700	1600	1600	1400	1200	—	
300	—	—	—	—	1600	1500	1500	1300	1200	—	250 and 500 V
400	—	—	—	—	1500	1500	1400	—	—	—	
500	—	—	—	—	1500	1400	—	—	—	—	
600	—	—	—	—	1500	1300	—	—	—	—	500 and 700 V
700	—	—	—	—	1300	—	—	—	—	—	
800	—	—	—	—	1250	—	—	—	—	—	

FIG. 7.2 Stator of a DC Shunt Motor (Courtesy Reliance Electric Company)

connected in series with the armature; they are used to improve commutation by opposing the armature reaction.

The armature is made up of laminations bolted together on a shaft, and having slots parallel to the shaft, in which formed coils are placed. A commutator is mounted on one end of the shaft, and the ends of the armature coils are connected to the bars of the commutator. Carbon brushes, mounted to the motor end frame, ride on the commutator bars, providing a means of connection to the armature coils. See Fig. 7.3.

The resistance of the shunt field is relatively high, since it is connected across the supply lines. The resistance is a matter of motor design, and the value must be obtained from the motor manufacturer.

Motor Torque. The motor field winding sets up a magnetic field, in which the armature is located. The armature windings and brushes are so arranged that, when the armature is energized, it becomes a magnet with its poles offset enough from the field poles to cause maximum attraction between field and armature, and so provide maximum turning torque. As soon as the armature starts to turn, the connections are changed through the commutator, so that the magnetic poles of the armature are set back enough to compensate for the movement of the armature. In other words, the commutator serves to change continually the portion of the armature winding which is energized, and to keep the north and south magnetic poles of the armature in a fixed position relative to the field poles. The armature therefore has a practically constant turning torque, which is proportional to the magnetic strength of the field and that of the armature. For any direct-current motor, then,

$$T = K_1 \phi_A \phi_F \tag{7.1}$$

where T = the torque of the motor
K_1 = proportionality constant
ϕ_A = the flux produced by the armature
ϕ_F = the flux produced by the field

Since the flux ϕ_A produced by the armature is created by the armature current I_A flowing through the armature conductor windings, and since

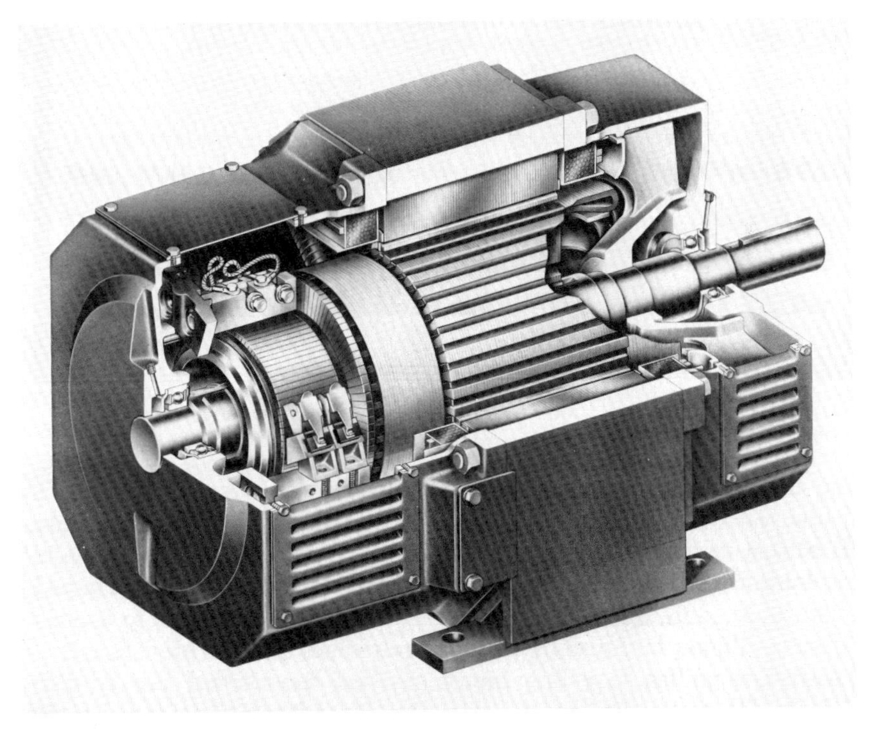

FIG. 7.3 Construction of a DC Shunt Motor (Courtesy Reliance Electric Company)

the armature magnetic circuit is customarily not operated in saturation, it is reasonable to say that

$$\phi_A = K_2 I_A$$

or that armature flux ϕ_A is directly proportional to armature current I_A. Equation 7.1 may then be written

$$T = K_1 K_2 I_A \phi_F$$

We can replace K_1K_2 with a simple constant K, and then

$$T = KI_A\phi_F \tag{7.2}$$

or the mcre common form

$$T = K\phi I_A \tag{7.3}$$

where it is understood that ϕ is the *field* flux, and the subscript $_F$ is dropped. Now if

$$I_A = I_A \text{ rated, and}$$

$$\phi = \phi_F \text{ rated}$$

we know that

$$T = T_{\text{rated}}$$

Therefore,

$$K = \frac{T_{\text{rated}}}{(I_{A\text{ rated}})(\phi_{\text{rated}})}$$

Substituting in equation 7.2,

$$T = \frac{T_{\text{rated}}}{(I_{A\text{ rated}})(\phi_{\text{rated}})}\phi I_A$$

or

$$\frac{T}{T_{\text{rated}}} = \frac{\phi}{\phi_{\text{rated}}} \times \frac{I_A}{I_{A\text{ rated}}}$$

In this form we have eliminated the constant K and need only concern ourselves with *ratio* of the actual conditions to *rated* or *normal* conditions.

We therefore customarily write

$$T' = \phi' I'_A$$

or

$$T' = F' I'_A \tag{7.4}$$

where T' is the torque in per-unit (or per cent) of full-rated torque, ϕ' (or F') is the field flux in per-unit of normal and I'_A is the armature current in per-unit of full-rated current.

Thus, if the field strength is weakened to half (50%) of its rated value and the armature current at a particular shaft load is 70% of its rated value, the torque is

$$T' = (0.5)(0.7) = 0.35, \text{ or } 35\%$$

of rated motor torque.

In equation 7.4 it is permissible to use armature current I'_A as a measure of the magnetic pull because the *armature* magnetic strength is proportional to the armature magnetomotive force (namely, armature current), that is, the armature magnetic circuit is not saturated. However, the motor field structure *is* usually operated to a considerable degree to saturation, and therefore the actual flux should be used in calculation instead of motor *field* current. This may be obtained from a field control curve of the particular motor, using the relation

$$\phi' = \frac{1}{S'} \text{ (which is true providing the armature voltage is held constant)}$$

where ϕ' is the per-unit field flux $= \dfrac{\phi}{\phi_{\text{rated}}}$

$$S' = \text{per-unit speed} = \frac{S}{S_{\text{rated}}}$$

A typical curve of this kind is shown in Fig. 7.4.

Motor Speed. When a motor armature is first energized, the only factor limiting the current is the ohmic resistance of the armature circuit. As soon as the armature begins to turn, its conductors start cutting through the magnetic field produced by the field winding, and by so doing generate in the armature conductors a voltage opposite in direction to the applied line voltage. This *countervoltage* is proportional to the speed S of the armature and to the strength of the field, or, if we work with the values in per cent of their normal or full-rated values, the countervoltage

$$E_a = KSF \tag{7.5}$$

where K is a constant. The voltage which forces current through the armature is the difference between line voltage E and countervoltage

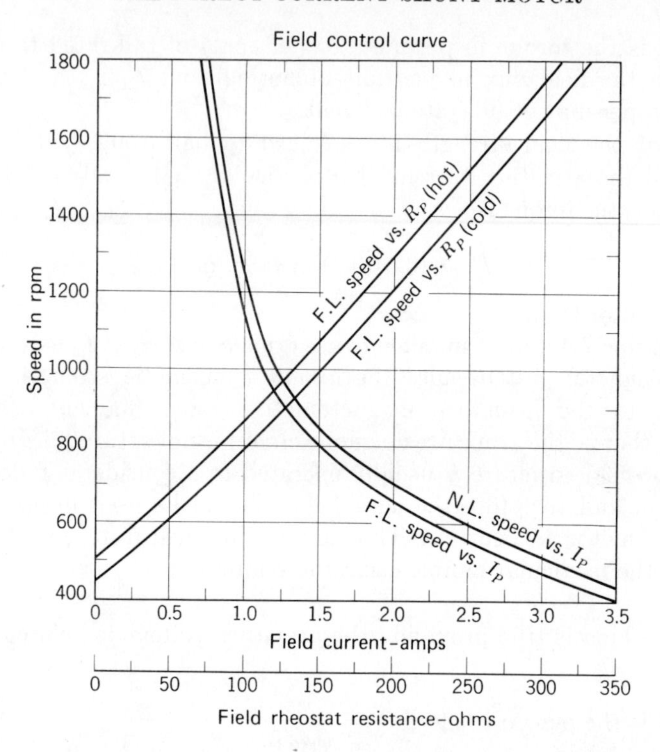

FIG. 7.4 Field Control Curves for a 20-hp Shunt-Wound DC Motor

E_a. Therefore, as the motor speed increases and the countervoltage rises, the armature current drops, and as it drops, the motor torque drops with it. The final speed of the motor can never quite reach the speed required to generate line voltage, because there would then be no armature current and no torque.

The final speed reached will be just enough below that value so that the difference between line voltage and countervoltage will send enough current through the armature to provide the torque necessary to turn the armature. This torque will depend on the load on the motor. The voltage $(E - E_a)$, which forces the current through the armature against the motor resistance R_m, is, by Ohm's law, equal to I_aR_m, and, therefore,

$$E - E_a = I_aR_m \tag{7.6}$$

or

$$E_a = E - I_aR_m$$

but

$$E_a = KSF \tag{7.5}$$

and, by substitution,

$$S = \frac{E - I_a R_m}{KF} \tag{7.7}$$

To determine the constant K, assume that $I_a = I_n$ or full load. Then by definition, S and F become 1.0, and

$$K = E - I_n R_m$$

$$S = \frac{E - I_a R_m}{F(E - I_n R_m)} \tag{7.8}$$

To determine the speed when there is external resistance in series with the armature, let R = the total resistance of the motor and external resistor. Then

$$S = \frac{E - I_a R}{F(E - I_n R_m)} \tag{7.9}$$

With equations 7.4 and 7.8 for the torque and speed, respectively, the procedures necessary to control the motor may be determined.

Starting. Inspection of equation 7.8 will show that the speed will vary with the motor voltage drop $I_a R_m$, and therefore it is necessary to make the armature resistance low to prevent wide speed changes with changes in load. A high motor resistance would also result in too high a resistance heat loss in the motor. If a motor having a resistance of $0.05E/I$ were connected directly to a supply line, the resulting inrush current would be 20 times the normal full-load motor current. The resulting torque would be 20 times normal, and the initial heating of the motor would be at the rate of 400 times normal. Probably neither motor, driven machine, nor power line would stand this punishment. The voltage across the motor must be reduced in some way, and the simplest and commonest way is by connecting a resistor in series with the armature. The resistor must allow more than full-load current to flow, so that there will be some excess current to accelerate the load. With large motors a usual value is 150% of normal, and with small motors as much as 300% may be allowed.

The ohms required may be determined by Ohm's law, from the equation

$$R = \frac{E}{I_s} \tag{7.10}$$

where R = the total ohms required in circuit
E = line voltage
I_s = desired starting current in amperes

From this total the motor resistance may be subtracted, but since this resistance is so low as to be negligible, R is commonly considered the value required in the controller.

Acceleration. When the motor starts to turn, it will accelerate in speed as long as its torque is greater than that required by the load. As the motor accelerates, its countervoltage rises, and its armature current drops, with a corresponding drop in torque. It is then necessary to reduce the resistance, as the motor speed increases, by cutting out, or short-circuiting, a step at a time, until finally all the resistance is out of circuit. The acceleration process is described in detail in Chapter 6.

A graphical method of designing the resistor for acceleration is shown in Fig. 7.5. Since the countervoltage developed is proportional to the speed, the ordinates represent both percentage of speed and percentage of countervoltage. The base line is percentage of full-load current. Assuming that 175% inrush is allowable, the total resistance, including motor and line, will be

$$R = \frac{E}{1.75I}$$

The point a represents the first inrush, and the motor will accelerate along a line drawn from a to 100, but if fully loaded will not accelerate beyond point b, as full torque requires 100% current. Therefore, at b a step of resistance must be cut out. Point c is selected to give an inrush equal to a, and a line is drawn from c to 100. This procedure is followed until a point is reached at which a line drawn to 100 will fall either directly on point h or very close to it.

The line gh represents the motor characteristic curve, and the line kh is the drop in speed between no load and full load, caused by the resistance drop in the armature and leads. The point h then represents the maximum speed to which the motor will accelerate under full load. It is only necessary to provide resistance to accelerate the motor to

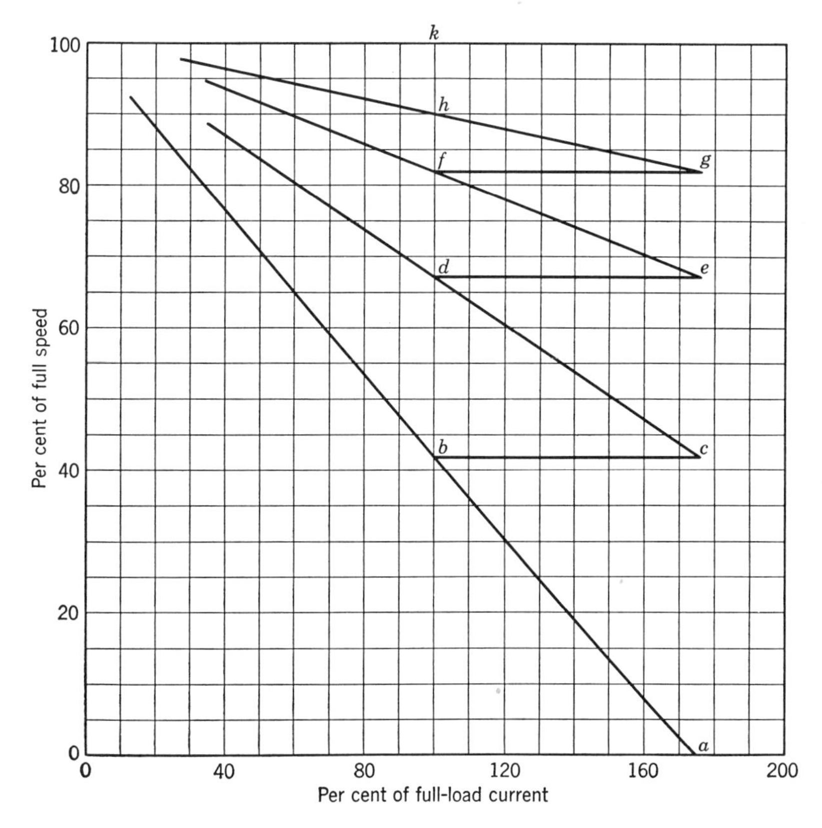

FIG. 7.5 Acceleration of a DC Shunt Motor

that point. If the last curve falls directly on point *h*, the design is correct, and with three steps the desired inrushes will be obtained. If it falls between *k* and *h*, the inrushes may be reduced slightly. If it falls between *f* and *h*, the inrushes must be increased or, if this is not permissible, another step will be necessary. In that event the inrushes would be reduced until the final accelerating curve passes through point *h*.

Each of the curves *ab*, *cd*, and *ef* represents a characteristic curve of the motor with a certain amount of resistance in series. The line *gh* is the curve of the motor alone plus line resistance, *ab* is the curve with all resistance in series, *cd* the curve with one step cut out, and *ef* the curve with two steps cut out. The line *gh* represents the motor performance with all the starting resistance cut out. Similarly *bd* is the drop across the first step of resistance, *df* that of the second step, and fh that of the third step. The ohms required are then

$$\text{First step} = \frac{bd}{bk} \times \frac{E}{1.75I}$$

$$\text{Second step} = \frac{df}{bk} \times \frac{E}{1.75I}$$

$$\text{Third step} = \frac{fh}{bk} \times \frac{E}{1.75I}$$

Under full load the motor will run at speed b with all resistance in circuit, at point d with one step cut out, at point f with two steps out, and at point h with all resistance out. If the load varies, the speed will vary also, always falling on one of the curves.

To design a resistance graphically in the manner described, it is necessary to know or to assume the motor resistance. The base current, or point at which the resistance steps are cut out, will be known. Then, if the inrush is known, the number of steps can be found, or, if the steps are known, the inrush can be found, and in either event the taper of the resistance steps can be determined.

The following example will illustrate a *nongraphical* method of calculating an accelerating resistor.

Assume a 50-hp 230-V motor having a full-load current of 180 amps, and assume that it is desired to limit the starting inrushes to 175% of full-load current, or 315 amps. The resistance steps are to be cut out when the inrush current has dropped to normal. The resistance of the motor and the leads to it may be assumed to be approximately 11% of E/I, or

$$\text{Motor resistance} = 0.11 \times \frac{230}{180} = 0.14 \text{ ohm}$$

The total resistance including the motor is

$$R = \frac{230}{315} = 0.73 \text{ ohm}$$

When the circuit is closed, an inrush of 315 amps occurs and the motor starts to accelerate. As it accelerates, it generates a countervoltage, so that the voltage across the resistance and the current from the line decrease. When the current has dropped to a normal full-load value, voltage across the resistance is

$$E_2 = 0.73 \times 180 = 131 \text{ V.}$$

At this point the first step of resistance is cut out. Since the voltage across the armature only cannot change instantaneously (speed cannot change instantaneously), the voltage across the resistor *remaining in the circuit* must also stay the same as the total voltage across the resistor was the instant before a resistance step was cut out. Only one thing can change, namely, the current which we desire to limit to 315 amps. The remaining resistance, to give a second inrush equal to the first, is

$$R_a = \frac{131}{315} = 0.418 \text{ ohm}$$

Further acceleration of the motor causes the current to drop to normal again, when

$$E_3 = 0.418 \times 180 = 75 \text{ V}$$

Proceeding in the same manner, the rest of the calculation is carried through:

$$R_b = \frac{75}{315} = 0.238 \text{ ohm}$$

$$E_4 = 0.238 \times 180 = 42.8 \text{ V}$$

$$R_c = \frac{42.8}{315} = 0.136 \text{ ohm}$$

This value is practically that of the motor resistance. The controller resistance steps may now be calculated:

$$R_1 = R - R_a = 0.73 - 0.418 = 0.312 \text{ ohm or } 53\% \text{ of the total}$$

$$R_2 = R_a - R_b = 0.418 - 0.238 = 0.180 \text{ ohm or } 30\% \text{ of the total}$$

$$R_3 = R_b - R_c = 0.238 - 0.136 = 0.102 \text{ ohm or } 17\% \text{ of the total}$$

$$\text{Allowance for motor and leads} = 0.136 \text{ ohm}$$

$$\text{Total} = 0.730 \text{ ohm}$$

Three steps of resistance are therefore necessary. If lower inrush peaks are desired, or if the steps are cut out at a current higher than normal, more steps will be required. In other words, the number of steps is inversely proportional to the ratio of the peak value and the base value.

In this discussion the acceleration curves of the motor have been assumed to be straight lines. Actually, they follow the shape of the motor characteristic curve, and bend slightly, so that the current peaks obtained in practice will not be quite so high as those calculated above.

The taper which has been determined in the example will be correct for any resistor of three steps having inrush peaks of 175% and base currents of 100%. That is, the total ohms required having been calculated, the first step will be 53% of the total, the second step 30%, and the last step 17%, regardless of the horsepower of the motor.

Taper tables may be worked out for any number of steps and for different values of inrush and base currents. The base current will generally be 100% for time-limit controllers, and 110 or 120% for current-limit controllers. The relays of current-limit controllers must be set to close above full-load current, to insure that the motor will have torque enough to accelerate. The starting torque required by some machines is higher than normal, and the inrush and base currents must then be high enough to allow the necessary torque. If the starting torque is low, as in a fan or a centrifugal pump, it is advantageous to reduce the inrushes, applying only as much torque as is necessary. This will give a smoother acceleration and will not apply so severe a shock to the machine. Also, it will reduce the power required for starting and will generally result in some saving in resistance material.

Figure 7.6 illustrates the acceleration of a motor driving a centrifugal pump. The load curve shows the torque required to drive the pump at any speed. At low speed the pump is not delivering much water, and the torque is low. As the speed increases, the torque increases also, until full torque is reached at full speed. The rise in the load curve at zero speed is caused by the addition of the torque required to overcome friction at starting. Obviously the application of 150% torque to this drive would be excessive and unnecessary. The acceleration curves have been laid out graphically to give increasing torque as the speed and load increase, and the base currents have been selected safely above the value required to insure acceleration. Since the average current carried by the resistor is less than it would be if the base currents were at normal or above, less resistance material will be required.

Reversing. Equation 7.3, for torque, shows that the motor may be reversed by reversing either the armature or the field but not both. It is customary to reverse the armature because the high inductance of the field windings makes them relatively hard to open and slow in response. Motors are often plugged for quick reversal, which means opening the armature circuit and immediately reconnecting it in the reverse direction before the motor has stopped.

Plugging. When a motor is plugged, or connected to the line in the reverse direction while still running full speed in the forward direction,

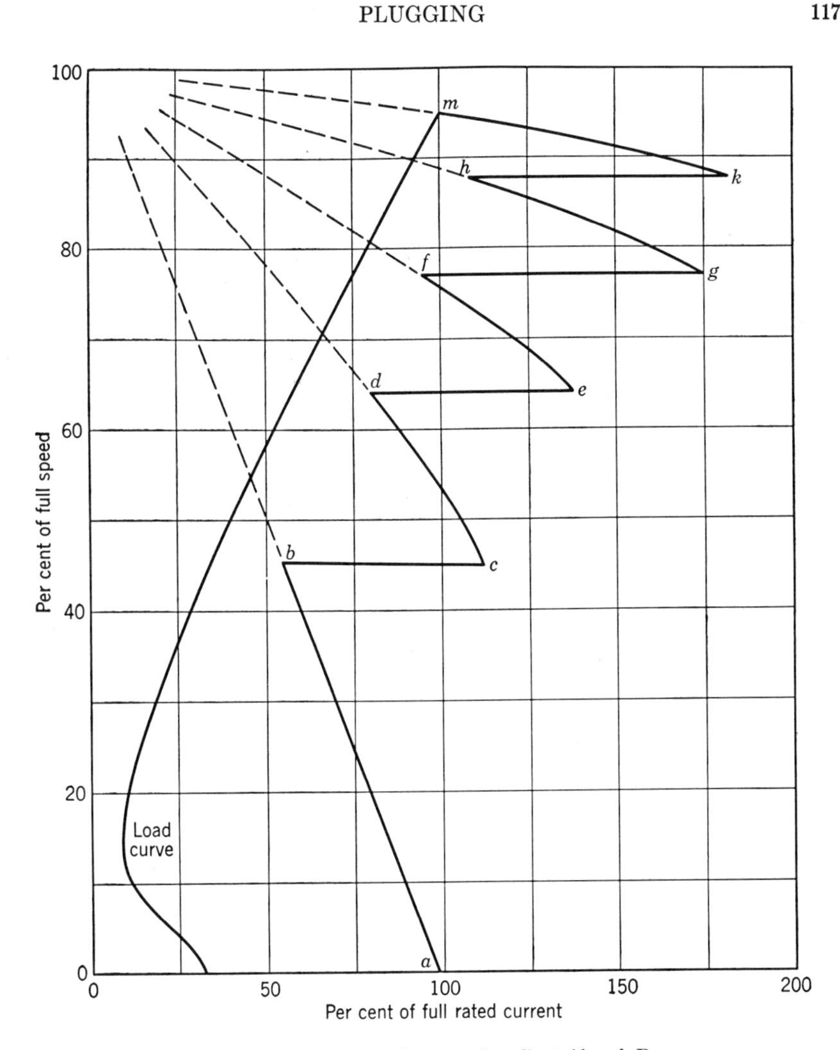

Fig. 7.6 Acceleration Curve of a Centrifugal Pump

the countervoltage of the armature is added to that of the line in forcing current through the armature and series resistance. The series resistance must be increased to limit the current to the same value obtained in starting from rest. This requires an additional step of resistance and a contactor to short-circuit it just before the motor reaches zero speed and reverses. The total resistance is

$$R_t = \frac{E + (E - I_a R_m)}{\text{Inrush current}}$$

A convenient formula for general use assumes about 90% of the double voltage available:

$$R_t = \frac{1.8E}{1.5I} = 1.2\frac{E}{I}$$

where E = line voltage and I = normal current.

The plugging step is determined by subtracting the value of the accelerating resistance from R_t.

If the motor has speed regulation by field weakening, and the control is arranged to strengthen the field during acceleration, the countervoltage will be increased in proportion to the field strength. For instance, for a 2-to-1 motor,

$$R_t = \frac{2.7E}{1.5I}$$

and for a 3-to-1 motor,

$$R_t = \frac{3.6E}{1.5I}$$

However, it is usually not essential to plug motors having speed control by field, and such controls may be arranged to prevent plugging. Some motors, particularly those for high-speed metal planers, are designed to stand plugging at weakened field.

Shunt Motor. Dynamic Braking. The purpose of dynamic braking (Fig. 7.7) is to obtain a quick stop. The armature is disconnected from the line, and a step of resistance is connected across it. The field remains on the line. Since the motor is turning, it acts as a generator and forces current around the loop formed by the armature and the braking re-

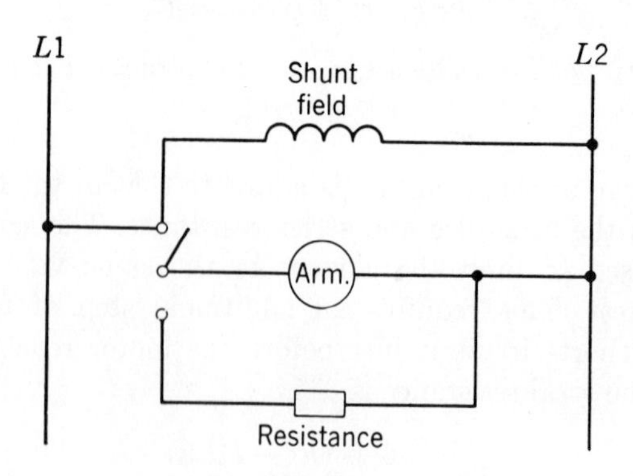

Fig. 7.7 Dynamic Braking Connections for a Shunt Motor

sistance. At the start of the braking period the armature voltage is $E - I_a R_m$, and the ohmic value of the braking resistance is

$$R = \frac{E - I_a R_m}{I}$$

The value of I is dependent on the severity of the braking desired. It is usually set at 150% of normal current, but where very quick stopping is desired, the current may be allowed to go as high as 300% of normal. As the motor slows down, the generated voltage falls, and consequently the braking current and braking effect are reduced. At zero speed there is no braking effect, and the final stopping of the motor is due to friction. For very quick stopping the braking resistance is sometimes cut out in two or more steps as the voltage falls, thus keeping the current high. It is not unusual to stop a large machine in 1 sec or less.

Shunt motors having speed regulation by field control will commutate heavy braking currents better if the field is at full strength. It is customary to provide means to strengthen the field during the braking period by short-circuiting all or a part of the field rheostat. If this is done, the braking resistance must be increased to compensate for the increased countervoltage of the motor. For a motor having 2-to-1 speed increase by field, the resistance would be doubled. In any event,

$$R = \frac{E - I_a R_m}{I} \times \text{Speed range by field}$$

Speed Regulation below Normal. Equation 7.8 will show that the speed of a shunt motor may be reduced below normal by reducing the voltage across the armature or by increasing the strength of the shunt field. Since the field is already across the line, there is no practical way to increase its strength. The voltage to the armature may be reduced by reducing the voltage of the generator supplying it, but the use of resistance in series with the armature to obtain speeds below normal is widespread because of the simplicity of the method.

Let E = line voltage
R_m = motor resistance
R = controller resistance
I = current
S_n = normal speed
S = reduced speed
E_n = countervoltage at normal speed
E_a = countervoltage at reduced speed

Then

$$E_n = E - IR_m$$

$$E_a = E - (IR_m + IR)$$

Since the shunt field strength is constant, the speed is proportional to the countervoltage, and

$$\frac{S}{S_n} = \frac{E_a}{E_n} = \frac{E - IR_m - IR}{E - IR_m}$$

$$\frac{S}{S_n}(E - IR_m) = E - IR_m - IR$$

$$IR = E - IR_m - (E - IR_m)\frac{S}{S_n}$$

$$IR = (E - IR_m)\left(1 - \frac{S}{S_n}\right)$$

Since S/S_n is the reduced speed in percentage of normal speed, $1 - S/S_n$ is the percentage speed reduction, and

$$R = \frac{(E - IR_m) \times \text{Per cent reduction}}{I} \tag{7.11}$$

The value R_m may be assumed to be approximately $0.05E/I$. The current I is not the full-load rated current of the motor but the actual current taken by the motor when running at reduced speed. The load characteristics of most machines will fall into one of two classes, either machine-type or fan-type load. Experience shows that with a machine-type load the current follows approximately the curve of Fig. 7.8, being

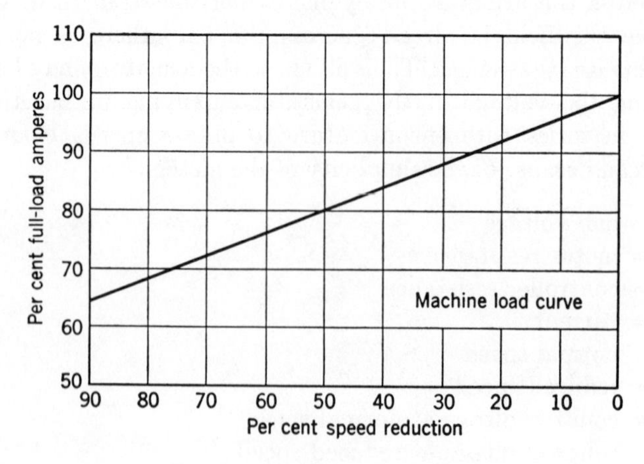

Fig. 7.8 Machine Load Curve

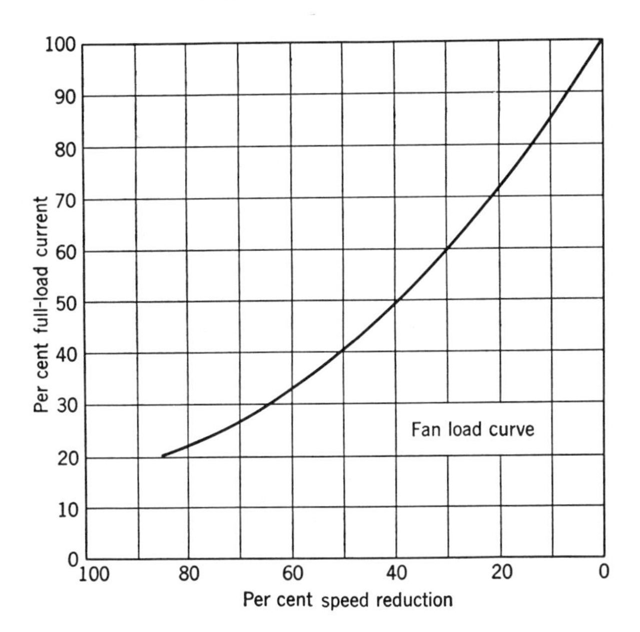

FIG. 7.9 Fan Load Curve

80% at 50% speed, and 65% at 10% speed. The machine takes less power to drive it at reduced speeds, because of the reduction in friction and windage losses. With a fan-type load, the power required to drive the load varies as the cube of the speed. In addition, the friction varies with the speed. The resultant curve of current versus speed is shown in Fig. 7.9. The two curves of fan and machine load characteristics will be applicable to the majority of installations, and the results obtained by employing them will be sufficiently accurate. Of course, if information about the actual load is available, calculations should be based on that load, and if accuracy is essential, data on the load should be obtained. There are two disadvantages in the use of series resistance for speed reduction, both of which grow more serious as the speed reduction is increased. The first is the fact that, if the load is variable, the speed will vary with it, and the second is the fact that considerable power is wasted in the resistor. From equation 7.11,

$$\text{Per cent reduction} = \frac{IR}{E - IR_m}$$

Since IR_m is so small as to be practically negligible, the speed reduction varies almost directly with the load current. Reference to the curve of Fig. 7.5 will show that this variation will not be serious with a small speed reduction but may be quite serious with a large one. If sufficient resistance is in circuit to give a small reduction at full load, so that the motor will run on curve *ef*, a 20% increase in load will cause a speed change from 81 to 77, or 5%. With a higher resistance and a greater reduction at full load, the motor will run on curve *cd*, and a 20% increase in load will cause a speed change from 67 to 60, or 10.5%. With still higher resistance, and the motor running on curve *ab*, a 20% increase in load would cause a speed change from 43 to 31, or 28%.

Because of this variation of speed with load, regulation by series resistance is usually limited to 50% speed reduction.

If the motor resistance is neglected, the speed reduction is proportional to the voltage drop across the controller resistor. At 50% reduction the voltage across the resistor is approximately equal to that across the motor, and since the current is the same in both, the power lost in the resistor is equal to the power used by the motor. At 75% reduction the power loss in the resistor is three times the power taken by the motor. This method of speed regulation is, therefore, very inefficient. It should be used where the required reduction is 50%, or preferably less, and where the slow speed is infrequently used.

Armature Shunt Resistance. Variation of speed with changes in load may be reduced by means of an armature shunting resistor. The resistor is connected in parallel with the armature, and its purpose is to increase the current flowing through the series resistor. If the current flowing through the shunt is high in proportion to that in the armature, changes in the armature current will have relatively little effect on the current in the series resistance. Consequently, the voltage across the series resistor will not change very much with changes in load, and the speed will remain more nearly constant. The armature shunt method is of particular value for speed reductions greater than 50% and may be

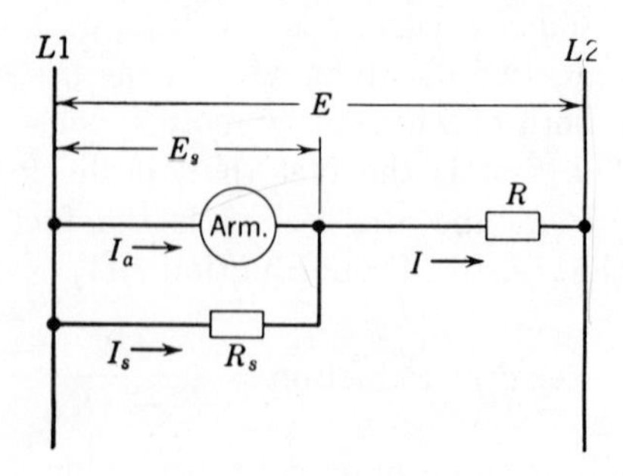

FIG. 7.10 Armature Shunt Connections for a Shunt Motor

safely used to obtain reductions up to 90% without danger of stalling. The method, however, is even more wasteful of energy than the straight series-resistance method.

When designing an armature shunt resistance, the series resistance must be known. The other known factors are the desired speed reduction and the load at the reduced speed. Referring to Fig. 7.10,

E = line voltage

I_a = the armature current required to drive the load. If the load is unknown, use the ampere rating of the motor

R = the series resistance (This usually will be the accelerating resistance but sometimes will be determined by the load at the speed required.)

R_a = the resistance of the armature. For speeds above 20% of normal speed, and for motors larger than 200 hp, this may be assumed to be zero. In other cases average values are

$$5 \text{ hp } R_a = 0.10E/I$$

$$50 \text{ hp } R_a = 0.05E/I$$

$$200 \text{ hp } R_a = 0.04E/I$$

E_a = the countervoltage of the armature at the desired speed. For constant field strength the countervoltage is directly proportional to the speed:

$$\text{Full speed countervoltage} = E - I_aR_a$$

The desired speed being known,

$$E_a = \frac{\text{Desired speed}}{\text{Normal speed}} \times (E - I_aR_a)$$

E_s = the voltage across the armature (and the armature shunt resistance). This is the sum of the countervoltage and the resistance drop in the armature,

$$E_s = E_a + I_aR_a$$

I = the current in the series resistance,

$$I = \frac{E - E_s}{R}$$

I_s = the current in the shunt,

$$I_s = I - I_a$$

R_s = the resistance of the shunt,

$$R_s = \frac{E_s}{I_s}$$

The current-carrying capacity of this resistor is easily determined, since the current is practically constant at the calculated value I_s.

Speed Regulation Above Normal. Equation 7.8 shows that speeds above normal may be obtained by increasing the armature voltage, or by weakening the strength of the field. The field is readily weakened by connecting resistors or rheostats in the field circuit. Standard industrial dc motors will permit a speed increase of up to 400% by field weakening (Table 7.1), but for anything beyond that range a specially designed motor is required. A range of about 6 to 1 by field weakening is considered to be the upper limit. This method of speed control is efficient, because the power lost in the field rheostat is negligible. The speed regulation at weakened field current is greater (more speed droop) than with full field.

Increasing the armature voltage above so-called rated 230 V or 240 V is acceptable on a modern industrial motor. Speed increase is obtained approximately in proportion to the increase in voltage. It is usually permissible to increase the voltage to 480 V (double speed). What this amounts to is that the motor is only being used to half its voltage capability when operated on a 230-V or 240-V line. Since the current (or torque) rating of the motor remains essentially constant at higher voltages, the horsepower capability of the motor increases with voltage.

Wide Speed Ranges. Where extremely wide speed ranges are required, they may be obtained by a combination of speeds below and above normal. The normal speed might, for instance, be reduced to 20% by a combination of series and shunt resistance, and increased to 4 times normal by field weakening. The whole range would then be 20 to 1. Adjustable voltage is often combined with speed increase by field weakening to give a wide speed range.

Overhauling Loads. As the load on a motor decreases, the speed increases, until, at zero load, there is no current in the armature, and the countervoltage is equal to the line voltage. If the load is negative, or overhauling, the countervoltage will be higher than the line voltage, and the motor will return power to the line. When the load is positive,

$$\text{Countervoltage} + \text{Resistance drop} = \text{Line voltage}$$

When the load is negative,

$$\text{Countervoltage} - \text{Resistance drop} = \text{Line voltage}$$

The increase in speed of a shunt motor, when the load is overhauling, is not very great, since the resistance drop is usually not more than 10%. However, if external resistance is inserted in the armature circuit, the speed will increase instead of decreasing.

Special Adjustable-Speed Motors. There are two types of dc motors whose speed can be adjusted by varying the reluctance of the field circuit instead of by changing the field current. In one construction the armature and the field poles are slightly tapered, and the armature is shifted along the axis of the shaft by means of a hand wheel. By thus shifting the armature, the air gap between the armature and the field poles is changed. With a small air gap the flux is greater and the motor speed is lower; with a large air gap the speed is higher.

The other motor of this type has hollow field poles with movable plungers, and the reluctance is varied by moving the plungers in and out. Comparatively few motors of this type are used.

Adjustable Voltage and Multivoltage Control. Motors are frequently controlled by varying the supply voltage to the armature, instead of having resistance in series with the armature. The advantages of this method are many, primarily:

1. Wide speed range control.
2. Speed not appreciably affected by changing load.
3. Less power wasted at low speed.
4. Ease of adaptability to sophisticated regulated drive systems.

Adjustable voltage may be obtained by varying the generator field on an M-G set supplied specifically for this purpose, or by varying the control to a static power conversion unit. Adjustable voltage controllers are discussed in detail in Chapter 9.

The use of two voltages is possible, and some installations have been made with a number of different voltages. Two voltages plus some resistor control plus motor field weakening for speeds above normal, can be used to provide speed ranges of 10 or 15 to 1. However, with the advent of many versions of adjustable voltage, this method is seldom used today.

Motor Protection. In addition to its principal function of controlling a motor, a controller usually includes devices to protect the motor and its driven machine. Protection against overload is included as a part of most controllers. Thermal overload relays are used to some extent for dc motors, but a more common practice is to use a magnetic relay which has a time-delay action on ordinary overloads but will trip out instantaneously if the motor stalls. The instantaneous trip feature is desirable because the stalled motor current may be as high as 20 to 40 times normal, and the motor might be burned out before a thermal relay would trip.

Shunt motors are often protected against loss of field, and against too rapid acceleration or deceleration, by field control. The methods are described in Chapter 8.

Manually Operated Controllers. Although seldom used, manually operated controllers for shunt motors are made in three principal forms: face-plate controllers, multiple-switch controllers, and drum controllers.

Face-Plate Type. Figure 7.11 shows a typical face-plate-type controller, and Fig. 7.12 is a diagram of it. Connection is made from one side of the line to a lever, which carries a contact brush, or shoe. The brush rides on a series of stationary contact buttons to which resistance is connected. When the lever is moved to the first button, the circuit is closed, and as the lever is moved on across the buttons, the resistance is gradually short-circuited. The last contact is connected directly to the motor. The lever is provided with a spring which will return it to the OFF position if it is released at any time during the starting period. When the lever has reached the final contact button,

FIG. 7.11 DC Face-Plate Starter

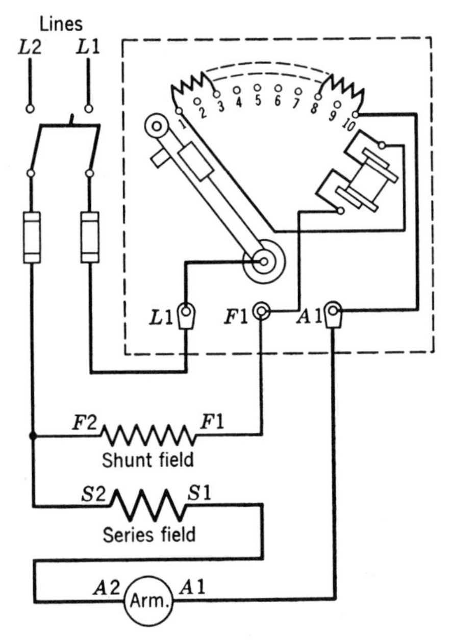

FIG. 7.12 Diagram of a DC Face-Plate Starter

it engages a pivoted armature, or keeper, which closes the magnetic circuit of a small magnet. The magnet holds the lever in the ON position as long as there is voltage on the line. In the event of voltage failure, the lever is released and returns to the OFF position. This avoids the danger of connecting the motor directly to the line when power returns.

When it is used with shunt motors, the common practice is to connect the release coil in series with the shunt field, so that any failure of the field circuit will release the lever and stop the motor. The shunt field is energized on the first contact button, insuring full field from the moment of starting, and a connection is also made from the first button to the frame of the release magnet so that the starting resistance will not remain in the field circuit when the motor is running. The use of the release coil in series with the field requires that the coil be selected to suit the current of the motor with which it is to be used. Usually, when these release magnets overheat or fail to hold properly, the trouble is that the field current is not what it was expected to be. The release coil may be connected directly across the line if desired. With small series motors, the coil is connected in series with the line, but above 10 amp a shunt coil connected across the line is used.

Drum-Type Controllers. The drum type of manually operated controller has a number of advantages over both the face-plate and the multiple-switch types. These advantages are principally due to the method of construction, which is radically different from that of either of the other types of controller.

The essential parts of a drum controller are

1. The case
2. The contact fingers
3. The cylinder
4. The blowout magnets
5. The driving mechanism

The case may be of iron, cast in one piece—a common construction for small drums—or it may consist of cast-iron end pieces joined by a back piece of boiler plate. In the second type, the back piece is welded to the end pieces. This is a flexible arrangement, since one design of end piece will do for drums of different lengths. A cover of sheet metal fits over the end pieces, to protect the sides and front of the drum. The cover may be provided with rubber gaskets to make it dust-tight. The wiring is brought into the drum through bushed holes, or conduit fittings in the back plate or in the end casting.

The fingers are the stationary contacts to which the supply lines, the motor, and the resistor are connected. Each finger consists of a copper tip riveted to a supporting channel of steel or brass. The finger is mounted so that it is pivoted at the back end and held in contact under the pressure of a spring. An adjustment is provided to permit changing the arc gap to compensate for wear. There are a number of methods of mounting the fingers. A common construction is shown in Fig. 7.13, where the fingers are mounted on a square steel shaft which extends the length of the drum. The shaft is insulated by being wrapped with an insulating material or covered with a tube of insulation. Over this insulation steel brackets are mounted, and on them the finger is supported. The brackets also serve to mount a small insulating board which prevents copper dust from falling on the shaft and eventually causing a short circuit.

The cylinder provides the moving contacts that make connections between various fingers. No wiring is brought to the cylinder. An insulated square shaft is generally used, on which cast-iron or brass sections are mounted. Rolled copper contact segments are screwed to the cylinder castings. These segments may be of any desired length and may be arranged to make contact with a finger at any desired position as the cylinder is rotated. The segments may be easily renewed when worn out.

One of the principal differences between the drum-type controller and

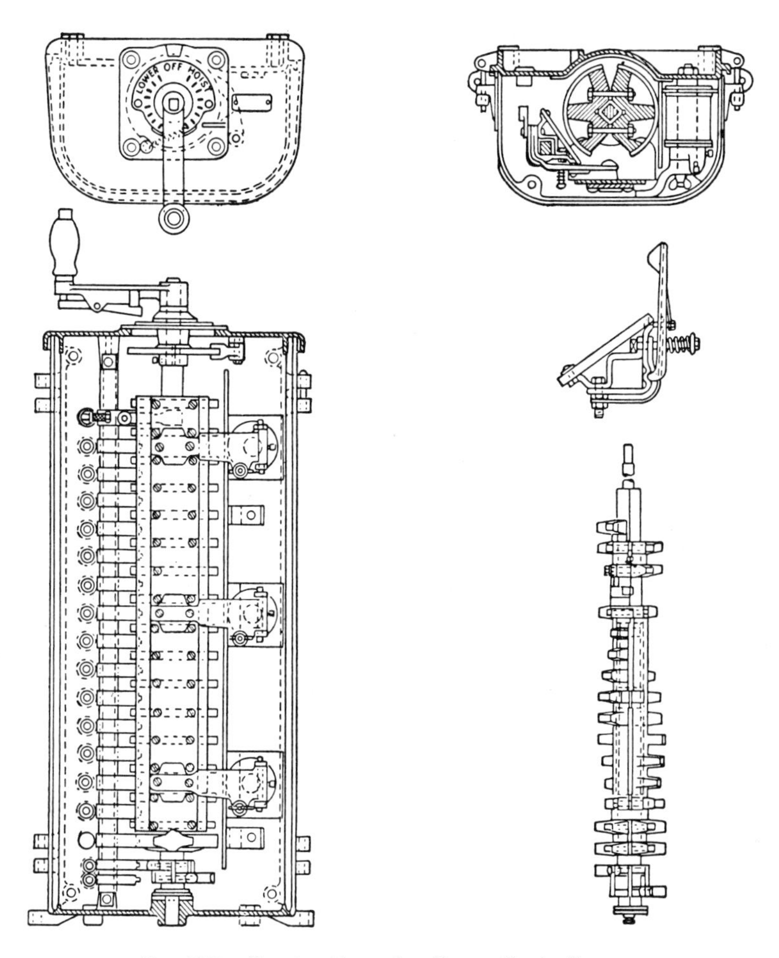

Fig. 7.13 Construction of a Drum Controller

the face-plate type is that a magnetic blowout can readily be incorporated in the drum structure but cannot be easily built into a face-plate controller. This is a decided advantage, since much heavier currents can be commutated with the aid of the blowout magnet. The blowout coil is made of copper wire or bar, the size and number of turns being determined by the motor current. The coil is wound on a steel core fastened to the drum case. The circuit for the magnet flux is through the core, through the drum case, then to the cylinder. From the cylinder the flux crosses the air gap in the path of the contact finger and completes its circuit through the core support to the core. Assuming that the drum cylinder has just been moved to the off position, and that there is an arc hanging between the end of the segment and the tip of the finger, the reaction between the magnetic field because of the arc and the magnetic field of the blowout coil will cause the arc to be deflected either up or down, depending on the direction of current in the arc. The arc, in deflecting, will strike the barrier either above or below the finger, and will be cooled and dissipated.

The simplest type of drive for a drum is the radial handle, once common on streetcars. It is broached to fit on the end of the shaft, or it may be keyed to the shaft. Straight-line motion of the operating lever is secured by gearing the lever to the shaft through a bevel gear. Rope drive is obtained by mounting a sheave wheel on the drum shaft. Any of these drives may be arranged to return to the OFF position when released or to remain in any selected position.

The self-returning action is accomplished by means of a coiled spring that is wound up as the drum lever is moved to a running position and unwinds when the lever is released. If spring return is not used, the lever is held in any desired position by a star wheel and pawl. The star wheel is a toothed cam keyed to the drum shaft. The pawl, or roller, is arranged to roll over the surface of the star wheel as the cylinder is moved, and it is held in contact with the surface by a tension spring. The spring is strong enough to move the cylinder, so that the cylinder always stops in a position where the roller is in one of the star-wheel notches. This is an important feature of the drum controller because it insures stopping the cylinder at definite positions which can be selected so that the fingers are squarely on the segments and making good contact. If it were not for the star wheel, the operator would be unable to feel the speed points and might pause during the operation in such a position that a finger would be making poor contact with a segment. This would produce heating or arcing between the finger and the segment, which would shorten the life of the contacts. The relation of the star wheel and the contacts can also be arranged so that the finger tips do not rest directly on top of one of the screws which hold the segment to the cylinder casting.

These advantages of the star wheel also apply to spring-return drums, and star wheels are sometimes used with that type of drive, but it is rather difficult to obtain a positive spring return and at the same time a star-wheel action that is strong enough to indicate definite positions. For this reason the star wheel is generally omitted with spring return.

Advantages of Drum Controllers. The drum controller has a number of advantages over the face-plate controller and the multiple-switch controller. In the first place, the drum has a better mechanical construction. It is stronger and able to stand more abuse than either of the other controllers. The construction of the drum also is favorable to good insulation, which is an important point.

Another advantage of the drum controller is that, owing to its construction and method of operation, heavy pressures can be maintained between contact surfaces without causing undue strains or making the operation difficult. If the contact pressure of a face-plate controller is increased very much, there will be a severe bending strain on the bearing shaft and a tendency to crack the slate at that point. Heavy pressure between the contacts will also increase the friction and make the operation of the starter difficult. These limitations do not apply to the drum construction.

Blowout magnets and arc barriers can readily be applied to the drum, as has been mentioned.

The drum structure lends itself more readily to complete enclosure than either of the other types. This is an important advantage because it affords safety to the operator and protection to the drum itself.

Complicated circuits, like those for reversing a motor or for handling both the primary and secondary windings of a slip-ring motor, are more easily handled in a drum controller than in any other manually operated controller. It is a simple matter to have the segments of the proper length so that contacts are made in any desired position. Sections of the cylinder may be insulated from each other, if necessary, by having separate castings for the two sections and mounting them on the insulated shaft separately.

The space required by a drum controller is usually less than that required by a face-plate or multiple-switch controller of the same rating and providing the same functions.

A drum controller is easier to operate than the other types of manual controllers, and the wide variety of drives for a drum is an advantage in adapting it to a particular service condition.

Auxiliary Functions. Contactors and relays are used in connection with drum controllers to obtain protective functions.

Low-voltage protection (Fig. 7.14), which is inherent in the face-plate controller, necessitates a contactor with the drum. The coil of the contactor is energized through a pair of auxiliary fingers in the drum, arranged to close the circuit in the OFF position of the drum only. When the contactor has closed, the pickup circuit in the drum is paralleled by interlocks on the contactor, so that the contactor will remain closed when the drum is moved to a running position. The main circuit to the motor is fed through both the contactor and the drum. If the voltage fails during running, the contactor will open, and it will be necessary to return the drum to the OFF position in order to energize the contactor coil again. It will be noted that with this arrangement the contactor, once closed, will remain closed unless the power supply is interrupted. All making and breaking of the current is done in the drum.

Since a contactor is specially adapted to the opening and closing of heavy currents, it is often desirable to use one for that purpose in connection with a drum, energizing the contactor coil on the first point of the drum instead of in the OFF position (Fig. 7.15). This does not give true low-voltage protection, since it only requires that the drum be brought to the first position after a voltage failure. If the drum is in that position when power fails, the motor will be restarted upon return of power. However, the arrangement has the advantage of using the contactor to open the circuit and the further advantages that the contactor coil is energized only when the motor is running. Both the low-voltage protective arrangement and the low-voltage release arrangement are common. The advantages of both may be obtained by the use of a relay in connection with the contactor, the relay being energized in the OFF position and the contactor in the first position of the drum.

Limit switch protection in two directions of travel may be secured

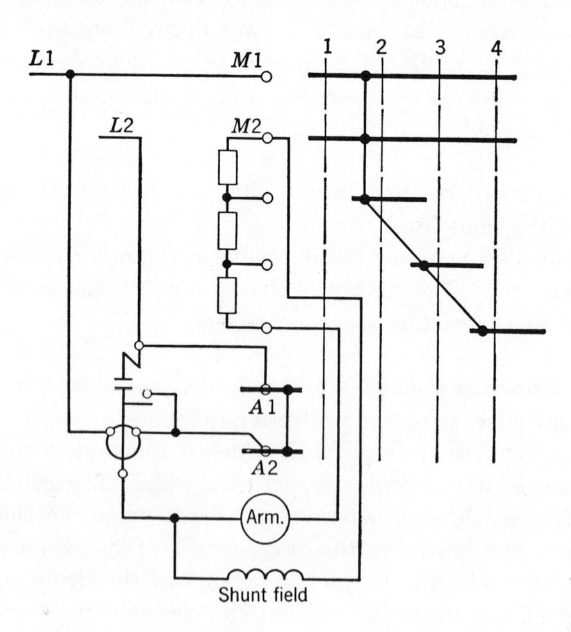

FIG. 7.14 Connections for a Drum Controller with Low-Voltage Protection

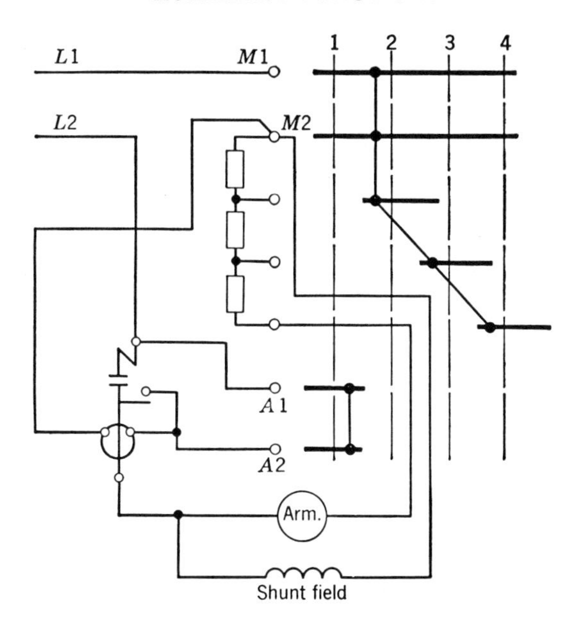

FIG. 7.15 Connections for a Drum Controller with a Line Contactor

by a single contactor in connection with a reversing drum controller. In the forward direction of travel the coil of the contactor is energized through the forward limit switch. When the switch trips, the circuit to the coil is broken and the contactor opens, stopping the motor. The drum may then be moved to the reverse position, when the contactor coil is energized again, this time through the reverse limit switch, and the motor will run in the reverse direction until that limit switch trips.

Drums for shunt motors are built in several forms, as follows.

1. Across-the-line, rated to 3 hp, 115 V, and 5 hp, 230 V.
2. Starting only, by armature circuit resistor.
3. Speed regulating, by armature resistor, to 50% speed reduction.
4. Starting only, by armature resistor, and speed regulating by field resistor.
5. Speed regulating, below normal by armature resistor, and above normal by field resistor.

All types are built either reversing or nonreversing. They can also be built with armature shunt for 90% speed reduction, and with dynamic braking circuits. All except type 1 are rated as shown in Table 7.2.

Continuous ratings apply wherever any point on the drum switch will be used for any period exceeding 5 min. Intermittent ratings should

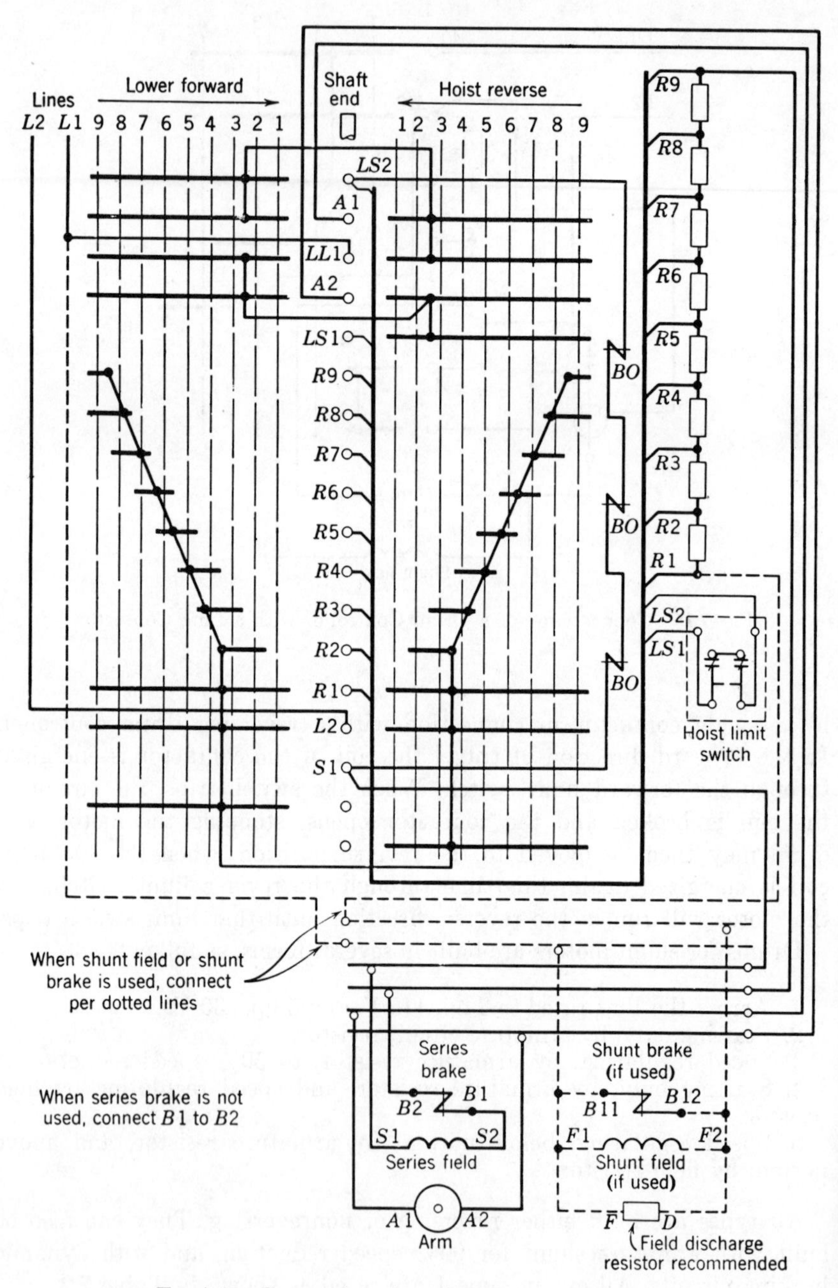

Fig. 7.16 Connections for a DC Drum Controller

be used for crane, hoist, or other duty where the running time is not over 50% of the total time, and the maximum running time is not more than 5 min.

The resistance material used with drum controllers for armature starting or speed regulation is made up of cast-iron grids or units mounted

TABLE 7.2

NEMA Drums for Shunt Motors

8-hour Ampere Rating	8-hour Ratings, Horsepower at 115 V	8-hour Ratings, Horsepower at 230 V	Intermittent Duty, Horsepower at 115 V	Intermittent Duty, Horsepower at 230 V
50	5	10	7½	15
100	10	25	15	30
150	20	40	25	50
300	40	75	50	100

separately from the drum. Resistors for connection in the shunt field circuit are sometimes mounted in the drum case, since they occupy a relatively small space.

Magnetic Controllers. Some small shunt motors, up to 1½ hp, 115 V, and 2 hp, 230 V, are designed so that they can safely be started by direct connection to the power supply. Magnetic controllers are made for that purpose, both in reversing and nonreversing forms. These controllers are very simple, being just a magnetic line contactor, or pair of reversing contactors, and an overload relay.

Controllers for larger motors are made in a number of forms to suit the requirements of the machines which they control. They may be any of the following.

1. Nonreversing, starting service.
2. Nonreversing, starting service, plus dynamic braking.
3. Nonreversing, starting service, with speed regulation by field control.
4. Nonreversing, starting and dynamic braking, with speed regulation by field control.
5. Reversing, with dynamic braking.
6. Reversing, with dynamic braking, with speed regulation by field control.

The control devices used for these controllers are:

A line contactor or a set of reversing contactors.
A set of one to four accelerating contactors.

An accelerating means, usually of the time-limit type.
An overload relay.
A dynamic braking contactor.
A field accelerating relay for speed-regulating controllers.
A field discharge resistor.

In addition, the controllers may include a field-failure relay, field decelerating relay, and control circuit fuses. The armature resistor, NEMA class 135 or 136, is made up of wire-wound units, or welded steel grids, and is contained in the case which mounts the control panel. Operating pushbuttons, or master controllers, are separately mounted. Standard ratings for these controllers are shown in Table 5.1.

Figure 7.17 shows a typical controller of the reversing type, with speed regulation by field control. The four reversing contactors as shown in the center of the panel are mechanically interlocked in pairs to avoid

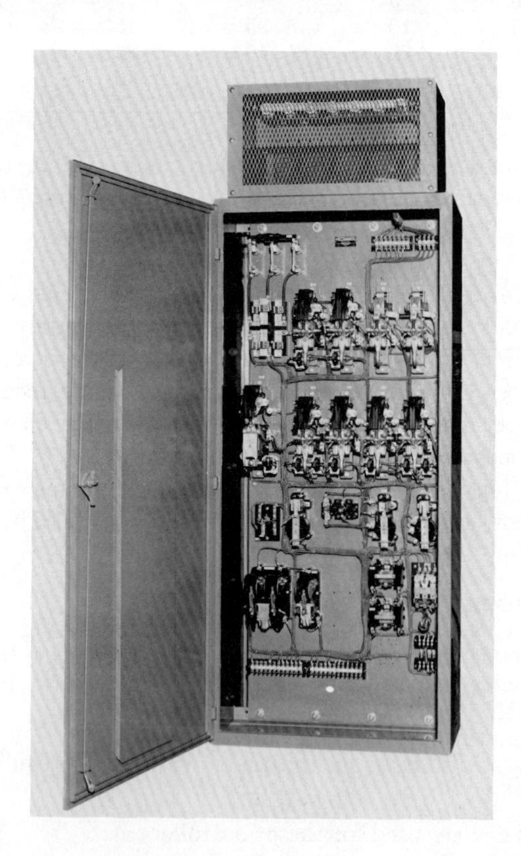

Fig. 7.17 Definite-Time-Limit Automatic Reversing Starter with Dynamic Braking

inadvertent line short circuit. The dynamic braking contactor is mounted to the left of the reversors at the extreme left of the panel.

Shunt motors are built in sizes much larger than those listed above for controllers, and, of course, it is possible to build magnetic controllers for them in any required size. For very large motors, whose current is above the rating of available contactors, magnetically operated circuit breakers are used as line-closing devices. The starting resistor for such motors is connected in parallel instead of series steps, so that the accelerating contactors each carry only a portion of the total current.

Application of Shunt Motors. Direct-current shunt motors are particularly applicable to those machines requiring a wide range of speeds, or a large number of operating speeds. The ability to obtain a quick stop by dynamic braking is also an advantage. Pumps, fans, some machine tools, printing presses, paper mills, elevators, and electric power shovels all use shunt motors. If a machine requires a constant speed at varying loads, a shunt motor will be selected. If rapid acceleration is desired, and relatively constant running speed, a compound motor should be used, and the series field might be cut out of circuit after the motor has accelerated. This is the usual practice for elevator installations.

Motors having speed adjustment by field control are sometimes larger and more costly than constant-speed motors, and these factors increase as the range of speed control is increased. Motors with high base speeds are smaller and less costly than slow-speed motors. For example, a 40 hp 1150 rpm motor weighs about 960 lb, whereas a 40 hp 650 rpm motor weighs about 1790 lb. A 7½ hp 500 to 2000 rpm motor weighs about 84% as much as a 7½ hp 400 to 1600 rpm motor.

Problems

1. Write an equation for the torque of a shunt motor.

2. Write an equation for the speed of a shunt motor.

3. A 25-hp 230-V dc shunt motor has a full-load current of 100 amp. Neglecting armature resistance, how many ohms will be required in the starting resistance to limit the starting torque to 200% of normal torque?

4. Again neglecting armature resistance, what would be the running speed of this motor when operating without resistance in its circuit and at half load?

5. Suppose that this motor were running with resistance in series with its armature limiting the voltage across the armature to 70% of line voltage, and suppose the motor is drawing 50 amp from the line; then what will the torque be in per cent of normal?

6. Under the conditions of problem 5, what will the speed be in per cent of normal?

7. A 50-hp 230-V motor has a full-load current rating of 180 amp, and its armature resistance is 5% of E/I. What is the armature resistance in ohms?

8. A 75-hp 230-V 268-amp motor is driving a centrifugal pump. The controller is designed for starting characteristics like those of Fig. 7.6. What is the ohmic value of the controller resistor?

9. If it is desired that the pump run at 45% speed with all resistance in circuit, what will be the controller resistance required?

10. A 35-hp 230-V 128-amp motor is equipped with a 3-step controller and a resistor which gives starting characteristics like Fig. 7.5. What is the ohmic value of the starting resistor?

11. If this motor is plugged, what current inrush will be obtained at the instant of plugging?

12. What is the ohmic value of the additional resistor step which must be added to hold the plugging inrush current to the same value as that obtained when starting from rest?

13. If the speed of this motor was increased 50% by field weakening, and the field rheostat was short-circuited during starting, what would be the ohmic value of the plugging resistor step?

14. A 10-hp 230-V 38-amp motor has an armature resistance of 10% of E/I. What is its countervoltage at full load, in per cent of the applied line voltage?

15. Calculate the ohms in a dynamic braking resistor which will limit the braking current to a maximum of 300% of the motor rating.

16. If the braking is to be done in two steps, each limited to 300% of motor rating, and the first step cut out when the braking current has dropped to 100%, how many ohms will be in each step?

17. What will be the approximate speed of the motor when the first step is cut out?

18. A 50-hp 550-V 75-amp motor has an armature resistance of 5% of E/I. How many ohms will be required in a series resistor to obtain 20% speed reduction below full-load speed, when the motor is operating at 70% load?

19. How many amperes must the resistor of problem 18 carry continuously?

20. Referring to Figs. 7.8 and 7.9, if a resistor is designed to produce 50% speed with a machine-type load, what speed will the same resistor produce with a fan-type load?

21. If the motor of problem 18 is used with a resistor designed to give 50% speed reduction with a machine-type load, how many additional ohms will be required to give the same speed reduction with a fan-type load?

22. Referring to Fig. 7.5, what per cent of no-load speed will be obtained on each of the four controller points if the motor is loaded to 140% of rated load?

23. If this motor is running with all resistance in circuit, how much will the speed change if the load changes from 140% to 90%?

24. If the motor is running with only the last step of resistance in circuit, how much will the speed change if the load changes from 140% to 90%?

25. A 40-hp 230-V 146-amp motor has a series resistor designed to give 50% speed reduction below normal full-load speed, when the motor is operating at full load. How many ohms are required in an armature shunt resistor which will give 80% speed reduction at full load?

26. With the series resistor and the armature shunt resistor in circuit, what will the motor speed be at zero load?

27. Under the conditions of problem 25, how many horsepower will the motor be delivering, and how many horsepower will be lost in the resistor?

28. What are the delivered horsepower and the lost horsepower under the conditions of problem 26?

29. A 25-hp 230-V 92-amp motor is running with a series resistor of 1.6 ohms and an armature shunt resistance of 6.8 ohms. The motor armature current is 20 amps. What is its speed?

30. What would be the speed of the motor if the load dropped to 10 amps?

31. What would be the ohmic value of an armature shunt resistor which would produce the speed of problem 29, with a load of 10 amps?

32. How many watts are lost in the resistor under the conditions of problem 29?

33. If the series resistance is reduced to 0.8 ohm, calculate the ohms in the armature shunt to produce the speed required by problem 29.

34. What is the wattage loss with the combination of series and shunt resistors of problem 33?

8

SHUNT FIELD RELAYS AND RHEOSTATS

When a shunt motor or compound motor is used to drive a machine, the handling of the shunt field introduces a number of factors into the control problem. Some of these will be present whenever a shunt field is involved; others may or may not need to be considered.

Discharge Path. When the shunt field circuit is opened, the high inductance of the field windings tends to oppose the opening of the circuit and to keep the current flowing. Opening the circuit abruptly may damage the insulation of the field windings. It is advisable, therefore, to supply a discharge path, so that the generated voltage may be limited to a safe value. With nonreverse controllers, it is customary to connect the shunt field behind the line contactor, so that when the circuit is opened a discharge path is provided through the armature of the motor and the starting resistance. With reversing controllers, the field cannot be connected in this manner, and a separate discharge path must be provided. The same is true of controllers having dynamic braking, for then the shunt field must be kept energized after the armature circuit has been opened.

In general, a discharge resistance is not essential on 115-V equipment but should be provided for 230-V motors rated at 7½ hp or higher, and for 550-V motors rated 5 hp or higher.

The recommended ohmic value for a discharge resistor is between one and three times the ohmic value of the shunt field. The discharge voltage will depend upon the resistance of the discharge path, so that, if a resistance of three times the field ohms is used for the discharge path, the generated voltage will be four times normal line voltage.

Where the wattage consumed by the discharge resistor is relatively low, it is good practice to connect the resistor directly across the line, behind the service knife switch. The resistor must then have continuous capacity for the current flowing in it. For larger equipments, a discharge resistor is connected into circuit by a discharge clip on the knife switch, so that it is not consuming energy when the field is connected to the line. The capacity may then be figured on the basis of handling the discharge current for approximately 15 sec.

If the field circuit may be opened at different points, as, for instance, by a set of reversing contactors, and also by a field knife switch, it is necessary to provide a discharge path to take care of each condition, or else to connect the discharge resistance permanently across the field itself.

Thyrite. Certain crystal materials have the characteristics of being good insulators at voltages below a critical value and good conductors at voltages above that value. One such material, developed by the General Electric Company under the name of Thyrite, is being used by control manufacturers as a field discharge resistor. Thyrite is a dark gray, dense, nonporous, ceramic compound, having mechanical characteristics somewhat similar to those of dry-process procelain. It does not burn or disintegrate under red heat, and it is not affected by moisture, oil, or gases. It conducts current without arcing and is not dependent on ionization, deionization, or breakdown of gas. The resistance follows a definite law, and the critical voltage may be determined from the equation

$$E = RI^{1-a}$$

where E = the applied voltage
I = the current in amperes
R = the resistance in ohms when 1 amp is flowing
a = a constant (approximately 0.76)

A typical block of this material for use as a 250-V discharge resistor is cylindrical in shape, $1\frac{5}{8}$ in. in diameter, and 1 in. long, with depressions in the ends to provide a conducting path of $\frac{1}{2}$ in. The equation for such a block is

$$E = 1350\, I^{0.24}$$

If the block is connected across a 250-V line, in parallel with a 1-amp field, the continuous current taken by the Thyrite will be approximately 0.001 amp. When the circuit is opened, and the field discharges through the Thyrite, the voltage will be limited to 1350 V, on the assumption that the circuit will be opened instantaneously. Actually the arc on the interrupting device will not be extinguished instantaneously, and the peak voltage will be lower than the theoretical value. The losses in the resistor will be 0.24 W, as against approximately 60 W for an ordinary resistance-type discharge unit. Another advantage of Thyrite is the fact that it may be connected across a field without affecting the characteristics of a rheostat connected in the field circuit.

Field-Failure Protection. With shunt or compound motors it is good practice to provide for shutdown in the event of failure of the shunt field circuit, as such failure means a runaway condition. Protection is obtained by means of a relay having its coil connected in series with the shunt field and its contacts in the stop circuit. Then, if the field fails at any time, the relay coil will be deenergized, and the relay in opening will shut down the motor. The coil of a field-failure relay must be able to carry the maximum field current continuously and at the same time must be able to close the relay at approximately 65% of the minimum field current. The 65% factor serves to take care of low-voltage conditions and of heating of the field and the relay coil. It is not difficult to design a relay to operate at these values, provided that field weakening is limited to a reasonable amount, but where the speed range by field is greater than 2 to 1, the current range may become so wide that it will be difficult to design a suitable relay. Under such conditions, it is generally satisfactory to assume that the field rheostat will be turned back sufficiently to permit picking up the field-failure relay when the motor is started. It is also generally satisfactory to connect the contacts of the field-failure relay in the maintaining circuit instead of in the pickup circuit. Since the field rheostat is short-circuited during the accelerating period, the relay will receive sufficient current to close as soon as the motor starts to accelerate.

When field-failure relays are used with series shunt motors having a series field of more than 40%, it is necessary to take special precaution to insure that the delay will remain closed during the acceleration period. During this period, the series field reacts upon the shunt field in such a manner as to greatly reduce, or even reverse, the current in the shunt field. If only a single-coil relay were used, the relay would drop out during the acceleration period and shut down the motor. One means of avoiding this is a double-coil relay, one coil being connected in series with the shunt field and the other in series with the armature accelerating resistance. The second coil then insures that the relay is held closed during the starting period. After acceleration, this coil is short-circuited.

Another method of obtaining the same result is to short-circuit the contacts of the field-failure relay by an interlock on the final accelerating contactor. With this arrangement, the field-failure relay may open during the accelerating period, but it will not shut down the motor.

After the acceleration is over, and the last accelerating contactor has closed, the relay is free to function as it normally should.

Economizer Relay. An economizer relay, or field-protective relay, is used to insert resistance in series with the shunt field whenever the motor is not running. It may be necessary to keep the shunt field energized when the motor is idle, for a number of reasons, and since the ventilation of the field is greatly reduced when the motor is not running, it is necessary to insert a resistance to prevent the field from overheating. The relay is simply a single-pole shunt-type relay having its coil connected in parallel with the coil of the main line contactor. The resistance may be designed to reduce the voltage across the field to one-half of the normal value, which will reduce the wattage in the field to one-fourth of the normal value.

Reversing. Wherever rapid reversing is required, it is necessary to reverse the armature circuit; but a number of drives require only emergency reversing, and for such drives it is often advantageous to reverse the shunt field instead of the armature. If the motors are large, the cost of the control may be reduced by omitting the large armature-reversing contactors and using the smaller reversing contactors required by the field. Where field reversal is employed, it is the usual practice to interlock the controller, so that it cannot be plugged and so that the field cannot be reversed until the motor has stopped. When reversing contactors are used, they are provided with a normally closed contact, which serves to set up a discharge circuit when the field is opened. Where remote control is not particularly required, the field is often reversed by a knife switch provided with discharge clip. This knife switch is usually interlocked with the main control circuit, so that, if it is opened when the motor is running, the main contactors will immediately be deenergized and the motor disconnected from the line.

Plugging. When a shunt motor is plugged under full field conditions, the countervoltage of the armature is added to the voltage of the line, and additional resistance, nearly equal to the accelerating resistance in ohms, must be used to limit the inrush current. If the motor has some speed increase by field weakening, the field will be strengthened when the motor is plugged and the countervoltage of the armature will be increased in proportion to the strengthening of the field. For example, assume that a motor having a speed increase of 2 to 1 by field is running at maximum speed and is plugged. As soon as the circuit is closed in the reverse direction, the accelerating relay will close and short-circuit the field rheostat. The field strength is then doubled and the armature countervoltage correspondingly doubled. The total voltage is therefore three times the line voltage; and, if the inrush current is to be limited to that obtained when starting from rest, resistance equal to three times the value of the starting resistance will have to be used. With motors of greater speed range by field, the resistance must be still higher in ohms.

Frequently plugging is not essential to the proper operation of the drive, and the extra resistance is only a safeguard in case of accidental plugging. Then it is usually more economical to arrange the control so that the motor cannot be plugged, and so avoid the use of additional resistance. The method of doing this when dynamic braking is employed has been explained. Where there is no dynamic braking relay, a normally closed voltage relay, having the coil connected across the armature, will serve the same purpose. The contacts of the relay are connected in series with the direction contactor coils and are bypassed by normally open interlocks on the direction contactors. It is not necessary that the motor be entirely stopped before the voltage relay closes to permit plugging. The relay is usually set so that the speed is low enough to insure that the countervoltage of the motor will not be excessive.

Speed Increase by Field Control. In order to increase the speed of a motor by weakening the field, resistance is inserted in series with the field, either by means of a rheostat or by means of a set of magnetic contactors. Contactors are used in connection with the control of large motors, where the field currents are high, and also where rapid and frequent commutation of the resistance is necessary.

When the field strength is reduced by the insertion of a step of resistance, the countervoltage generated by the motor is also reduced, and the difference between line voltage and the new countervoltage causes an inrush of current to the armature. If the field strength is reduced 10%, and the countervoltage correspondingly 10%, the inrush which occurs will be in the neighborhood of 100% of normal full-load current. This is on the assumption that the inrush is limited only by the resistance of the armature and leads to it, and that this resistance is approximately 10% of E/I. Actually the inrush current will be less than this value, because of the inductance of the field windings which prevents the field current from changing instantly.

The amount of resistance which may be inserted in the field at any one time is limited by the characteristics of the motor. A motor with a slow field will permit of a greater field weakening without excessive inrushes. If the field characteristic is such that the field current changes rapidly, a greater number of steps will be necessary. The number of steps is influenced also by the amount of current inrush which may be permissible from the standpoint of line disturbances and also from that of good motor commutation. Other factors to be considered are the effect of the inrush on the machine which is being driven and the extent to which smooth acceleration is necessary. Adjustable-voltage-control equipment for large motors having a speed increase by field of 2 to 1 generally requires about 10 steps as minimum. Automatic self-starters for control of continuous-running mill rolls, or similar machinery, may be safely accelerated with approximately 4 steps. Plate-type field rheostats have between 30 and 60 steps. The large number of steps is necessary because of the limited wattage which can be handled by any one step of a device of this kind, and it is an advantage because of the finely graduated control that is obtained.

Accelerating Relay. In order to obtain full torque for acceleration up to full field speed, it is necessary to short-circuit the field rheostat during the armature acceleration period. This is generally accomplished by means of a full field relay, the coil of which is energizing during armature acceleration and the contacts of which short-circuit the field rheostat. If the field rheostat is left in a set position and the motor started from rest with some resistance in the field circuit, it is necessary to use a full field relay.

If the speed increase is greater than 25%, it is also advisable to use an accelerating relay of the vibrating type as shown by Fig. 5.9. This relay acts to insert the resistance in the field circuit and short-circuit it again when the inrush reaches a predetermined limit. The coil of the accelerating relay is in the armature circuit and energized by the armature current. The contacts of the relay-short-circuit either all the field rheostat or a portion of it. After the motor has been accelerated to full field speed, the full field relay opens, inserting the field resistor, and an inrush immediately occurs in the armature circuit. This inrush closes the accelerating relay, which short-circuits the field rheostat again. In the meantime, the motor has accelerated somewhat above full field speed. When the rheostat is short-circuited, the inrush current drops and the accelerating relay opens, again inserting the field resistor. A second inrush then occurs, and this action goes on until the motor has been accelerated up to full speed. The acceleration obtained is shown by the curve A of Fig. 8.1.

It is possible to combine the functions of the full field relay and the accelerating relay by having a second coil on the accelerating relay. This coil is connected in series with the last step of armature-accelerating resistance, and it is of sufficient strength to close the accelerating relay during the time of acceleration to normal speed. When the armature resistor is short-circuited, the starting coil is also short-circuited, and the accelerating relay is then controlled by the running coil only and vibrates in the manner described.

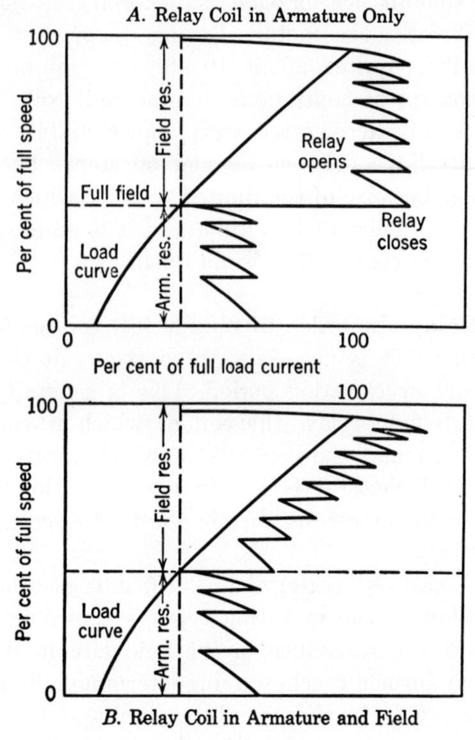

FIG. 8.1 Field Acceleration Curves

The currents at which the accelerating relay will open and close can be determined by adjustments on the relay. Since these values are fixed, once adjustment has been made, the relay will always open at the same current, and consequently it is necessary that this current be above the full-load running current in order for the motor to accelerate. For this reason the torque which is obtained during the first period of field acceleration is higher than necessary. For the majority of drives this does not do any particular harm and is perfectly satisfactory. However, if the driven machine has a high inertia, and particularly if it is belt-driven, the excess torque will be objectionable.

Curve *B* (Fig. 8.1) shows a method by which the accelerating relay may be made to vibrate in such a manner that the torque obtained will follow the torque required by the load. With this arrangement, one coil of the relay is connected in the usual manner in series with the armature, and the other coil is connected in series with the field. At the full field speed the armature current is low and the field current is high; at the weak field speed the armature current is high and the field current is low. Consequently, at full speed the field coil is strong, and a relatively low armature current will close the relay. As the field is weakened, the effect of the field coil constantly decreases, so that more and more armature current is required to cause the relay to close. By properly proportioning the coils, an acceleration curve as shown in Fig. 8.1 may be obtained.

Decelerating Relay. If the field is suddenly strengthened by removing resistance, the countervoltage of the motor will be increased and the current flowing from the line will be decreased. If the deceleration is too rapid, the armature current may even be reversed. It is possible to control the deceleration by means of a relay handling one coil in the armature circuit, and a second coil connected across the line. During normal operation these coils are arranged to oppose each other. The contacts of the relay are normally closed and are arranged so that in opening they will insert a step of resistance to weaken the field. As long as the deceleration is not too rapid, the decelerating relay remains closed. However, if the rheostat is moved too rapidly, so that the current in the armature reverses, then the two coils of the relay are accumulative in effect and cause the relay to open. Resistance is then inserted in series with the field to weaken it again. The relay vibrates in the same manner as the relay used for accelerating.

Miscellaneous Relays. The diagram of Fig. 8.2 shows all the devices which have been described, including the accelerating relay *FA*, decelerating relay *FD*, field-failure relay *FL*, and field-protective relay *FP*. It also shows a method of obtaining a desired speed in one direction of travel and a different speed in the other direction, both being adjustable. This control feature is of value for such machines as planers, where it is desired to cut at slow speed and to return at high speed. Two rheostats are used in series. The relay *C*, which has one normally open and one normally closed contact, is arranged to short-circuit one rheostat when energized and the other when deenergized. The relay is then energized in one direction of travel and deenergized in the other direction.

The diagram also includes a relay *FF*, which is used to short-circuit the field rheostats and to provide full field for dynamic braking. It is generally desirable to strengthen the field during the braking period in order to increase the braking torque and also to improve the commutation of the motor when handling heavy braking currents. When the master is moved to the OFF position, the line and direction contactors

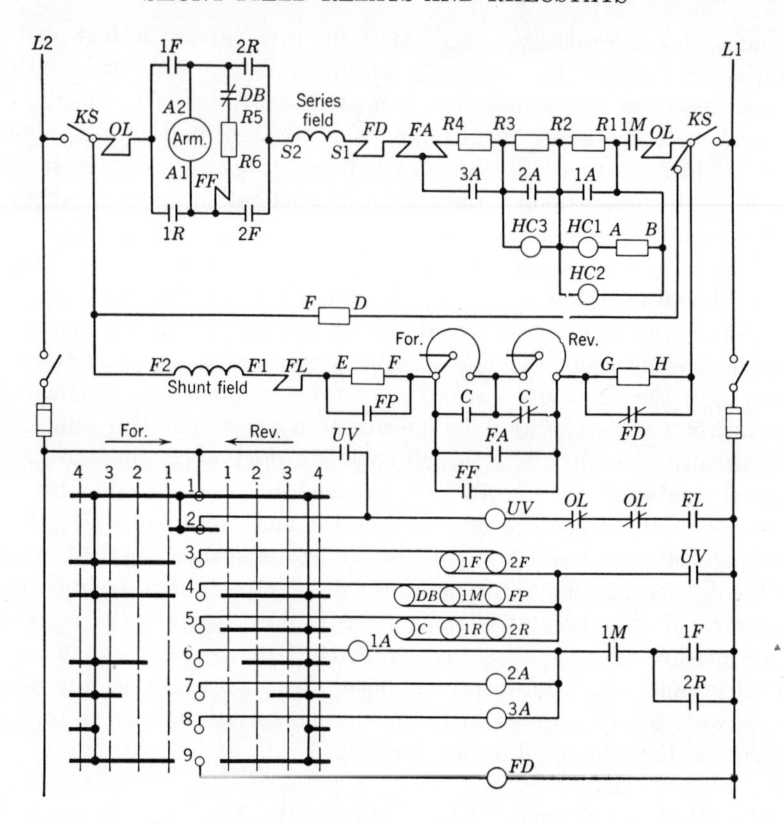

FIG. 8.2 Controller with Various Field Relays

are open, and the dynamic braking contactor *DB* is closed. Current then flows through the armature, the braking resistor, and the coil of the relay *FF*. The relay closes and short-circuits the field rheostat. The relay remains closed until the current in the dynamic circuit reaches a low value, unless the motor is restarted, in which case the dynamic loop would be opened. The relay is usually interlocked with the rest of the control so that the motor cannot be restarted until the dynamic current has fallen to a value low enough to permit *FF* to open. In other words, the motor cannot be restarted, or plugged, until it has practically come to a stop. This feature is not shown in the diagram. It is secured by using a normally closed contact on relay *FF*, connected in series with the direction contactor coils, and bypassed by normally open interlock contacts on the forward and reverse direction contactors.

Field Rheostat Construction. Field rheostats serve to add resistance in field circuits of shunt motors for the purpose of speed adjustment.

They are usually manually operated but also can be provided with a motor drive for remote operation. Basically a rheostat is just a variable resistor with operating knob and mounting suitable for use in operator's control consoles. Consideration must be given to the dissipation of the heat so that temperature limitations of the rheostat or the surrounding air are not exceeded.

There are several basic constructions used for rheostats. One type consists of winding resistance wire as a toroid on a ring-shaped base of ceramic material. The wire turns are protected by a vitreous enamel coating, except for one edge which is bared to provide for a brush contact. This design has as many steps as there are turns and usually has uniform resistance per turn. However, the wire size can be changed in order to provide a resistance taper from one end of the rheostat to the other. A wide range of sizes is available up to 1000 W. Rheostats can be mounted in tandem and driven from a single knob.

Another construction, known as plate type, consists of separate resistance steps connected to a ring of contact segments near the center of a base plate. The base can be of insulating material or of metal which is coated to provide insulation. The latter has better heat-dissipating properties. Resistor steps are usually made from reflexed resistance wire which is attached to the base with good thermal contact. In this design the resistance value can be varied step-by-step; however, the maximum wattage capacity of every step is equal since equal areas are available for each step of resistance. Larger diameter plates provide greater wattage dissipation, and standard sizes run from about 6 in. to 15 in. in diameter. As in the first type, tandem mountings are used for greater capacity.

A third type of rheostat construction that should be mentioned is the so-called face-plate type. In this design the contact segments and the commutating arm and brush mechanism are mounted on an insulating panel, and a resistor compartment is located behind this panel. The resistors are wired to terminals located in back of the segments. These separate resistors provide flexibility as to resistance and wattage rating, and the space provided for them can be made suitable for the capacity needed. This construction is often used for large, heavy-duty requirements.

Rheostat Design. A motor field rheostat is usually designed to give either equal speed changes per step, or equal percentage changes per step. A typical example of each design is presented to illustrate the method. Figure 8.3 is a curve giving field current against speed for a 500-hp 250-V motor having a speed range by field of 150 to 450 rpm. The rheostat to be designed has 150 steps. Experience has shown that

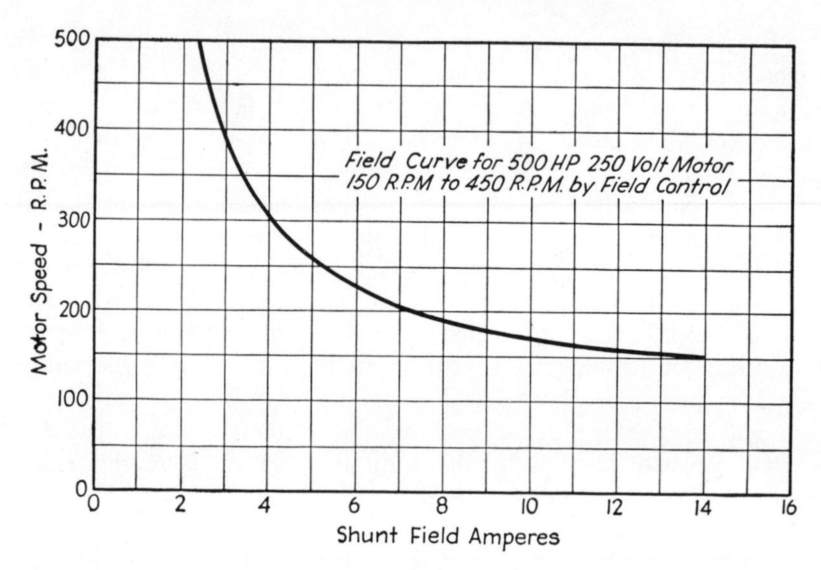

FIG. 8.3 Shunt Motor Field Curve

it is not necessary to calculate each step, which would be a very long process, but that it is sufficiently accurate to divide the rheostat into a number of groups of steps. In this case fifteen groups of ten steps each have been used. The design calculations of Table 8.1 are for a rheostat giving equal speed changes per step. The step numbers are given in the first column. The speed change at each step is readily calculated by dividing the total speed change by the number of groups of steps; that is, 300 divided by 15 gives 20 rpm per group.

These data are listed in column 2. The field amperes to give the required speed at each point are then read from the field curve.

The total ohms of column 4 are obtained by dividing the line voltage by the field current at each point. The value obtained for point zero is the resistance of the field itself, in this case 15.7 ohm. The rheostat ohms, in the next column, are obtained by subtracting the value of the field resistance from the total ohms of column 4. Column 6 gives the resistance in each group. The wattage which must be dissipated by each group is calculated from columns 3 and 6. When the ohms per group and the wattage per group are known, the resistor material required for the group may readily be selected from the units or cast grids available, whose resistance and current-carrying capacities will be known. Each group will be divided into ten equal steps. As a check it is desirable to calculate the wattage on the first step of the group, particularly at the high-current end of the rehostat, to be sure that equal division of the resistance over the ten steps will not overload the first step. If that step has ample capacity, the rest of the group will be on the safe side.

With the rheostat arm at any setting the current at the last step cut in will be at the full capacity of that step, but all other steps up to that point will be working below full capacity. Only one step can be operating at full capacity at any one time. It follows that the step which is at full capacity is not subjected to as much heat from the rest of the rheostat as it would be if the entire resistor were operating at full rating. The continuous rating of resistor material is based on a number of units bunched together, and so helping to heat each other. It is possible, therefore, to rate the resistor material of a field rheostat approximately 15% higher than its regular continuous rating.

The calculations of Table 8.2 are for a rheostat based on the same field but designed for equal percentage speed increases. The rheostat is again divided into fifteen blocks of ten steps each. It is then necessary to find a factor by which the low speed (150 rpm) can be multiplied

TABLE 8.1

RHEOSTAT DESIGN

500 hp 250 V—150–450 rpm by Field—150 Steps
Equal Speed Increase per Step

Rheostat Step	Motor rpm	Field Amperes	Total Ohms	Rheostat Ohms	Group Ohms	Group Watts
0	150	15.9	15.7	0	0	0
11	170	10.2	24.5	8.8	8.8	915
21	190	8.0	31.3	15.6	6.8	436
31	210	6.75	37.0	21.3	5.7	269
41	230	5.9	42.4	24.7	5.4	188
51	250	5.2	48.1	32.4	5.7	155
61	270	4.65	53.8	38.1	5.7	123
71	290	4.25	58.8	43.1	5.0	90
81	310	3.97	63.0	47.3	4.2	66
91	330	3.70	67.6	51.9	4.6	63
101	350	3.50	71.5	55.8	3.9	48
111	370	3.28	76.3	60.6	4.8	52
121	390	3.08	81.2	65.5	4.9	47
131	410	2.90	86.2	70.5	5.0	42
141	430	2.80	89.3	73.6	3.1	25
151	450	2.70	92.6	76.9	3.3	24

TABLE 8.2

RHEOSTAT DESIGN

250 hp 250 V—150–450 rpm by Field—150 Steps
Equal Percentage Speed Increase per Step

Rheostat Step	Factor	Motor rpm	Field Amperes	Total Ohms	Rheostat Ohms	Group Ohms	Group Watts
0	—	150	15.9	15.7	0	0	0
11	1.076	161	12.0	20.8	5.1	5.1	750
21	1.157	173	9.7	25.8	10.0	5.0	470
31	1.245	186	8.35	29.9	14.2	4.1	287
41	1.339	201	7.2	34.7	19.0	4.8	250
51	1.442	216	6.6	39.2	23.5	4.5	195
61	1.551	232	5.9	43.7	28.0	4.5	157
71	1.668	250	5.2	49.6	33.9	5.9	159
81	1.795	269	4.7	55	39.3	5.4	120
91	1.931	289	4.3	60	44.3	5.0	93
101	2.056	307	3.9	64	48.3	4.0	61
111	2.235	335	3.63	69	53.3	5.0	66
121	2.404	360	3.4	76	60.3	7.0	81
131	2.586	386	3.1	83	67.3	7.0	67
141	2.79	418	2.9	87	71.3	4.0	34
151	3.00	450	2.7	92.6	76.9	5.6	41

fifteen times to give the high speed (450 rpm). This factor is the fifteenth root of 3, or 1.076. The speed at the beginning of the second block is then 150 multiplied by 1.076, or 161. At the beginning of the third block the speed is 150 multiplied by $(1.076)^2$, or 173, and in this manner the speed at each point is calculated. The speeds required having been determined, the currents are read from the curve, and the rest of the design is calculated in the same manner as has been described for Table 8.1. The irregularities in the tables of group ohms and group watts are caused by slight errors in plotting and in reading the field current curve.

Any desired stepping of a rheostat can be worked out by the method described. Both the design giving equal speed increments and that giving equal percentage increments result in an unequal distribution of wattage over the rheostat. With a face-plate rheostat this does not make much difference, since the resistance material is separate, and any required amount may be used for any step. However, if the amount of material per step were fixed, as it is in a plate-type rheostat, these designs could not be used unless each step of the plate had capacity for the maximum step wattage. The use of equal percentage speed increments per step gives a more nearly balanced condition, but in order to get anything like equal wattage per step the resistance would have to be tapered much more steeply. The speed increases at the low-speed end of the rheostat would have to be smaller, and those at the high-speed end would be larger. For this reason plate-type rheostats may not always work out well for motor field control.

Potentiometer Rheostats. Potentiometer rheostats are connected directly across the line, and the field of the motor is connected across a portion of the rheostat.

Referring to Fig. 8.4,

E = line volts
R = total rheostat resistance
r = resistance of the field
i = amperes in the field
I = amperes in the portion of the rheostat which is in series with the field
x = resistance of that portion of the rheostat which is in series with the field

Then

$$R - x = \sqrt{Rr + \left(\frac{E - iR}{2i}\right)^2} - \left(\frac{E - iR}{2i}\right)$$

$$I = \frac{E}{\dfrac{r(R - x)}{r + R - x} + x}$$

The equations give the value of the resistance x in terms of quantities which are known (E and r) or which can be assumed (R and i). The

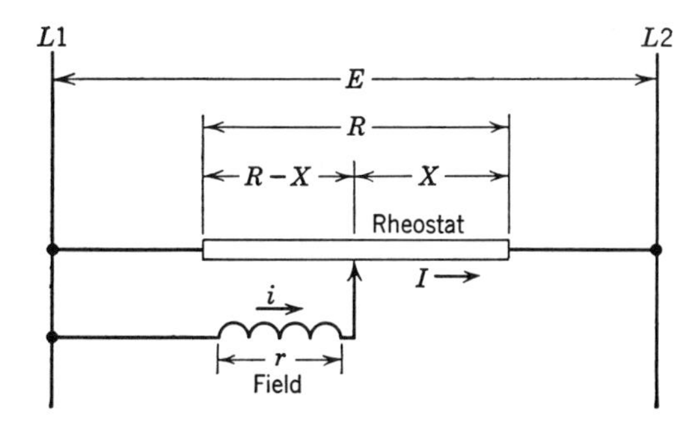

FIG. 8.4 Potentiometer Rheostat Connections

total rheostat resistance R should be made as high as is possible within the limits of the resistor materials used, to keep the watts of the rheostat low. A good value for R is approximately $4r$.

Instead of calculating each step of the rheostat, it is simpler to calculate a few points and plot curves of field amperes and rheostat amperes against the position of the rheostat arm, x. The other steps can then be read from the curve and be proportioned to give approximately equal current increments per step.

Problems

1. The speed of a dc motor is to be varied by field control. If the voltage is 250 V and the desired range of speed control is obtained by varying the field current from 5 amp to 2.5 amp:

(*a*) What is the resistance of the motor field?
(*b*) What total resistance is required in the field rheostat?
(*c*) What is the current taper of the rheostat?

2. The resistance of the shunt field of a 230-V motor is 55 ohms. The field current at maximum speed is 1.2 amp. If the field rheostat is made up of resistor units having a continuous rating of 50 W each, how many will be required?

3. If units of various continuous ratings are available, and if the rheostat has 10 steps of equal ohmic value, what will be the continuous wattage of the unit used for the first step of the rheostat?

4. What will be the ohmic value of the unit used for the last step of the rheostat?

5. Referring to Fig. 8.3, calculate a rheostat of 100 steps in 10 blocks, to give equal speed increases per step.

6. Referring to Fig. 8.3, calculate a rheostat of 100 steps in 10 blocks, to give equal percentage speed increases per step.

7. A 230-V motor field having a resistance of 150 ohm is connected to a 600-ohm potentiometer rheostat at a point where the field is in parallel with one-third of the rheostat. What field current is obtained?

8. What will be the relative cost of the resistor material in the potentiometer rheostat of problem 7, compared with the cost of a straight series resistor that will give the same motor speed?

9. A 230-V shunt motor has a maximum field current of 7 amp, and the field current at twice base speed is 2 amp. Calculate a rheostat of 100 steps in 10 blocks, to give equal speed increases per step.

10. A 550-V shunt motor has a shunt field resistance of 100 ohms, and the resistance required in a rheostat to give a desired speed range is 300 ohms. Calculate a 150-step rheostat, using blocks of 10 steps each, and giving equal speed increases per step.

9

DIRECT-CURRENT ADJUSTABLE SPEED DRIVES

Basic Principles. The speed of a shunt-wound dc motor is, for all practical purposes, proportional to the motor countervoltage. The motor terminal voltage is the sum of the countervoltage and the armature-resistance voltage drop, the last being a relatively low value. The motor then will run at any desired speed below rated base speed by applying a suitable voltage to the armature while maintaining full voltage on the shunt field. When the motor must be operated from the main plant dc power supply, it is obviously impractical to change the voltage of the supply lines, since that would affect the operation of every other motor connected to the lines. To vary the supply voltage directly at the motor, it is necessary to utilize some type of power conversion unit supplying only the motor, or motors, of the particular drive. The applied voltage may then be varied by adjusting a control current or voltage such as the strength of the generator field.

The motor armature is connected directly to the conversion unit, and the motor may be started and controlled without accelerating resistors or contactors. Speeds above base speed may be obtained by weakening the motor shunt field. Reversing is achieved by controlling the conversion unit to reverse the polarity of the voltage applied to the motor armature.

The basic functional units of an adjustable voltage drive system are, then:

1. A controllable power conversion unit with a dc adjustable voltage output.
2. A means to energize or connect the conversion unit.
3. A dc motor.
4. A dc supply to excite the motor fields.
5. A means to control the power conversion unit (and sometimes also the motor field strength).

Power Conversion. Almost all power utilized by industry is brought into the plant through an electrical ac distribution system. In the United States this is invariably a three-phase, 60-Hz system. Only on mobile units such as diesel locomotives or earth movers, or at locations remote from all power lines such as pumping stations, is the energy conversion from fuel to electricity an integral part of the local drive system.

Since power is normally distributed at constant frequency and constant voltage, and since dc adjustable speed systems require direct current and adjustable voltage, a fully controllable power conversion unit is of primary importance. Historically, an M-G set utilizing an ac motor to drive a dc generator, with control of the generator field, has provided this function. On this type of unit, rectification from ac to dc is performed by the commutator of the dc generator by electromechanical means. The magnitude of the voltage is controlled by manipulation of the generator field excitation.

Conversion of ac to dc can also be accomplished by static rectifier devices which allow current to flow in one direction, but not in the opposite direction. The magnitude of the dc voltage may then be controlled by gating the portion of each ac cycle that current is allowed to flow. The gating function can be performed with external devices, such as saturable reactors, or internally in controlled rectifiers through special device design and firing techniques. Semiconductors with the ability to block current in both directions until triggered by a firing signal are known as *thyristors*, described in Fig. 9.1.

Power conversion of this type is completely static, and no electromechanical devices are required between the power source and the terminals of the drive motor.

Characteristics. Since the controlling means operate in the pilot or control circuits of the conversion unit, the control can be either very simple or sophisticated as demanded by the application. Dynamic performance can be optimized since no mechanical switching is required.

Large contactors are not required, although a single main circuit contactor is often used. Armature accelerating and regulating resistors are eliminated. Power losses resulting from added armature circuit resistance are also eliminated.

Acceleration and deceleration are smooth without current or torque peaks. This is important for continuous processing such as paper-making machines, printing, or metal finishing. Braking torque by regeneration is smoothly applied for slowdown and stopping.

By suitable design of the power conversion unit it is possible to automatically obtain special speed-torque characteristics. The electric shovel

Power Semiconductor Properties

The rectifying element in each thyristor and diode is a small, thin wafer of semiconductor material which has been doped in manufacturing. (Doping is the process of adding impurities to achieve a desired characteristic.) The wafer in a diode is made up of two layers. One is doped to make what is called *p*-type material and the other is doped to make *n*-type material. The surface between these two layers is the rectifying junction. A thyristor wafer has four or more layers with *p*-type and *n*-type material alternated to produce multiple rectifying junctions.

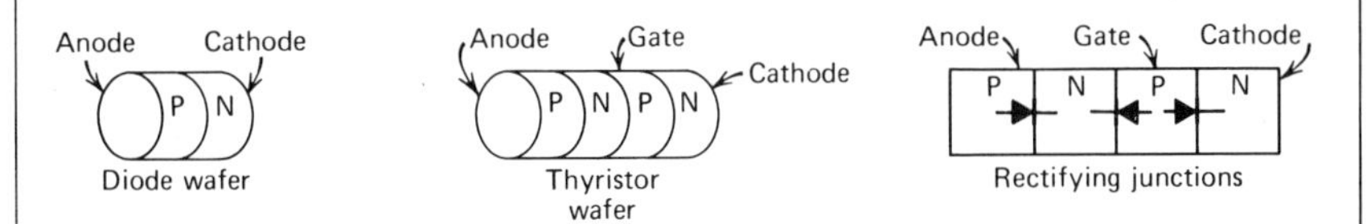

Applying a positive voltage to the thyristor gate when the anode is positive and the cathode is negative causes current flow from gate to cathode. The gate current upsets the balance of electrical charges within the wafer, which negates the rectifying characteristic of the center junction. Current can then flow from anode to cathode as long as the unbalance of charges remains.

Fig. 9.1 Semiconductor Devices

is a good example since it must slow down and stall when the shovel strikes an obstruction requiring more than a safe torque to move it. When the operator backs the shovel away, the speed should come back to normal. A three-field generator having a self-excited shunt field, a differentially connected series field, and a separately excited shunt field will meet these qualifications.

Groups of motors such as those in an annealing furnace, run-out tables, or range drives can be connected to a single adjustable voltage system and their speeds adjusted simultaneously under the control of one regulator.

Most significantly, the dc adjustable voltage system offers a means of obtaining wide speed ranges for many drives in a single plant where only alternating current is available.

M-G Set Conversion. When an M-G set is utilized for power conversion, the drive motor is matched to the available voltage and frequency of the supply line. A means is provided to connect this drive to the power system and to start and stop the M-G set. The dc generator is controlled by adjusting the generator field strength. The control system can range from a simple, manually operated rheostat to any of the more sophisticated schemes outlined in Chapter 10.

Multiple fields are frequently used on generators to increase the flexibility of this type of drive. For example, drives which must operate over a wide speed range may be limited by the residual voltage. A field of limited ampere turns can be connected with its polarity the reverse of the main shunt field, to assure enough negative ampere turns to bring the armature voltage to zero.

A distinct advantage of a dc generator, as a power conversion unit, is its ability to accept regenerative current. Since the generator is electrically bidirectional, current (because of motoring or overhauling loads) can flow in either direction depending on whether the load is motoring or overhauling. It is possible to reverse the polarity of the armature merely by reversing the polarity of the shunt field.

Since the dc generator is a rotating machine, it involves wear, maintenance, and noise. Physical configuration and weight are also limited by basic design fundamentals. These are obvious disadvantages as compared with the static conversion units.

Single Motor M-G Set Drive. Figure 9.2 is a diagram of a single-motor, adjustable-voltage controller using manually operated rheostats and arranged for low-voltage starting and speed regulating. The motor armature is connected to the generator armature through a magnetic contactor *M* and an overload relay *OL*.

In the OFF position of the controller the contactor *M* is deenergized and open. The generator rheostat is disconnected by an interlock contact on *M*.

To start up the drive the ac drive motor is energized, bringing the generator and exciter up to speed, and the exciter voltage is adjusted to its rated value by means of its field rheostat. These machines continue to run as long as the drive is in use. The start button is pressed and the contactor is energized, provided the generator rheostat is in the weak field position with the interlock contact on the rheostat closed. When *M* closes, its main contact closes the circuit between the generator armature and the motor armature, but little or no current flows because the generator field is short-circuited. One interlock contact of *M* provides a maintaining circuit, and the other connects the generator field rheostat into the circuit. The motor is now accelerated to the desired speed by strengthening the generator field.

A potentiometer rheostat is used for the generator rather than a series rheostat, so that the initial generator field current may be zero. Speeds

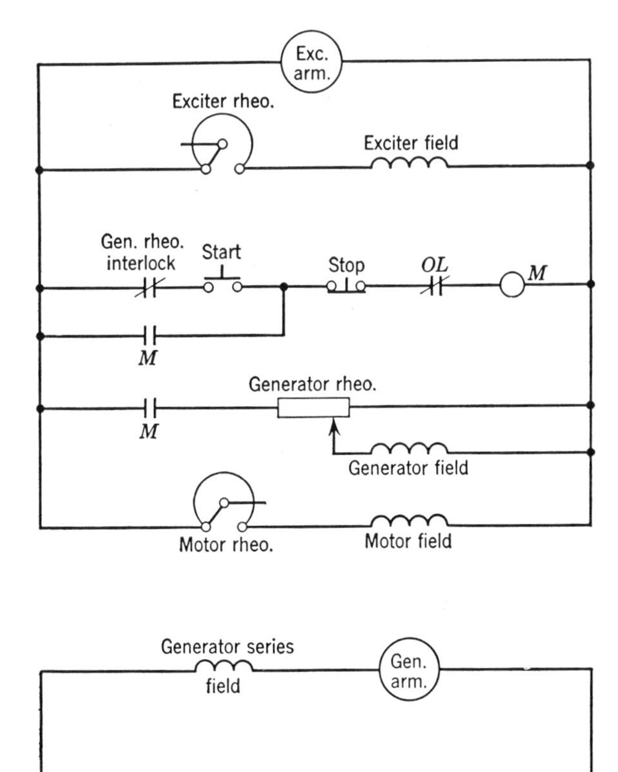

FIG. 9.2 Single-Motor Adjustable-Voltage Controller with Manually Operated Rheostats

above base speed are obtained by weakening the motor field. The motor is stopped by pressing the stop button, which deenergizes M. It is impossible to restart until the generator rheostat is returned to the weak field position to reclose the interlock switch.

In this very simple form of controller there is no provision for protection against too rapid acceleration or decleration. The generator rheostat and the motor field rheostat could be assembled as one unit, driven by a pilot motor. With this added feature the speed of the pilot motor would control the rate of acceleration or deceleration. The drive also could be remotely operated, since control devices for the rheostat pilot motor, as well as the start-stop pushbuttons for the main motor, could be located at a remote operator's station. In addition the pilot motor could be connected to return the rheostat automatically to its minimum position each time the drive is stopped.

Static Rectifiers. Devices with the ability to pass current when the voltage is of a given polarity and block the current when the voltage polarity is reversed are called diode rectifiers. Barrier surfaces have been created from many materials such as copper oxide, selenium, germanium, and silicon. A similar characteristic is exhibited by the gas-filled tube. Figure 9.3 represents the performance of a diode in an electrical circuit. When the anode of the rectifier is positive with respect to the cathode, the rectifier is said to be in the conducting state. At this time it presents a very small impedance to the flow of current and allows the supply voltage to appear across the load. When the cathode of the rectifier is positive with respect to the anode, the rectifier presents a very high impedance to current flow so that essentially no voltage ap-

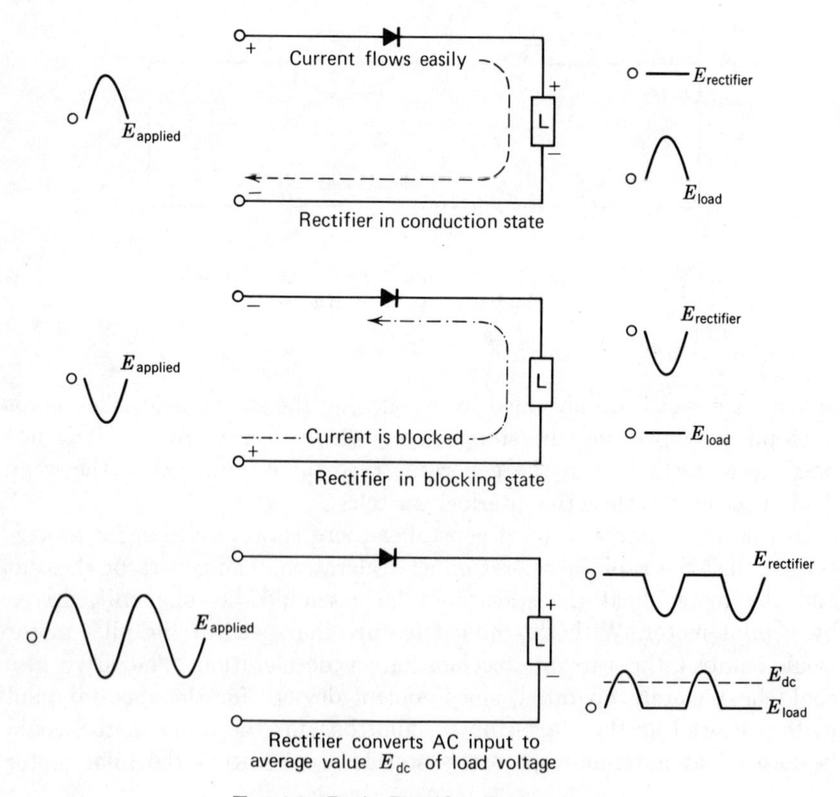

FIG. 9.3 Basic Rectifier Performance

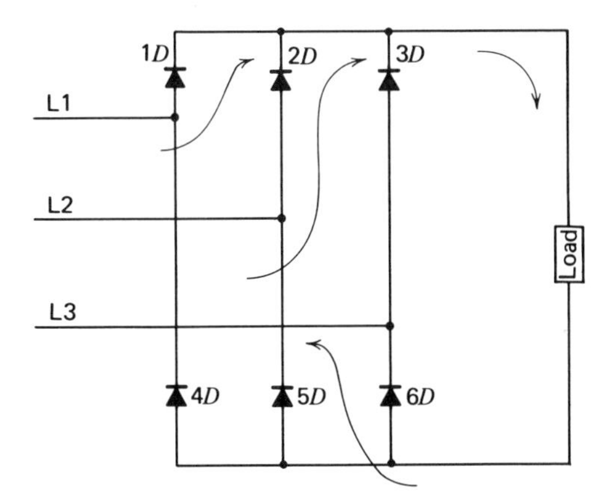

FIG. 9.4 Three-Phase Rectifier

pears across the load. During this time the rectifier is in the blocking or nonconducting state.

By combining six diodes in a three-phase bridge, the dc voltage will be produced with only a very small 360-Hz ripple. The path of conventional current at any instant depends upon the relationship of the three line voltages to each other at that instant. Current flows from the most positive line through a rectifier, the load, and another rectifier, to the most negative line. For example, in Fig. 9.4, at the instant when line *L*1 is the most positive with respect to line *L*3, current flows from *L*1 through diode 1*D*, the motor, and diode 6*D* to line *L*3. If the voltage between *L*2 and *L*3 is increasing while the voltage between *L*1 and *L*3 is decreasing, line *L*2 soon becomes more positive than line *L*1. At that moment current transfers from *L*1 to *L*2. Diode 1*D* stops conducting, and the new current path is established through diode 2*D*. A complete sequence of similar line current transfers to other rectifiers causes each of the six possible paths to conduct for a period of each ac cycle. These current transfers result in 120 electrical degrees conduction in each rectifier, as shown in Fig. 9.5, and provides the maximum dc power output that can be obtained from a three-phase ac line with a three-phase rectifier.

Controlled Rectifiers. Some of the more sophisticated rectifying devices such as ignitrons, thyratrons, and thyristors not only have the ability to block voltage in the reverse direction, but will also block the flow of current with forward voltage until a pilot signal initiates the flow of current from anode to cathode. This pilot circuit is normally called the gate, and the signal, the gating signal. Figure 9.6 illustrates

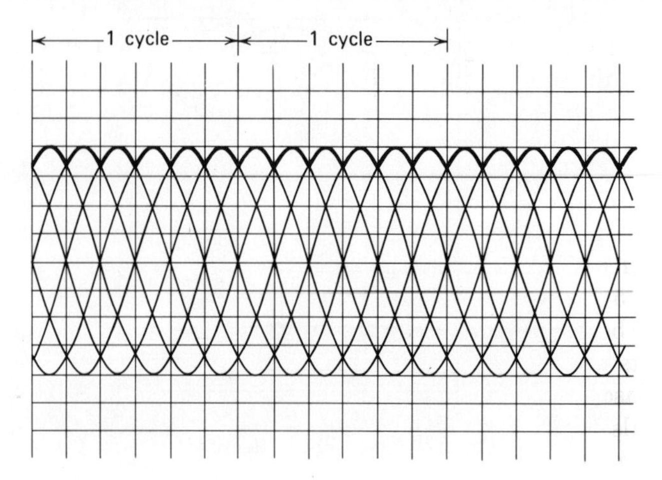

Fig. 9.5 Voltage Output of Three-Phase Full-Wave Rectifier

this characteristic in a single-phase circuit. By controlling the point in the cycle when the gate voltage is applied, it is possible to adjust the average value of direct-current through the load. If the gate voltage either is never applied or is applied when the anode is negative with respect to the cathode, the controlled rectifier blocks current during the positive

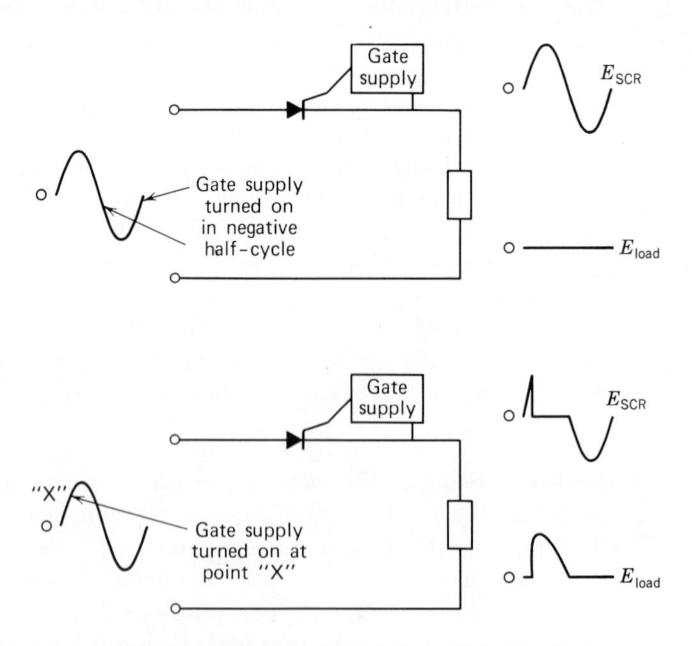

Fig. 9.6 Performance of Controlled Rectifier

and negative half-cycles and the average load voltage is zero. If the gate voltage is applied at point "X," current will flow from that point in time until the input voltage sine wave goes to zero. If the gate signal is continually applied, the controlled rectifier will perform exactly as a conventional half-wave rectifier.

Static Excitation. To reduce the number of moving parts and rotating machines, ac line power is rectified statically for constant voltage, dc excitation of control and fields. Where the power requirements are less than about 5 kW, a single phase bridge is used. If there are a number of fields to be excited from constant dc voltage, or if the control power requirements are high, a three-phase, diode rectifier is used (Fig. 9.4), In either case, transformers may be used between the power lines and the terminals of the rectifier to electrically isolate the control and field circuits from the main power sytem. In additon, taps on the transformer allow adjustment of the dc voltage to correspond with field and control component ratings.

The generator fields must operate to swing the armature voltage through its full adjustable voltage range. The generator field is therefore supplied from a separate rectifier bridge in which diodes have been replaced by controllable rectifiers, normally thyristors. Figure 9.7 illus-

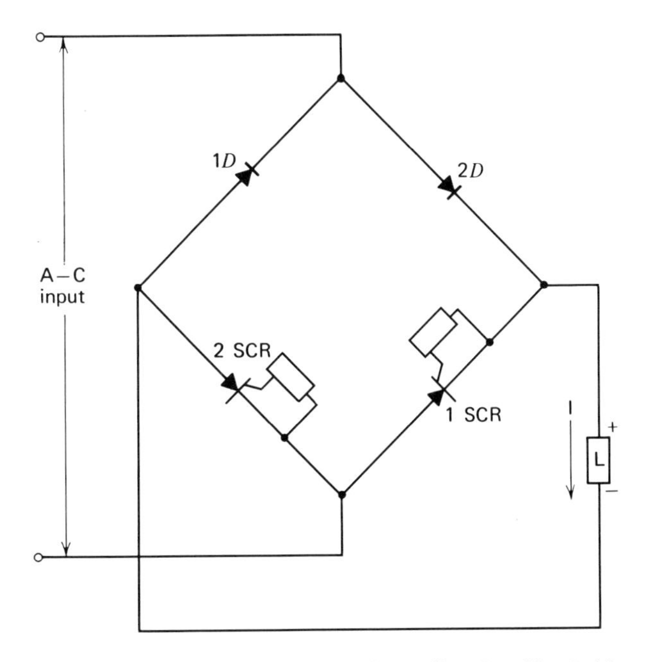

FIG. 9.7 Single-Phase, Full-Wave, Controlled Rectifier Bridge

trates a single-phase, full-wave bridge capable of supplying field current from zero to rated excitation. Similarly, where the motors must be operated at higher than base speed, the fields for those motors are each supplied from a separate controllable bridge.

Since the field supply is an amplifier, a pilot duty potentiometer, or similar low-level signal source, is used to establish the reference. The output voltage of the generator is compared with the reference voltage, and the error initiates action to correct the output voltage as described in Chapter 10.

Semiconductors of the thyristor type have extremely high efficiency, fast switching times, and ability to block significant voltages. The capability of the thyristor has encouraged the rapid development of static power conversion units for dc drives.

Static Drives. The simplest three-phase, static drive is illustrated by Fig. 9.8. The basic power conversion circuit consists of three diodes and three controlled rectifiers. If the controlled rectifiers are gated full on, the dc output will be exactly the same as a three-phase, diode bridge (Fig. 9.5).

To control motor speed, however, it is necessary to adjust the dc voltage. Since normal current transfer from rectifier to rectifier provides maximum output voltage, the output can be reduced by delaying the point in the ac cycle where each rectifier starts to conduct. The portion of each cycle during which the rectifier is allowed to pass current, measured in electrical degrees, is called the conduction angle.

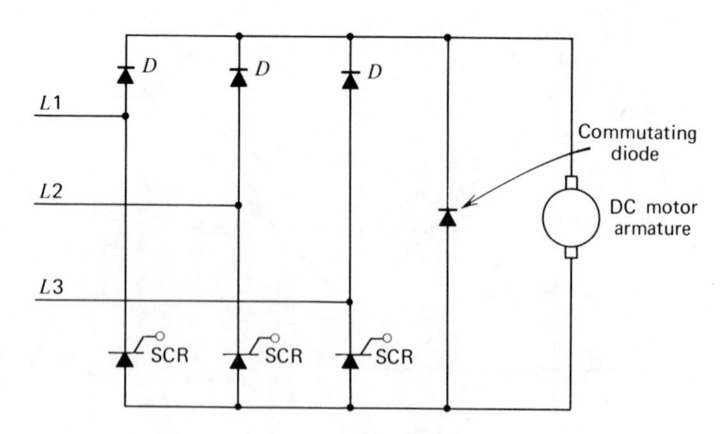

FIG. 9.8 Three-Phase, Single Convertor

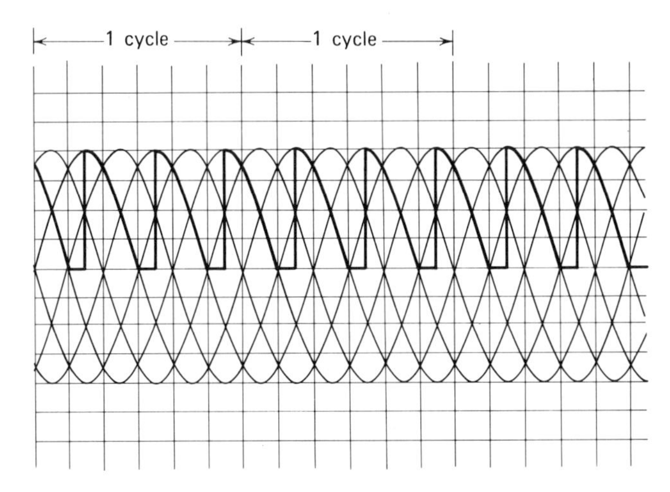

FIG. 9.9 Three-Phase Single Convertor Operating at Rated Load, One-Half Speed

Delaying the start of conduction by 60 electrical degrees reduces average output voltage by shifting the 120-degree conduction angle to a lower-voltage portion of the input ac wave. A delay or phase shift of more than 60 degrees reduces the output voltage by a greater amount, since the conduction angle then becomes less than 120 degrees. Controlled rectifiers need only be provided in three of the six legs of the three-phase bridge to obtain full range control of output voltage. The dc armature voltage wave form at one-half speed and rated load is illustrated in Fig. 9.9. Each controlled rectifier is gated in sequence, and in this illustration at 90 degrees, allowing a 90-degree conduction angle.

Commutation. The transfer of current from one rectifier to another as voltage relations change is called commutation. When a bridge output is controlled by phase-shifting the conduction starting point, the point of commutation must shift also. A phase shift of greater than 60 degrees means that the forward line voltage on a conducting rectifier goes to zero before the next rectifier in sequence starts to conduct. Consequently, in a controlled rectifier bridge, there are three times in each cycle when there is no path for line current.

Even though current in the supply line decreases to zero, inductance in the motor armature circuit tries to keep the motor current flowing. This current will attempt to find a path through one diode and one controlled rectifier so that the controlled rectifier may not turn off and

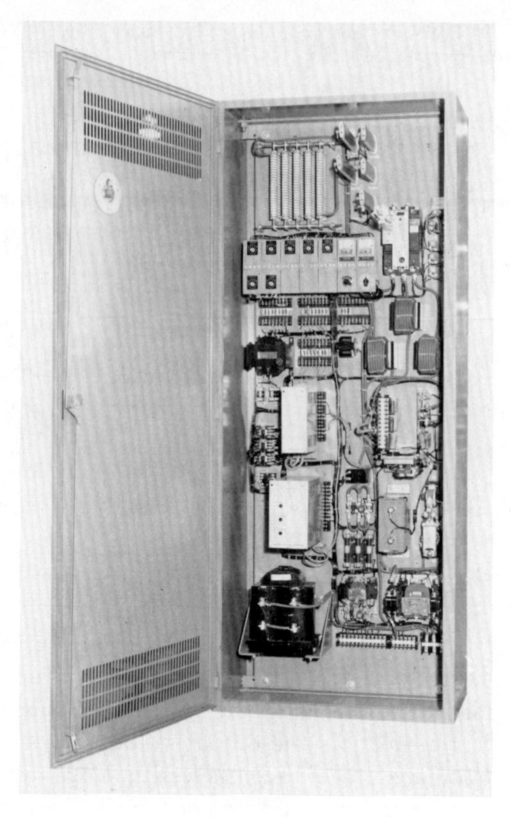

FIG. 9.10 Cutler-Hammer Thyristor Convertor and Control for DC Drive

cause a fault. By introducing a commutating diode (Fig. 9.8), the armature current will not attempt to flow through the controlled rectifiers and diodes in series, but rather will transfer to the commutating diode. If the start of conduction is phase-shifted more than 60 degrees, the commutating diode commutates current away from each controlled rectifier in sequence, conducting three times in each complete cycle. The commutating diode, therefore, reduces the length of time the controlled rectifiers and bridge diodes have to carry current. Second, by forcing positive commutation of the controlled rectifiers in sequence, additional reactance can be used in the armature circuit with assurance that each thyristor will shut off at zero line voltage and "shoot-through" faults will not occur.

Reactance in series with the motor armature is generally desirable to improve the form factor of the direct current and, thereby, reduce the heating in the motor. The reactance will also keep the conversion unit in continuous current down to a lower voltage level.

Convertor-Inverter. Utilizing a three-phase bridge with controlled rectifiers in each leg as shown in Fig. 9.11 provides for conversion of alternating-current to direct-current with six current pulses in each cycle as each of the six controlled rectifiers is gated in sequence. For a given set of armature parameters the form factor will be improved over the bridge with three controlled legs since the ripple frequency will be doubled to 360 Hz.

This type of bridge is also capable of inverting direct-current to alternating-current by firing the six controlled rectifiers in proper sequence and returning power to the line. For example, if a motor supplied from this type of conversion unit is driving a winch hoisting a load, the conversion unit will operate as a convertor supplying current and voltage to that motor in the raise direction. As the convertor unit is phased back to the point where the output of the convertor is less than the torque required to raise the load, the drum will reverse and the load will begin to move in the lowering direction. Although the torque, and thereby the current in the motor, will continue in the same direction, the reversal of rotation will reverse the polarity of the dc voltage and the motor will now function as a generator. The voltage of the generator will tend to rise above the dc output of the convertor. By proper phase control of the controlled rectifiers, the output of this dc machine can be inverted to the line frequency in the proper sequence to return the power to the ac lines.

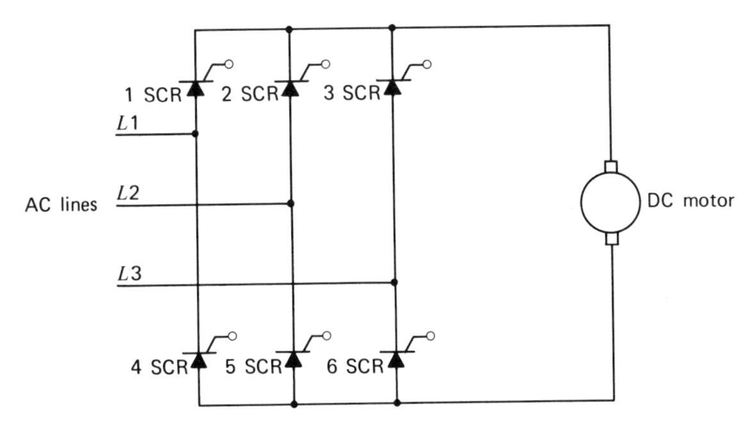

Fig. 9.11 Inverting, Nonreversing Convertor

Reversing Static Drives. Two convertor-inverter units are connected in a back-to-back arrangement as shown in Fig. 9.12 to provide a full reversing, regenerative drive. Under normal motoring operation, Unit A will function as a normal convertor. If, under some conditions of operation, the motor is overhauled by the load, Unit B will act as an inverter, returning power to the line. Similarly, Unit B can be gated as a convertor supplying the motor with reverse voltage to drive it as a motor in the reverse direction. If under these conditions the motor is overhauled by the load, Unit A will act as the inverter and return power to the line. Functionally, this configuration of static conversion units will provide all the advantages of a dc generator with added advantages of no moving parts, long life, low maintenance, and high efficiency.

When connecting convertor-inverter units in a back-to-back configuration, logic circuitry must be added to the firing units to prevent short circuits through the dc connections to the convertor. For example, in Fig. 9.12, if 1SCR and 8SCR are ever allowerd to fire simultaneously, there will be a short circuit from line $L1$ to $L2$. The simplest design technique is to create a dwell period between Convertor A and Convertor B during which both units are turned off. For dynamic response of the total system this period must be minimum time. Most designs use a current-measuring device in the output of each convertor to check for zero current. The logic circuitry establishes a turn-on signal for the desired mode of operation. The signal from the current-measuring device will not allow a converter to turn on if current is still flowing through the complementary unit.

In the design of a dual-convertor drive system, careful consideration must be given to both the ac and the dc levels. If, under any condition such as an overhauling load, the motor is driven by the load, it performs as a generator. The convertor must then operate in an inverting mode. Current is returned to the ac power system by fabricating a simulated sine wave from the dc power generated by the dc drive. The logic system is obligated to key the inverter output as near as possible to an in-phase relationship with the ac line. Sufficient voltage margin must exist between the ac power system and the inverted alternating current returning from the drive to allow current to flow from the drive to the line. Consideration must, therefore, be given to the ac line voltage level, gear-in speed of the drive, and maximum operating dc voltage under regenerative conditions.

During the inverting mode, current must commutate from one thyristor to the next in sequence. If a line transient causes a dip in the power supply voltage, the differential between the supply and the inverting

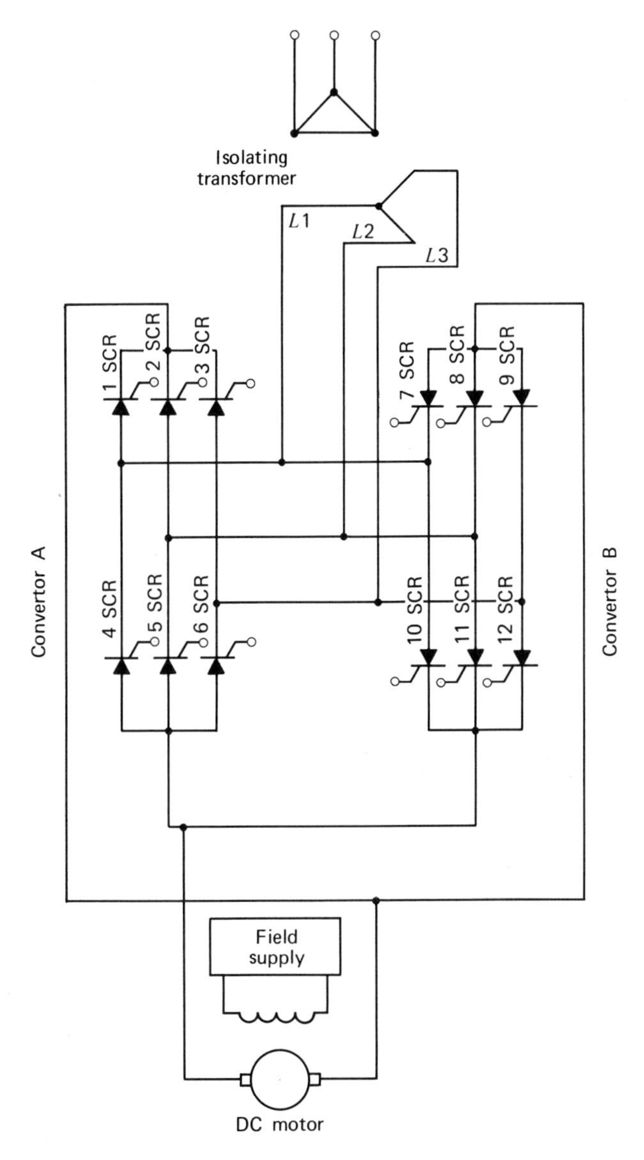

FIG. 9.12 Reversing, Regenerative Dual Convertor

voltage may disappear and cause a misfire on the particular thyristor which should fire at that time. The discontinuity in the firing will cause the next thyristor to gate full on and cause a fault condition in the convertor. The design of the dual convertor should therefore include current-limiting or current-interrupting devices for protection against this contingency. In the application of dual convertors the stability of the overall power system is a part of the engineering analysis.

Thyristor Protection and Special Considerations. Since all thyristors are designed for minimum size and maximum efficiency, they inherently have little mass and little ability to absorb thermal deviations. The critical criterion is the temperature of the rectifying junction. Depending on the cooling medium and the size of the wafer, the junction temperature usually reaches steady state in a few seconds. Most conversion units utilize forced air cooling. Temperature-sensing devices are frequently mounted on the heat sink assemblies to shut down the drive on overtemperature because of excessive loading or failure of the air supply.

Thyristors are also susceptible to failure from voltage transients. Resistance-capacitance networks may be connected around each semiconductor and across the incoming lines to absorb voltage peaks. Snubber devices, which are usually selenium rectifier stacks with predetermined breakdown voltage, are connected across the incoming lines to clip the voltage peaks. Power semiconductors frequently have a rate of change of voltage rating (dv/dt). The resistance-capacitance networks will decrease the slope of the voltage wave to prevent failure from this phenomenon.

Rate of change of current (di/dt) must also be limited to allow uniform firing over the entire surface of the wafer. Reactance in the ac lines and proper shaping of the firing pulse provide this type of protection.

Since all controlled rectifiers utilize only a portion of the incoming sine wave, the distorted wave form reflects back into the power system. In addition, commutation of the thyristors causes momentary overlap of the current flow between complementary thyristors and creates a line-to-line short circuit for a few microseconds. This phenomenum is known as *line notching.* If the thyristor bridge is connected directly to the incoming ac lines, the voltage wave at the terminals of the conversion unit will literally "notch" to zero during the transient. Good design, therefore, dictates that reactance be added between the ac source and the rectifier bridge, in the form either of series reactors or of a transformer. If the conversion unit includes reactance between its line terminals and the rectifier bridge, the line notching will be reduced by the ratio of the convertor impedance to the line impedance. Isolation transformers are also frequently used to provide a buffer between the convertor and the power lines. In addition to adding impedance, the isolating transformer prevents faults resulting from multiple grounds on the ac power system and the dc drive.

Current surges or transients will destroy power semiconductors. These high-current conditions may result from extreme overloads, short circuits, or misfiring of the thyristors. High-speed circuit breakers and fast-trip fuses have been developed specifically for protection of semiconductors. These protecting devices may be connected in various combinations, depending on the design of the convertor. Usually, however, they are expected to clear the circuit only on catastrophic failures. Under overload conditions or anticipated high-current surges, current-measuring devices are used to delay or halt the firing of the thyristors until the overcurrent condition disappears. Reactance is incorporated in circuits involving power semiconductors to limit the rate of rise in current and to allow time to adjust the gating circuits.

On high-horsepower drives, standardized conversion modules are paralleled to obtain adequate capacity for the motor requirements. On heavy-duty applications, such as rolling mills, the drive may be rated several thousand horsepower and require the paralleling of several hundred conversion units (Fig. 9.13). Reactors are used to balance load between modules.

Although the conversion unit utilizing six thyristors inherently provides a better form factor than the three-thyristor bridge, low-level output from either will be discontinuous current. The motor under these conditions will see current pulses. Reactors in the dc loop will store energy to keep the convertor in continuous current to lower output levels, particularly when used in conjunction with a commutating diode. At normal operating speeds, reactance will improve the form factor and reduce motor heating. On major, high-kilowatt drives, special input transformer connections are sometimes utilized to create a full twelve-phase drive for optimum form factor at the motor.

Chopper Drive. Since the thyristor is basically a high-speed switch, the switching characteristic has been used to create adjustable-speed dc drives, illustrated in Fig. 9.14. A constant-voltage rectifier or a dc bus is used as the incoming supply. When thyristor 1SCR is gated on, the dc motor is, in effect, connected to the constant-voltage bus. As the motor speed rises above the reference level, thyristor 2SCR is turned on, shutting off 1SCR, and the motor momentarily receives the energy from power stored in inductor L. As soon as the motor speed drops below

FIG. 9.13 General Electric Co. Thyristor Drive for Reversing Hot Strip Mill

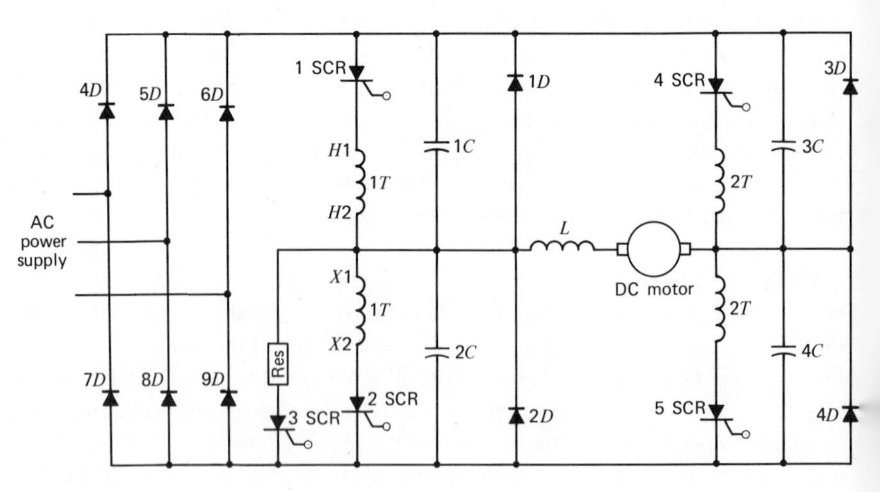

FIG. 9.14 Direct-Current Adjustable-Speed Chopper Drive

the reference level, 1SCR is turned on, turning off 2SCR, to supply power again to the armature from the bus.

The techniques used in the chopper drive are exactly the same as one phase of a McMurray-Bedford inverter as described in Chapter 18. The commutation of 1SCR and 2SCR is achieved by discharge of capacitors of 1*C* and 2*C* through transformer 1*T*. While 1SCR is firing, capacitor 2*C* will charge to nearly bus voltage. When 2SCR is gated on, it shorts capacitor 2*C* and the discharge current transient through windings *X*1–*X*2 of 1*T* causes winding *H*1–*H*2 to drive the anode voltage of 1SCR to zero.

Static dynamic braking is added by gating 3SCR whenever energy must be dissipated into the dynamic braking resistor. Since it is impossible to return current to the line through the rectifier bridge, power generated by overhauling loads or quick slowdown must be dissipated in the dynamic braking resistor.

Diodes 1*D* and 2*D* provide a free wheeling path during the transient switching conditions. Thyristor 4SCR and 5SCR, with the attendant capacitors, diodes, and transformer, provide a full reversing function.

In some horsepower ranges the chopper drive does not utilize the capabilities of the thyristors as completely as the phase-shifted thyristor convertors. However, it has added advantages. It can be used on a dc bus to create an adjustable-speed dc drive. There is no pollution of the ac lines, since a three-phase rectifier is used to convert ac to dc. The form factor of the power delivered to the dc motor is excellent. This type of drive is particularly desirable on applications such as adjustable-voltage hoists, because power supplies to cranes are subject to wide voltage deviations. Power dissipated during regeneration is minimal.

Problems

1. Assume a dc adjustable-speed drive and motor with the following parameters:

Rated terminal voltage—250 V
Armature loop resistance—0.2 ohm
Rated armature current—100 amp
Constant torque load at motor rating
Rated speed at rated voltage and load—1800 rpm

What will be the motor speed for each of the following conditions?

(*a*) Reduce armature current to 50 amp.
(*b*) Reduce terminal voltage to 135 V.
(*c*) Reduce terminal voltage to 20 V.
(*d*) Add 0.8 ohm in armature circuit.

2. Draw the circuit for a three-phase full-wave diode rectifier and sketch the voltage wave it will superimpose on its load.

3. Describe "natural commutation."

4. Draw the power circuit for a dc motor supplied from a dc adjustable voltage generator including main contactor, overload, and knife switch.

(*a*) Indicate polarity of voltage across motor and generator and direction of current flow with a motoring load.

(*b*) Indicate polarity of current flow for overhauling load.

5. Describe the phenomena of "line notching" as related to thyristor drives.

6. Sketch the power circuit for a three-phase full-wave static drive utilizing six thyristors.

7. Sketch a triode-type thyristor, identifying the wafers and barrier layers.

10

AUTOMATIC REGULATING SYSTEMS

Closed-Loop Regulators. Most regulating systems used in industrial motor-control applications are of the closed-loop type. A closed-loop regulator is an amplifying system which operates from an error or difference between a reference quantity and an actual resulting quantity, in a manner tending to keep the error as small as possible.

Motor-Control Applications. The need for automatic regulating systems in electric motor control has developed largely because of changes in manufacturing methods. Formerly, single operations were performed on materials being processed, and the materials were moved from one machine to another until finally completed. This method required that the raw materials be cut into suitable sizes and handled as individual pieces throughout the processing.

In keeping with improved manufacturing methods, the natural trend was to eliminate as much as possible the handling of materials between process operations. This has led to processing materials in continuous form, as coils of strip or sheet and rolls of wire or rod. When materials are handled in continuous form and pass through a series of processing operations, control problems are encountered that often require automatic regulating systems for their satisfactory solutions. The several machines in a continuous processing line are usually driven by individual motors, the motors receiving their power from one or more adjustable-speed systems. Individual motor drives are usually required because the length of the line makes shafting impractical, but a more important reason is the requirement for flexibility between process sections. This would be extremely difficult to provide by mechanical means.

The proper coordination of a number of motors driving separate machines, which process a continuous length of material, requires precise regulation of voltage, frequency, speed, current, tension, or position. Often the linear speed in one section of a machine must be held constant, and this becomes the master section of the group. Other sections must be synchronized with it so that the material does not pile up between sections, or become torn because it gets too tight. Sometimes the tension in the material entering a machine is critical, and this requires tension regulation. Other machines may require a slack loop in the material, and a regulator is used to maintain the loop at its proper form. Lateral location of the moving material is often important, and position regulators are used to compensate for irregularities in the edge of the material and other variables. Current regulators are usually used with helper motors and cause these motors to operate at constant torque output.

Continuous-processing methods require handling of the materials in roll, coil, or bobbin form. These coils must be wound under conditions of carefully controlled tension, for process reasons and to prevent damage to the coil in handling. Constant-tension winding is commonly done not only in the continuous-processing operations, but also in preparatory processes when getting the material ready for the final process or for sale. Properly wound coils or rolls will not telescope in handling and will not be damaged under normal storage or handling conditions.

Regulators used in industry must meet a wide range of performance requirements, and very often the type of regulator used is determined by the requirement. Beyond the ability to do the job satisfactorily is the need for dependability. This is especially true in the case of high-speed processing lines, where the failure of one part or function may shut down the entire line and where down time is expensive in terms of lost production.

Basic Regulating Systems. The basic closed-loop regulating system is shown in Fig. 10.1 The regulated quantity is the end result and can be either electrical or mechanical in nature. For instance, the regulated quantity can be the output voltage of a generator, the rotational speed of a shaft, or the position of a strip of material being processed. The regulating system usually furnishes only a small portion of the total power involved, and this is often in the form of shunt field excitation for a motor or generator. The regulated quantity has some principal power source, and this can also be in the form of electric or mechanical power.

The basic regulator has a source of power sufficient to operate the regulating devices and to supply the control power for regulating purposes.

The reference can be mechanical, as, for example, a calibrated spring, but the customary reference used is a known voltage or frequency. This reference should be stable, and where precise results are required, it is carefully regulated so that it does not fluctuate with line voltage, load,

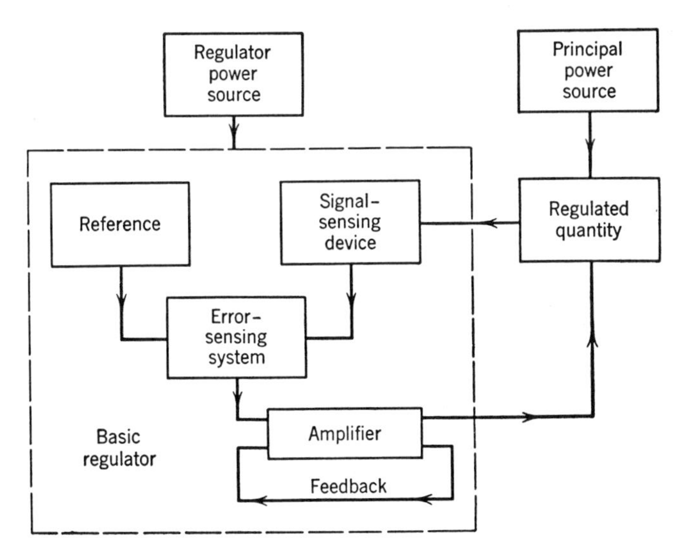

FIG. 10.1 Basic Closed-Loop Regulating System

and temperature changes. The reference is used to preselect the performance level required of the regulated quantity.

The signal-sensing device is a means of measuring the actual performance level of the regulated quantity. It must be capable of measuring the thing to be regulated and of producing a signal which bears a fixed relationship to the measurement. The signal is usually a voltage that is proportional to the level of the regulated quantity.

The error-sensing system is that portion of the regulator where the comparison of the actual signal is made with the preselected reference. The reference represents the wanted result, and the signal represents the actual result. Any difference resulting from the comparison is the error signal used to drive the amplifier system. The polarity, or sense, of the error is in the direction to produce a correction that tends to reduce the error.

The purpose of the amplifier is to enlarge the relatively weak signal from the sensing system to the power level needed to control the performance of the regulated quantity effectively. The power gain in the amplifying system, which is the ratio of the regulator output power to the signal power, is dictated by the sensitivity required or, in other words, by the allowable error. The principal difference in regulating systems is in the type of amplifier used.

Circuits are usually employed which take off a portion of the amplifier output and feed it back into the input end or at some intermediate stage of the amplifier. If the feedback is additive, it is called regeneration, and if subtractive, it is degeneration. In general, feedback is used to stabilize the system, and the usual arrangement is to feed back transients, rather than to use steady-state feedback. Stabilizing circuits are frequently called antihunt circuits, and their purpose is to prevent oscillations of the regulated quantity. In transient feedback antihunt circuits, the sense of the transient is in opposition to the change taking place in the amplifier output.

Sensing Devices. As pointed out previously, the function of the sensing device is to measure the actual performance of the quantity that is being regulated. An axiom of the control industry states: "If you can measure it, you can control it." There are a great many devices that can be used for this purpose, and some of the more common ones are shown in Fig. 10.2.

The simplest device for converting mechanical motion into a signal is the rheostat, but because of the relatively short life of the sliding

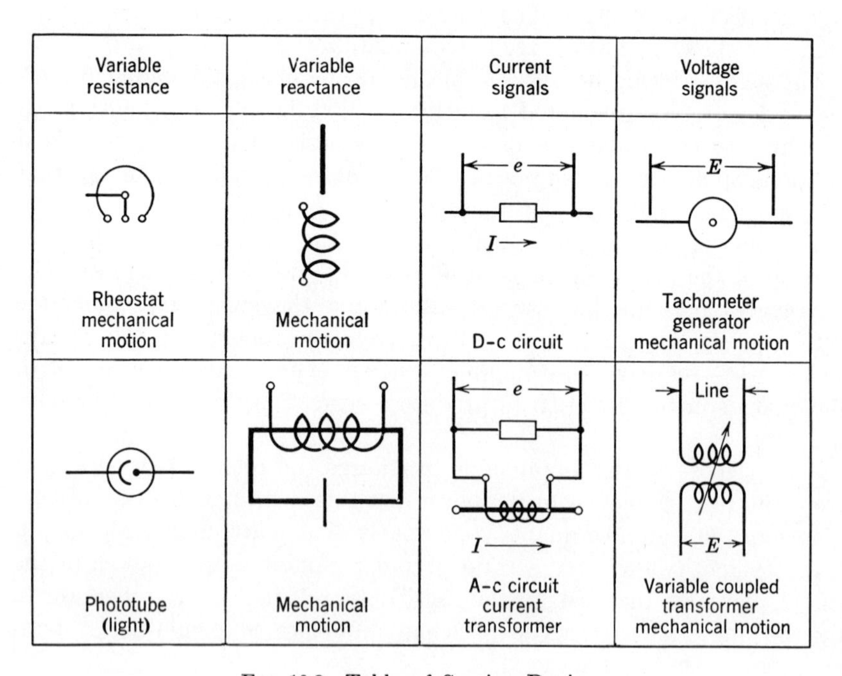

FIG. 10.2 Table of Sensing Devices

contact it has limited use. Signals from mechanical motions, and without objectionable wear, can be obtained from variable-reactance devices. This type is shown in the ordinary coil and plunger form, where the impedance varies with plunger position, and the biopolar reactor structure, where the impedance is varied by the introduction of an iron member between the poles. The mechanical motion is usually linear, but this is not a necessity.

High-level ac signals can be obtained from mechanical motions by using the variable-coupled transformer sensing device. The primary and secondary coils can be stationary, and the mechanical motion varies the coupling between them by changing the magnetic path characteristics. The movable iron structure can be designed so that very little or no force is produced by the magnetic circuit, thus making it an extremely acute sensing device that can be operated from delicate mechanical apparatus.

Signals from rotating devices are usually produced by small tachometer generators. These generators are of the permanent-magnet field type and have voltage outputs that are linear with the rotational speed, over a wide speed range. They are used in both ac and dc outputs. The relatively high signal level obtained is advantageous where precise control of speed is a requirement.

Current regulators need signals which are proportional to the current flow in power circuits. In dc circuits the signal e is produced across a resistor which carries the current being regulated. This is commonly used in tension systems, and the resistor can be a winding of one of the electric machines, rather than an additional voltage drop introduced for this purpose alone. An additional transducer for measuring direct current consists of a pair of iron cores excited by an ac source and encircled by the primary dc conductor. As long as the iron in the toroids is not saturated, the ampere turns of the ac signal winding must equal the ampere turns created by the direct current. The signal voltage produced across a load resistor in series with the ac winding will, therefore, directly represent the primary direct current. The ac sensing device uses a current transformer having a resistor load across its secondary terminals. The voltage e produced across the resistor is proportional to the load current.

The phototube is used as a variable resistor, to signal changes in light falling upon it. This sensing device is used on positioning regulator equipment. The device or material to be positioned cuts into or varies the beam of light that is directed on the phototube, and the variable-resistance characteristic of the phototube acts as the signal to the sensing system.

Error-Sensing Systems. The function of the error-sensing system is to compare the signal from the sensing device with the reference and to transmit the error to the input of the amplifying apparatus. The two generally used methods are voltage comparison and magnetic comparison. Each method has noticeable advantages and disadvantages which must be taken into account when selecting one to use for a specific application.

Voltage Comparison Circuit. In this circuit the voltage signal from the sensing device is electrically compared with the reference voltage, and any difference voltage is directly applied to the regulator or amplifier input. In Fig. 10.3, the error voltage e is shown applied to a coil on a magnetic structure, but it can be applied to a field of a rotating regulator, or to a gating element of an electronic amplifier.

The circuit makes efficient use of materials, especially where magnetic structures are used in the amplifier input. All of the copper in the input winding is active in producing ampere turns from the error voltage. The circuit is the most effective type to use for simple regulating systems where a single control or error voltage is employed.

The principal disadvantage of the circuit is the necessity of making the reference circuit and the signal circuit electrically common. This means that the two circuits must be actually connected together, so that the voltage comparison can be accomplished. Some applications require that additional signals be fed into the sensing system and enter into the comparison. The circuit does not lend itself readily to the use of a multiplicity of signals, since these must also be made electrically common with the principal voltage comparison circuit. Sometimes, where a magnetic structure is involved in the input, an additional coil is used

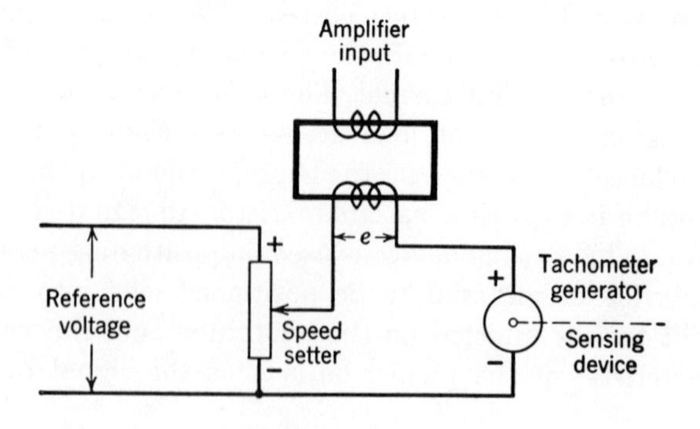

Fig. 10.3 Voltage-Comparison Sensing System

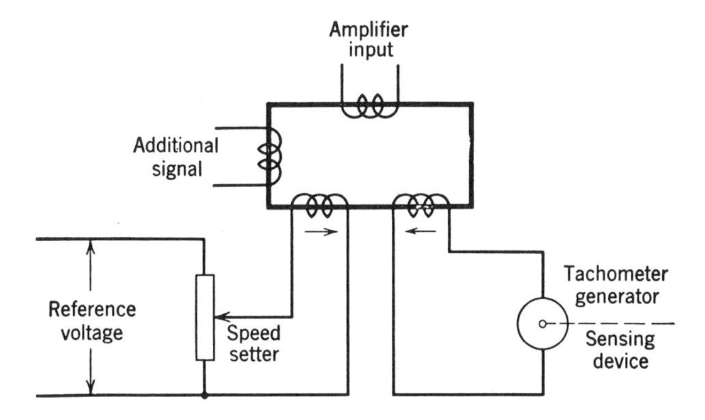

Fig. 10.4 Magnetic-Comparison Sensing System

to feed in an isolated signal. Whatever signal is fed in, in this manner, is opposed by the voltage comparison circuit, with the net result that the new signal biases the regulator and forces it to operate at a somewhat different error voltage. If the isolated signal were exceedingly strong, it might bias the regulator to some extreme value, which would result in damage to one of the coils because of coil heating.

Magnetic Comparison Circuit. It is necessary to use some form of magnetic structure in the regulator input stage if magnetic comparison is to be used. This type of circuit is inherently present when rotating regulators and magnetic-amplifier regulators are used. It can also be accomplished with an electronic regulator if a saturable core reactor or pilot magnetic amplifier with minimal time constants is used for the input.

This circuit offers a great deal of flexibility because a variety of isolated signals can be introduced through separate coils. The signal sent in to the regulator from the sensing system is the result of the algebraic sum of the separate signals. In the circuit (Fig. 10.4) the reference is established from the reference voltage and, through the speed setter, is connected to a coil on the magnetic structure. The sensing device shown—a tachometer generator used to measure rotational speed—is connected to another coil. The design of these coils is such that the ampere turns balance each other at any steady-state condition. In operation the ampere turns of the reference coil and the signal coil are in opposition to each other. Any difference in the ampere turns of these two coils shows up as resultant ampere turns which are active on the magnetic structure and produce the error signal fed into the input of the regulator. Any number of additional signals can be fed in through isolated coil circuits on the magnetic structure, and the resultant ampere turns of all of the coils is the signal to the regulator.

The principal disadvantage of the magnetic comparison circuit is that it does not use material efficiently in the input circuit. The major portion of the reference and signal energy is dissipated in heating in the coils, rather than in producing effective ampere turns in the input unit. This inefficiency can be overcome by increasing the size of the magnetic structure and coils of the input unit appreciably. Another method is to accept the performance obtained from a smaller magnetic input unit and regain the circuit amplification required by designing higher gain into the amplifying system.

Regulating System Stability. The closed-loop regulator action sometimes tends to overshoot when making corrections and may set the system into oscillation about the desired value. Such oscillation or hunting is damped out by the use of stabilizing or antihunt circuits. The design of system stability into regulating systems is the most involved part of the entire problem. The problem goes beyond the regulator itself, including time constants of the electric machinery and mechanical inertia of rotating elements in the electric machinery and the process machinery. Methods of calculation of stability problems have been derived, but in all of them complete knowledge of the mechanical and electrical factors external to the regulator is required.

The designer is confronted by the time constants of all parts of the closed-loop regulating system. In many cases the time constant of the regulated quantity is large in relation to all other time constants in the loop. When this ratio is large, the problem of stabilizing the system becomes easier. In general, the time constant of the regulator should be kept low; in other words, the regulator response should be fast.

Within the regulator, the gain and response time, in combination, are important factors. The regulator must have adjustments of sufficient range to adapt it to a variety of field applications. When it is used where high gain is necessary, the speed of response must provide adequate machine performance.

Regulators used for industrial control applications usually have antihunt circuits that employ transient feedback signals. These feedback signals may be picked up at the regulator output and fed back into the input, or they may feed back around just a portion of the amplifier. Transient signals can be obtained from a dc output by feeding the output signals back through capacitors. The capacitors block out steady-state voltages but allow changes in voltage to be fed back. Another feedback method uses a transformer with its primary winding excited from the output circuit. The transformer design must avoid saturation of the iron when excited from the maximum output. Under steady-state conditions there is no voltage induced in the transformer secondary winding. However, when the regulator output changes, voltages are induced in the secondary and are fed back.

The sense of the transient feedback signals is in opposition to the change that is taking place. This damps out the tendency of the system to oscillate. The amount of antihunt feedback signal is usually made adjustable. Typical antihunt circuits are shown in the following paragraphs which deal with complete regulating systems.

Voltage Regulators. Automatic regulators are often used to maintain the output voltage of generators, alternators, and static power supplies within certain required limits, for industrial control applications. The generating equipment used for power purposes in mills and factories is subjected to wide variations in loads and temperature, both of which can cause the voltage to fluctuate enough to be objectionable.

Individual processes may frequently require closely regulated voltage, where an individual power supply provides all or a portion of the process. Typical examples are electroplating, some high-temperature ovens, high-cycle tools, automatic welding, and excitation generators used on continuous processing lines. The high-cycle alternators used to supply hand tools, and for high-speed resistance welding, inherently have relatively poor load regulation. This imposes heavy duty on the regulating apparatus, often requiring that the alternator excitation current be doubled to compensate for the sudden application of full load. The regulator response must be fast, to prevent an objectional dip in voltage when the load is applied.

Since dc motor speed is proportional to the countervoltage, voltage regulators are frequently used to control motor speed in low accuracy systems. The basic equation for dc motor performance is

$$E = V - IR$$

where E = countervoltage
V = terminal voltage
I = armature current
R = armature resistance

In the basic equation voltage drop, IR, attributed to resistance, is normally 5 to 10% of rated voltage and insignificant at high speeds. However, when speed is reduced by lowering armature voltage, the voltage fluctuation caused by load change is appreciable. To reduce this effect, an *IR* compensation circuit may be added in the voltage regulator which senses load current and modifies the reference signal in proportion to the armature current, regardless of motor speed. Where there is no means available to measure speed directly, the voltage regulator is a reasonable substitute for a speed regulator.

Speed Regulators. When precise control of rotational or linear speed is a process requirement, an automatic speed regulator is used. Applications such as paper-making machines or metal coating lines must operate at carefully regulated speeds, since speed variations will cause deviations in the product or coating thickness. The measured variable is actual speed. The speed transducer is a dc tachometer producing voltage proportional to speed or a pulse tachometer producing a pulse frequency proportional to speed. Since the actual speed is measured and compared with a reference, the closed-loop regulator completely encompasses all causes of speed variations such as changes in load, temperature, field, and line voltage.

An analog speed regulator is in reality a voltage regulator and operates to hold the dc tachometer generator voltage constant in accordance with the reference representing the desired speed. The output of the regulator is fed into some element which is capable of changing the speed. This element is usually the power conversion unit that supplies power to the motor.

Digital speed regulators are used for control of either ac or dc drives, particularly when the drive must run for long periods without drift. The speed-measuring transducer is a pulse generator which produces pulses in proportion to the speed. The reference is an oscillator, either of the electrically driven tuning fork variety or, on more sophisticated applications, of the crystal type. The simplest digital speed regulators are merely add-subtract counters, which, over a programmed period of time, count the number of pulses from the reference and the number from the feedback. The analog regulator reference is adjusted by an increment corresponding with the error. More sophisticated regulators not only adjust for the number of pulses over or under the reference chain, but also continually monitor the phase displacement between concurrent pulses and adjust the speed to maintain exact phase relationships.

Since oscillators can be built to accuracies of one part in several thousand, this type of regulator can regulate speed to 0.001% or better. In addition, long-term drift can be essentially eliminated. Since the input signal is in digital form, the control system can readily be programmed from computers or similar pulse-generating devices. Speed-matching of consecutive drives in a process is easily achieved by utilizing a pulse tachometer on the preceding machine as the reference for subsequent machines. By adding or subtracting pulses from the reference chain, it is possible to provide a specified speed ratio between drives.

Position Regulators. Moving strips of material, particularly in continuous processing lines, sometimes require positioning regulators for process reasons or to guide them laterally in passing through the machines. Some applications, such as printing, require precise coordination of a moving strip of material with the machine. This is sometimes called register control, referring to the requirement that the material and the machine must operate in register. Color printing, where several colors are applied, each in a separate printing unit, requires careful control of register. Register control usually employs marks or dots on the moving strip as the measuring medium, and a light system with photocells takes regulating signals from the register marks. The regulating system operates on some mechanical means that is capable of shifting the strip to maintain the required register.

When a moving strip of material, under tension, must be guided laterally, the regulating signal is picked up at the edge of the strip by means of a light and photocell system. The regulator can shift the strip by moving rolls over which it travels. When the signal is taken from one edge of the strip, regulation is about a point where the light beam is partially intercepted by the strip. Some systems use a light beam at each edge of the strip, and the automatic regulator action keeps the amount of intercept of the two beams equal.

Some processes require a slack loop in the continuously moving strip of material. Certain cloth finishing ranges handle the cloth in completely relaxed form, which is attained by allowing it to hang freely in a festoon between process machines. In coating processes, such as paper coating, the material is not allowed to ride on rolls until the surface coating has dried. The material may be supported on air jets and allowed to form a slack loop before entering the next process. Steel-mill process lines require slack-loop control at certain slitters, in process tanks, and at other places.

When slack loops are used in continuous processing, the regulating problem involved is to synchronize the speed of one side of the loop with that of the other side, to maintain the length of the loop constant. The sensing device must be able to measure the length of the loop and send an appropriate signal into the sensing system. If the strip is of magnetic material, a variable-inductance or variable-coupled transformer can serve as the sensing device. In nonmagnetic materials the sensing

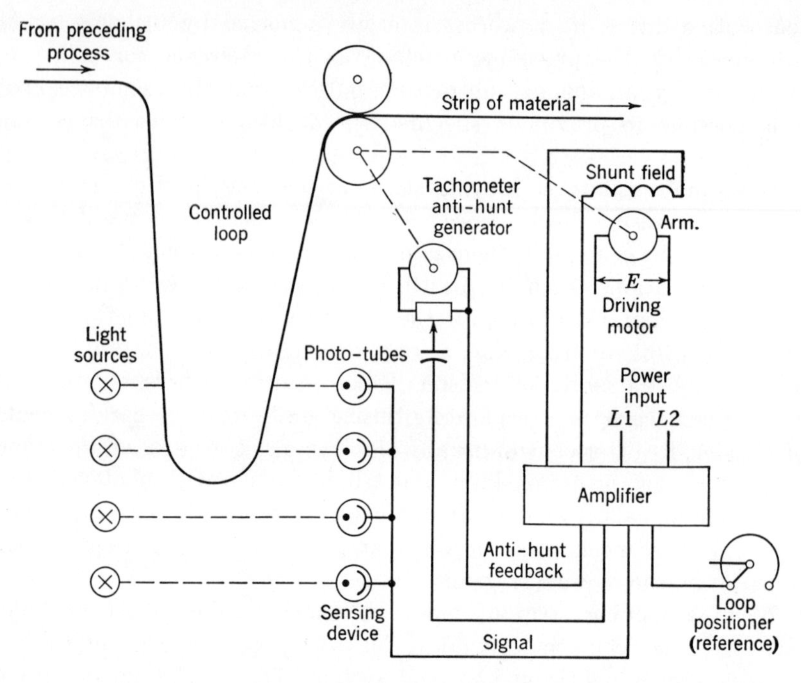

FIG. 10.5 Loop Regulator

device is usually one or more photocell systems. The regulator maintains the loop length in agreement with some preselected reference.

A slack-loop regulating system is shown in elementary form in Fig. 10.5 The strip of material is fed up to the loop from a process. It can be fed from rolls or might fall naturally from a supporting system of air jets. The next motor-driven process in line pulls the material from the loop at a synchronized speed, so as to maintain the loop at substantially constant length. The motor takes armature power from some source such as E, which is usually an adjustable-voltage system.

It is advantageous to use a wide zone of regulation for the bottom of the loop, because this desensitizes the mechanical system, resulting in less tendency to hunt. Figure 10.5 establishes a wide regulation zone by using four light sources and photocells. The combined effect of impedance change of the four photocells is the signal that is compared with the reference. The reference, in this case, selects the desired loop position in the light system. The output of the amplifier supplies excitation for the driving-motor shunt field. Any change in loop length changes the amount of light intercepted, resulting in a signal change from the photocells. The signal change is amplified and applies a correction to the motor shunt field, in a direction tending to restore the preselected loop length.

Sometimes the process or space available will not permit a deep loop or festoon as illustrated in Fig. 10.5. A long shallow loop is a more critical mechanical system than a deep loop and, for that reason, usually requires effective antihunt methods in order to obtain stability. One such method is to measure speed changes with a tachometer generator, and feed the transient tachometer voltage changes into the amplifier in degenerative sense. A capacitor in the feedback circuit blocks out steady-state signals from the tachometer generator. A potentiometer across the tachometer output permits adjustment of the strength of the antihunt signal. This signal should be as small as possible, to avoid objectionable interference with changes in the basic process speed. This type of feedback is particularly effective on applications where high mechanical inertias are involved and where there is a tendency to oscillate at low frequency.

Constant Tension Regulation. Long strips of material, such as paper, cloth, steel, and others are usually wound at constant tension. This is important for handling and, in many cases, prevents damage of the material, both in handling and in storage. Considerable attention is give to holding the winding tension uniform all the way from the inner core to the outside diameter. Where the tension cannot be held uniform for one reason or another, it is customary to taper it, the turns being somewhat tighter at the core than they are at the outside.

The tension systems employed are roughly classified as surface winding and mandrel winding. Surface winders make use of winding drums which are driven so that their surface speed is the same as the linear speed of the material being processed. The roll of material is supported in such a manner that it rests in contact with the winding drum, and the winding effort is transmitted to the roll by the surface contact between the winding drum and the outside surface of the roll. Since the surface speed of the winding drum does not need to change as the roll diameter changes, and since the driving effort of the drum can be constant, the driving motor operates at constant torque and constant speed for a given linear speed of the strip.

In mandrel winding the mandrel on which the roll of material is wound is coupled to the driving motor. When a roll is started, the strip is usually secured to the mandrel, and this firm connection permits winding at much greater tension than can be obtained from surface winding.

Basic Constant Tension Relations. In Fig. 10.6 the strip of material is fed up to the reel from a pair of tension rolls. These rolls can be

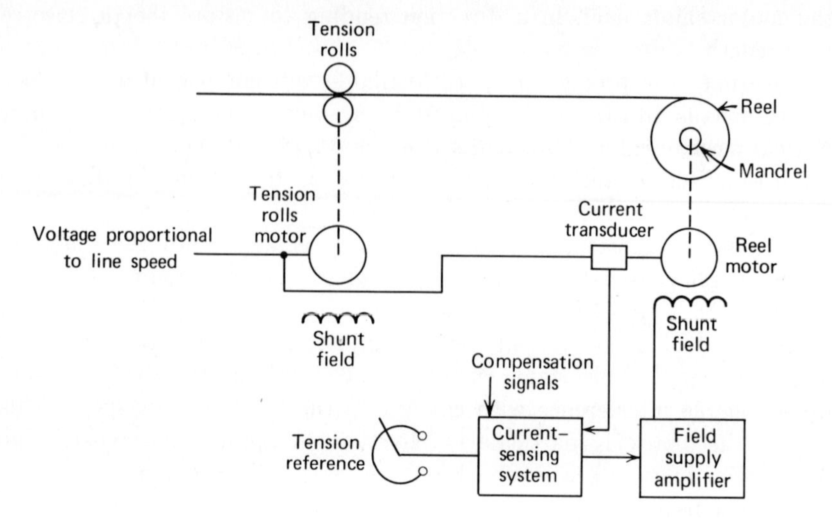

FIG. 10.6 Constant Tension Regulation by Motor Field Control

a portion of a processing machine, the only requirement being that some machine member must have a firm grip on the material to produce back tension.

The reel motor is directly coupled to the mandrel of the reel and is shown as receiving its power from adjustable-voltage lines. The reel motor can have a separate adjustable-voltage system, but the voltage applied to the motor should be proportional to the linear speed of the strip of material.

Because the length of turn varies with the diameter of the roll, the speed of the reel motor must vary inversely with the diameter of the roll, if the linear speed of the strip is constant. For a tension P in pounds and a linear speed S in feet per minute, the horsepower in the strip is

$$\text{Horsepower} = \frac{PS}{33{,}000}$$

Since the horsepower is in no way related to the diameter of the roll, it follows that the reel drive is constant horsepower.

The reel motor must be able to deliver constant-horsepower output to the reel over a speed range determined by the ratio of the outside diameter of the roll to the mandrel diameter. This requires an adjustable-speed constant-horsepower rated motor. The speed range of the motor should be somewhat greater than the diameter ratio of the coil, to allow some range at each extreme for regulating purposes.

With the constant-horsepower output, the reel-motor power input EI is essentially constant. The voltage E is fixed by the linear speed of the process, so the product EI can be held constant for constant tension by regulating I to a constant value.

The motor torque must vary directly with the diameter of the roll, and this relationship is

$$\text{Motor torque} = \frac{PD}{2} = K_1 FI$$

where D = the diameter of the roll
K_1 = a motor design constant
F = the motor field flux

Also

$$\text{Motor speed} = \frac{S}{\pi D} = K_2 \frac{E}{F}$$

where K_2 is a motor design constant.

The adjustable-speed motor performance for mandrel winding is shown by Fig. 10.7. The horsepower and armature input watts are held constant

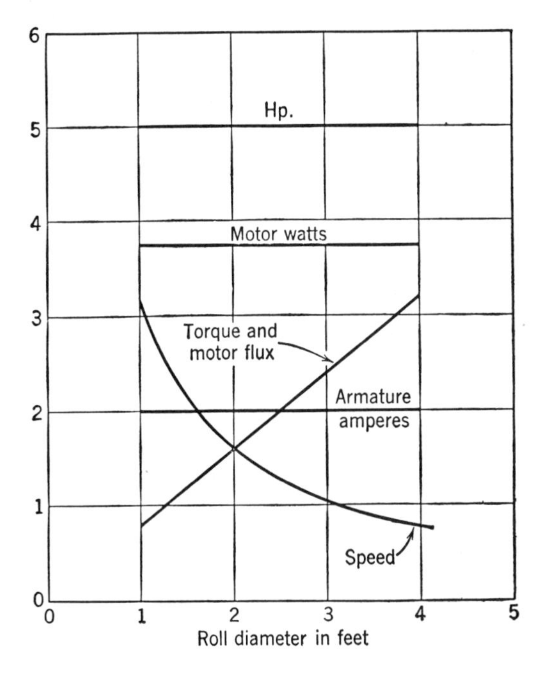

FIG. 10.7 Mandrel-Winding Motor Performance Curves

by holding the armature amperes constant. The motor flux must vary directly with the roll diameter. The motor torque and speed can be kept in the proper relation for constant tension by simply changing the motor field flux as the diameter of the roll changes. If the armature current, I, is to be a direct indication of tension, the following are the requirements.

1. The applied voltage must be a direct representation of the measured speed of the strip.
2. Mechanical losses must be minimal as compared with the power utilized for tension. These include inefficiencies of gears, bearing friction, machine losses, process energy dissipated, and power required to accelerate mass of moving parts.
3. Compensation must be provided for voltage drop caused by copper or brush resistance in the armature circuit and similar electrical losses.
4. Motor field flux must be properly adjusted in proportion to roll diameter buildup regardless of line speed and tension setting.

Constant-Tension Control Systems. The commonly used constant tension system shown in Fig. 10.6 makes use of an automatic regulator to supply the shunt field excitation of the reel motor. The tension-setting device establishes a constant voltage reference which is fed into the sensing system of the regulator. The quality of this reference signal is extremely important, since in any regulating system the total regulator performance can be no better than the established reference. The regulating signal is produced in a current transducer proportional to the reel motor armature current. In the sensing system any difference between the signal and the reference voltage shows up as an error which is fed into the amplifier system and results in a correction of the shunt field current of the reel motor.

Assuming stable operation at the desired tension, with the material being wound at constant linear speed, the roll will steadily increase in diameter. As the roll diameter increases, the tension tends to increase resulting in a slightly greater armature current. This increases the signal voltage fed into the sensing system and sends an error signal into the amplifier calling for more shunt field excitation. This greater field flux tends to slow down the motor and to increase its torque, in keeping with the increase in roll diameter. Thus by holding the armature current constant, the motor field flux, speed, and torque are in a proper balance with the roll diameter at all times to produce constant tension. With this type of tension regulator the speed range of the motor by adjustment of the motor field must be at least as great as the buildup of the coil. If additional material is wound on the reel after reaching a diameter corresponding to full field of the motor, this type of regulator cannot slow down the motor further, and the tension will increase. The result will be the undesirable condition of a coil wound tighter on the outside than on the inside.

CEMF Regulator. With the advent of a separate adjustable voltage power supply for the reel motor and faster response characteristics in the power supply, a more sophisticated system, the counter-electromotive-force regulator, has come into general use. This regulator, illustrated by Fig. 10.8, compares the armature current with a current or tension reference. Current excursions causing an error are immediatly corrected by adjustment of the armature voltage power supply. This, in turn, causes a change in reel speed.

Simultaneously voltages representing the tension rolls motor speed and reel surface speed are compared. The error between these voltages initiates adjustment of the reel motor field to eliminate the error. With this type of regulator the time constants in the current-sensing system are short, providing excellent dynamic response of the system. Conversely, the motor field should only be adjusted at a slow rate corresponding with buildup of the coil. The speed-sensing system is designed with longer

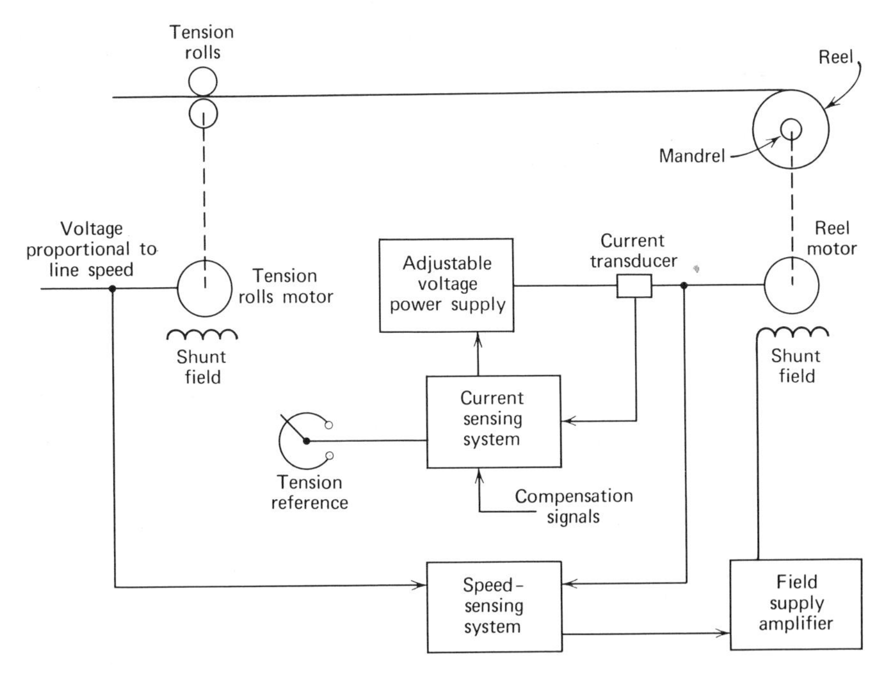

FIG. 10.8 Tension Regulation by Control of Counter-electromotive Force

time constants, adequate for these conditions, and providing a stable high-performance system. If the coil diameter is extended beyond the range within which constant tension can be provided by motor field adjustment, the current regulating system will cause the armature voltage to decrease while holding constant armature current. This will reduce the tension on the outside of the coil which is the preferable condition if constant tension cannot be maintained.

Compensation circuits or signals may be used to adjust for mechanical or electrical losses. The most common concept is WR^2 compensation. If the linear speed of the material is changed during the winding operation, additional horsepower must be added for acceleration or subtracted for deceleration, to overcome the effects of mechanical inertia. The usual method employed is to bias the regulator in a direction to add to or subtract from the reference the fixed increment of armature current necessary to accelerate or decelerate the mass.

Although Figs. 10.6 and 10.8 illustrate a constant-tension winder taking material from a process, it is sometimes necessary to feed material at constant tension into a process. This is referred to as unwinding or payoff. During unwinding the process pulls the material from the roll. The unwinder motor operates as a generator returning power into the adjustable-voltage system. Constant tension holdback is obtained by regulating constant current in the unwind motor armature. The control equipment is much the same as is used for winders. Since generating action rather than motoring action is controlled, the sense of the amplifier output and the polarity of the inertia compensation signals are the reverse of the winder equipment.

Regulator Evolution. The predecessor to the automatic closed-loop feedback system was the regulating system in which the operator performed many of the basic functions. Typically, the operator observed the process directly or by watching instruments, and adjusted a rheostat to bring the controlled variable to the desired value. It is therefore logical that the original automatic regulators were of the electromechanical type utilizing significant power for the regulating action.

Since closed-loop feedback systems truly represent the "brain" of the control system necessary for maximizing production in processing of material, it is also logical that the most modern electronic techniques should be used. In a relatively few years there has been a transition through electromechanical devices, rotating machines, magnetic amplifiers, tube-type devices, transistors, and integrated circuits. Semiconductor, solid-state control devices in combination with static conversion equipment offer many advantages in reliability, longevity, maintainability, speed of response, and overall systems performance.

Solenoid Operated Regulators. In this general classification are included those regulators which convert electric signals into mechanical motions, the motions being capable of changing the electrical characteristics of circuits of the controlled machines.

The reference is usually mechanical and in the form of a spring and linkages, which, in combination provide a fixed mechanical bias for the complete throw of the regulator mechanism. The signal is a voltage from the regulated quantity. This signal voltage furnishes the power to operate a solenoid having a movable plunger. The solenoid pull on the plunger balances the pull of the reference spring when the regulator is in operation. The solenoid is designed for a pull characteristic to match the reference spring characteristics, so that a small change in solenoid ampere turns can swing the regulator over its entire range. The solenoids usually have straight-line plunger motions, but some regulators use rotary-motion solenoids or torque motors to set up the operating forces.

The precision of this type of regulator is intimately related to the efficiency of the mechanical system employed to translate the sensing signal into a regulating quantity. The regulators built by the different manufacturers differ mainly in the types of mechanical motions used, to eliminate friction and to employ the signal forces to the best advantage. The antihunting means is also mechanical and operates to damp oscillations of the mechanical system.

The control output of this type of regulator is a variable resistor or rheostat, which is connected in series with a shunt field of the machine to be regulated, so that it can automatically change the machine excitation. Similarly, the output can be ac with variable reactance or variable coupling transformers.

Figure 10.9 shows, in simplified form, an Allis Chalmers rocking contact regulator connected to maintain constant the output voltage of a self-excited dc generator. The generator output voltage is the signal applied to the solenoid coil. The pull of the solenoid is opposed by the reference spring, and any resulting motion rocks the contact sector. The plunger system transmits motion to the dashpot piston through an elastic member. This permits fast initial corrections, but its damping action tends to limit the amount of regulator movement for a given disturbance.

If the generator output voltage drops for any reason, the pull of the

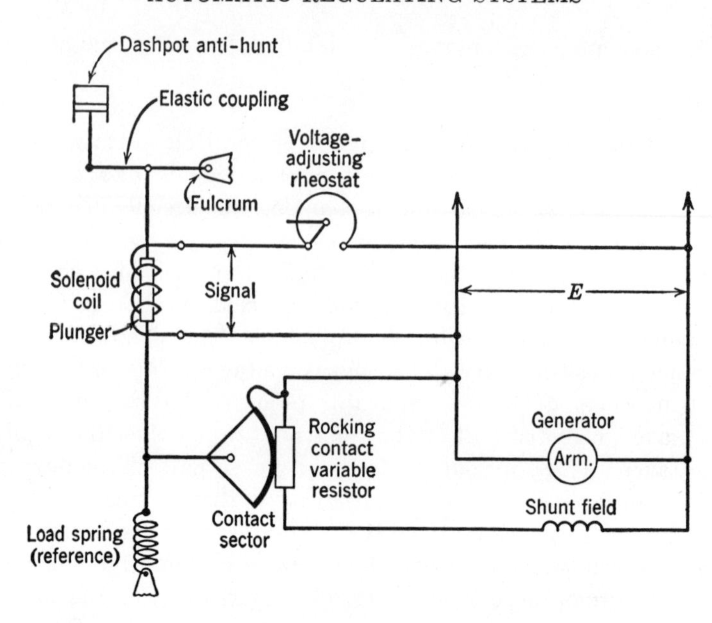

FIG. 10.9 Solenoid-Operated Regulator Connected to Maintain Constant Voltage on a Generator

solenoid decreases somewhat allowing the spring to pull the plunger down and rocking the contact sector downward. The contact sector commutates the resistor in the direction to reduce the resistance in the shunt field circuit. The increased field excitation restores the generator output voltage to the preselected value. The voltage-adjusting rheostat is connected in the signal circuit in series with the solenoid coil. Inserting resistance in this rheostat requires an increase in generator voltage, to supply the necessary solenoid ampere turns to balance the regulator.

Rotating Regulators. Under this general classification are several specific types, which differ in the detailed principle of operation of the amplifying elements. They are alike in the broad sense that the amplifying elements are rotating dc electric machines. These rotating amplifiers, or generators, are usually driven from the ac power lines by induction motors. The reference and signal power is applied to low-energy shunt field windings, and the amplifying generator armature output is the regulator output used for control purposes. The power gain is the slope of the curve of regulator output watts plotted against signal watts.

The difference in the amplifying generators sold under various trade names are, in general, in the circuits employed to increase the power gain.

Special attention is given to obtaining fast rates of response, because the regulators must have low time constants to give stable operation in many applications. One method of getting high gain from a dc generator is to supply most of its excitation from a self-excited field circuit, whose resistance is adjusted to the critical value that just allows excitation to be supported. Low power signals, applied to other fields on the generator, can then readily change the generator output over its entire range. Another method is to operate two dc generators in two stages. These can be built as two stages in one machine, or in two machines.

The Reliance VSA is a two-stage amplifier, consisting of two separately excited dc generators that are driven at constant speed by an induction motor. A simplified circuit of this regulator is shown in Fig. 10.10, where it acts to maintain the speed of a motor constant at some present value. The motor takes its armature power from a dc adjustable-voltage generator, and its field is excited from constant-voltage lines. The shunt field of the adjustable-voltage generator receives its excitation from the regulator output. The sensing device is a tachometer generator, which is driven by the controlled motor. The speed selector furnishes a reference taken from a reliable reference voltage source.

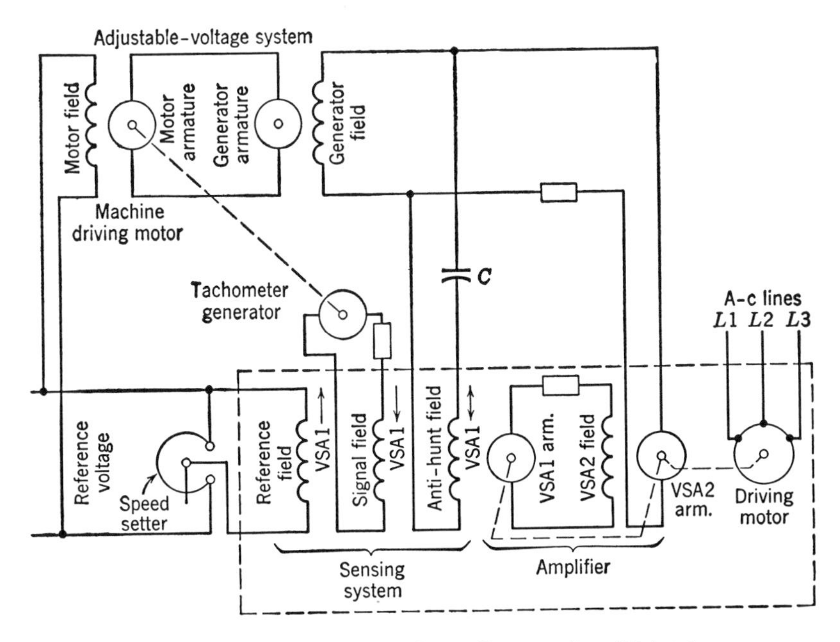

Fig. 10.10 Rotating Regulator Connected to Maintain Constant Speed on a Motor

The amplifying generators are driven by the ac motor, and therefore the principal power involved within the regulator comes from the ac lines. The input stage is generator VSA1, and it is fitted with a number of shunt fields coils. These shunt field coils, whose ampere turns combine algebraically to excite the machine, are the sensing system of the regulator. When used in this form they operate as a magnetic comparison system.

The speed setter determines the speed level by presetting the reference-field ampere turns. The sense of the reference field is to increase the regulator output. This strengthens the adjustable-voltage generator field, raises its armature voltage, and increases the speed of the controlled motor. The reference field operates in opposition to the signal field.

The signal field takes its signal from the tachometer generator. When operating normally, the ampere turns of the signal and reference field almost balance, and any small difference is the error that is amplified through the regulator. The steady-state operating point is automatically attained when the net difference ampere turns just produce the required regular output.

The voltage produced at the armature of the input unit, VSA1, as a result of the error ampere turns, is applied to the shunt field of the second-stage generator, VSA2. The armature output of VSA2 furnishes the shunt field excitation for the adjustable-voltage generator. The generator furnishes power for the controlled motor and thus completes the closed-loop regulating system.

The antihunt field in Fig. 10.10 is connected to feed transients from the regulator output, back into the input. The capacitor C prevents steady-state feedback. The sense of the antihunt field is always in opposition to the change taking place.

Vacuum Tube Regulator. Electronic regulators which employ vacuum and gaseous tubes to amplify error signals have the disadvantage of using components with unpredictable life, but when compared with other types, they have many advantages of superior performance.

The error signal, set up in the sensing system, is applied to a grid circuit of a high-vacuum tube in the input stage. The input tube characteristic requires very low signal power, which enables the regulator to operate from low-energy signals. The signal input can be from a high-impedance circuit. Stages of amplification are used as required to get the necessary overall gain for the application.

The output stage usually uses thyratron tubes connected in single-phase or polyphase circuits, supplying direct current for motor-armature or shunt field circuits. A wide range of output kilowatt ratings is available because of the wide selection of thyratrons available and the various circuit connections that are used.

These regulators can have extremely low time constants, because the electronic components are inherently fast. The low-energy grid circuits can be designed for high rates of response. When moderate rates of response are required, saturable-core reactors are used in phase-shift circuits to control the output thyratrons. When extremely fast regulator action is required, other phase-shift circuits are used that eliminate the time constant of the saturable-core reactor.

An electronic regulator built to the requirements of mill-type control is shown in Fig. 10.11. This uses industrial-type components, with special care given to adequate electrical creepage and clearance distances. The upper panel mounts the three thyratrons which operate in the three-phase

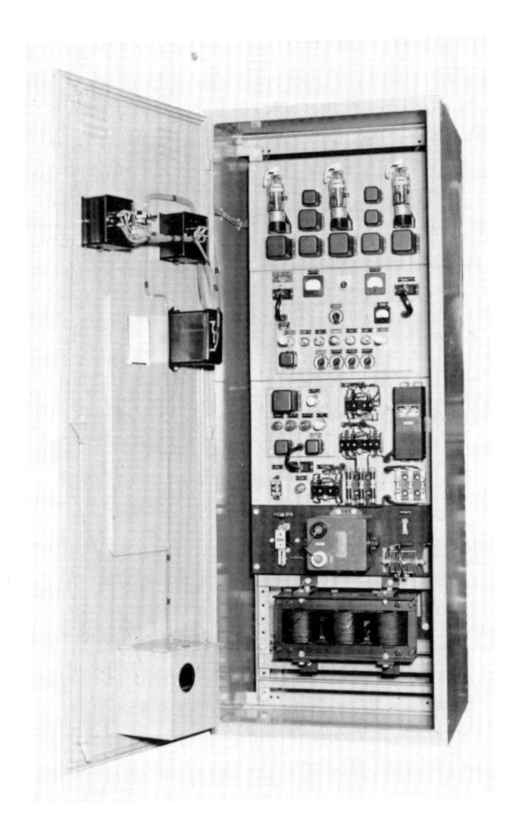

FIG. 10.11 Electronic Regulator Assembly, Constructed to Meet Heavy-Duty Mill Requirements

power-supply circuit of the regulator. Directly below the thyratron panel is the amplifier panel. The circuits of this panel can be disconnected by use of the plug connectors shown. The amplifier panel can then be removed for inspection and servicing. The operating conditions are indicated by the instruments. The remainder of the panel devices consist of a removable dc supply panel, constant-voltage transformer, anode transformer, and miscellaneous contactors and relays.

Magnetic Amplifier Regulator. Magnetic amplifiers (Fig. 10.12) are a combination of saturable reactors and diodes with the power circuit connected in such a way that the load current assists in the saturation of the reactor core. The core materials of nickel-iron alloys have low magnetizing force requirements with resulting steep saturation curves.

With the core unsaturated, the ac load windings present maximum impedance in the ac line. Control coils wound on this same magnetic structure can be excited from direct current to saturate the iron. As soon as saturation begins, load current starts to flow, and for the balance of the half-cycle the combination of the two will totally saturate the iron. The resulting minimum impedance in the load winding causes the voltage drop across the reactor to fall to a minimum, and almost the entire voltage will appear across the load. The gating action of the saturable reactor is very similar to that obtained from a controlled rectifier.

Diodes are connected in series with the power windings to polarize

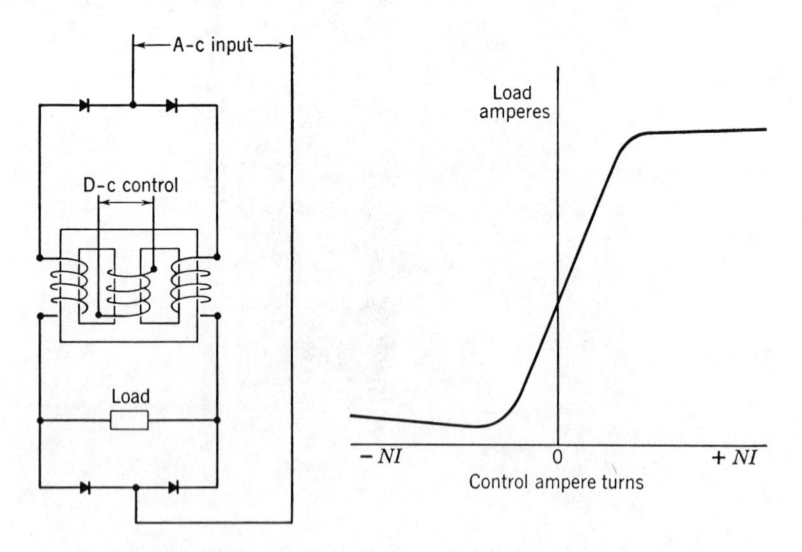

Fig. 10.12 Full-Wave Bridge-Connected Self-Saturable Reactor Circuit

the load current and create the condition of self-saturation. In addition, rectifiers are connected in conventional bridge arrangements to convert the output to direct current for most industrial applications.

The efficient use of self-saturable reactors requires that the load resistance be matched to the reactors. If the unit is to be rated for continuous duty, the load resistance selected will limit the reactor temperature to the allowable rise. Additional resistance may be added to reduce the time constant of the load circuit. The response time, usually expressed as time required to obtain 63% of the ultimate change in output, is predominantly a function of the L/R ratio of the control coil circuits. Since the power requirements of the control windings are very low, adding resistance and thereby additional power is usually not significant.

Multistage Magnetic Amplifier Regulator. Since magnetic amplifier output can be controlled by excitation of low wattage control windings, these devices are logically control system elements. If an entire regulator is to be built from magnetic amplifiers, the desired gain and time constants may not be achieved with a single unit.

The tension system in Fig. 10.13 uses a three-stage magnetic amplifier regulator. The maximum roll-diameter change ratio is 3/1, and the adjustable-speed constant-horsepower-rated driving motor selected has a shunt-field-control speed range of 4/1. In compensating for changes in roll diameter, the regulator must vary the motor shunt field over a range of 250/50 V, for the 4/1 speed range. Constant tension throughout the coil is obtained by regulating the motor armature current to a preselected constant value.

The three magnetic amplifiers are cascaded in the order of input, intermediate, and output stages. Only the essential control-coil circuits are shown for each stage. It is understood that each stage has load-coil windings and the associated rectifiers that combine into self-saturable reactor circuits, operating from some ac input. The output of the first stage supplies the control power for coil $2C$ of the intermediate stage. The output of the intermediate stage supplies the control power for coil $3C$ of the output stage. The power from the output stage supplies the winder-motor shunt field excitation.

The power amplification, in progression through the stages, is illustrated by the performance curves. In each case these are plotted between control-coil ampere turns and output watts of the unit involved. On these curves is indicated the approximate excursion, or portion of curve used, necessary to produce the steady-state swing of 250/50 V of the motor shunt field. The ratio of the output-watts excursion to the control-watts excursion is the power gain of the stage. The ratio of the

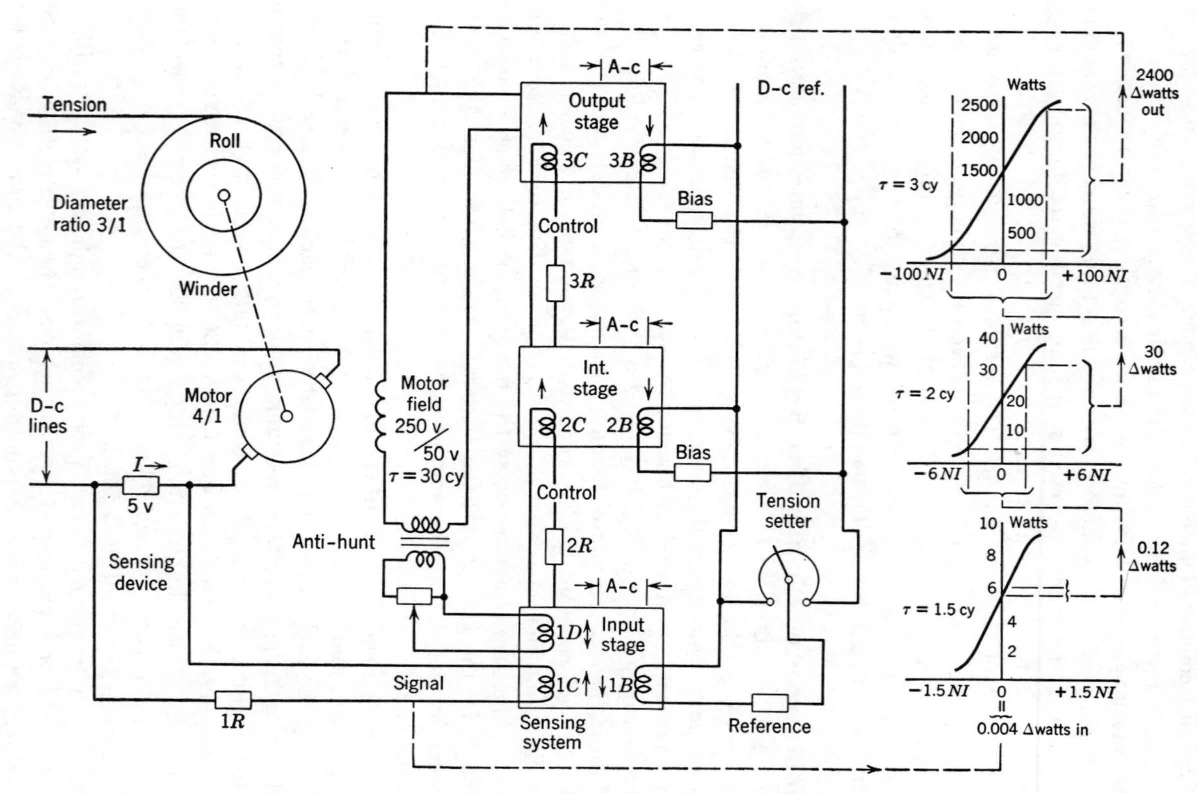

FIG. 10.13 Three-Stage Magnetic-Amplifier Regulator

output-stage output-watts excursion to the input stage control watts is the regulator power gain.

The sensing device is a resistor in the motor armature circuit. This armature supply is usually an adjustable-voltage system which also supplies other motors involved in the process. At the maximum tension setting the armature current I produces a 5-V signal. This signal voltage is proportionately less for lower tension settings. Series windings in the motor can be used in place of the dropping resistor, providing they produce a sufficiently high signal voltage.

The circuit in Fig. 10.13 employs magnetic comparison in the sensing system. The tension-setter rheostat preselects the ampere turns of the reference coil $1B$ which acts to reduce the stage output. The ampere turns of the control coil $1C$ are in opposition to the reference and act to increase the stage output. They automatically find an operating point where the resultant ampere turns of these two bucking coils produce sufficient output to adjust the motor field excitation to the roll-diameter requirement. The $1R$ resistor sets up a favorable ratio of inductance to resistance in the control-coil circuit, to obtain the needed speed of response of the input stage.

The bias coils of the intermediate and output stages are connected to reduce the stage output. The sense of the control coils is to increase the stage output.

The antihunt winding $1D$ in the input stage is connected to feed transient output changes back into the input stage. The antihunt transformer has its primary winding connected in series with the motor shunt field. Rapid changes in shunt field current induce voltages in the secondary winding of the transformer. A portion of the induced voltage, as selected by a potentiometer rheostat, is fed into coil $1D$. The sense of coil $1D$ can be either to increase or to decrease the regulator output. It is always in the direction to oppose the change that is taking place in the motor field current.

The action of the regulator as the roll diameter increases has been described in the paragraphs under constant-tension control system.

Solid-State Regulators. The need for infinite life, low power, high gain, fast response, low cost, control elements dictated the development of closed-loop feedback regulators utilizing semiconductor devices. The semiconductor elements operate throughout their control range in accordance with the input signal to perform the same functions as their predecessors, the vacuum tubes. The control elements may be discrete components arranged on a printed circuit board, or integrated circuits in which combinations of circuitry, active elements, and passive elements are produced by deposition on a substrate. Frequently the regulator unit boards include a combination of discrete components and integrated circuit chips utilizing the optimum features of each.

A preferred construction for regulators is unitized printed circuit boards (Fig. 10.14). The design of each printed circuit board encompasses one or more basic regulator functions. A complete feedback control system regulator is generally a combination of these individual, functional, printed circuit boards combined in a manner which will provide the basic regulating function required by the application. These application requirements are regulation of speed, voltage, position, current, and tension.

A fundamental element of each of these types of regulators is a mixing, weighting, and comparing of the various incoming signals. The signal comparison may be electrical or magnetic, but the ultimate output expected of the board is an amplified signal representing the error between the reference and feedback signals. The name assigned to this particular unit may be comparator, operational amplifier, mixing amplifier, or other similar nomenclature. The output of this unit works directly into a power amplifier which may be the field supply of a rotating generator or the firing circuit of a static power conversion unit.

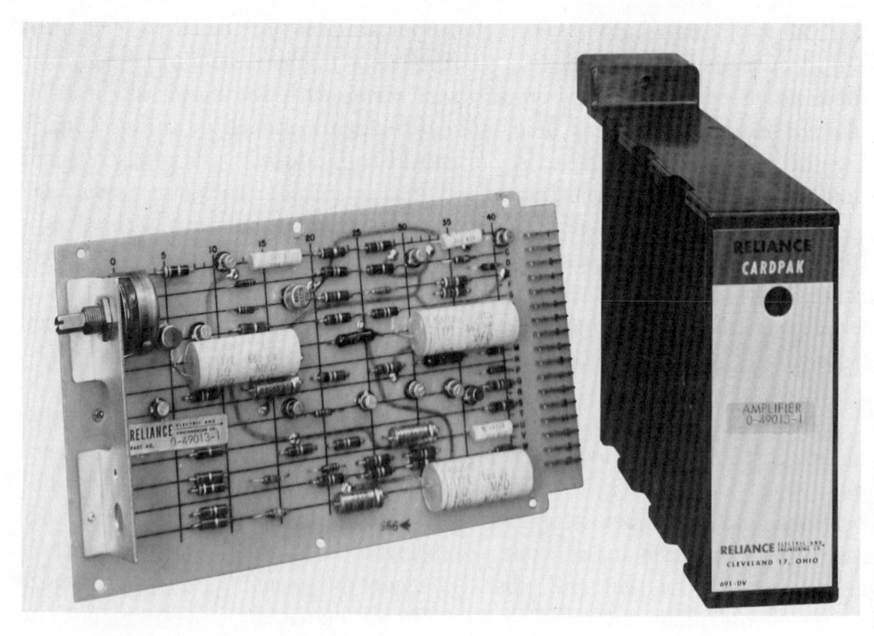

Fig. 10.14 Reliance Electric Co. Cardpak Unit for Analog Regulator

Since the equipment or machinery frequently cannot tolerate maximum torque or current, an additional printed circuit board may be added to provide a current-limiting function. The ceiling is established with semiconductor devices of known parameters so that if the load current exceeds the prescribed limit, there will be a clipping or limiting action by the control elements to modify the reference of the basic regulator.

Controlled rate of acceleration or deceleration of machines or processes is generally desirable. Therefore, an additional board provides a time reference or ramp function as an input to the basic sensing system. Functionally the circuitry on this board is interposed between the reference-setting device and the signal comparison system to program the rate at which the reference end of the regulator can increase or decrease. Frequently this rate signal is controlled as a function of the charging of a capacitor with a transistor regulating the charging current to a constant value. The ultimate performance is exactly the same as using a motor-operated potentiometer as the reference-setting device.

Motor-operated rheostats are used in this manner on applications requiring exceptionally long times, unique nonlinear characteristics, or memory of reference value on loss of power.

When adjustable speed drives require utilization of both the voltage range of the conversion unit and the field-weakening range of the motor, the control is expected to bring the motor up to base speed by increasing armature voltage and then to weaken the field to the final desired speed. To provide this function a crossover board is required which holds the motor field supply at full rated value until the power conversion unit has reached rated output.

Since semiconductor devices required unique voltages, and since high quality regulators require regulated voltages as a basic reference, a power supply board is used to convert available line power to the desired values. By the use of transformers or voltage-dividing networks, proper voltages are established. Diode rectifiers convert to direct current, and resistance-capacitance filter networks establish a ripple-free output. Voltage-rated devices such as zener diodes are used to regulate the output voltage within close tolerances.

If signal isolation is desired, a small toroid with multiple windings may be connected as a magnetic amplifier to mix the incoming signals. This toroid with its diodes can also be mounted on a printed circuit board (Fig. 10.15). In addition to signal isolation, the magnetic comparison allows selection of control windings on the toroid which match the impedance of each incoming signal. If the system has low gain requirements, this pilot magnetic amplifier may function as the basic regulator with its output directly controlling the power conversion unit. If more

FIG. 10.15 Cutler-Hammer mPAC II Speed Regulator. Side of Enclosure Removed

gain is desirable, additional amplification can be provided with semiconductor circuitry.

With a well-designed family of solid-state modules, high-performance regulators can be assembled to fulfill all reasonable application needs. Careful attention must be given to gain, response, and stability to achieve optimum performance. Printed circuit boards are normally assembled in some type of enclosure for mechanical protection. Printed circuit boards are designed for ease of maintenance by either designing the entire board to plug into a receptacle or with quick-change connectors.

Problems

A. The constant-tension regulating system, shown in Fig. 10.6, has the following assigned data:

Maximum linear speed = S = 1000 ft/min
Tension = P = 330 lb
Adjustable-voltage lines, 230-V dc
Diameter of mandrel = 1 ft
Diameter of full roll = 4 ft
Gear ratio between armature and mandrel = 6/1
Mechanical efficiency of winder = 80%

1. What horsepower is required to wind the strip?

2. What is the motor speed at the start of a roll?

3. What is the motor speed at a full roll?

4. What is the motor torque at the start of a roll?

5. What is the motor torque at a full roll?

6. What type of motor and what horsepower rating should be selected?

7. What speed range, by motor shunt field control, is required of the motor to allow for automatic tension regulation at the roll-diameter extremes?

8. With 230 V supplied to the motor armature, what armature current produces 330 lb tension?

9. How many ohms are required in the sensing-device resistor to produce a signal of $e = 5$ V, to operate the automatic tension regulator?

B. Assume a mechanical efficiency of 40% for the mandrel winder of problem *A*. The regulator accuracy is the same in both cases.

1. What is the loss (friction and windage) horsepower?

2. What motor horsepower is required?

3. State the ratio of the per cent of the motor horsepower that produces useful tension in the high-efficiency drive to that in the low-efficiency drive.

4. The regulator holds the total horsepower to a constant value. Why then is it important that the variable losses be as low as possible?

C. The mandrel winder of problem *A* has a sensing-device resistor which produces a 5-V signal for maximum tension. This signal must change 0.1 V to swing the regulator from the empty mandrel to the full roll conditions.

1. What per cent error, from constant armature current, does this produce at the maximum tension setting?

2. When preset for one-fifth of maximum tension, the sensing device will produce a 1-V signal. What is the per cent error in this case?

3. What conclusion can be made relative to defining the signal level when specifying the regulator accuracy?

D. A constant-speed application takes a regulating signal from a tachometer-generator sensing device. The regulator requires a signal change of 0.1 V to produce the necessary corrections.

1. If the allowable error at maximum speed is 1%, what voltage must the tachometer generator produce at the maximum speed?

2. What tachometer voltage is required at the maximum speed to reduce the allowable error to 0.5%?

3. What conclusion can be made relative to the magnitude of the signal level where a choice exists?

11

THE DIRECT-CURRENT SERIES MOTOR

Series motors are those having the main field connected in series with the armature, and so carrying the load current. They may be provided with a relatively light shunt field also, to prevent excess speed under light loads; they are then called series-shunt motors. The relative strength of the two fields may be anything up to 50% for each. Since the load current passes through the series field, the field strength will vary with the load, and the speed will decrease on heavy load and increase on light load.

Construction. The general construction of a series-wound motor is similar to that of a shunt-wound motor, except that the field windings are of heavy wire or strap, for connection in series with the armature.

The frame is of heavy steel construction and may be split horizontally and hinged, so that the top half may be swung back for easy access to the armature and the bearings. Armatures are relatively long and of small diameter to reduce their inertia, which helps in quick starting, stopping, and reversal. In the motor shown in Fig. 11.1, the field coils are wound in a sealed steel box, into which the field pole fits, the bottom of the box being held between the pole and the motor frame. The commutator and brush construction is similar to that of a shunt motor. The motors are made in several types according to the degree of protection desired. Totally enclosed motors have a solid frame, without openings. Open motors have openings in the top and the bottom halves of the frame. These are covered with perforated metal or louvers. Protected self-ventilated motors have a solid top frame and openings in the bottom frame.

Series-wound motors are manufactured in the same frames as shunt-wound motors. There are two basic configurations, one the *industrial* dc motor and one the *mill* motor. The vital dimensions of the mill motor have been standardized by the Association of Iron and Steel Engineers (AISE), so there is interchangeability from one manufacturer to another.

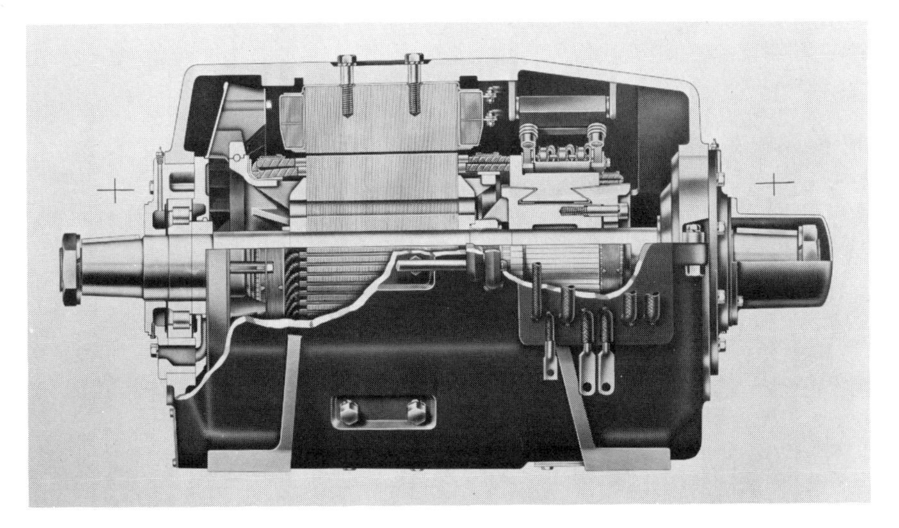

FIG. 11.1 Westinghouse Type-MC Series-Wound Mill Motor

Selected parts reprinted from "D-C Mill Motor Standards," *Iron and Steel Engineer,* September 1968, for mill motors follow. Note that shunt and compound motor ratings are also shown.

> The purpose of this standard is to define the capability of the dc mill motors and the mounting dimensions. The motors are for heavy duty steel mill service. This standard supersedes the previous Standard No. 1 AISE dc mill motors.
>
> The scope of this standard covers series, shunt, and compound wound motors of two standard enclosure types: TENV (totally enclosed nonventilated), and TEFV (totally enclosed forced ventilated from a separate source); these motors shall be suitable for use outdoors.
>
> The standard mounting position shall be horizontal.
>
> The TENV and TEFV enclosures shall be convertible by addition or removal of suitable covers.
>
> Frames shall be horizontally split in such a manner to allow removal of the armature by a straight vertical lift after the top frame half is opened or removed.
>
> The voltage, horsepower, speed and time rating shall be as listed in *Fig. 11.2.*
>
> The performance standards of motors conforming to this standard are based on operation from a direct current source of supply such as a generator or battery and ratings are based on 230 V, but motors shall be suitable for operation on voltages up to 500 V. Ratings do not necessarily apply when appreciable ripple is present. Maximum running torque at voltages above 230 V are reduced.
>
> These standard mill motors in frames 802 through 818 shall be

AISE FRAME SIZE	TOTALLY ENCLOSED 1 HR OR FORCED VENTILATED CONTINUOUS					TENV SERIES MOTORS 30 MINUTE		TENV 30 PERCENT TIME ON DUTY CYCLE (3)						AIR REQUIREMENTS FOR CONTINUOUS FORCED VENTILATED RATINGS			MAXIMUM STARTING TORQUE (LB - FT)			MAXIMUM RUNNING TORQUE ON 230 VOLTS (LB-FT)			MAXIMUM ARMATURE WR^2 (4) ($LB-FT^2$)	MAXIMUM SAFE SPEED RPM
	HP	RPM				HP	RPM	SERIES		COMPOUND		SHUNT		CFM	STATIC PRESS. AT INLET (INCHES H_2O)									
		SERIES	COMPOUND	STRAIGHT SHUNT	ADJUSTABLE SPEED (2)			HP	RPM	HP	RPM	HP	RPM		AIR IN DRIVE END	AIR IN COMM END	SERIES	COMPOUND	SHUNT	SERIES	COMPOUND	SHUNT		
802A (1)	5	900	1025	1025	1025/2050	6.5	750	5.5	840	5	1080	5	1130	110	3/4	1/2	145	115	92	116	90	75	6	3600
802B (1)	7 1/2	800	900	900	900/1800	10.0	675	8	780	7.5	950	7.5	1000	110	3/4	1/2	245	198	158	196	154	130	6	3600
802C (1)	10	800	900	900	900/1800	13.5	675	10	800	9.5	940	9.0	1000	160	1	1/2	330	263	175	262	205	160	6	3600
803	15	725	800	800	800/2000	19	620	15	725	14.5	840	14	880	200	1 1/4	1/2	545	445	295	440	345	265	12	3300
804	20	650	725	725	725/1800	26	580	20	650	18.5	775	17	800	250	1 1/4	1/2	810	650	435	650	505	390	30	3000
806	30	575	650	650	650/1950	39	500	30	575	28.5	690	25	715	335	1 1/2	3/4	1370	1100	725	1100	855	650	50	2600
808	50	525	575	575	575/1725	65	450	40	570	37.5	625	35	630	425	1 1/2	3/4	2500	2050	1370	2000	1600	1220	90	2300
810	70	500	550	550	550/1650	90	440	60	550	52.5	615	45	600	525	1 3/4	1	3700	3000	2000	2950	2330	1800	145	2200
812	100	475	515	515	515/1300	135	420	85	525	75	580	60	565	750	2	1 1/2	5500	4600	3060	4430	3600	2750	220	1900
814	150	460	500	500	500/1250	200	400	115	515	110	565	85	560	900	2 1/4	1	8550	7100	4725	6850	5550	4250	400	1700
816	200	450	480	480	480/1200	265	400	150	500	140	540	110	535	1200	2 1/2	1 1/4	11,700	9800	6550	9300	7650	5900	600	1600
818	250	410	435	435	435/1100	325	360	185	485	165	490	130	470	1600	3	1 1/2	16,000	13,600	9050	12,800	10,600	8150	1100	1500

(1) FRAME SIZE 802 IS ASSIGNED THREE RATINGS. MOUNTING DIMENSIONS ARE THE SAME FOR EACH RATING, BUT THE ELECTRICAL DESIGNS WILL BE DIFFERENT.

(2) A LIGHT STABILIZING SERIES FIELD MAY BE USED AS REQUIRED TO OBTAIN THESE SPEED RANGES.

(3) CONTINUOUSLY REPEATED DUTY CYCLES OF 5 MINUTES DURATION WITH LOAD ON FOR 1 1/2 MINUTES, POWER OFF FOR 3 1/2 MINUTES, WITH SHUNT FIELDS CONTINUOUSLY EXCITED.

(4) LIMITING MAXIMUM VALUES, NOT TO BE USED FOR APPLICATION CALCULATIONS.

AISE FRAME SIZE	TOTALLY ENCLOSED				PROTECTED SELF-VENTILATED								ADJUSTABLE SPEED		ENCLOSED FORCE-VENTILATED SERIES, COMPOUND, SHUNT AND STABILIZED SHUNT WOUND		MAXIMUM STARTING TORQUE (LB-FT)			MAXIMUM RUNNING TORQUE ON 230 VOLTS (LB-FT)		
	1 HOUR, 75 DEG. C RISE				CONTINUOUS, 75 DEG. C RISE				1 HOUR, 75 DEG. C RISE				SHUNT OR STABILIZED SHUNT WOUND		THE CONTINUOUS, 75 DEG. C RISE RATINGS OF ENCLOSED FORCE-VENTILATED MOTORS ARE EQUAL TO THE 1 HOUR, 75 DEG. C RISE TOTALLY ENCLOSED RATINGS WHEN THE REQUIRED VOLUME OF VENTILATING AIR FOR SUCH RATINGS IS SUPPLIED TO THE MOTORS.							
	HP	RPM			HP	RPM			HP	RPM			1 HOUR, 75 DEG. C. ENCLOSED CONT., 75 DEG. C., SELF-VENT.									
		SERIES	COMPOUND	SHUNT		SERIES	COMPOUND	SHUNT		SERIES	COMPOUND	SHUNT	HP	RPM	REQUIRED AIR CFM	STATIC PRESSURE AT MOTOR AIR INLET (INCHES H_2O)	SERIES	COMPOUND	SHUNT	SERIES	COMPOUND	SHUNT
620	275	370	390	390	275	370	390	390	350	350	370	390	275	390/975	2000	1 3/4	19,500	16,650	13,320	15,600	12,950	11,100
622	375	340	360	360	375	340	360	360	475	320	340	360	375	360/1080	2700	1 3/4	29,000	24,600	19,660	23,200	19,110	16,380
624	500	320	340	340	500	320	340	340	625	300	320	340	500	340/1020	3500	1 3/4	41,100	34,740	27,790	32,800	27,020	23,160

FIG. 11.2 AISE Standardized DC Mill Motor Ratings—230 Volts

suitable for operation on adjustable voltage (up to 460 V dc) rectified power supplies (such as silicon controlled rectifiers) the equivalent of three-phase, 60-Hz, 6 controlled legs (360-Hz ripple frequency) with a maximum of 400 V ac (RMS line to line) applied to the rectifier bridge.

The shunt field voltage rating shall be 230 V.

The rated temperature rise of all windings shall be 75°C rise by thermometer or 110°C rise by resistance.

The motor shall be capable of being operated in either direction of rotation.

The maximum values of armature WR^2 shall be as shown in *Fig. 11.2.*

The following minimum information shall be given on all nameplates:

A. Manufacturer's type and frame designation.
B. Horsepower output at base speed.
C. Time rating at base speed.
D. Temperature rise at base speed.
E. Rpm at full load base speed.
F. Voltage.
G. Full load amperes.
H. Winding—straight shunt, stabilized shunt, compound, or series.
J. Enclosure.

Motor Torque. The general equation for dc motor torque

$$\text{Torque} = \text{Armature current} \times \text{Field strength}$$

applies to the series motor. The field current and field strength change whenever the armature current changes. Since the field coils are wound on iron cores, which are subject to saturation, the field strength does not vary directly with the field current but follows a saturation curve, like that of Fig. 11.3. The design of motors is sufficiently standardized so that a typical curve will apply to most of them accurately enough for most controller calculations. The curve of Fig. 11.3, compiled from data on a number of motors and checked against many others, has been found satisfactory for general use.

If a shunt motor were started with an initial inrush current of 1.5 times its rated current, the starting torque would also be 1.5 times normal. With a series motor, if saturation were not present, the field strength would also be 1.5 times normal and the torque would be 2.25 times normal. The curve of Fig. 11.3 shows the actual field strength to be 1.11 times normal, and the torque would actually be 1.5 $\times$ 1.11 or 1.67 times normal. These figures illustrate one of the advantages of the series motor, as the high torque results in rapid starting and reversing.

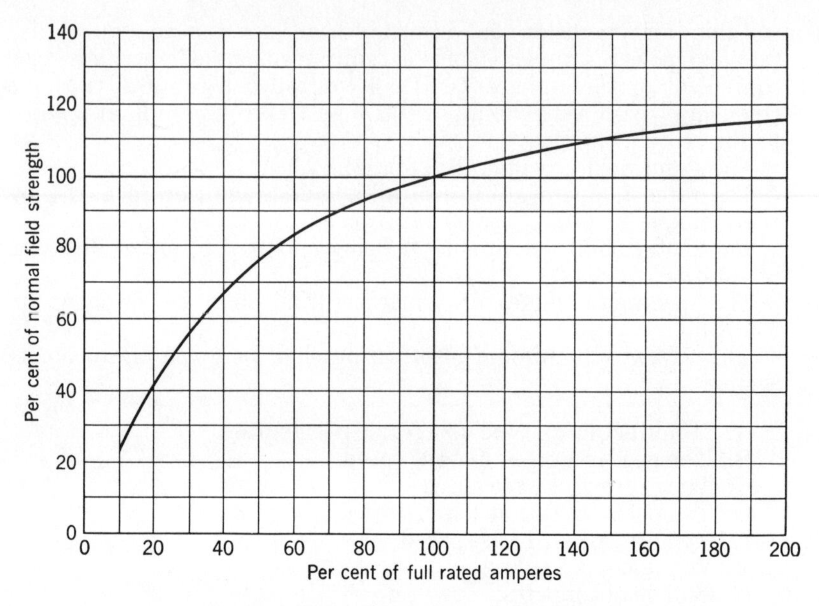

FIG. 11.3 Field Curve for a Series Motor

Motor Speed. The general equation for speed discussed in Chapter 7 also applies to the series motor:

$$S = K\frac{E - IaRa}{\phi} \tag{11.1}$$

where S = speed
K = proportionality constant
E = applied voltage
Ia = armature current
Ra = armature circuit resistance
ϕ = motor field flux

The proportionality constant may again be figured from rated or normal conditions, since by definition the speed is then the *rated* speed:

$$S_{\text{rated}} = K\frac{E_{\text{rated}} - I_{A\ \text{rated}}R_m}{\phi_{\text{rated}}}$$

where

$$R_M = \text{Motor armature resistance}$$

$$K = \frac{S_{\text{rated}}\phi_{\text{rated}}}{(E_{\text{rated}} - I_{A\ \text{rated}}R_m)}$$

Also

$$K = \frac{S\phi}{(E - I_a R_a)}$$

from equation 11.1. Therefore

$$\frac{S_{\text{rated}}\phi_{\text{rated}}}{(E_{\text{rated}} - I_{A\ \text{rated}}R_m)} = \frac{S\phi}{(E - I_A R_A)}$$

or

$$\frac{S}{S_{\text{rated}}} = \frac{(E - I_A R_A)/(E_{\text{rated}} - I_{\text{rated}}R_m)}{\phi/\phi_{\text{rated}}} = \frac{E_{cv}/E_{cv\ \text{rated}}}{\phi/\phi_{\text{rated}}} \tag{11.2}$$

This may be written in per-unit terms:

$$S' = \frac{E'_{cv}}{\phi'} \tag{11.3}$$

where S' = speed in per unit of rated speed
E'_{cv} = countervoltage in per-unit of rated countervoltage
ϕ' = field flux in per-unit of rated flux

Example: A 230-V 25-hp dc series motor is rated 100 amp. Find the speed at half-rated current with 20% E/I series resistance.

Assume the motor resistance is

$$\frac{.05E}{I} = .05\,\frac{230}{100}$$

$$\begin{aligned}
E_{cv\ \text{rated}} &= E_{\text{rated}} - I_{A\ \text{rated}}R_m \\
&= 230 - 100\left(.05\,\frac{230}{100}\right) \\
&= 230 - .05(230) = .95(230) \\
E_{cv} &= E - I_A R_A \\
R_A &= R_m + \text{Series resistance} \\
&= .05\,\frac{230}{100} + .2\,\frac{230}{100} = .25\,\frac{230}{100} \\
E_{cv} &= 230 - (50)\left(.25\,\frac{230}{100}\right) = 230 - .125(230) \\
&= .875(230)
\end{aligned}$$

The flux from curve Fig. 11.3 at half-load (50% current) is 77%, or $\phi' = .77$.

$$S' = \frac{.875(230)/.95(230)}{.77} = \frac{.875/.95}{.77} = 1.20$$

or 120% of rated speed.

Since ϕ will vary with the load, a series motor will not accelerate to a constant stable speed like a shunt motor, but will accelerate to a speed determined by the load. With light loads and corresponding low currents the speed may be excessively high, even runaway. *The motor must not be applied to drives where the load may at any time fall below a safe value, unless some arrangement is made in the control to prevent a runaway.* The inherent characteristic of running slowly under heavy load and fast under light load is exactly what is desired for many applications, particularly cranes and hoists.

Acceleration of Series Motors. Series motors are commonly accelerated by means of series resistance, just as shunt motors are, but the

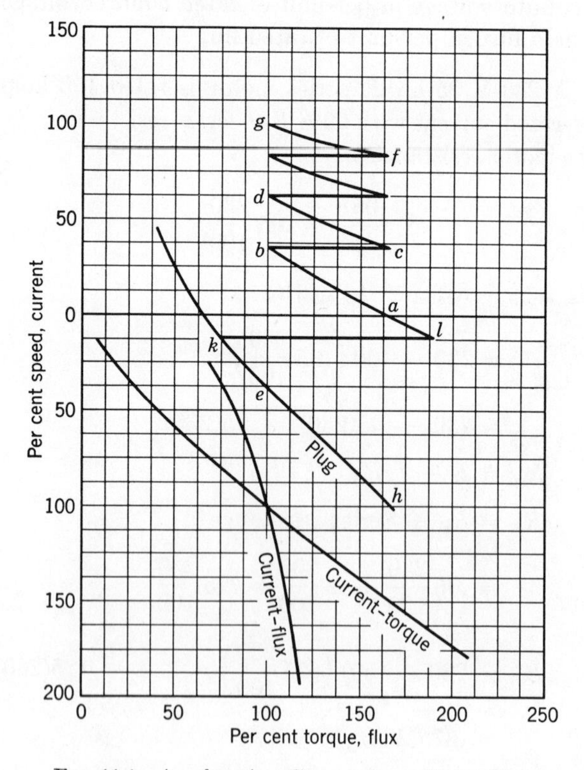

FIG. 11.4 Acceleration Curves for a Series Motor

starting curves are different owing to the different field characteristics. Figure 11.4 shows the starting of a series motor using a resistance calculated to give 150% of current for the inrush peaks. The accelerating curves are plotted against torque. The current-torque curve shows that, for 150% current, the torque is approximately 167%. The motor first accelerates along the curve *ab*, and at *b* the first resistance step is cut out. This gives the second current inrush *c*, and the motor accelerates along the curve *cd* until the second step is cut out. These curves do not pass through 100% speed at zero current, as in a shunt motor, but show that the speed keeps increasing as the load decreases. In plotting the curves, the resistance of the motor and leads has been assumed as 11%.

From equation 11.2,

$$S' = \frac{E - I_A R_A}{E_{\text{rated}} - I_{A\ \text{rated}} R_m} \times \frac{1}{\phi'}$$

Since E = line voltage, or rated voltage,

$$S' = \frac{E_{\text{rated}} - I_A R_A}{E_{\text{rated}} - I_{A\ \text{rated}} R_m} \times \frac{1}{\phi'}$$

$$= \frac{E_{\text{rated}}\left(1 - \dfrac{I_A R_A}{E_{\text{rated}}}\right)}{E_{\text{rated}}\left(1 - \dfrac{I_{A\ \text{rated}} R_m}{E_{\text{rated}}}\right)} \times \frac{1}{\phi'}$$

Now

$$R_A = \text{say, } .15\,\frac{E_{\text{rated}}}{I_{\text{rated}}}$$

or some R'_A (per unit value). Then

$$S' = \frac{1 - \dfrac{I_A R'_A \dfrac{E_{\text{rated}}}{I_{\text{rated}}}}{E_{\text{rated}}}}{1 - \dfrac{I_{A\ \text{rated}}(0.11)\dfrac{E_{\text{rated}}}{I_{A\ \text{rated}}}}{E_{\text{rated}}}} \times \frac{1}{\phi'}$$

$$= \frac{1 - \dfrac{I_A}{I_{A\ \text{rated}}}\dfrac{E_{\text{rated}}}{E_{\text{rated}}} R'_A}{1 - 0.11\,\dfrac{I_{A\ \text{rated}}}{I_{A\ \text{rated}}}\dfrac{E_{\text{rated}}}{E_{\text{rated}}}} \times \frac{1}{\phi'}$$

$$= \frac{1 - I'_A R'_A}{(1 - 0.11)} \times \frac{1}{\phi'} \tag{11.4}$$

where I'_A = the current in per-unit of full-load current
R'_A = the resistance in the circuit expressed in per unit of

$$\frac{\text{Rated voltage}}{\text{Rated current}}$$

The value of the steps of resistance may be obtained by determining the distance between the curves on the line bg, just as with the shunt motor.

Probably the best method of graphical solution is first to plot the speed-current curve of the motor alone and then assume a value for the resistance in the last step of the controller. Applying this value in the above equation, plot the curve for that step. In the same manner, assume values for the other steps and plot those curves. From the point where each curve crosses the full-load line, draw a straight line vertically until it intersects the next curve. If equal peaks are not obtained, select new values for the resistance and try again. If the curves are very nearly correct, it will be possible to draw them in closely enough freehand, following the shape of the nearest calculated curve.

This is, admittedly, a cut-and-try method; it is easier to calculate the resistance values by the nongraphical method and then plot the curves if desired. The speed-torque curve is generally of more interest than the speed-current curve. The curve of the motor alone can always be obtained from the motor manufacturer, and such curves usually give the losses at any load. The motor resistance and the voltage drop across the motor at full load can be calculated from the losses. Average values are $0.23E/I$ for motors up to 10 hp, $0.14E/I$ for motors up to 50 hp, and $0.11E/I$ for the larger motors. If no speed-torque curve for a motor

TABLE 11.1

RESISTOR DESIGN FOR SERIES MOTORS

Accelerating Steps	Accelerating Peak in Per Cent of Full Load	Plugging Peak in Per Cent of Full Load	Motor Resistance, Per Cent E/I
1	185	150	23
2	159	150	14
3	147	150	11
4	133	150	10
5	133	150	8

is available, one can be plotted from the average flux curve of Fig. 11.3. Table 11.1 gives the values of inrush current and resistor taper for series motors. The accelerating steps do not include the plugging step.

The total resistance for plugging is obtained by dividing the plugging voltage by the plugging inrush current and subtracting the motor resistance:

$$R_t = \frac{E \times 1.8}{I_p} - R_m$$

The total resistance for accelerating is

$$R_{\text{acc}} = \frac{E}{I_{\text{acc}}} - R_m$$

To be exact, the value of E in these two equations should be multiplied by a factor to cover the increase in voltage caused by the increase of the field strength above normal, which will depend on the values of I_p and I_{acc}. If these current values were 150% of normal, the factor would be about 1.11, and the increase in the resistance values would just about equal the value of R_m. The general practice is to neglect both the increase in field strength and R_m, since they practically balance each other.

The resistance of the plugging step above is

$$R_p = R_t - R_{\text{acc}}$$

The accelerating resistance may be divided as shown by Table 11.2. Figure 11.4 is a typical speed-torque curve plotted from these values.

TABLE 11.2

RESISTOR TAPER FOR SERIES MOTORS

	Per Cent of Total Accelerating Resistance				
No. Steps	Step 1	Step 2	Step 3	Step 4	Step 5
1	100	—	—	—	—
2	66	34	..	....	..
3	43	33	24	....	..
4	32.5	26	23	18.5	..
5	24	22	20	18	16

The equations are

$$S = \frac{E - RI_a}{\text{Flux } (E - R_m I_n)}, \text{ and } T = \text{Flux} \times I_a$$

where S = speed in percentage of full-load speed
T = torque in percentage of full-load torque
E = line voltage = 100%
R = total resistance including the motor in percentage of E/I_n
I_a = current in percentage of full-load current I_n
R_m = resistance of the motor in percentage of E/I_n
Flux = field strength from field curve

Reversing and Plugging. The series motor may be reversed by reversing either the armature or the series field, but not both. The usual practice is to reverse the armature. Series motors are often plugged by connecting the armature for reversed direction while the motor is running in the forward direction. This is done to get a quick stop or, more frequently, a quick reversal. When the motor is plugged, the countervoltage of the armature is added to the line voltage, and it is necessary to provide an additional step of resistance in the controller to limit the inrush current to a safe value. The practice of plugging series motors is so general that it is customary either to provide a plugging resistance step or, when it is known that plugging is not desired, to arrange the control so that the operator cannot plug. It is assumed that he will try to do so.

Figure 11.4 includes a curve which shows the plugging of a series motor. Additional resistance, measured by the line *eb*, has been added to limit the current inrush to 150%. When the motor is plugged, it decelerates along the line *hk* to the point *k*, at which point the plugging step of resistance is short-circuited and the motor continues to decelerate along the curve *la* to *a*. At this point it has come to rest, and from there on it accelerates in the opposite direction. It is possible to delay short-circuiting of the plugging resistance until the current has dropped to the point *k*, because it is certain that the motor will continue to decelerate to zero speed, and then reverse. The curve shows that at zero speed the current is below point *k*.

Dynamic Braking. Dynamic braking may be obtained with a series motor, as shown in Fig. 11.5, but the connections become a little complicated. If the motor were simply disconnected from the line and shunted by the step of resistance, no braking would be obtained, because the current would flow through the field in the wrong direction, and the

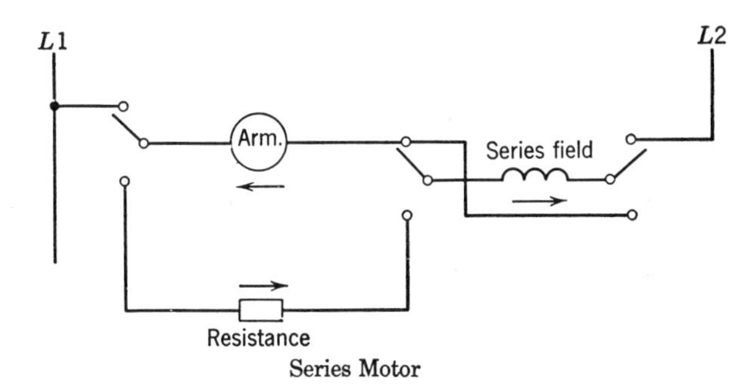

FIG. 11.5 Dynamic Braking of a Series Motor

field would be demagnetized. Assuming that, with the motor running, the current is flowing from $L1$ to $L2$, the countervoltage of the armature is in the opposing direction, as shown by the arrow. Current is flowing in the field from $L1$ to $L2$. When the motor is disconnected from the line and connected for braking, it will be evident that the field must be connected in the reverse direction in order that the current will flow through it in the same direction as it was flowing with the motor on the line.

Sometimes, for dynamic braking, the series field is connected directly across the line in series with a resistance. It thus becomes temporarily a shunt field. This method is wasteful of energy since a high percentage of full-load current must be supplied to the field.

Since dynamic braking is present only when the motor is turning, a mechanical brake of some sort is required to hold the load after stopping it. Such brakes are usually released electrically and applied mechanically by springs. When they are used with series motors, the usual practice is to equip the brake with a series coil and connect it in series with the motor. This arrangement insures that the brake will release only when current is flowing through the motor and motor torque is available to hold or move the load.

Speeds below Normal. The speed of a series motor may be reduced below normal by means of series resistance, and also by armature shunt resistance, and the results are the same as with a shunt motor, except that in calculating the speeds the current in the series field must always be considered.

Series Motor. Armature Shunt. The method followed in designing an armature shunt resistance for a series motor differs from that for

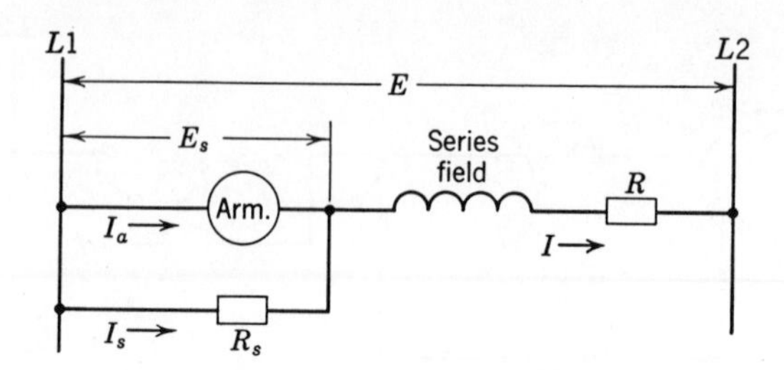

FIG. 11.6 Series Motor with Armature Shunt

a shunt motor, in that the effect of the series field must be considered. The field current will be the sum of the currents through the armature and the shunt, so that, to obtain normal full-load torque, the armature current will be less than full-load value. The torque desired at the reduced speed will be known, but neither the armature current nor the field current can be readily determined. Direct calculation for the shunt resistance becomes very involved, if not impossible. The simpler method is to assume a value for the field strength and calculate the speed obtained at the desired torque. If we refer to Fig. 11.6, and use all values in percentage of normal:

E = line voltage—known
R = series resistance plus field resistance—known
R_m = armature resistance—known
T = torque required—known
F = field strength—assumed
I_f = the current in the series field. This is determined by the assumed value of field strength
I_a = the current in the armature $= T/F$
I_s = the current in the shunt $= I_f - I_a$
E_s = the voltage across the armature (and the shunt) $= E - I_f R$
E_a = the countervoltage of the motor $= E_s - I_a R_m$

$$S = \text{the speed} = \frac{E_a}{F(E - I_a R_m)}$$

The value of F which will give the desired speed having, by a few trials, been determined, the shunt resistance is

$$R_s = \frac{E_s}{I_s}$$

Speeds above Normal. It is possible to increase the speed of a series motor above normal by shunting the series field and so reducing the current through it, but the method is not very practical because of the low resistance of the field and so is infrequently used.

Overhauling Loads. It may seem something of a paradox that the series motor, which will run away if too lightly loaded, should be the motor most universally used on cranes to handle overhauling loads. One reason has been mentioned, the fact that its speed characteristics are exactly those desired for hoisting loads. The high torque available for rapid acceleration is also desirable. Still another reason for its popularity is the fact that the heavy windings of the series fields are much less liable to damage in service than the fine wire windings of the shunt motor.

The series motor can readily be made to handle overhauling loads safely by suitable arrangement of the control. If a series motor were connected for dynamic braking, as shown in Fig. 11.5, and were then driven by an overhauling load, the motor would generate current, dissipating energy in the resistance, and act as a brake on the load. Almost any desired speed could be obtained by varying the resistance. The speed would, however, vary considerably for different loads, and a light load could not be driven down. To overcome these objections, a circuit like Fig. 13.1 is used. Here the motor armature and field are connected in parallel, each with resistance in series, and the motor characteristics are like those of a shunt machine. The armature circuit resistance is used to prevent excessive starting inrushes, and is cut out in several steps as the motor acclerates. The minimum resistance in the field circuit must be high enough to limit the field current to about normal, and additional resistance is then cut in, in several steps to weaken the field and increase the motor speed. On the hoisting side of the controller the circuits are the same as for a straight nonreversing control. A controller of this kind is called a dynamic-lowering hoist controller, and is described in detail in Chapter 13.

Motor Protection. Controllers for series motors usually include magnetic overload relays, which have a time-delay action at ordinary overloads but trip instantaneously if the motor stalls. It is not necessary to guard against loss of the series field, since the heavy construction makes this unlikely, and the circuit is such that opening of the field could hardly happen without the armature being disconnected also. Controllers for series-shunt motors often include a relay to protect against loss of the shunt field. Ordinarily no other protection is provided for a series motor.

Manually Operated Controllers. Manually operated face-plate controllers are used to some extent with small series motors, both for starting and for speed regulation. Their construction is the same as that of similar starters for shunt motors.

Multiple-switch starters are seldom, if ever, used for series motors, because the starter is inherently one which takes time to operate, whereas the motor is particularly adaptable to machines which must be started frequently and rapidly.

Drum Controllers. Historically, drum controllers have been widely used for the control of series motors, particularly on street-railway cars and on cranes. Their general construction is the same as that of the drums used for controlling shunt motors.

The reversing and nonreversing drums used for crane trolley and bridge motions are arranged for 50% speed reduction by resistance in the armature circuit. If greater speed reduction is desired, an armature shunt is used, and 90% reduction may be obtained in this way. The drum ratings are the same as for shunt motors.

Dynamic lowering drum controllers are for use with hoists. They are employed with series motors, and the circuits are so arranged that, when the drum cylinder is moved in the hoist direction, the motor is brought up to speed gradually by reducing the resistance in the circuit in the usual manner. The drum is arranged so that, when the cylinder is moved to the "lower" position, the motor, when overhauled by the load, acts as a generator and forces current through a resistor. If the speed increases, the voltage on the terminals of the motor also increases, forcing more current through the resistor. The speed is thus automatically kept at a safe value with the controller on any given point, but by moving the controller to another point the speed can be changed by increasing or decreasing the value of the resistance through which the generated current must flow.

Figure 11.7 shows in an elementary manner how the above effects are secured. It will be noted that in the OFF position a resistance is connected around the armature and field of the motor. The residual magnetism causes the motor to build up as a series generator, and current is forced through the resistance until the inertia of the motor and the falling load is overcome.

In the lowering direction, the series field of the motor is connected across the line in series with a resistor, and another resistor is connected across the armature. This gives the effect of a shunt-wound motor, since the strength of the field is independent of the armature current. An

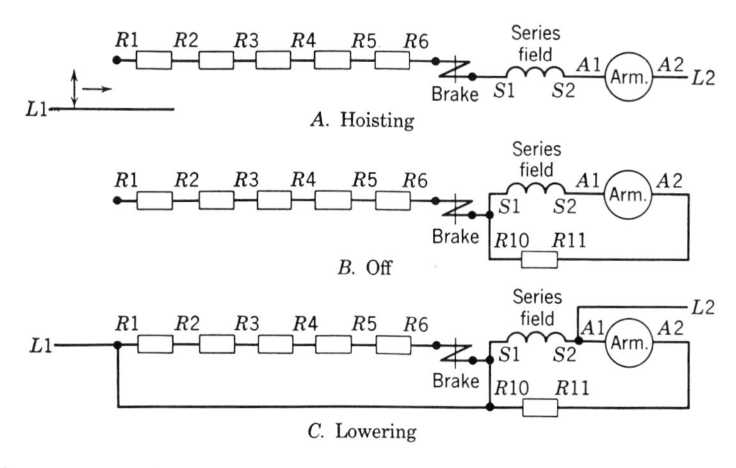

FIG. 11.7 Connections for a Dynamic-Lowering Drum Controller

inspection of Fig. 11.7 will show that, if the load is not heavy enough to overhaul the motor, current will flow from the line through the armature and fields of the motor in such a direction as to drive the load downward. These drums also are rated the same as those for shunt motors.

In more common use today are combinations of "cam masters" or pilot duty drum switches together with contactors mounted on the control panel. The action is substantially the same as the power drum controller. Contacts on the master pickup and drop out contactors on the panel by energizing and deenergizing the corresponding contactor coils. With this arrangement only small control currents pass through the master circuits, and the power switching is done entirely on the panel.

Magnetic Controllers. Magnetic controllers for series motors are generally used on cranes, coal and ore bridges, and for auxiliary machinery in the processing of steel and other metals. They are made in several types as follows:

Nonreversing, plain starting duty.
Nonreversing, starting and regulating duty.
Nonreversing, starting and regulating duty, with dynamic braking.
Reversing, plugging.
Reversing, with dynamic braking.
Reversing, dynamic lowering. These are described in detail in a separate chapter.

TABLE 11.3

NEMA Ratings for Steel-Mill Auxiliary Controllers

			For Continuous Duty			For Intermittent Duty	
				Minimum Number of Acelerating Contactors			
Size of Contactor	8-hour Contactor Rating (amp)	Contactor Mill Rating (amp)	Hp Purpose (230 V)	Mill Motors	General-Purpose Motors	Hp Rating (230 V)	Minimum Number of Accelerating Contactors
3	100	133	25	2	2	35	2
4	150	200	40	2	2	55	2
5	300	400	75	2	3	110	2
6	600	800	150	3	4	225	2 or 3
7	900	1200	225	3	5	330	3
8	1350	1800	350	4	5	500	4
9	2500	3350	600	5	6	1000	5

Note: The number of accelerating contactors does not include the plugging contactor.

A typical controller of the reversing, dynamic-braking type would include the devices listed below. These devices were in the past assembled on a single panel of slate or inpregnated asbestos panel board. Modern construction utilizes unitized contactors with integral insulated moldings and bases mounted directly on a steel panel. The panel is ordinarily mounted on an angle-iron supporting frame or enclosed in a sheet steel enclosure.

1 double-pole single-throw main disconnect knife switch
1 double-pole single-throw control circuit knife switch

TABLE 11.4

NEMA Ratings for Crane Controllers

Size of Contactor	8-hour Contactor Rating (amp)	Contactor Crane Rating (amp)	Hp Rating (230 V)	Minimum Number of Accelerating Contactors Exclusive of Plugging Contactor
2	50	67	15	3
3	100	133	35	3
4	150	200	55	3
5	300	400	110	3
6	600	800	225	4
7	900	1200	330	4
8	1350	1800	500	4

2 fuses for the control circuit
2 magnetic overload relays
1 set of reversing contactors, mechanically interlocked
1 negative line, or main, contactor
1 set of accelerating contactors
1 spring-closed dynamic-braking contactor
1 undervoltage relay
1 set of accelerating devices (series or time relays)
1 plugging contactor

For separate installation there would be a master controller, a set of resistor material, and possibly limit switches, or other accessory devices. The main or motor circuits of these controllers are much alike, but the control circuits vary widely with the application, depending on the functions required. Figure 11.8 is typical of the control schemes used for mill auxiliary machines. It is for a motor driving a machine through an eccentric, each operation of the motor causing the wheel of the drive to make one revolution and moving the eccentric rod forward and back. Such a machine might be used to push steel billets into a furnace. In the OFF position of the master the relay *UV* closes. When

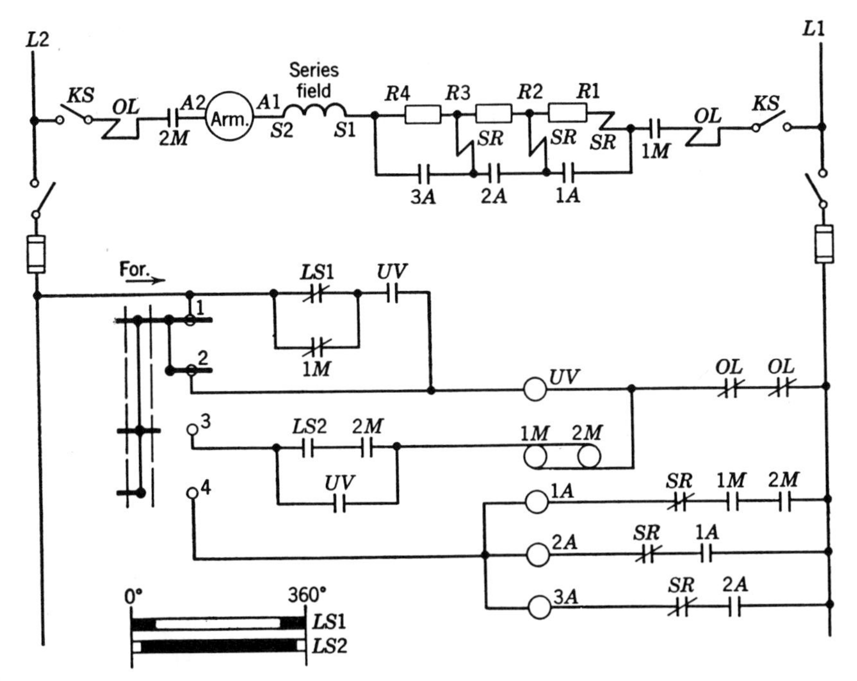

FIG. 11.8 Single-Cycle Controller for a Steel-Mill Auxiliary Drive

the master is moved to the forward position, contactors 1*M* and 2*M* are energized through a contact of *UV*. The motor starts and accelerates to full speed under the control of the series relays. The limit switch contacts *LS*1 and *LS*2 are those of a rotating cam-type switch (see Fig. 4.12), geared to the driven wheel. As soon as the wheel rotates a small amount, contact *LS*2 closes. Shortly afterward contact *LS*1 opens, deenergizing *UV*, which opens. The motor continues to run until *LS*2 opens, and then it stops. Just before this happens, contact *LS*1 recloses, but when the motor stops, it cannot restart until the operator returns the master to the OFF position and recloses *UV*. The machine then makes one complete cycle for each operation of the master. In case of trouble the operator can stop the motor at any point in the cycle. The normally closed interlock contacts on 1*M* permit him to reset *UV* if he makes an emergency stop at a point in the cycle where *LS*1 is open.

Crane-Protective Panels. When drum controllers are used for one or more of the motors of a crane, it is usual practice to obtain low voltage and overload protection by means of crane-protective panel, mounting the apparatus listed below.

1 double-pole isolating knife switch, with provision for padlocking in the open position
2 single-pole (or one double-pole) contactors arranged to open both sides of the line
1 set of automatic-reset overload relays, one in the positive circuit to each motor, and one in the common return to the other side of the line
2 pilot lights with fuses
2 control-circuit fuses

TABLE 11.5

RATINGS FOR CONTACTORS USED ON CRANE-PROTECTIVE PANELS

Size of Contactor	8-Hour Rating (amp)	Total Horsepower of All Motors (230 V)	Greatest Horsepower Required by Any Single Crane Motion
3	100	55	35
4	150	80	55
5	300	160	110
6	600	320	225
7	900	480	330
8	1350	725	500

For motors of other voltage ratings, the 8-hour rating of the main-line contactors should not be less than 50% of the combined ½-hr or 1-hr rating of the motors, nor less than 75% of the ½-hr or 1-hr rating of the largest individual motor.

Sometimes magnetic controllers are also connected behind a crane-protective panel, and the disconnect knife switch, overload relays, and undervoltage relay are then omitted from the individual controllers. With that arrangement the main line contactor should not be of smaller rating than that of the largest contactor on any of the individual controllers.

The line contactors of the crane-protective panel are operated by a two-button momentary-contact pushbutton station, marked RESET and STOP. The circuit is arranged so that the contactors cannot be held closed under overload.

For crane bridges driven by two motors the arrangement shown in Figs. 11.9 and 11.10 will meet all the requirements of good braking. This method was developed and patented by Herman Wilson and Charles Ritchie, two electrical engineers in one of the steel mills, and is known as the Wilson-Ritchie method of braking. The armature of one motor is handled by the reversing contactors in the conventional way, but the field of the other motor is reversed. Four spring-closed, magnetically released contactors are used for the braking circuit. These are standard contactors without any additional mechanism. All of them are open when the motor is running and closed during braking.

Referring to Fig. 11.9, which shows the basic scheme, and assuming that the bridge is moving forward, current is flowing from $L2$ to $L1$, through armature 1 in the direction from $A2$ and $A1$, and through armature 2 in the direction from $A12$ to $A11$. Current flows through field 1 from $S2$ to $S1$, and through field 2 from $S12$ to $S11$. In stopping, all direction contactors open, and all braking contactors close. The countervoltage of armature 1 causes current to flow through the armature from $A1$ to $A2$, and through field 2 from $S12$ to $S11$. Similarly, the countervoltage of armature 2 causes current to flow from $A11$ to $A12$, and through field 1 from $S2$ to $S1$. Both fields are therefore energized in the right direction, and each provides a braking field for the armature of the other motor. It is evident that this must be so for either direction of travel, since current always flows in the same direction through armature 1 and field 2, which form one braking circuit, and since armature 2 and field 1 are always reversed together. The controller of Fig. 11.9 therefore gives a simple and positive braking method, either for normal stopping or when power fails.

Since the speed of a crane bridge varies widely, being low on short runs and high on long runs, the available countervoltage, and corresponding braking effect, will also vary widely. It is usually desirable to provide several steps of braking resistance if the braking is to be used for normal or service stopping. Figure 11.10 shows the usual arrangement, with three steps. The contactors which short-circuit these steps are controlled either from the operating master or from a separate master or foot switch having three positions. For stopping from high speed, point 1 is used, on which contactors 3*DB*, 4*DB*, 13*DB*, and 14*DB* are open and all braking resistance is in circuit. This resistance will allow approximately 125% of normal current to flow. On the second point, contactors 3*DB* and 13*DB* are closed, and the remaining resistance will allow approximately the same current if the motor is at normal speed. On the third point, contactors 4*DB* and 14*DB* close, and the remaining resistor is designed to allow sufficient current for good braking from half-speed. Series relays are connected in the braking circuits, controlling 4*DB* and 14*DB*, so that, if an attempt is made to close these contactors at too high a speed, the relays will operate to prevent it. On power failure the circuit is the same as it is on the second, or normal speed, braking position. If power failed with the bridge at high speed, the braking would be more severe than might be satisfactory for service stopping but not too severe for an emergency condition.

The Wilson-Ritchie scheme has been successfully applied to a number

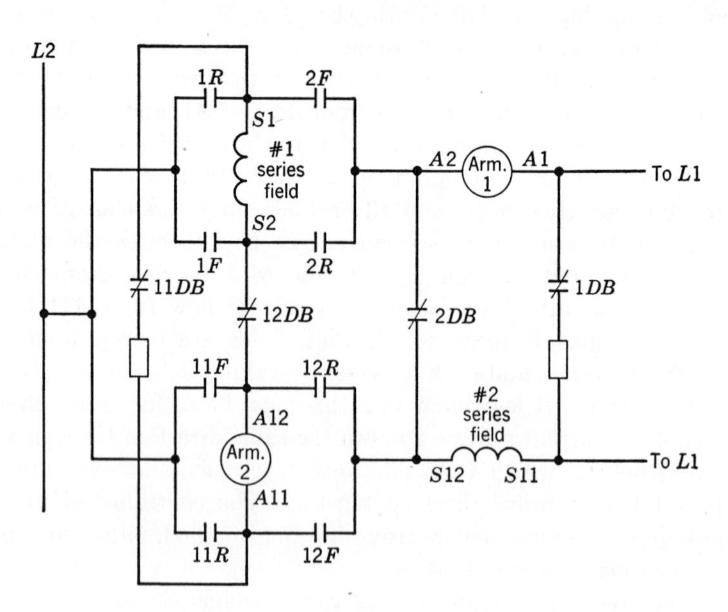

FIG. 11.9 Single-Step Wilson-Ritchie Braking Method

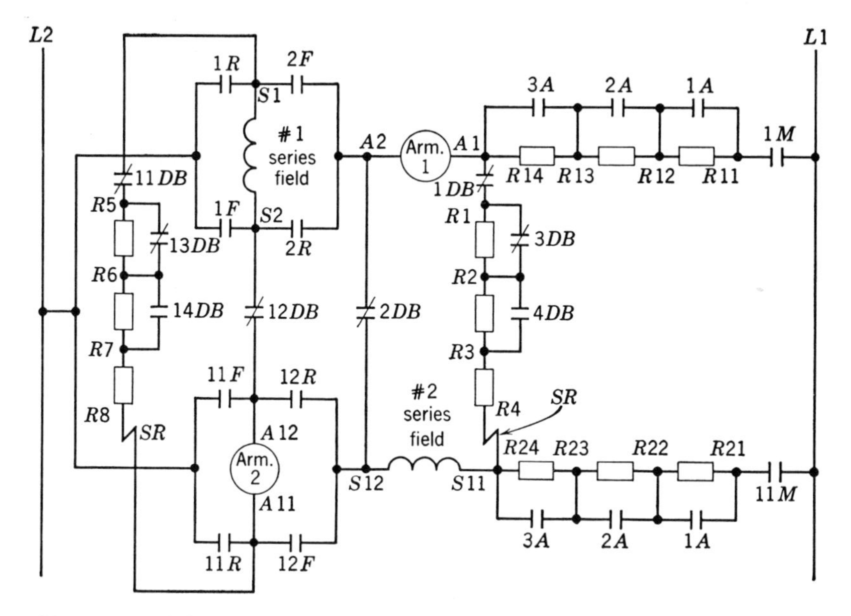

Fig. 11.10 Wilson-Ritchie Braking Method with Graduated Braking Torque

of cranes. Some interesting tests were made at one installation, where there were three cranes on one runway, each equipped with this control. It was therefore possible to operate the center crane in either direction, or to allow it to drift, or to remain at rest for some time, and then in each case push it with one of the other cranes. It was found that braking always occurred, even after the crane had been allowed to stand at rest for some time. There was always sufficient residual magnetism in the fields to start a current flow, and the braking effect, once started, built up rapidly. With this in mind, the scheme has been applied to large coal and ore bridges, which are liable to be moved on their tracks by a strong wind. If that happened, dynamic braking would apply to limit the speed. To the best of the author's knowledge, no other control scheme will accomplish this result.

Application of Series Motors. Series motors are applicable to drives requiring high starting torque, fast reversing, and a motor as sturdy in construction as is available. These characteristics make the series motor ideal for such applications in steel mills, as roll tables, screw-downs, furnaces pushers, manipulators, and many others. The motor is ideal for cranes, hoists, coal- and ore-handling bucket hoists, and cargo winches, because of its inherent characteristic of running fast with light loads and slowly with heavy loads. Its ready adaptability to dynamic braking and dynamic lowering is a further advantage on this service. Its speed and torque characteristics have made it ideal for street-railway cars and other motor-driven vehicles.

Series-shunt motors are applied to drives when it is desired to limit the speed with light load. Crane bridges and trolleys may use series-shunt motors for that reason. Car dumpers, which may require a high starting and accelerating torque, may have some portion of their cycle in which the load is very light, or even overhauling, and some shunt field is desirable to limit the speed there. Many machines formerly driven by series motors with constant potential dc controls described in these chapters are now equipped with adjustable voltage drives (usually shunt wound motors), described in Chapter 7.

Problems

Note. Use field curve Fig. 11.3 in calculations.

1. Write an equation for the torque of a series motor.

2. Write an equation for the speed of a series motor.

3. A series motor is driving a load which causes it to take 110% of rated full-load current. Approximately how much current would an equivalent shunt motor require to supply the same torque?

4. The torque required to drive a certain machine is 5000 lb-ft. The inrush current when starting the driving motor is limited to 175% of normal full-load current. What is the starting torque if a shunt motor is used? What is the starting torque if a series motor is used?

5. The resistance of the armature and field of a series motor is $0.11E/I$. If the full-load speed of the motor is 1000 rpm, what is the speed at half-load?

6. A motor is provided with both series and shunt fields, so proportioned that at full load the series field provides 80% of the field strength. If the resistance of the armature and series field is $0.11E/I$, what will be the speed at half-load?

7. Plot the speed-torque curves for acceleration of a series motor with three steps of resistance, using the data given in Table 11.1.

8. If the resistance of a motor armature and field is $0.11E/I$, what is the motor countervoltage when running at full load?

9. What is the total voltage if the motor is plugged?

10. If this is a 100-hp 230-V 375-amp motor, what will be the ohmic value of a resistor which will limit the starting current to 150% of normal when the motor is plugged?

11. A motor is provided with both series and shunt fields, so proportioned that the series field provides 70% of the field strength. What will be the speed if the shunt field circuit is opened?

12. A 20-hp 230-V 80-amp motor has an armature and field resistance of $0.14E/I$ and is provided with a dynamic-braking resistor of 1.8 ohms. What initial braking torque will be obtained?

13. A 30-hp 230-V 110-amp motor is used with a series resistor of 1.3 ohms. Calculate the value of an armature shunt resistance which will produce a speed of 30% of full-load speed, with a torque of 50% of full rated torque. Assume armature resistance of $0.10E/I$, and field resistance of $0.04E/I$.

14. A 40-hp 230-V 146-amp motor is used with a series resistor of 1.0 ohms, and an armature shunt resistor of 2.0 ohms. What will be the speed of this motor in per cent of normal speed when the armature current is 50 amp, if the resistance of the armature and the field are neglected?

15. What percent of full-rated torque will the motor be delivering?

16. Draw an elementary diagram of a single-cycle controller like that of Fig. 11.8, except arranged for reversing in emergency. The reversing operation may be continuous and not under the control of the limit switch.

17. Draw an elementary diagram for a controller for a 50-hp 230-V 190-amp series motor, the controller to have the following characteristics:

Reversing
Three steps of accelerating resistance
Inductive-timed acceleration like Fig. 6.8
Armature shunt on the first point in each direction
Field shunt on the last point forward only
Master controller with 5 speed points in each direction
Low-voltage protection and overload protection

18. Calculate the ohms required in each step of the accelerating resistance. using the data of Tables 11.1 and 11.2.

19. Calculate the ohms required in the armature shunt step to give a speed of 50% of full-load speed, when the motor is running at zero load.

20. Calculate the ohms required in the field-shunting resistor to give a speed of 115% of full-load speed, when the motor is running at 100% load.

21. A small series motor is started by a controller having a single step of resistance, the accelerating contactor being set to close at 100% of rated current. If the controller is changed to use a five-step resistor, with the same contactor setting, how much will the average accelerating torque be reduced? Use data of Table 11.1.

22. A 75-hp 230-V 280-amp series motor has a resistance of 11% of E/I. Using the data of Tables 11.1 and 11.2, calculate the ohms in each step of a five-step resistor which will give the following performance:

Current on first point when plugging, 100% of rated current.
Current on second point when plugging, 150% of rated current.
Current on third point starting from rest—see table.

23. A motor is provided with both series and shunt fields, the shunt field providing 30% of the field strength. The motor is belted to the machine which it drives. What speed will the motor reach if the belt breaks?

24. A mill has a series-shunt motor rated 100 hp, 230 V, and 375 amp. It has a shunt field which provides 50% of the field strength. How many ohms must be used in series with the motor to obtain 50% speed at full-rated load?

12

TWO-MOTOR AND MULTIPLE-MOTOR DRIVES

It frequently happens that two motors, instead of one motor of twice the size, may be used to definite advantage for a drive. The motors are usually connected mechanically, so that any change in speed or direction of rotation of one is accompanied by a corresponding change in the other. The mechanical connection may be accomplished by direct coupling, or by gearing, or it may be through track and wheels in a car or a crane bridge.

Space conditions sometimes influences the choice of two motors, since it may be easier to find room for two small motors than for one large one. If the single motor were of an odd size it might be advantageous to select two smaller ones in order to make them duplicates of other machines in the plant, and so eliminate the necessity of carrying a new line of repair parts. Where emergency operation of a machine is essential, as in hot-metal-handling equipment, it is advantageous to have two or more motors. In the event of motor burnout the damaged machine may be dismantled, and the equipment operated temporarily by the remaining motor or motors. Some two-motor drives have been installed with the idea of obtaining more rapid acceleration and reversal, because the moment of inertia of two small motors is less than that of one large one. There is some difference of opinion as to whether this condition actually obtains in practice, but undoubtedly many two-motor drives have been installed for that reason.

When using two motors mechanically connected together, the principle difficulties are to obtain equal division of the load when starting and when running, to avoid circulating currents when plugging, to commutate highly inductive currents when the motors are in series.

Division of Running Load. Motors which are supposedly exact duplicates may vary enough in their characteristics to cause serious unbalancing of the load; this is particularly true of shunt motors. Slight imperfections in manufacture, unequal heating of the field coils, differences in the length of leads, or any one of a number of other causes may result in unbalancing the load. The effect is most marked in shunt motors because of their flat speed-current characteristic.

The speed of a shunt motor changes very little over a wide range of load current and, conversely, if the motor is forced, by mechanical connection, to run at a speed only slightly different from its natural speed, wide variation in current will result. With a series motor the speed changes more rapidly with changes in load, and consequently a slightly incorrect speed will not cause a very great unbalancing of load. The unbalancing obtained with compound motors will be somewhere between that of the shunt and the series motors, depending upon the percentage of compounding.

Calculations based on some typical motor curves show that, when two shunt motors having a difference of 5% in full-load speed are mechanically coupled, one may take twice as much current as the other. When series motors having the same speed difference are coupled, the load of one might be approximately 15% higher than that of the other. It will be evident that the control equipment is not the factor which causes unbalancing of the running load. However, the blame is often laid to the control equipment, and control engineers are called upon to correct the trouble. For that reason a brief description of a method of balancing, applicable to shunt or compound motors in parallel, is given here.

Balancing Shunt Motors

1. Set the motor brushes at neutral.

2. Check the speed at full field. If the speed is different, increase the speed of the slow motor by taking out shims of the main field poles. If there are no shims on the main field poles, or if all have been taken out, the speed of the fast motor should be decreased by adding shims to the main field poles. The thickness of the shims should be not more than $\frac{1}{32}$ in.

3. Check the speed of the motors with weak field maximum speed, at no load. If necessary, disconnect the motors to obtain no load. The brushes are to remain on neutral. If this speed differs more than 1%, adjust the field resistor to make the speed the same.

4. Check the speed at full load with weakened field. If the speed differs, shift the brushes of the fast motor in the direction of rotation. The brushes should not be shifted more than one bar at the maximum. If this is not sufficient, return the brush to neutral and increase the

air gap on the interpoles of the fast motor. First take out the shims of the two adjacent poles. If this does not suffice, shift the brushes, not more than one bar, in the direction of rotation on the fast motor. If still more adjustment is required, return the brush to neutral and take out shims of the other two interpoles on the fast motor, and shift brushes, if necessary. If the removal of the shims from the interpoles slows down the motor from which they are removed to such an extent that it becomes slower than the other motor, the other motor then becomes the fast motor, and the brushes on it should be shifted, not more than one bar, in the direction of rotation. The shims removed at any one time should be not more than $\frac{1}{32}$-in. thick.

5. If the motors do not parallel and equalize the load on the intermediate steps, adjust the field resistor.

Notes. The fast motor is the one that takes the greater load; and vice versa, the slow motor is the one that takes the lesser load.

The motors should be adjusted so that difference in current taken by them is not more than 5% of full-load current. This may mean that on light loads one of the motors may take more than twice as much current as the other, in extreme instances. This is considered satisfactory because the load on either of the motors will be well within its rating.

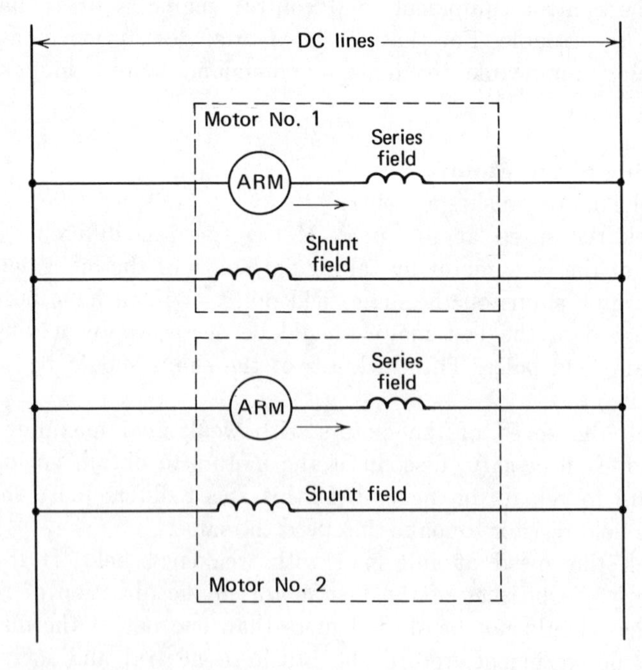

FIG. 12.1 Compound Motors Mechanically Coupled Connected for Load Sharing

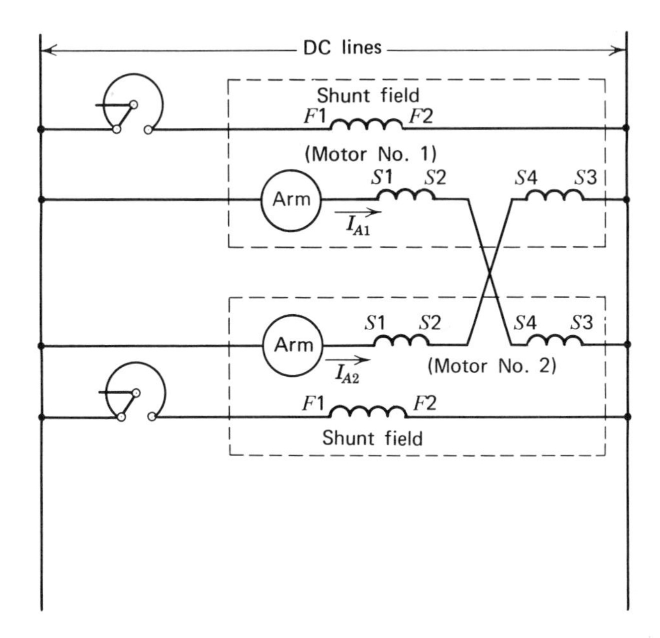

FIG. 12.2 Load Sharing with Two Series Fields

Two compound motors may be caused to share the load by combinations of series field connections. One such arrangement might be as shown in Fig. 12.1.

Another method requires two series fields on each motor. This makes the motor more expensive but has the advantage that, *at equal loads,* the motors behave essentially as straight shunt motors. See Fig. 12.2.

Current flowing in the direction of the arrows as indicated in Fig. 12.2 causes a field flux which *aids* the flux from the shunt field in its own series field $S1$–$S2$ but bucks or is differential to the field flux in the other motor. When the motor armature currents are equal (equal load), the fields buck each other and the motor behaves essentially as a straight shunt motor. If motor 1 trys to draw more current, its field will be strengthened (causing it to tend to slow down allowing the torque to shift to the other motor) at the same time motor 2 field strength is weakened (causing it to tend to speed up and supply more of the total torque).

Starting Motors in Parallel. Shunt motors, or slightly compounded motors, may be started as shown in Fig. 12.3*a,* using a common starting resistor and one set of accelerating contactors. A common set of reversing

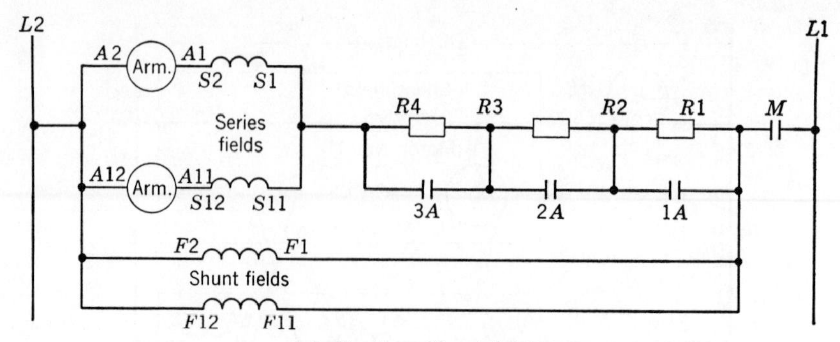

(*a*) Starting Shunt Motors

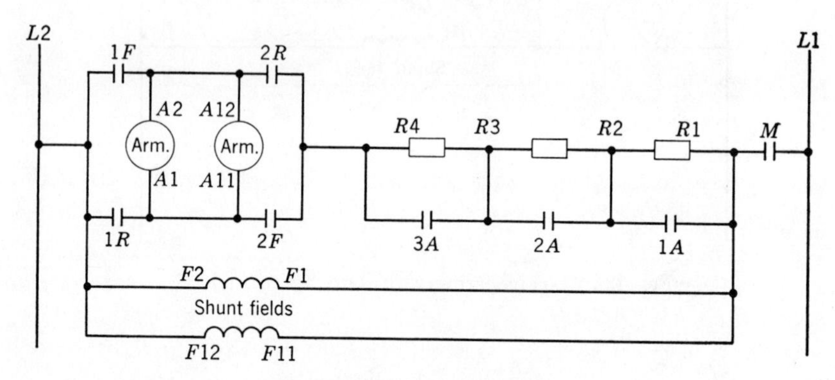

(*b*) Reversing Shunt Motors

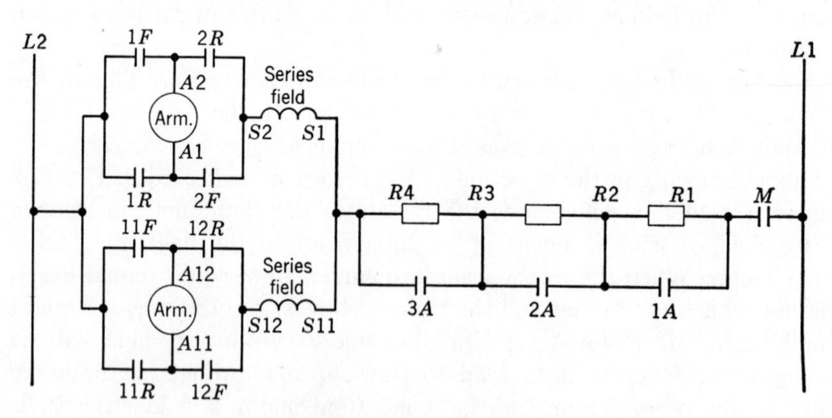

(*c*) Starting Series Motors, or Reversing without Plugging

Fig. 12.3 Starting Motors in Parallel

contactors may be used for shunt motors, as shown in Fig. 12.3*b*. With compound motors two sets of reverse contactors will be required, in order that the series fields may receive current in the same direction and that the current of each armature may pass through its own series field. If a single set of reversing contactors were used, the series fields would have to be connected in parallel, outside of the reversing contactors. The current would then divide through the fields in inverse ratio to their resistances. Since the resistance of a series field is very low, it is difficult to balance two of them exactly and to maintain this balance whether the motors are hot or cold. If the series field is only a small percentage of the shunt field strength, this objection may not be very serious, but two sets of direction contactors make a better arrangement, and with heavily compounded motors or series motors they become a necessity. The connections shown in Fig. 12.3*c*, therefore, are satisfactory for starting series motors or for reversing them if plugging is not required.

Since each motor develops full horsepower, and since the armature current of both passes through the starting resistor, it is evident that the resistor, and the control contactors, must be designed on the basis of the total horsepower of both motors. If one motor is cut out of service, the starting currents will remain the same and the second motor will receive double its normal starting peaks. This may be satisfactory in emergency, but where such a condition is likely to be of frequent occurrence it is customary to provide some arrangement for changing the resistance when one motor is cut out. Figure 12.4 shows an arrangement for starting a number of motors together. In this case it was frequently desired to start different combinations of motors. Each motor was provided with its own resistor, which could be paralleled with any other resistor by closing a four-pole knife switch. The knife switches used had two clips for each blade, since this was necessary to break up the circuit properly. When starting any combination of motors, the knife switches for those motors were closed, thus providing the correct resistance for the combination and, at the same time, permitting the use of a single common set of accelerating contactors. One pole of each knife switch disconnected the motor, so that it was impossible to set up an incorrect resistor combination.

The connections shown in Figs. 12.3 and 12.4 are also satisfactory for speed regulation by armature resistor. With speed regulation by field resistor it is customary to provide separate rheostats in each motor field. However, for certain types of drives—cloth calender trains, for example—a single rheostat is used to control all the fields in parallel. Correction is then applied by a small vernier rheostat in each field, except that of the leading motor.

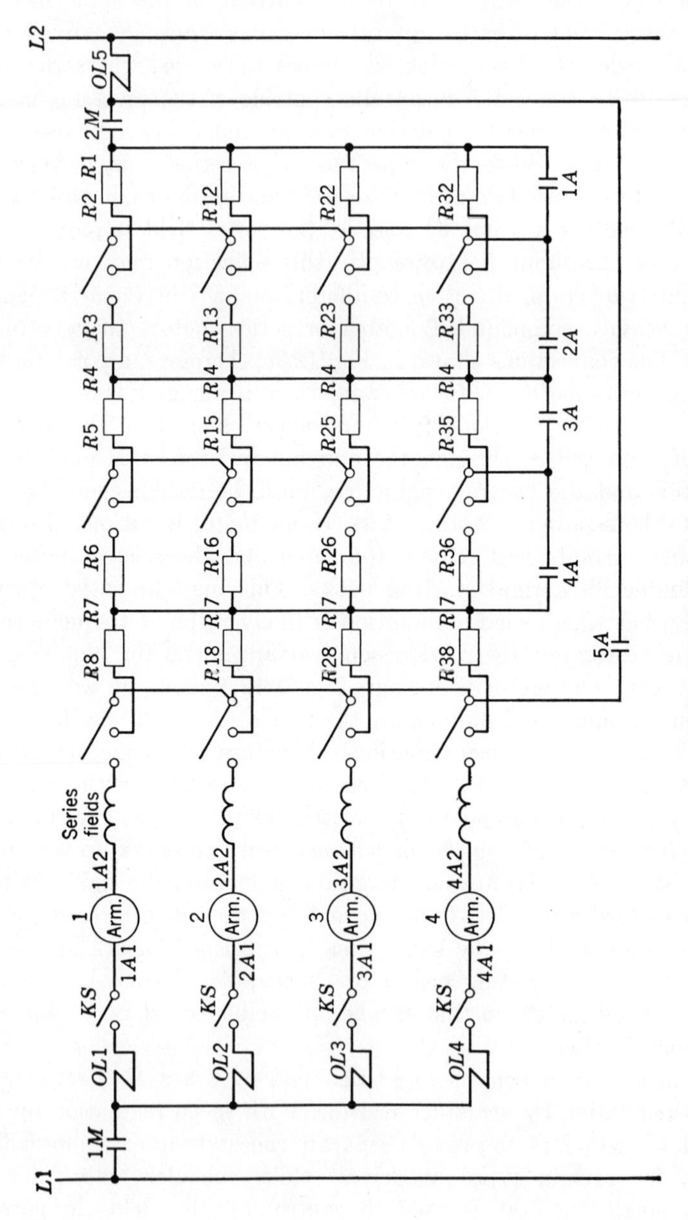

FIG. 12.4 Starting Shunt Motors in Parallel

Plugging Motors in Parallel. When series motors in parallel are to be plugged, none of the control schemes discussed so far will be satisfactory. The series fields cannot be simply connected in parallel, as each machine must feed its own field to insure proper division of starting and running current. The connections shown in Fig. 12.3*c* will cause serious trouble. There is certain to be some slight difference in the resistance of the two series fields. As long as the motors are running forward, they will tend to balance the load, since any increase in load on one motor will cause an increase of current through the corresponding series field, and that motor will tend to slow down, allowing the other motor to take a greater share of the load. However, when the motors are plugged the conditions are different. The countervoltage of the armatures, which has been opposing the line voltage, is then added to the line voltage. Since there is certain to be some difference in the field characteristics, there is also certain to be some difference in the countervoltage developed by the two armatures. The armatures and fields form a closed loop of very low resistance, and any difference in countervoltage will force current around the loop. Because of the low resistance of the loop, a small voltage will cause a considerable current flow. Even if the initial current flow is very small, the result is disastrous, because the circulating current strengthens the field of the motor whose voltage is already high and weakens the field of the motor whose voltage is low. Consequently, the voltage of the strong motor increases still more, while that of the weak motor decreases further. The result is an increasingly greater difference in voltage and heavier circulating current, and the effect builds up until the weak motor is not generating any voltage. It then acts as a short circuit across the other motor.

In order to avoid the effect just described, it is necessary to insert resistance into the motor loop during the plugging period. One method of doing this is to have entirely separate plugging and accelerating contactors for each motor, as shown in Fig. 12.5*a*. This is really having two separate controllers operated from one master. No circulating currents can occur, and the scheme is highly satisfactory for starting, reversing, and plugging. Since each motor has its own contactors and its own resistor, it is a very simple matter to cut out one motor in case of trouble. In order to insure smooth acceleration by closing the corresponding accelerators of each motor at the same time, double-pole accelerators may be used, or single-pole contactors may be mechanically tied together in pairs. Both methods have certain advantages. With double-pole contactors, the number of coils will be smaller and the control circuit simpler. The controller with single-pole contactors will be some-

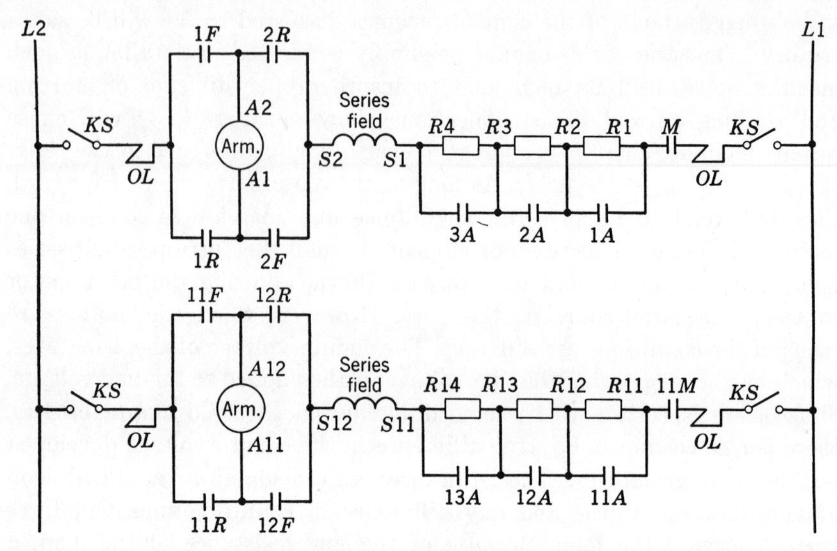

(*a*) Separate Accelerating Contactors

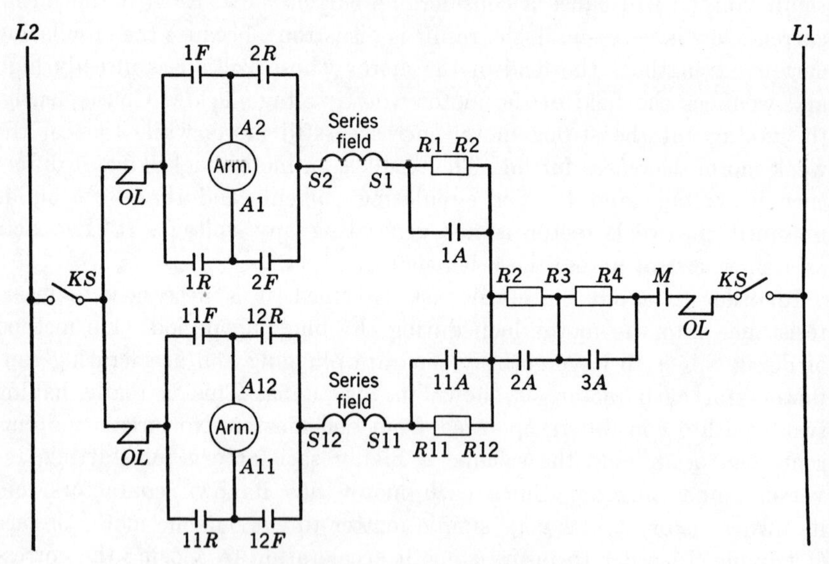

(*b*) Common Accelerating Contactors

FIG. 12.5 Plugging Series Motors in Parallel

what more flexible, and the contactors will duplicate those for single-motor controllers.

Another method of arranging a plugging controller is shown in Fig. 12.5*b*. Here there are separate plugging contactors and resistance in the loop circuit, to limit the circulating currents, but the rest of the accelerating contactors are common to the two motors. The only advantage of this arrangement is lower cost, and this is somewhat offset by the fact that the common contactors must have capacity for the current of the two motors. If it happened that the motor size was such that the same size of contactor was required for either one or two motors, then this arrangement would result in a worthwhile saving, but where the common contactors must be larger, very little saving results.

Series Motors in Series. When series motors are connected in series, the armatures are in series inside the reversing contactors and the fields are in series outside the reversing contactors. The purpose of this arrangement is to limit the speed of the motors to one-half of its normal value. For some drives the acceleration appears to be more rapid if the motors are required to reach only half-speed, and also there is less tendency to reach undesirable speeds on light loads. The current flowing through the circuit is limited to that of one motor; and since the voltage across each motor is half-line voltage, it follows that the total horsepower of the drive is that of one motor. The contactors are selected and the resistance designed on that basis, with the exception that the severity of the arcing must be taken into account. The inductance of the two series fields in series is greater than that of a single field, and this causes the arcing to be more severe when the circuit is opened by the contactors. It is therefore advisable not to use contactors to the limit of their rating, but to allow some margin of capacity. The life of the arcing contacts will be increased by so doing. In plugging, the countervoltage of two motors in series will rise higher than that of a single motor and, coupled with the increased inductance of the fields, will cause severe arcing. The ohmic value of the plugging resistance should be high enough to compensate for the higher countervoltage. The usual practice is to have approximately 50% more ohms than in a single motor.

Frequently a standard single-motor controller may be satisfactory to handle two motors in series, but the capacity of the contactors should be considered and the plugging resistance increased.

Series Motors—Dynamic Lowering. Two series motors are often used to drive the hoist motion of a large crane or of a bucket hoist. The characteristics of the series motors are such as to insure proper division of the load during hoisting. If one motor should attempt to take more load than the other, its field strength would be increased and it would tend to slow down. The other motor would tend to speed up until the load was equalized. During dynamic lowering, however, the opposite takes place, and, if one of the motors takes the greater part of the load, it tends to slow down and increase the unbalancing still more.

A satisfactory means of insuring a balance during lowering is a control scheme similar to that shown in Fig. 12.6, in which the series fields of the motors are cross-connected during lowering. In the hoisting direc-

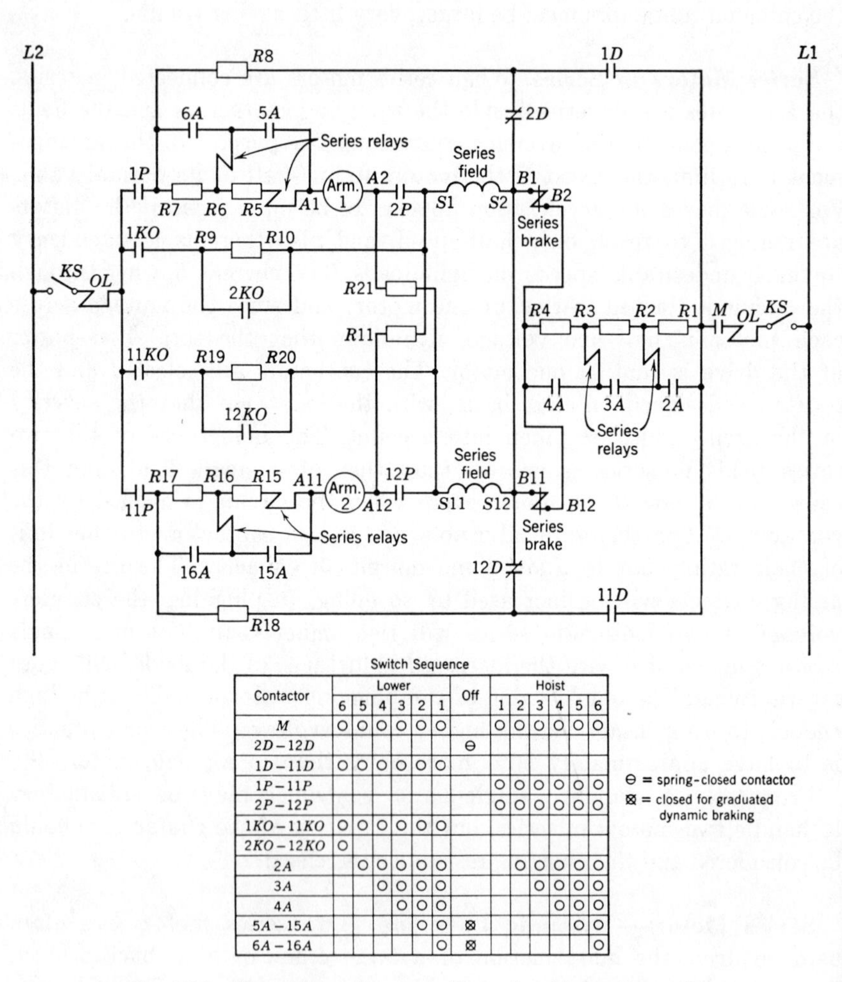

Switch Sequence

Contactor	Lower 6	Lower 5	Lower 4	Lower 3	Lower 2	Lower 1	Off	Hoist 1	Hoist 2	Hoist 3	Hoist 4	Hoist 5	Hoist 6
M	O	O	O	O	O	O		O	O	O	O	O	O
2*D* – 12*D*							⊖						
1*D* – 11*D*	O	O	O	O	O	O							
1*P* – 11*P*								O	O	O	O	O	O
2*P* – 12*P*								O	O	O	O	O	O
1*KO* – 11*KO*	O	O	O	O	O	O							
2*KO* – 12*KO*	O												
2*A*		O	O	O	O	O			O	O	O	O	O
3*A*			O	O	O	O				O	O	O	O
4*A*				O	O	O					O	O	O
5*A* – 15*A*					O	O	⊠					O	O
6*A* – 16*A*						O	⊠						O

Fig. 12.6 Two-Motor Dynamic-Lowering Hoist Controller

tion, contactors, 1*P*, 2*P*, 11*P*, 12*P* are closed and each motor armature is in series with its own field. The connections are then the same as they would be with two separate controllers. In the lowering direction contactors 1*KO*, 11*KO*, 1*D*, and 11*D* are closed. The armature of motor 1 is now connected in a dynamic-braking loop circuit including the field of motor 2, and the armature of motor 2 is similarly connected to the field of motor 1. Any tendency of one motor to take more current will increase the field strength of the other motor and so equalize the load again.

This control scheme, which has been used on many bucket hoists and on some large cranes, insures proper division of the load. However, there are other considerations which often outweigh this one when determining crane hoist control. Most of the cranes large enough to require a two-motor hoist are ladle cranes, and here it is essential to be able to move the hoist with one motor in the event of trouble. The cross-connected scheme does not lend itself readily to single-motor operation; consequently the majority of ladle-crane controllers consist simply of two entirely separate controllers operated from a single master and provided with knife switches to cut out either motor in case of trouble. With such controllers it is possible to secure fairly satisfactory division of load by carefully balancing the motors on each step of the control during lowering, and it is necessary to accept some unbalancing in order to secure the flexibility of the separate control scheme.

Motors Not Mechanically Coupled. Some machines require two or more motors which are not mechanically tied together, or which have a somewhat flexible mechanical tie. In such installations the problem is not one of division of load but often one of speed synchronization or position synchronization. For example, a machine for making paper consists of a number of separate sections, each driven by its own motor, and the only tie between the sections is the piece of paper under process. The paper is, of course, too weak to hold the sections to the same speed, and so some sort of accurate speed synchronization is necessary. The same is true of a multistand continuous steel rolling mill.

For these applications it is desired to run the motors at a controlled constant speed over a wide speed range. This dictates the choice of shunt wound or lightly compounded motors with the armatures connected in parallel to a single source of adjustable-voltage direct current (thyristor or M-G set) as shown in Fig. 12.7. At the same time it is desired to have the motors run at approximately the same speed when "threading up" (before the steel is pulled through the line). Once the steel is in and tension is established, however, the motors become mechanically con-

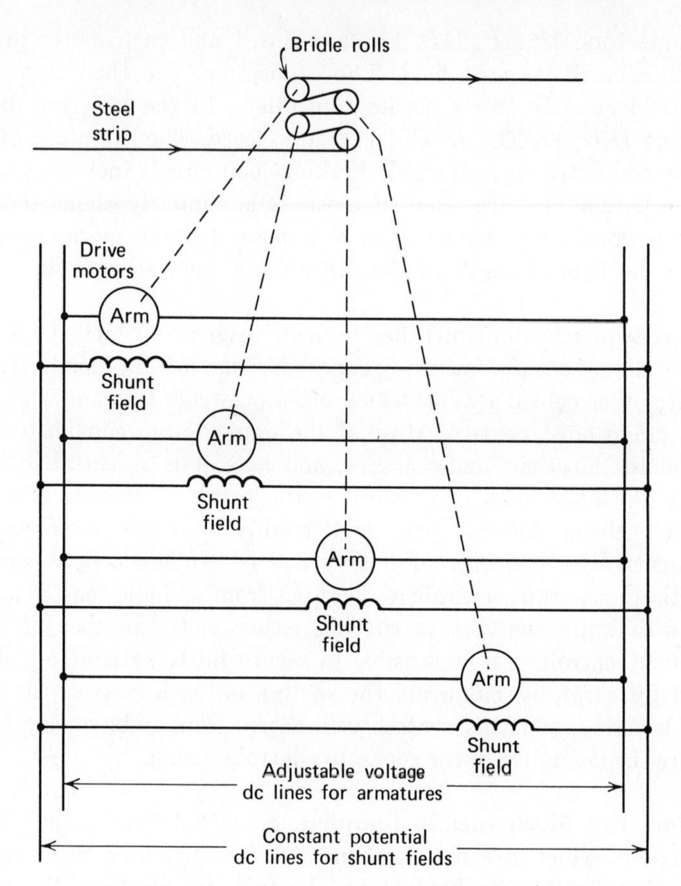

Fig. 12.7 Four Motor-Driven Rolls for Propelling Steel Strip

nected together by virtue of the strip wrapped around the bridle rolls. It is therefore important to obtain load sharing even though sophisticated regulated control is used for the group. This load sharing is usually accomplished by a combination of a permanent resistor in series with each armature and/or a small amount of series field on each motor, and a vernier field rheostat on each motor, as shown in Fig. 12.8.

Most of the speed-synchronizing methods are quite complicated and usually peculiar to one particular process. They are really a part of the problem of applying automatic control to a definite manufacturing process. Some are described in Chapter 10.

The bridge motion of an ore or coal bridge is a good example of position synchronization. The cross-member of such a bridge is mounted on large king pins, so that the ends of the bridge may move out of line with each other without doing any harm. It is advantageous to be able to move one end of the bridge without moving the other, for such purposes as loading or moving freight cars. A large bridge may permit a misalignment of 40 ft before the danger point is reached. The control generally consists of separate drum- or magnetic-type controllers for motors on each end of the bridge. A skew limit switch is used to prevent misalignment beyond the danger point. This switch is mounted

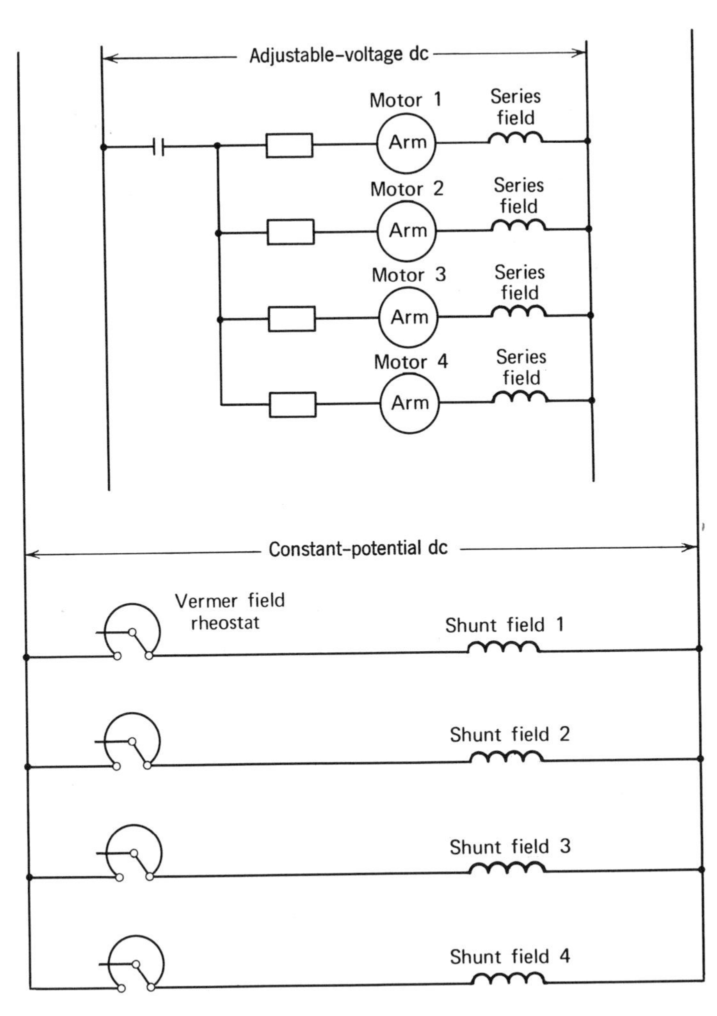

Fig. 12.8 Load Sharing on Multimotor "Bridle"

at the top of one of the bridge legs and is operated by a cam on the cross-member of the bridge. As the end of the bridge moves, the cross-member turns on the king pin, and when the danger point is reached the cam trips the skew limit switch, stopping further motion in that direction. Generally a second contact of the limit switch is used to slow down before final stop occurs. If the bridge is moving with both motors energized, and one side gets too far ahead, the skew switch will operate and slow down the leading motor. If further misalignment occurs, the leading motor will be stopped.

Series-Parallel Control. Two motors mechanically connected may be controlled by the series-parallel method. With this arrangement the motors are first connected to the line in series and with resistance in circuit. The resistance is then gradually cut out until the motors are across the line in series. The next step is to reinsert the resistance and connect the motors in parallel. Then the resistance is again cut out step by step until the motors are across the line in parallel. With the motors in series across the line, half-speed is obtained; with the motors in parallel, full speed. Since the half-speed is obtained without power loss in the resistance, the scheme is of advantage where half-speed is frequently required. A further power saving is obtained where the accelerating period is long, as the motors may be accelerated in series with half of the current that would be required if they were in parallel. The series-parallel control also gives a smooth, even acceleration.

The principal application of this type of control is to moving cars. In operating city trolley cars, for instance, a considerable power saving is effected by having two running speeds which are obtained without losses in resistors. A relatively long acceleration period is required, and a smooth acceleration is desirable. Similar requirements are found in controlling electric locomotives, interurban cars, larry cars in steel mills and coke plants, and ore bridges.

The most important feature of a series-parallel control scheme is the method of transition from series to parallel. It is desirable that the transition be rapid, smooth, and without loss of speed or torque. In order to obtain these results, additional control material is required so that the series-parallel controller is relatively complicated and expensive. Three types of transition are common. They are known as open-circuit transition, shunt transition, and bridging transition.

Open-Circuit Transition. This method is illustrated in Fig. 12.9. The motors are first connected in series with resistance in circuit. The resistance is then cut out until the motors are across the line in series

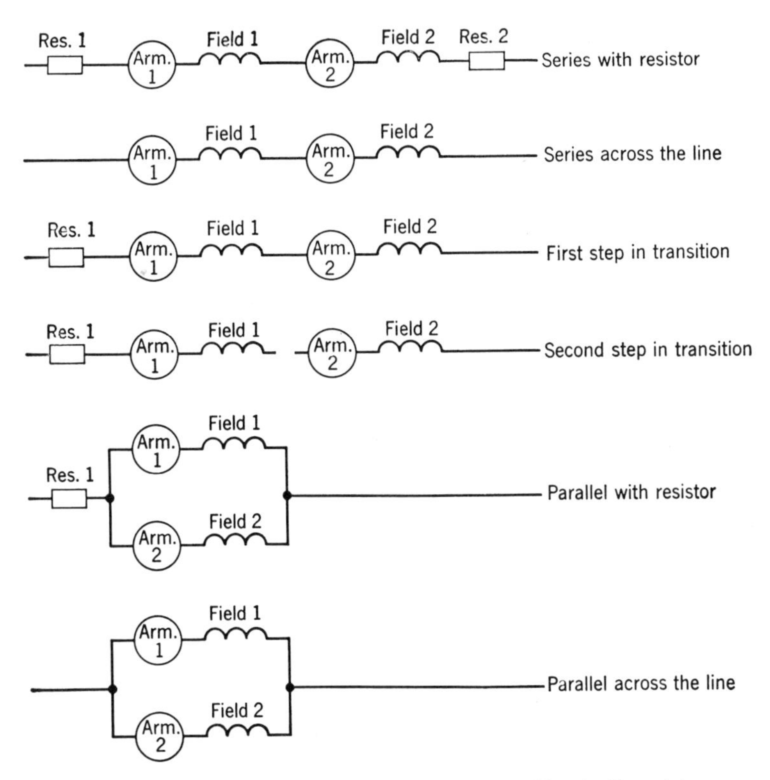

Fig. 12.9 Series-Parallel Controller with Open-Circuit Transition

and running at half-speed. At the transition point the motor circuit is opened and the motors are momentarily disconnected. They are then reconnected in parallel and with the resistance back in circuit. The resistance is then cut out until the motors are across the line in parallel and at full speed. This method is simple and involves the minimum complication of the controller, but the opening of the circuit is undesirable, as it causes a loss of speed and torque during the change-over period. The arcing which occurs when the circuit is opened shortens the life of the controller contacts. For these reasons the open-circuit method is limited to small motors.

Shunt Transition. The shunt-transition method shown in Fig. 12.10 provides a simple means of avoiding an open circuit during the change from series to parallel. The motors are brought up to half-speed in the same manner as for open-circuit transition. At the transition point resistance is introduced into the circuit, and then one of the motors is

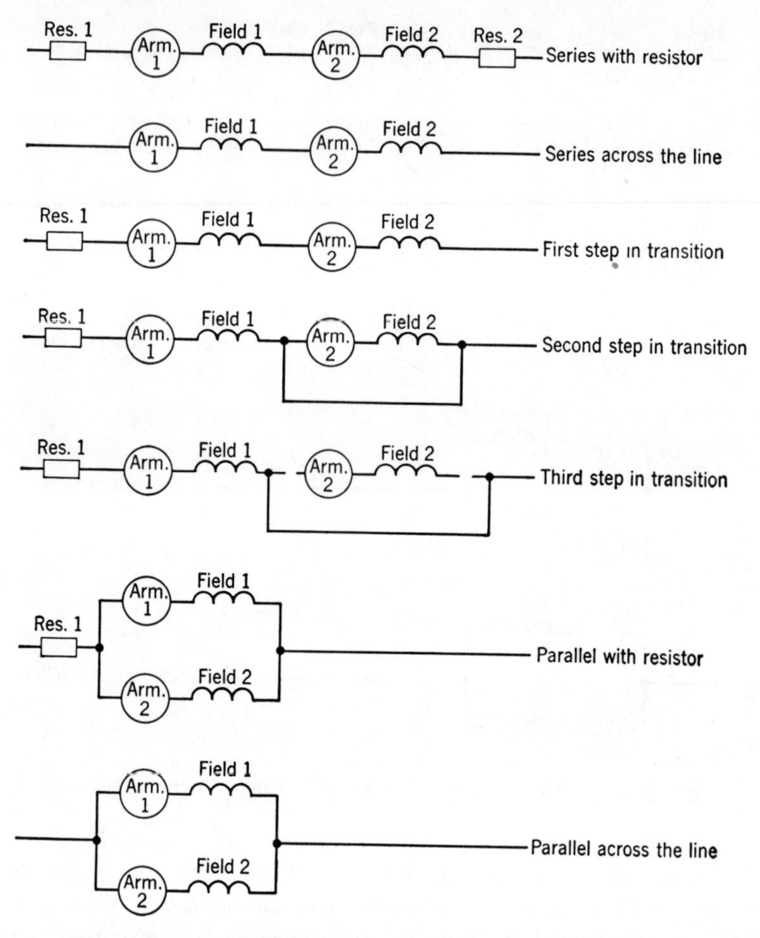

FIG. 12.10 Series-Parallel Controller with Shunt Transition

short-circuited. The resistance compensates for the countervoltage which the short-circuited motor has been developing and prevents an excessive inrush of current. The short-circuited motor is now disconnected and reconnected in parallel with the other motor, after which the resistance is cut out of circuit until the motors are across the line in parallel and at full speed. With this method the circuit is never broken, and at least one motor is always producing driving torque. The total torque, however, is reduced during the transition. The most serious objections to the method are that the short-circuited motor receives a heavy overload for a short period, and that the sudden application of this overload and the sudden change in torque produce severe shocks to the driven machinery.

Bridging Transition. The bridging method is shown in Fig. 12.11. With this arrangement the motors are brought up to half-speed in series in the usual manner. Resistance is then connected in parallel with each motor, so that the circuit is equivalent to that of a Wheatstone bridge. If the resistances are of equal value, no current will flow in the bridging circuit. When this circuit is opened, the motors are connected in parallel. The resistance is then cut out until the motors are up to full speed. By adjusting the resistance to suitable values, the torque may be kept up during the transition or may even be increased. Increasing the current through the resistors will increase the motor torque when the bridge is opened.

The bridging method gives a smooth, even acceleration without loss

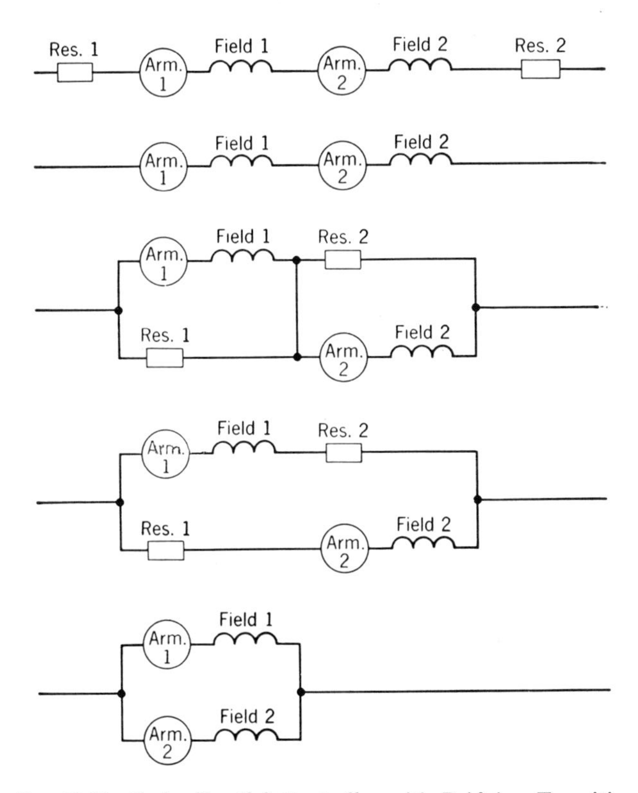

Fig. 12.11 Series-Parallel Controller with Bridging Transition

of speed or torque and without shock to the driven machine. It is, therefore, the most extensively used method, although it requires a more complicated and expensive controller.

Types of Series-Parallel Controllers. Series-parallel controllers may be either of the drum type or of the magnetic-contactor type. Since their greatest field of application is to cars, where space is usually limited, drums are widely used. However, the advantages of magnetic control in general apply to series-parallel control also, and where space conditions will permit, magnetic control is desirable.

Figure 12.12 illustrates the connections of a magnetic-contactor controller; it includes a sequence table, showing the contactors which are closed on each point of the master controller. On the first point forward, the direction contactors 1*F*, 2*F*, 3*F*, and 4*F* are closed, and the series contactor is 1*S*. The motors are then connected in series with all the resistance in circuit. On the second point, contactors 2*A* and 3*A* are closed, and a portion of the resistance is short-circuited. On the next point, contactors 4*A*, 14*A*, 5*A*, and 15*A* are closed, short-circuiting all the resistance except the step *R*6–*R*7. This step is cut out on the fourth point, by closing contacts 6*A*, and the motors are then across the line, in series, and at half-speed.

The transition starts on the fourth point and is completed on the fifth point. As soon as contactor 6*A* closes, it energizes a relay which opens the circuit to the series contactor 1*S* and to the resistance contactors 2*A*, 3*A*, 4*A*, 5*A*, 14*A*, and 15*A*. This does not affect the operation in any way, since the circuit is still complete through 6*A*, but it does prepare the resistance circuit for the next step in the transition. When the master is moved to the fifth point, the parallel contactors 1*P* and 2*P* are closed, thus setting up the bridging circuit. Two relays are also energized at this time. One of these is used to connect the coils of contactors 4*A* and 14*A* together, and also the coils of 5*A* and 15*A*. This is done to insure equal acceleration of the motors in parallel, by cutting out the resistance steps in each motor circuit at exactly the same time. The second relay opens the circuit to 6*A*, thus opening the bridge and connecting the motors in parallel, with resistance in each motor circuit. On the sixth point the resistance contactors 4*A* and 14*A* are closed and then 5*A* and 15*A*, and the motors are in parallel across the line at full speed.

The operation in the reverse direction is the same, except that the direction contactors 1*R*, 2*R*, 3*R*, and 4*R* are closed instead of the forward-direction contactors.

In the controller just described, six speeds in each direction are ob-

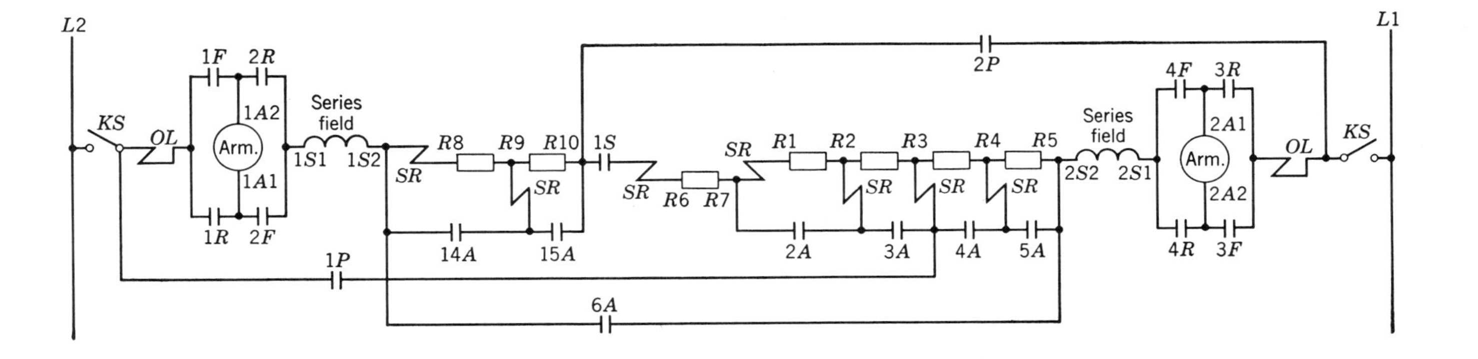

Contactor	Forward						Off	Reverse					
	6	5	4	3	2	1		1	2	3	4	5	6
$1F-2F-3F-4F$	O	O	O	O	O	O							
$1R-2R-3R-4R$								O	O	O	O	O	O
$1S$			O	O	O	O		O	O	O	O		
$1P-2P$	O	O										O	O
$2A$			O	O	O				O	O	O		
$3A$			O	O	O				O	O	O		
$4A$	O		O	O						O	O		O
$14A$	O		O	O						O	O		O
$5A$	O		O								O		O
$15A$	O		O								O		O
$6A$			O								O		

FIG. 12.12 Connections for a Series-Parallel Magnetic Controller

tained. This is the maximum number generally employed, but the same controller could give a number of additional speeds if desired. It would be necessary only to change the master so that each resistance contactor would be energized on a separate point. The maximum number of speeds in each direction would be as follows.

Speed	*Series*	*Speed*	*Series*	*Speed*	*Parallel*
1	1*S*	5	14*A*		
2	2*A*	6	5*A*	9	1*P*–2*P*
3	3*A*	7	15*A*	10	4*A*–14*A*
4	4*A*	8	6*A*	11	5*A*–15*A*

The resistance contactors are under the control of series relays, and they are so interlocked as to insure their closing in the proper sequence. The circuits are also arranged so that the master may be moved rapidly to the full-speed point, and the contactors will still go through the proper sequence to start the motors in series, effect the transition, and bring them up to speed in parallel.

The resistance steps *R*3–*R*4–*R*5 and *R*8–*R*9–*R*10 are the accelerating resistance, and should limit the current inrush to 150%. The total of these four steps is then $E \div 1.50 = 0.66E/I$. The steps *R*1–*R*2–*R*3 are plugging steps and should limit the current to 150% when the controller is plugged. If plugged from parallel to series, the countervoltage of both motors must be added to the line voltage. If we assume the countervoltage of each motor to be 80%,

$$E + 0.8E + 0.8E = 2.6E$$

$$\text{Total resistance} = \frac{2.6E}{1.5I} = 1.72E/I$$

$$R1\text{–}R2\text{–}R3 = 1.72E/I - 0.66E/I = 1.06E/I$$

The step *R*6–*R*7 is of low resistance and is used to prevent a short circuit if an arc should hang on contactor 1*S* until contactors 1*P* and 2*P* close. When the plug steps and the accelerating steps are tapered in the usual manner, the following table is obtained.

Step	*Ohms in per cent of E/I*	*Step*	*Ohms in per cent of E/I*
*R*1–*R*2	64	*R*8–*R*9	22
*R*2–*R*3	42	*R*9–*R*10	11
*R*3–*R*4	22	*R*6–*R*7	10
*R*4–*R*5	11		

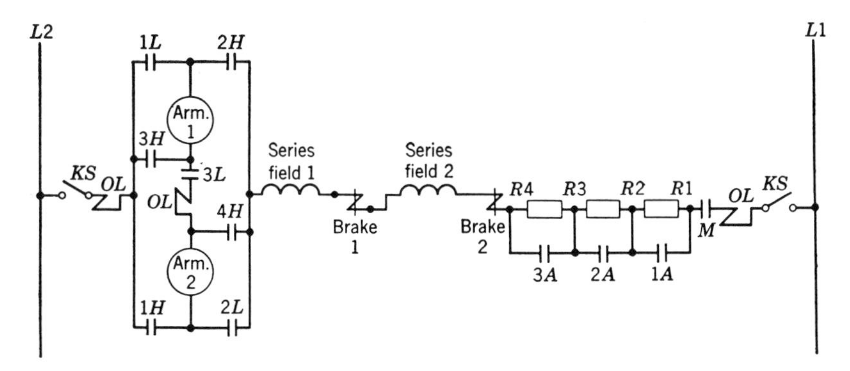

FIG. 12.13 Connections for a Simplified Series-Parallel Controller

Simplified Series Parallel. It is not always necessary to have full series-parallel control. Sometimes a controller giving series operation or parallel operation, but without transition, will meet the requirements of the drive. The controller can then be considerably simplified. A controller of this type might be used in connection with a screw-down on a rolling mill. The characteristics of the service require a slow speed at full torque in the lowering direction, and a high speed in the up direction. This can be obtained by connecting the motors in parallel when hoisting, and in series when lowering, and the ratio of speeds would be 2 to 1. If the controller were further simplified by connecting the series fields permanently in series and the armatures in series parallel, the hoisting speed would be 155% of the lowering speed, which might be sufficient to meet the requirements of the installation. Figure 12.13 shows the connections.

Problems

1. Three 100-hp 230-V 350-amp 480-rpm series motors are connected in series to drive a machine. Using the equations of the series motor, calculate the speed of the combination.

2. Calculate the total horsepower which will be delivered by the three motors of problem 1.

3. Draw an elementary diagram for the main and control circuits of a nonreversing controller for the motors of problem 1, including the following:

Main and control knife switches
2 overload relays
2 line contactors
3 accelerating contactors
Current-limit acceleration
Low-voltage relay
Four-speed master

4. Draw an elementary diagram for the main and control circuits of a reversing controller for the motors of problem 1, including the following:

Main and control knife switches
2 overload relays
1 line contactor
4 reversing contactors
3 accelerating contactors
Inductive-timed acceleration
Low-voltage relay
Four-speed master

5. Calculate the ohmic value of a starting resistor for the motors of problem 1, to give an initial starting current of 150% of rated current, and assuming the resistance of each motor as 11% of E/I.

6. Calculate the ohmic values of a series speed-regulating resistor for the motors of problem 1, to give 50% speed at full load.

7. Calculate the wattage to be dissipated by the resistor of problem 6. How does this compare with the wattage to be dissipated by a similar resistor for a single 100-hp motor?

8. Draw an elementary diagram for the control circuits of the arrangement shown in Fig. 12.3*c*, using a four-speed master controller, and condenser-timed acceleration.

9. Calculate the ohmic value and wattage capacity of a speed-regulating resistor for a controller like Fig. 12.3*c*, for two 35-hp 230-V 140-amp motors. The speed is to be reduced to 50% of rated speed, when running at 75% of rated load.

10. Draw an elementary diagram of the control circuits of Fig. 12.5*b*, using a four-speed master controller and inductive timing without a separate inductor.

11. Calculate the ohmic value of each step of a plugging resistor for two 25-hp 230-V 100-amp motors, using a controller like Fig. 12.5*b* and the data of Tables 11.1 and 11.2.

12. Two 50-hp 230-V 200-amp series motors are used to drive a crane trolley, using a controller like Fig. 12.5*b* except having two steps of resistor for plugging, and three steps of resistor for accelerating. Draw the elementary diagram of the main circuits only.

13. If the plugging inrush current is to be 70% of rated current on the first point, and 150% on the second point, calculate the ohms in all steps of the resistor, using the data of Tables 11.1 and 11.2.

14. Draw an elementary diagram of the main circuits only, of a reversing, plugging controller for four series motors, connected two in series, and the two pairs in parallel.

15. Two 50-hp 230-V 200-amp series motors are connected in a series-parallel arrangement like Fig. 12.12. If the resistance of each motor is 11% of E/I, calculate the ohmic values of the resistor steps of the controller.

16. Using the data of problem 15, calculate the speed which will be obtained on each point of the controller, when the motors are fully loaded.

17. Draw an elementary diagram of the control circuits of the controller shown in Fig. 12.13, using a reversing master arranged to give the maximum number of speeds for hoisting and lowering. Omit any means of delaying the accelerating contactors.

13

DYNAMIC-LOWERING HOIST CONTROLLERS

Characteristics Required. Dynamic lowering control is used in connection with series motors, or compound motors having a predominating series field, to operate crane hoists, winches, or other drives where the load may be overhauling. The requirements for hoisting are that it should be possible to take up a slack cable without undue jerking, and to accelerate and hoist both light and heavy loads smoothly, with a choice of several speeds. The requirements for lowering are that it should be possible to lower both light and heavy loads safely, with a choice of several speeds, and to lower an empty hook at a high speed, although the load under that condition would not be overhauling.

General Principles. The connections for the hoisting direction are the same as those of a full reverse controller; that is, the armature and the series field are connected in series. If this connection were used in the lowering direction, the motor would not retard an overhauling load but would soon reach a runaway speed. By connecting the armature and the field in parallel, with resistance in each circuit, characteristics approximating those of a shunt motor are obtained. With a light load the motor drives the hook down, taking a relatively heavy current from the line. As the load is increased, the current taken from the line becomes less, since less torque is required to drive the load. When the load is heavy enough to overcome the friction of the drive, it begins to overhaul the motor, which then acts as a generator and retards the load. The speed is varied by changing the amount of resistance in the braking loop or in the series field circuit.

Dynamic lowering controllers, as built by different control manufacturers, vary somewhat in detail, but the underlying principles are the same and the control schemes similar. Both current-limit acceleration and time-limit acceleration are used, with the trend decidedly toward time limit. The controller described below is of the inductive time-limit type. Several manufacturers have almost identical control schemes, with the exception of the method of acceleration. Since the purpose here is to describe the dynamic lowering principles, one control will serve as well as another.

Ltl Acceleration. The Ltl method of obtaining time for acceleration depends upon the fact that the current resulting when a voltage is applied to a circuit having inductance will not instantly reach its final value, but will build up to that value at a rate dependent upon the amount of inductance in the circuit. Also, if the voltage is removed and the circuit is closed, the current will not instantly drop to zero, but will reach zero after a time, again dependent upon the amount of inductance in the circuit.

In applying this principle, accelerating contactors having two coils are used. The upper coil is the closing coil and is connected to full line voltage. The lower coil is a holding-out coil which is so designed that it will prevent the contactor from closing until the current in the coil reaches a low value. The structure of the lower portion of the contactor is such that the holding-out coil operates on a magnetic circuit which is practically closed; that is, there is a very small air gap. This circuit is highly inductive. Consequently, if voltage is applied to the holding-out coil and then removed and the coil short-circuited, current will continue to flow in the coil for a period of time, and the contactor will remain open until this current has died down to very nearly zero. The timing of the contactor may be adjusted by increasing or decreasing the air gap in the magnetic circuit of the hold-out coil, which changes the value of the current at which the contactor will close. A coarse adjustment of this time is made by adding or removing steel shims which are from a part of the magnetic circuit. A fine adjustment is obtained by turning a knurled nut, which also varies the air gap. Ordinarily, sufficient adjustment can be obtained by means of the knurled nut, and it is seldom necessary to change the shims from the adjustment made at the factory.

Referring to Fig. 13.1, since the motor current obtained when contactor *LT* is closed is not high, no time delay is used on that contactor. The holding coil of contactor 1*A* is connected across the resistance step *R*2–*R*3, and is energized by the voltage drop across that step. When *LT* closes, the holding coil is short-circuited but because of its inductive effect it does not permit contactor 1*A* to close until a definite time has elapsed. In a similar manner the closing of contactor 2*A* is timed by its holding coil, which is connected across resistance step *R*2–*R*4, and is short-circuited by the closing of contactor 1*A*.

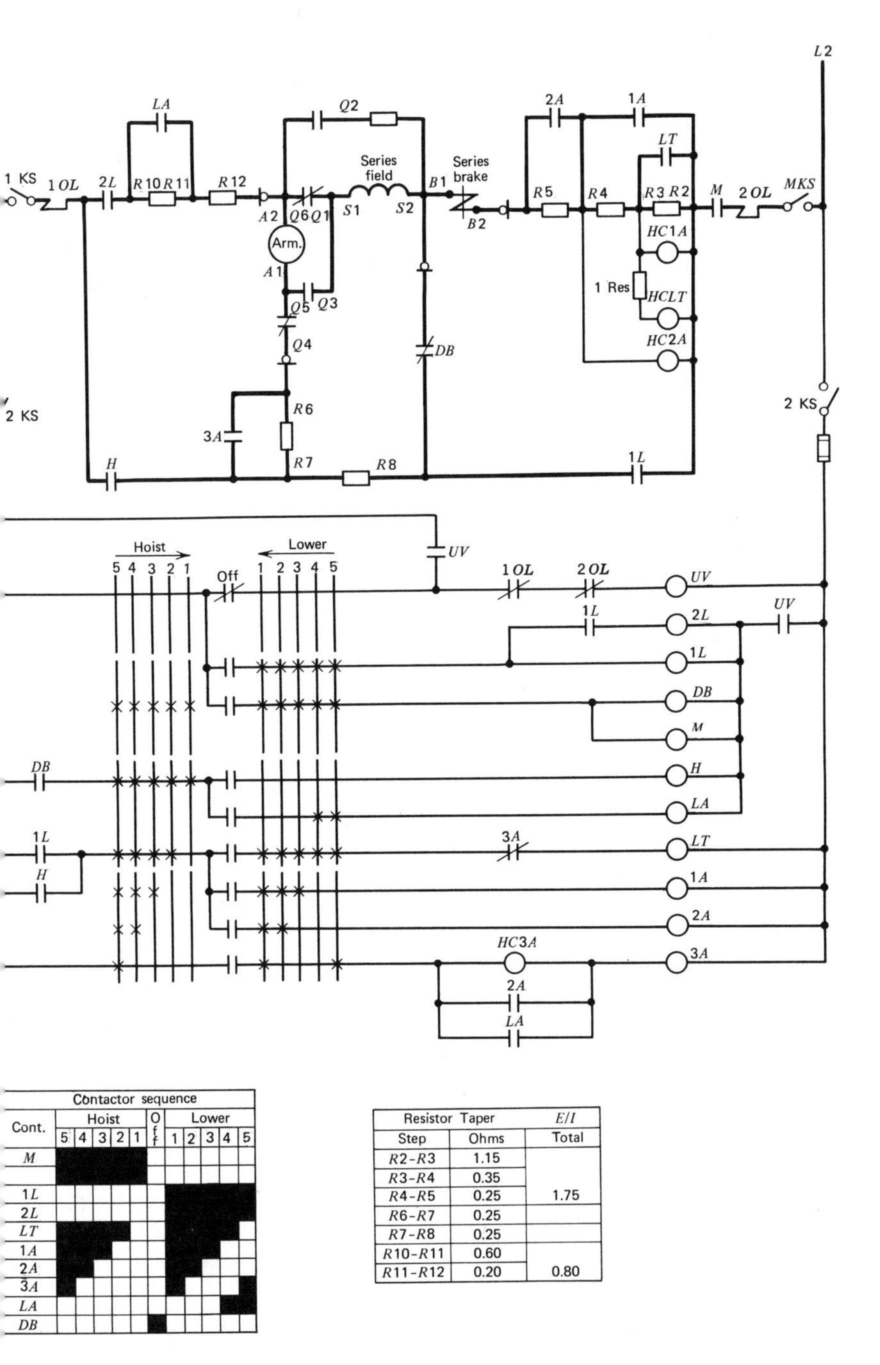

Cont.	Hoist 5	Hoist 4	Hoist 3	Hoist 2	Hoist 1	Off	Lower 1	Lower 2	Lower 3	Lower 4	Lower 5
M	■	■	■	■	■						
	■	■	■	■	■						
1*L*							■	■	■	■	■
2*L*							■	■	■	■	■
LT	■	■	■	■			■	■	■	■	
1*A*	■	■	■				■	■	■		
2*A*	■	■					■	■			
3*A*	■						■				■
LA										■	■
DB						■					

Contactor sequence

Step	Ohms	*E/I* Total
*R*2–*R*3	1.15	
*R*3–*R*4	0.35	
*R*4–*R*5	0.25	1.75
*R*6–*R*7	0.25	
*R*7–*R*8	0.25	
*R*10–*R*11	0.60	
*R*11–*R*12	0.20	0.80

Resistor Taper

Fig. 13.1 Standard Dynamic-Lowering Hoist Control

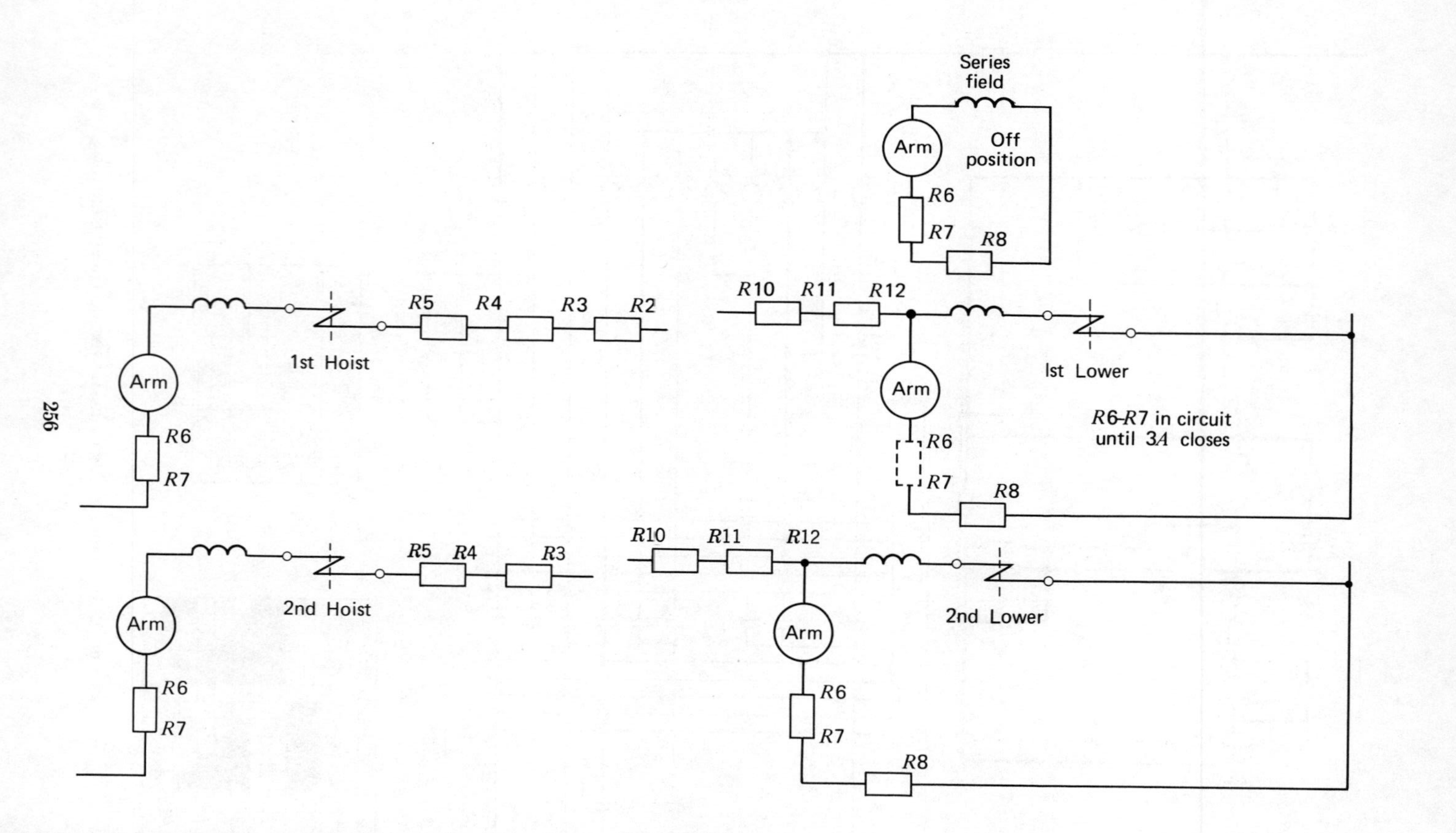
Series field
Arm
Off position
R6
R7
R8
Arm
R6
R7
R5
R4
R3
R2
1st Hoist
R10
R11
R12
Ist Lower
Arm
R6–R7 in circuit until 3A closes
R6
R7
R8
Arm
R6
R7
R5
R4
R3
2nd Hoist
R10
R11
R12
2nd Lower
Arm
R6
R7
R8

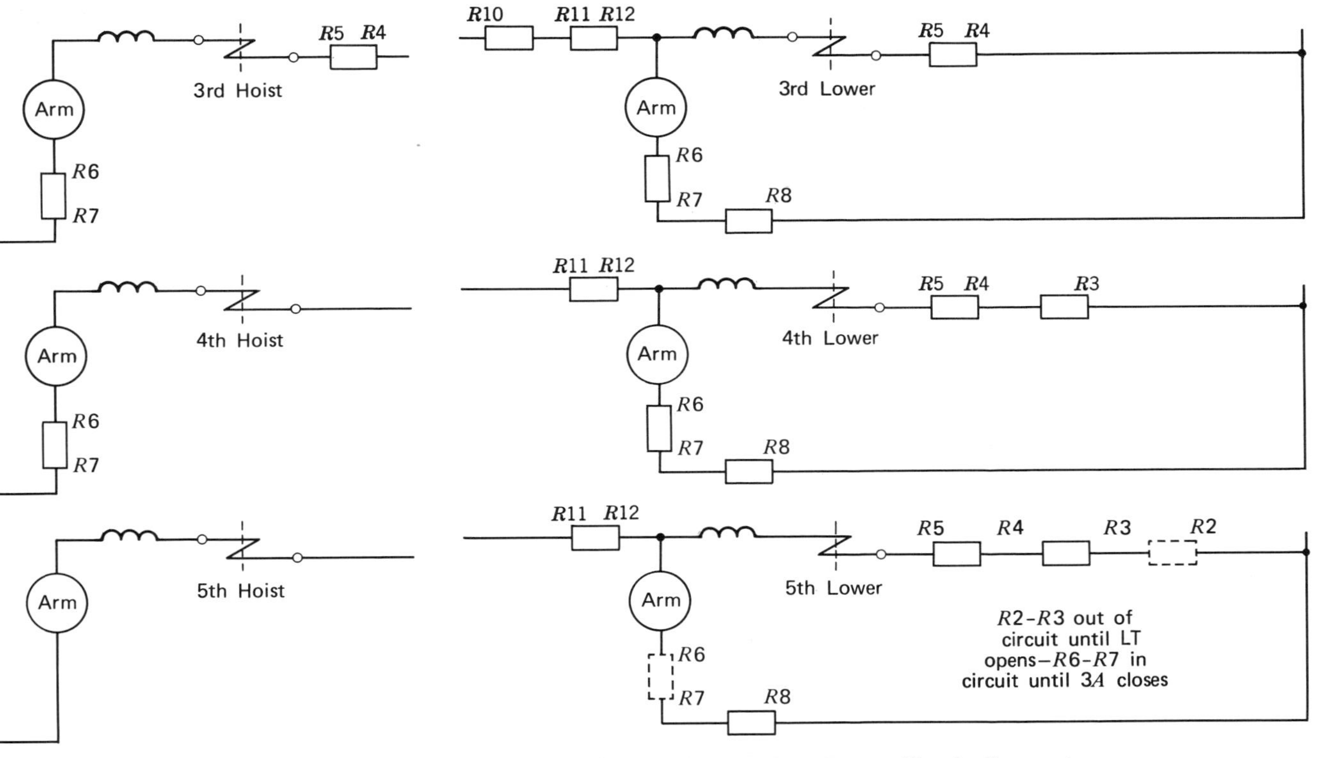

FIG. 13.2 Dynamic-Lowering Hoist Control. Step-by-Step Power Circuit Connections

Contactor $3A$ must be delayed in closing, both in hoisting, and on the last point lowering. To accompish this the holding coil $HC3A$ is connected in series with the closing coil of $3A$. With this arrangement the contactor will remain open until the holding coil is short-circuited, and then, after a time delay, it will close. In hoisting, the holding coil is short-circuited by interlocks on contactor $2A$. On the last point in the lowering direction, the holding coil is short-circuited by an interlock on contactor LA.

Hoisting (Figs. 13.1 and 13.3). Moving the master to the first point hoisting energizes the coils of contactors M, DB, and H. Since the coil of H is interlocked behind DB, the braking loop around the armature must be opened before power is applied to the motor. The closing of the contactors H and M connects the motor to the line through the starting resistance steps $R2$–$R5$ and $R6$–$R7$. Since the motor armature

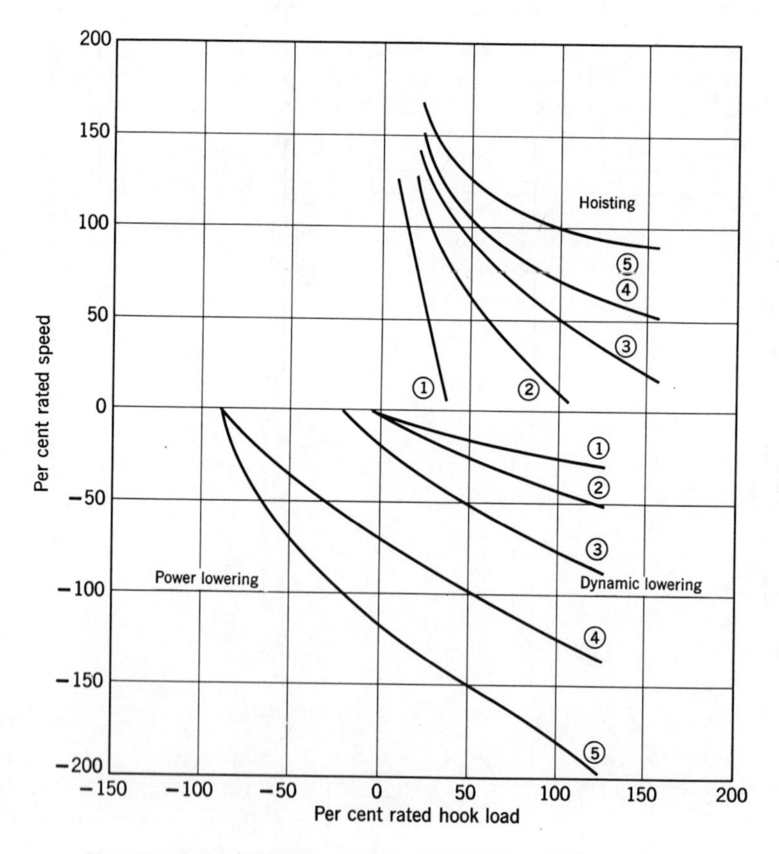

Fig. 13.3 Speed-Torque Curves for DC Series Motor with Dynamic-Lowering Hoist Control

and field are in series, the motor starts and operates as a series motor, exerting approximately 33% torque on the first point. The motor current also flows through the series brake coil, releasing the brake.

On the second point hoisting, the contactor *LT* is energized, short-circuiting resistor step *R*2–*R*3 and increasing the current and torque to approximately 100% in case the motor has not started on the first point. The first point is a slack cable point and the second point is the slowest full-load hoisting speed. Since the current obtained when *LT* is closed is not excessive, no time delay is introduced in the closing of this contactor.

On the third point of the master, contactor 1*A* is closed, short-circuiting resistor step *R*3–*R*4 which gives an increased hoisting speed. On the fourth point the contactor 2*A* is closed, and on the fifth point the contactor 3*A*. Each of these points gives additional hoisting torque and speed. On the last point the resistor is entirely short-circuited and the motor is connected directly to the line.

Accelerating Time. If the operator moves the master directly to the last point, the contactor *LT* will close at once since there is no time delay in connection with it. The contactors 1*A*, 2*A*, and 3*A* will close in a definite time under control of the inductive time-limit acceleration. When the controller leaves the factory, this total time is adjusted to be approximately 2½ sec.

In order to obtain time delay on the resistor contactors 1*A*, 2*A*, and 3*A*, it is necessary to energize their holdout coils before the closing coils are energized. Consequently, in hoisting, contactor *M* is closed first and then the contactor *H*. This applies power to the motor and causes current to flow through the starting resistor, which in turn energizes the holdout coils of the accelerators. The closing coils are connected behind interlocks on the contactor *H*, so that they are not energized until the contactor is closed and after the holdout coils have been energized.

Lowering. In the lowering direction the motor is connected to operate somewhat as a shunt motor; that is, the field is in parallel with the armature. On the first point the contactors 1*L*, 2*L*, *DB*, and *M* and all of the accelerating contactors are energized. Current flows from line *L*1 through the resistor *R*10–*R*12 to the junction of the armature and series field. The brake will release but the starting torque obtained is low, and if there is no load on the hook the motor may not have sufficient torque on this point to start the empty hook down. On the second point, contactor 3*A* is deenergized and opens, inserting resistance step *R*6–*R*7 in series with the armature. This does not increase the starting torque, but it does increase the lowering speed with a full load. On the third point, contactor *2A* is deenergized and opens, inserting resistance step *R4–R5* in series with the field. This weakens the field and increases the speed on all loads. The starting torque on the first three points is low. On the fourth point contactor *LA* is energized, cutting out resistor step *R10–R11*, thus increasing the starting torque. Contactor *1A* is deenergized and opens, inserting resistance step *R3–R4* to weaken the field further and increase speed at all loads. On the fifth point, contactor *3A* is energized, cutting out resistance step *R6–R7* and the interlock on *3A* opens to deenergize the coil on contactor *LT*, opening *LT* and inserting resistance step *R2–R3* in series with the field, thus again increasing speed at all loads.

Decelerating Time. If the master has been moved over to the last point or next to the last point in the lowering direction and is then moved back towards the OFF position to slow down, time for decelerating is obtained on the resistor contactors.

Since the slowest lowering speed is obtained with the accelerating contactors closed, it is desirable that they close immediately without any time delay when the master is moved to the first point lowering. This is accomplished by energizing their closing coils on this point before power has been applied to the motor. It will be noted that the closing coils are energized through an interlock on the *1L* contactor. It is necessary that *2L* be closed before power is applied to the motor and before the resistor contactor hold coils are energized. Consequently, the closing coils are energized first, and the resistor contactors closed immediately on the first point lowering. Figure 13.4 is a photograph of a size 5 dynamic lowering hoist controller.

Plugging. The holdout coil for the *LT* contactor is not used as an Ltl coil. Instead it is wired in series with a resistor unit 1RES and thus provides a voltage responsive holdout for the *LT* contactor. The coil is effective to hold out *LT* contactor only during plugging from lower to hoist.

Power-Limit Stop. It is considered good practice to use a limit switch to stop the hoist motor in case of overtravel in the hoisting direction, and to use a switch which will open the motor circuit directly (Fig. 13.5), rather than a pilot switch which would open the contactor coils. The contacts marked *Q* in Fig. 13.1 represent such a switch.

The switching mechanism is a double-pole, double-throw, cam-operated, quick-make, and quick-break device. It is reset to the normal

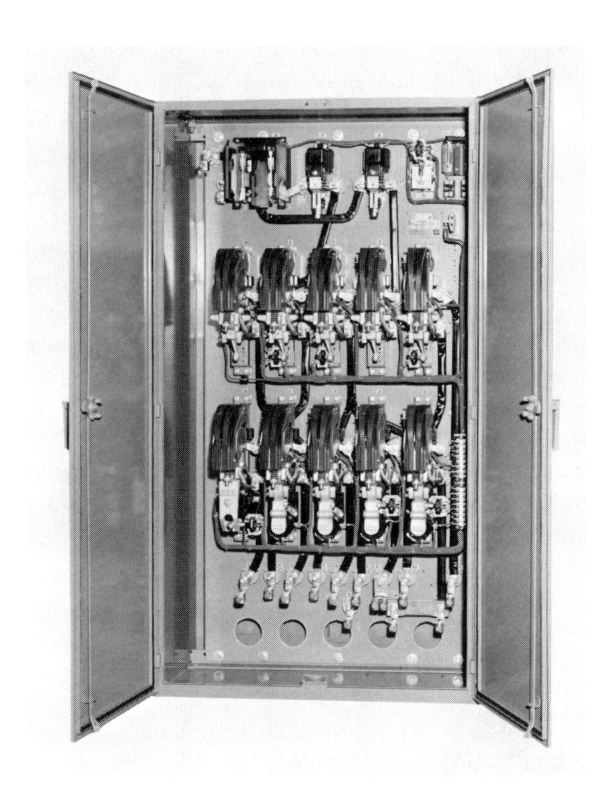

FIG. 13.4 Dynamic-Lowering Hoist Control

FIG. 13.5 Power Limit Stop (with cover removed)

position by a weight. The crane hoist cable runs through a slot in the weight, and when the crane hook is raised too high the hook block will strike the weight and lift it, tripping the switch.

As long as the hoist is in a safe position, contacts $Q1$-$Q6$ and $Q4$-$Q5$ are closed and contacts $Q2$-$Q6$ and $Q3$-$Q5$ are open. When the hoist overtravels and the switch is tripped, contacts $Q1$-$Q6$ and $Q4$-$Q5$ open to cut off power. At the same time contacts $Q2$-$Q6$ and $Q3$-$Q5$ close to set up a dynamic braking circuit and stop the motor quickly. When the limit switch is tripped, the motor cannot be energized in the hoisting direction but can still be energized in the lowering direction, which permits moving out of the danger zone. The limit switch will reset when the hoist has reached a safe position.

Hoisting. Operating of power limit stop in the hoisting direction disconnects line power from the motor and reconnects the motor as a self-excited generator to provide dynamic braking. (The limit stop contacts are marked $Q1$ thru $Q6$ in Fig. 13.1.) The total stopping torque obtained is motor dynamic braking *plus the electric brake torque.* Motor torque varies with speed, and is shown in Fig. 13.6.

Plugging. If the operator "plugs" the master switch to a lowering position just as the limit stop trips, the connection set up is a motor shunt with the motor driving down. Note that the *brake is held released.* If the master switch is in the fifth point, the motor torque obtained is approximately the same as dynamic braking torque. (See Fig. 13.6.)

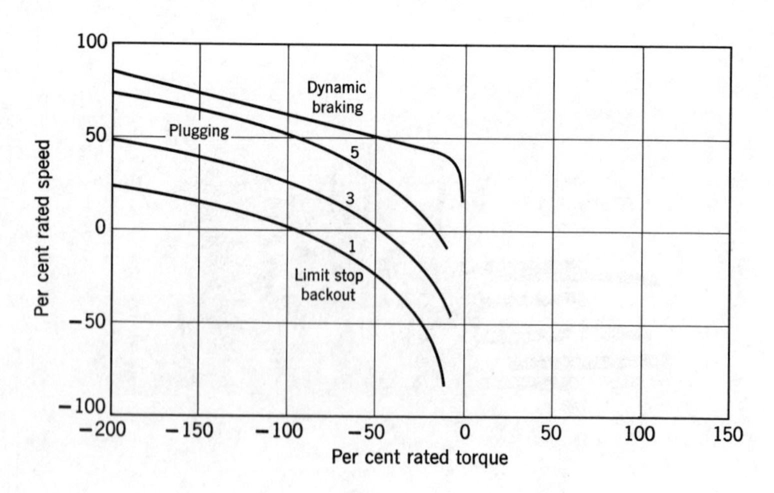

Fig. 13.6 Stopping Speed-Torque Curves

Note also that the torque falls off rapidly with speed, approaching 20%. Under these conditions, the net retarding torque is reduced by the amount of the brake torque, and is quite low as the motor slows down. On most hoists, this does not create a problem, but on a low-headroom crane where the upper sheave is close to the limit-stop trip point, it could result in a two-blocked hoist.

Backing Out. When lowering out of a tripped limit stop, the motor connection is the same as that described in plugging, above. The speed torque curves obtained are shown in Fig. 13.6. *No retarding torque* is obtained, but the load is *driven down.* On most cranes there is no danger of runaway conditions, because the limit stop resets in a short distance, before acceleration can occur. However, if the limit stop fails to reset for any reason, such as a fouled cable, a free-fall situation would develop quite rapidly, since the only retarding force is the friction of the drive.

Another potentially dangerous situation occurs on an extremely low-speed, fully loaded hook. The time required to back out of the limit stop is much longer than on a high-speed hook, allowing the motor to accelerate to a much higher speed.

When the limit stop does reset, if the operator has the master switch on a low-speed point, an extremely high, sudden torque is applied to slow down the hoist machine. If not dangerous, this is certainly undesirable abuse to equipment.

Limit-Stop Protective Circuits (Recommended Hoist Control Options)

Antiplugging. Antiplugging uses a lockout coil on the $1L$ contactor to prevent plugging from hoist to lower. This necessitates changing the $1L$ contactor from a shunt type to an Ltl type as described above. This contactor would have two holdout coils, connected as shown in Fig. 13.7. The coil designated *HC2L-A* would prevent plugging immediately prior to entry into the limit stop, and the coil marked *HC2L-B* prevents plugging after the limit stop is tripped.

Armature Shunt Backout. Armature shunt backout provides slow-speed lowering during limit-stop backout. Armature shunt backout requires the addition of a fifth collector rail on the crane and an armature shunt resistor and contactor as shown in Fig. 13.8. To operate the armature shunt contactor, and to keep the control on the slow-speed point, latch relay *LR* is used.

The close coil of the relay is connected across the *AS* power contacts. This permits the close coil to monitor the limit-stop contact $Q1$–$Q6$,

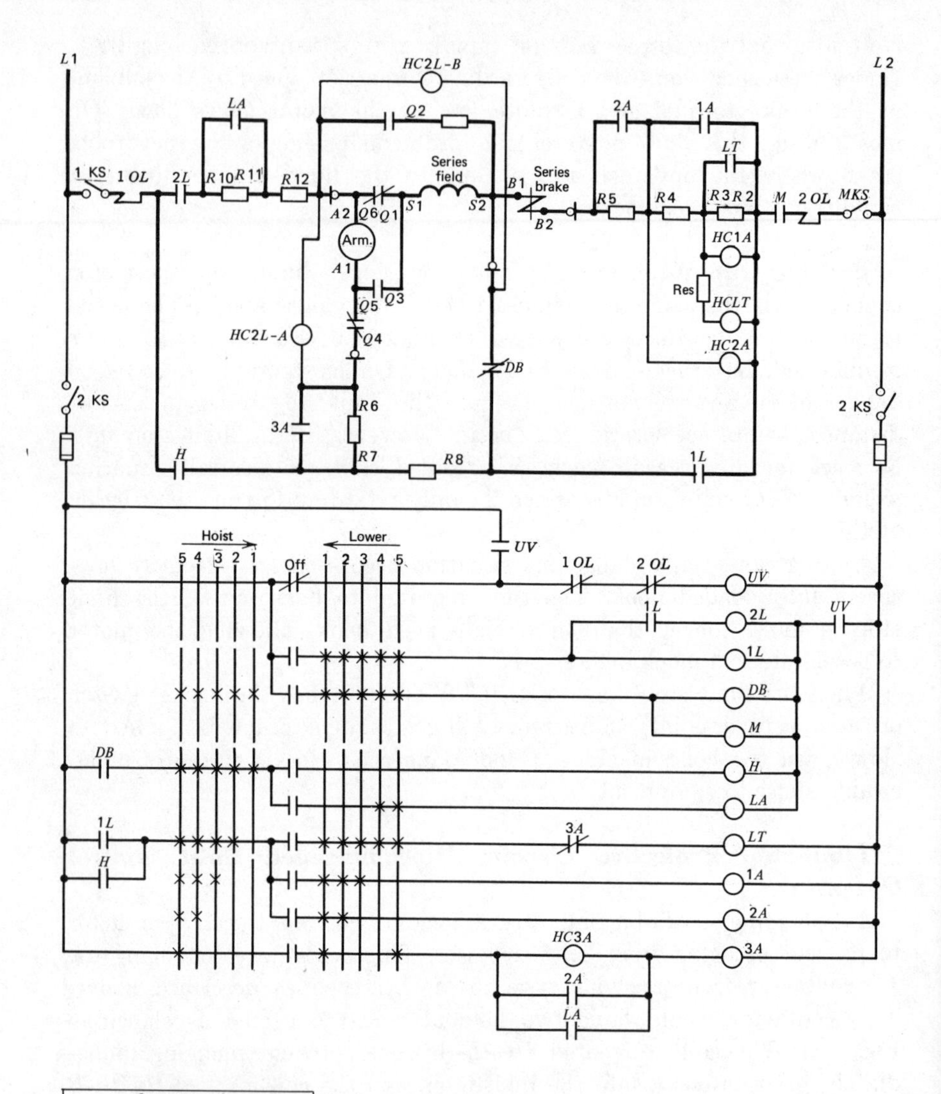

<table>
<tr><th colspan="12">Contactor sequence</th></tr>
<tr><th rowspan="2">Cont.</th><th colspan="5">Hoist</th><th rowspan="2">Off</th><th colspan="5">Lower</th></tr>
<tr><th>5</th><th></th><th>3</th><th>2</th><th>1</th><th>1</th><th>2</th><th>3</th><th>4</th><th>5</th></tr>
<tr><td>M</td><td>■</td><td>■</td><td>■</td><td>■</td><td>■</td><td></td><td></td><td></td><td></td><td></td><td></td></tr>
<tr><td>H</td><td>■</td><td>■</td><td>■</td><td>■</td><td>■</td><td></td><td></td><td></td><td></td><td></td><td></td></tr>
<tr><td>1L</td><td></td><td></td><td></td><td></td><td></td><td></td><td>■</td><td>■</td><td>■</td><td>■</td><td>■</td></tr>
<tr><td>2L</td><td></td><td></td><td></td><td></td><td></td><td></td><td>■</td><td>■</td><td>■</td><td>■</td><td>■</td></tr>
<tr><td>LT</td><td>■</td><td>■</td><td>■</td><td>■</td><td></td><td></td><td>■</td><td>■</td><td>■</td><td>■</td><td></td></tr>
<tr><td>1A</td><td>■</td><td>■</td><td>■</td><td></td><td></td><td></td><td>■</td><td>■</td><td>■</td><td></td><td></td></tr>
<tr><td>2A</td><td>■</td><td>■</td><td></td><td></td><td></td><td></td><td>■</td><td>■</td><td></td><td></td><td></td></tr>
<tr><td>3A</td><td>■</td><td></td><td></td><td></td><td></td><td></td><td>■</td><td></td><td></td><td></td><td>■</td></tr>
<tr><td>LA</td><td></td><td></td><td></td><td></td><td></td><td></td><td></td><td></td><td></td><td>■</td><td>■</td></tr>
<tr><td>DB</td><td></td><td></td><td></td><td></td><td></td><td>■</td><td></td><td></td><td></td><td></td><td></td></tr>
</table>

Resistor taper E/I		
Step	Ohms	Total
R2–R3	1.15	
R3–R4	0.35	
R4–R5	0.25	1.75
R6–R7	0.25	
R7–R8	0.25	
R10–R11	0.60	
R11–R12	0.20	0.80

Fig. 13.7 Hoist Control with Antiplugging Relay

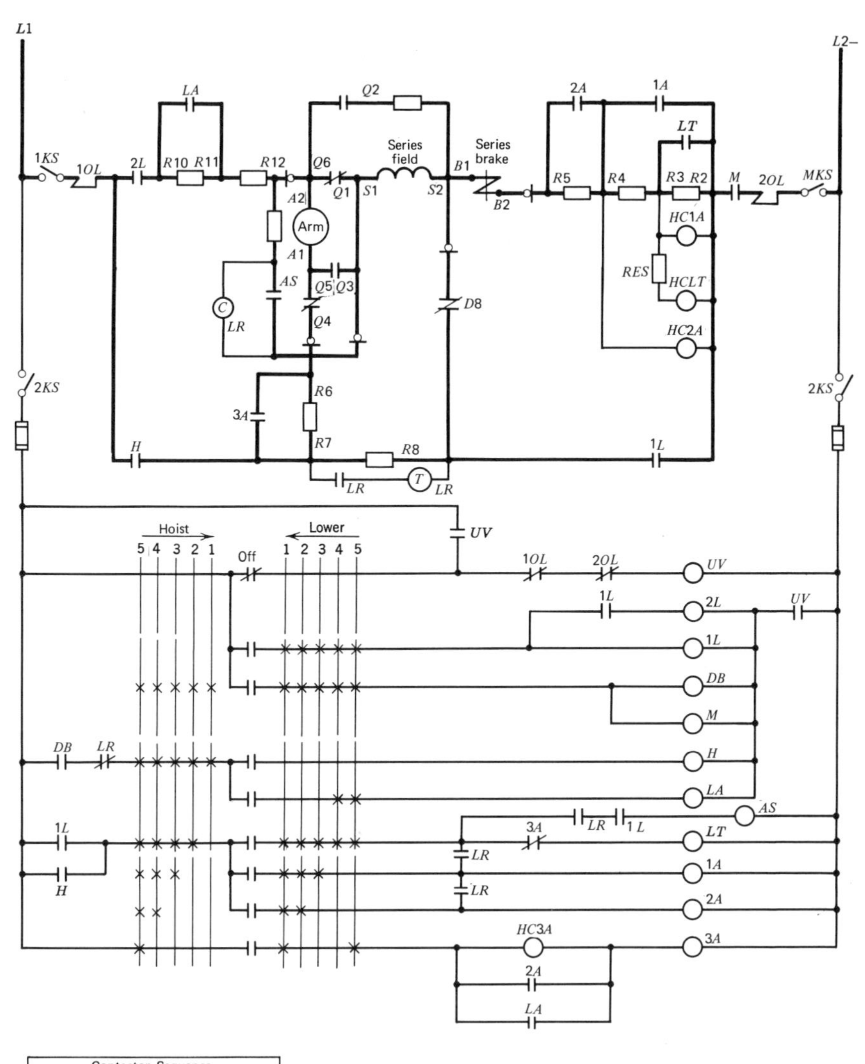

Cont.	Hoist 5	4	3	2	1	Off	Lower 1	2	3	4	5
M	■	■	■	■	■						
H	■	■	■	■	■						
1L							■	■	■	■	■
2L							■	■	■	■	■
LT	■	■	■	■			■	■	■	■	
1A	■	■	■				■	■	■		
2A	■	■					■	■			
3A	■						■				■
LA										■	■
DB						■					

Contactor Sequence

Resistor Taper Step	Ohms	E/I Total
R2–R3	1.15	
R3–R4	0.35	
R4–R5	0.25	1.75
R6–R7	0.25	
R7–R8	0.25	
R10–R11	0.60	
R11–R12	0.20	0.80

Fig. 13.8 Dynamic-Lowering Control with Armature Shunt and Latching Relay

which is a closed contact when the limit stop is not tripped. When the limit stop trips, the close coil sees motor armature voltage, which will usually operate the latched relay *LR* and close its normally open contacts and open its normally closed contacts. If the armature voltage is not high enough to trip the relay while the motor is hoisting, the voltage will be high enough to close the relay as soon as the lowering connection is made.

In the closed position the latched relay will hold the accelerators in the first point lowering condition and set up the armature shunt control circuit.

The hoist can now be lowered out at slow speed as indicated on the speed-torque curve in Fig. 13.9. The operator is limited to this one slow speed, regardless of master switch position. When the limit stop resets, a current will flow through the resistor *R*7–*R*8. The resulting voltage drop across *R*7–*R*8 will energize the trip coil on the latch relay and flip *LR* to its TRIP position, allowing the control to function normally. As can be seen on the speed-torque curves, a practically "bumpless" transfer will occur.

Undervoltage Relay. The diagram also shows the connections for the low-voltage protection relay which is supplied as standard on all types of controllers. Relay *UV* is energized in the OFF position of the

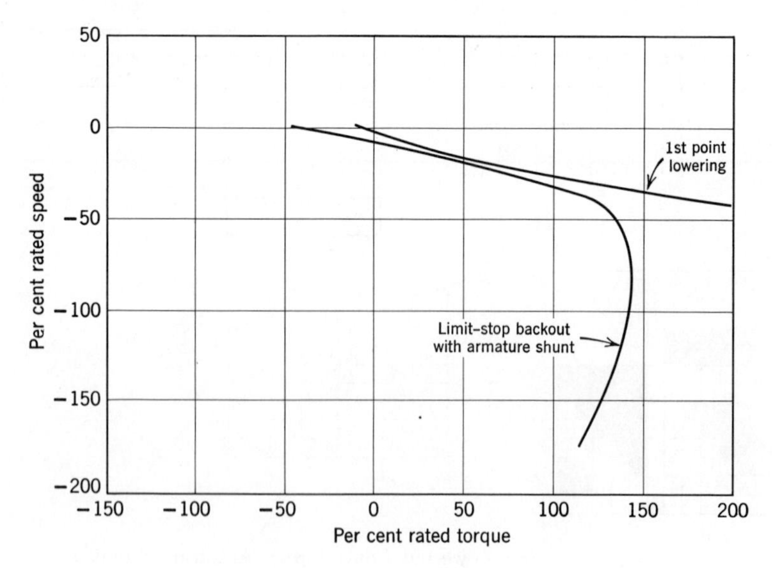

FIG. 13.9 Speed-Torque Curve with Armature Shunt

Hoisting

Arm

Res

0.08Ω 0.03Ω 0.01Ω

Master Switch Position	Motor ohms	Resistor ohms	Total ohms	Stalled Currents
1st	.12	2.00	2.12	.472
2nd	.12	.85	.97	1.03
3rd	.12	.50	.62	1.61
4th	.12	.25	.37	2.70
5th	.12	0.0	.12	

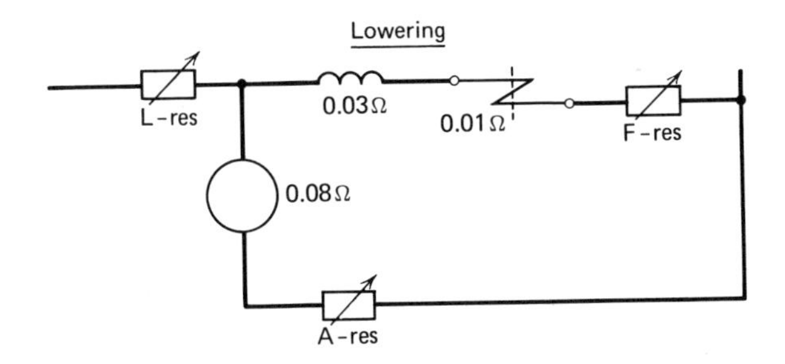

Master Switch Position	*L*-Res	Total Branch *A*-Res	Total Branch *F*-Res	Stall Current through Brake	
1st initial	.80	.58	.04	1.20	3*A* open
1st final	.80	.33	.04	1.15	3*A* closed
2nd	.80	.58	.04	1.20	
3rd	.80	.58	.29	.71	
4th	.20	.58	.64	.93	
5th initial	.20	.58	.64	.93	3*A* open
5th intermediate	.20	.33	.64	.81	3*A* closed
5th final	.20	.33	1.79	.42	*LT* open

Step	Resistor Taper %*E*/*I*	*CL*162%*I*
*R*2–*R*3	115	40
*R*3–*R*4	35	49
*R*4–*R*5	25	58
*R*6–*R*7	25	55
*R*7–*R*8	25	55
*R*10–*R*11	60	48
*R*11–*R*12	20	64

Fig. 13.10 Summary of Total Circuit Resistance and Currents

master. Once closed, the relay provides a feed for itself through one of its own contacts. The other contact provides a circuit for the coils of line contactors *M*, *1L*, *2L*, *DB*, *H*, and *LA*. When the master is moved to any running position, the relay coil is energized only through its own contact. If the relay opens, on account of voltage failure or the tripping of an overload relay, all contactors are opened and the equipment stopped. To restart, it is necessary to return the master to the OFF position and energize the relay again.

Calculation of Horsepower. The horsepower required to operate a crane hoist may be calculated as outlined below, and it is advisable to check the actual requirement against the rating of the motor actually used. If the hoist is overmotored, which is not uncommon practice, it may be necessary to increase the ohmic value of the resistor steps which determine the full-load lowering speed, as a design based on full-rated horsepower may not permit fast enough lowering speeds.

The following information is generally obtainable from the crane builder.

W = the rated load in pounds
S = calculated full-load hoisting speed in feet per minute
G = the gear reduction between the motor and the winding drum
D = the pitch diameter of the winding drum in feet
R = the rope reduction between drum and hook
Eff = the estimated overall efficiency of the hoist
The make and type of hoist motor
The ½-hr rating of the motor
The rpm at rated load

The horsepower required for hoisting can be calculated from the equation

$$\text{Horsepower} = \frac{W \times S}{33{,}000 \times \text{Efficiency}}$$

It often happens that S is not definitely known, and then the horsepower is calculated from the torque required. The torque in pound-feet at the motor shaft is

$$T = \frac{W \times D}{2 \times G \times R \times \text{Efficiency}}$$

When the torque is known, the motor speed can be determined from the torque-speed curve. The horsepower can usually be obtained from the motor curve or can be calculated from the motor speed and torque.

$$\text{Horsepower} = \frac{T \times RPM \times 2\pi}{33{,}000}$$

Resistor Design. The resistor must be designed to meet both hoisting and lowering conditions. Step $R2$–$R3$ is designed to give low torque for slack cable take-up. Steps $R3$–$R4$, $R4$–$R5$, and $R6$–$R7$, in series, limit the accelerating current in hoisting. Considered with the motor resistance, they should limit the accelerating current peaks to approximately 150% of full-load motor current.

Resistance $R10$–$R11$–$R12$ determines the starting, or kickoff, current in the lowering direction. For satisfactory operation this current should be approximately 125% of full load, so that the resistance step should be approximately $0.8E/I$.

The resistance of the dynamic step $R7$–$R8$ is $0.25E/I$. A number of conditions affect the value of this step. In lowering, the step is in series with the motor armature; and with a light load, the ratio of armature current to field current is dependent on the value of the resistance. An equal distribution of the current will give the maximum torque. It is also important that sufficient current flow through the field circuit to insure that the series brake will operate. About 40% of full-load current is required to lift the brake. In the last position lowering, steps $R6$–$R7$ and $R7$–$R8$ are in series with the armature, and if the operator moves the master lever so fast that the brake does not lift until the last point is reached, these resistances determine the inrush. Step $R6$–$R7$ is cut out as the motor accelerates, and $R7$–$R8$ should be high enough to limit the current to a safe value.

When the master is moved to the OFF position, these three steps establish a loop around the armature and field, serving to limit the initial braking current to a safe value.

In the last position, lowering step $R7$–$R8$ is in series with the armature, across the line, and the countervoltage of the armature is dependent on the voltage drop across the resistance. With an empty hook the voltage drop will reduce the speed of the motor, but with an overhauling load it will add to the armature voltage and increase the motor speed. It is desirable to have the empty hook speed relatively high; therefore, from this standpoint, the resistance step should be as low as possible. The value used is a compromise which best meets all the conditions desired.

Calculation of Speed-Torque Curves. The designer of a dynamic-lowering controller is interested in the speeds which are obtained in each point of the controller, in the lowering direction, and under varying load conditions. Sometimes these speeds are definitely specified by the purchaser or fixed by the requirements of the installation; in any event they must be within the limits of safety and good practice. A speed-torque curve, plotted for each point of the controller, will give the desired information. A method of calculating and plotting such curves follows.

All dynamic-lowering circuits when connected for lowering take the form shown in Fig. 13.2, in which the armature circuit and the series field circuit are connected in parallel. Either the armature or the field or both may have a resistance in series. The resistance in the line is generally in circuit on the slow speeds and cut out on high speeds. It will be understood that this circuit is typical. In an actual controller, any of these resistances may consist of two or more steps, but the circuit obtained on any given point in the lowering direction can always be reduced to a simple circuit similar to that shown. With this arrangement, and with a light load, the motor will drive the load down, exerting a torque in the lowering direction. With an overhauling load the motor will be driven as a generator and will exert a retarding torque. The countervoltage of the motor will depend on the speed and the field strength. The torque will depend on the field strength and the current in the armature. The circuit may be solved by Kirchhoff's laws. as follows:

$$\text{Armature resistance} = 0.08E/I \text{ (assumed)}$$

$$\text{Field resistance} = 0.03E/I \text{ (assumed)}$$

$$\text{Resistance in line circuit} = 0.80E/I \text{ (assumed)}$$

$$\text{Resistance in field circuit} = 0.26E/I \text{ (assumed)}$$

$$\text{Resistance in armature circuit} = 0.57E/I \text{ (assumed)}$$

E = line voltage = 1.0
E_a = countervoltage of armature as compared to line voltage
I_a = current in armature as compared to full-load current
I_f = current in field as compared to full-load current
I_l = current in the line as compared to full-load current
I_n = full-load current = 1.0
S = speed as compared to rated or full-load speed
T = torque as compared to rated or full-load torque
F = field strength or flux as compared to full field strength

Then, R_a = total resistance in the armature circuit = $0.65E/I$
R_f = total resistance in the field circuit = $0.29E/I$
R_l = total resistance in the line circuit = $0.80E/I$

The equation for speed is

$$S = \frac{E - I_a R}{F(E - I_m R_m)} = \frac{E_a}{F \times 0.89}$$

The equation for torque is

$$T = F \times I_a$$

By Kirchhoff's laws,

$$I_l = I_a + I_f \tag{13.1}$$

$$I_fR_f + I_lR_l = E - 1.0 \tag{13.2}$$

$$I_fR_f - I_aR_a = E_a \tag{13.3}$$

$$S = \frac{E_a}{0.89F} \tag{13.4}$$

$$T = F \times I_a \tag{13.5}$$

To calculate points on the speed-torque curve, assume values for I_f. To approximate the range of such assumed values, two points are readily calculated. These are zero speed, and zero torque.

At zero speed, $E_a = 0$.

From equation 13.1,

$$I_l = \frac{E}{R_l + \dfrac{R_a \times R_f}{R_a + R_f}} = \frac{1.0}{0.80 + 0.20} = 1.0$$

From equation 13.2,

$$I_f = \frac{E - I_lR_l}{R_f} = \frac{1.0 - 0.80}{0.29} = 0.69$$

At zero torque, $I_a = 0$. From equation 13.1,

$$I_f = I_l = \frac{E}{R_l + R_f} = \frac{1.0}{1.09} = 0.917$$

Now, having some idea of what the values of I_f will be, assume other values and prepare a table as shown below. The field strength is read from the curve of Fig. 11.3, and the other values are calculated from the equations indicated in the table headings.

I_f	I_fR_f	I_lR_l(13.1)	I_l	I_a(13.2)	I_aR_a	E_a(13.3)	F	S(13.4)	T(13.5)
0.69	0.20	0.80	1.00	+0.31	+0.20	0.00	0.883	0.00	+0.274
0.917	0.266	0.734	0.917	0.00	0.00	0.266	0.973	0.307	0.00
1.20	0.348	0.652	0.815	−0.385	−0.25	0.598	1.052	0.638	−0.405
1.50	0.435	0.565	0.707	−0.793	−0.515	0.950	1.11	0.962	−0.88

The speed-torque curve may now be plotted. Figure 13.11 shows the curve obtained in this example. If all of the line resistance is cut out,

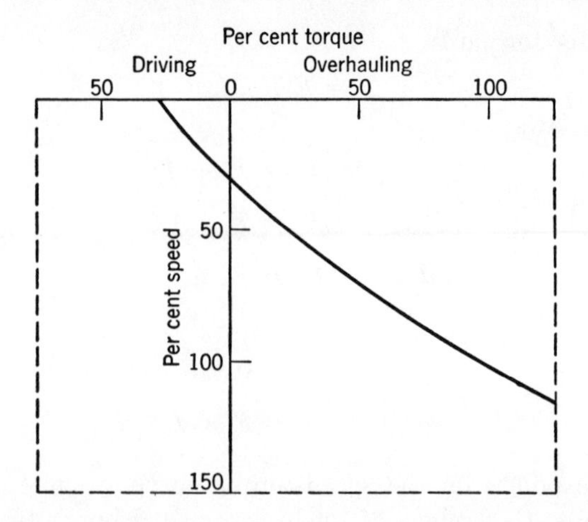

FIG. 13.11 Speed-Torque Curve of a Dynamic-Lowering Hoist

the field strength will be constant, and the curve will become a straight line. Decreasing the resistance in the armature circuit will make the curve more flat; that is, the speed will not vary so much with changes in load. Decreasing the resistance in the field circuit will decrease the speed in all loads. When estimating the resistance of the motor armature and series field, it may be assumed that about 70% of the motor resistance is in the armature, and 30% in the field. If the circuit includes a series brake, it will be in the field circuit, and the brake resistance should be included in the value of the resistance of that circuit. Since the resistance of a series brake coil is very low, it will not make much difference in the calculations. An average value is 0.01 E/I.

Calculation of Speed-Torque Curves—Alternate Method. It is possible to calculate the data for speed-torque curves directly in rpm and pound-feet. It is necessary to know the resistance of the motor armature and series field. If these values are not known, the motor resistance can be calculated from the motor efficiency curve, and it may be assumed that 70% of the motor resistance is in the armature and 30% in the field. A speed-torque curve of the motor is also necessary.

From the equations for the speed and torque of a series motor, and with constants added so that the equations are correct for actual values (not percentages),

$$KF = \frac{T}{I_a} \tag{13.6}$$

$$K_1F = \frac{E - I_aR_m}{S(E - I_nR_m)} \tag{13.7}$$

where F = field strength in any desired units
T = torque in pound-feet
I_a = armature current in amperes
E = line voltage in volts
R_m = motor resistance in ohms
S = speed in rpm
I_f = motor field current in amperes
I_m = motor current in amperes
I_n = motor full-load current in amperes
K, K_1, K_2 = constants

Since $E - I_nR_m$ is also a constant, and since $E - I_aR_m$ is the countervoltage, equation 13.7 may be written,

$$K_2F = \frac{E_a}{S} \tag{13.8}$$

It will be evident that, with a series motor operating normally,

$$I_f = I_a$$

The first step in the calculation is to assume a set of values for I_f (which will also be those of I_m and I_a), and from the motor curve read the corresponding values of T and S.

With T and I_a known, KF can be calculated from equation 13.6. Since E, I_a, and R_m are known, the countervoltage E_a can be determined. Then, with S known, K_2F can be calculated. A table listing these values would be headed as follows:

I_f	T	KF	I_aR_m	E_a	S	K_2F

It is now possible to plot curves for T/I_a and E_a/S, against I_f, and to use these curves for the calculation of the dynamic-lowering circuit.

The lowering circuit consists of three branches. One branch consists of the motor armature and a resistor in series with it. The current in this branch is I_a, and its total resistance (armature and external resistor) is R_a. A second branch consists of the motor field and a resistor in series with it. The current in this branch is I_f, and the total resistance (field and external resistor) is R_f. These two branches are connected in parallel. In series with the combination is the third branch which is a resistor R_l. The current in this branch is I_l.

The next step is to set up a table with headings as follows:

I_f	I_fR_f	I_lR_l	I_l	I_a	I_aR_a	E_a	$\frac{E_a}{S}$	S	$\frac{T}{I_a}$	T

Then assume values of I_f, and calculate the other values.

$$I_f = \text{assumed}$$

$$R_f = \text{known}$$

$$R_l = \text{known}$$

The voltage across $R_l = E - I_f R_f$. Call this E_1.

$$I_l = \frac{E_1}{R_l}$$

$$I_a = I_l - I_f$$

$$R_a = \text{known}$$

$$E_a = I_f R_f - I_a R_a$$

$$\frac{E_a}{S} = \text{read from curve}$$

$$S = \frac{E_a}{E_a/S}$$

$$\frac{T}{I_a} = \text{read from curve}$$

$$T = I_a \times \frac{T}{I_a}$$

The speed-torque curve may now be plotted from the values of S and T.

The method described works equally well for series and for series-shunt motors. It will be evident that if the same values of I_f are used in preparing the preliminary table and the final table, it is not necessary to plot the curves of I_f vs. T/I_a and I_f vs. E_a/S.

Accuracy of the Method. There will be slight errors in the results of these calculations because the calculations are based on the motor curve which is correct for normal series motor operation and the same current in armature and field. Under the conditions of dynamic lowering, the motor and field currents are not the same, and the motor is sometimes acting as a generator. Cross-magnetizing armature reaction will affect the field strength, tending toward demagnetization. Currents flowing in armature coils which are being commutated also affect the field strength. The net effect of these factors is to decrease the field strength when

TABLE 13.1

Effect of Varying Resistance Steps

Hoisting

Step	Increasing Ohmic Value	Decreasing Ohmic Value
*R*2–*R*3 slack cable step	Speed on the first point will be reduced. Current taken from the line on the first point will be reduced. If the ohms are made too high the brake will be slow in releasing.	Speed, torque and current from the line will all be increased.
*R*3–*R*4 *R*4–*R*5 *R*6–*R*7	These steps are designed for equal current peaks of approximately 150 % of normal full load current. Adjustment for hoisting should not be necessary. Increasing the ohms of any step will decrease the speed and torque on that point, and will also decrease the current peak at that point. The peak occurring when the step is cut out will be increased. If these steps are out of adjustment enough to cause very unequal peaks during acceleration, the operation will be jerky, and there will be strains on the ropes and gears.	Decreasing the ohms on any point will increase the current peak on that point, and decrease the peak occurring when the step is cut out of circuit.
*R*7–*R*8 dynamic	Not used in hoisting.	
*R*10–*R*12 kickoff	Not used in hoisting.	

Lowering

Step	Increasing Ohmic Value	Decreasing Ohmic Value
*R*2–*R*3	Increasing the ohmic value will increase the speed on the last point.	Speed on the last point will be decreased.
*R*3–*R*4	Speed on the fourth point will be increased.	Speed on the fourth point will be decreased.
*R*4–*R*5	Speed on the third point will be increased.	Speed on the third point will be decreased.
*R*6–*R*7	Speed on the second point with overhauling load will be increased. Starting torque on the second point will be decreased. In the OFF position dynamic braking will be less effective.	Speed on second point with overhauling load will be decreased. Starting torque on the second point will be increased. OFF position dynamic braking will be more effective.
*R*7–*R*8	Dynamic braking in OFF position will be decreased. The torque on starting will be decreased. The torque on all points lowering on power will be decreased. Light hook lowering speed will be decreased. Full-load lowering speed will be increased. If made too high, the proper ratio of full-load and light-load speeds will not be maintained, and the crane may not operate at its best efficiency.	Dynamic braking in OFF position will be increased. Torque on starting will be increased. Torque on all points lowering on power will be increased. Light hook lowering speed will be increased. If made too low, the brake may be slow in releasing. The accelerating inrush on the last point may be too high. Also sparking of the commutator may occur during dynamic braking in the OFF position.
*R*10–*R*12	Speed and torque will be decreased. Current taken from the line on starting will be decreased.	Speed and torque will be increased. Current taken from the line on starting will be increased.

the armature current is high. In general, if the field current is higher than 50% of normal, and if the armature current under this condition is not more than 100% of normal, the curves will be quite accurate. If the field current is higher than 50%, the armature current may be correspondingly higher without affecting the accuracy of the curves.

Adjustments. As an aid to the understanding of the controller described, the results which may be obtained by adjusting the various resistance steps are given in Table 13.1.

Proper Operating Functions. The following functions should be obtained when a controller is operating properly.

1. Automatic time-limit acceleration for hoisting.
2. Automatic time-limit deceleration for stopping from any lowering speed. This keeps down objectionable current peaks, particularly when checking from high speed.
3. Automatic time-limit acceleration for kickoff on last point lowering. This insures fast kickoff and prevents excessive peaks to the armature on kickoff.
4. Accurate speed control for both light and heavy loads either hoisting or lowering.
5. Slow speed on all loads either hoisting or lowering insures accurate spotting.
6. 100% of full-load current through brake when power is applied on the first and second points lowering insures quick releasing of the brake.
7. Sufficient current through brake on all lowering points to insure lifting of brake.
8. Loaded hook speed 190% of full-load hoisting speed. Light hook speed 125% of full load hoisting speed.

Problems

1. Using the following values, calculate and plot a speed-torque curve similar to Fig. 13.11, for a dynamic-lowering hoist controller:

Armature resistance	$= 0.08E/I$
Field resistance	$= 0.03E/I$
Resistance in line circuit	$= 0.80E/I$
Resistance in field circuit	$= 1.50E/I$
Resistance in armature circuit	$= 0.25E/I$

2. Calculate and plot a similar curve for the same circuit, but with the line-circuit resistance reduced to zero.

3. A bucket hoist has the following duty cycles:

Closing the bucket	6 sec at 40 hp
Hoisting	10 sec at 80 hp
Opening the bucket	3 sec at 30 hp
Lowering	10 sec at 45 hp
Resting	16 sec at 0 hp

If the rate of cooling of the motor at standstill is one-third of the rate when running, what is the rms horsepower required of the motor?

4. Assuming that it requires 150% of full-load current to supply the desired initial starting torque, how many ohms will be required in the accelerating resistor of a controller for the hoist of problem 3? The power is supplied at 230 V.

5. The bucket of a bucket hoist weighs 5000 lb, and the load in it weighs 10,000 lb. The hoisting speed is 150 ft/min, and the efficiency of the hoist is 80%. What is the horsepower for which the controller must be designed?

6. A hoist has the following characteristics:

Weight of bucket	2000 lb
Weight of load	6000 lb
Diameter of winding drum	2 ft
Efficiency of the hoist	80%
Gear reduction	5 to 1
Rope reduction	2 to 1

What is the torque on the motor shaft when hoisting full load?

7. What is the torque on the motor shaft when lowering the empty bucket?

8. A 50-hp 230-V 200-amp hoist motor has a controller with the circuits of Fig. 13.2. If the controller is in the OFF position, but with the brake released, and if the load is lowering at a speed which causes 100 amp to flow in the motor armature, what is the lowering speed, in per cent of full-load hoisting speed? The resistor values are:

	Per Cent of E/I
*R*6–*R*7	19
*R*7–*R*8	14
Armature	8
Field	3
Field curve	Fig. 11.3

9. What is the torque at the motor shaft, in per cent of full-rated torque?

10. A small crane is to be equipped with a controller like that of Fig. 13.1, except that it is to be simplified by the omission of 1*A*, 4*A*, the limit-stop contacts *Q*1–*Q*4 inclusive, and relay *CR*. The master is to have three speeds in each direction. Make an elementary diagram for the controller.

14

ALTERNATING-CURRENT CONTACTORS AND RELAYS

Alternating-Current Magnets. One difference between ac contactors and dc contactors is in the design of the operating magnet. With the dc contactor, any heating of the iron frame arises from copper loss in the coil. With the ac contactor, heating of the frame comes from iron loss in the frame itself. The resistance of the coil is relatively low, and the copper loss is small. In order to reduce the iron losses, ac-contactor magnets are made of thin laminations bolted together. The laminations must be bolted tightly together to prevent humming when the contactor is operated, but care must be exercised in locating the holding bolts, since they are solid and will be subject to heating. The bolts must be located at points of low flux density.

With a dc contactor, the current in the coil and consequently the flux available for closing the contactor are dependent upon the line voltage and the coil resistance. In the ac contactor, the current is dependent upon these factors and in addition upon the reluctance of the circuit, the frequency of supply, and the number of turns in the coil. Alternating-current contactors are quite sensitive to slight changes in voltage and frequency and are likely to overheat if the voltage is higher or the frequency lower than normal. On the other hand, if the voltage is lower than it should be or the frequency higher, the contactor may not have sufficient closing pull and will become noisy because it will start to open each time the voltage passes through zero.

There are many types of ac magnets in general use for contactors and relays. Two basic classifications can be drawn on whether the motion of the magnet armature is rotary or linear. The first has a pivot point about which the armature moves in an arc. The moving contacts may have a rotary motion similar to the armature or this rotation may be translated into a linear motion for the contacts.

Most ac contactors and relays are designed with bridging contacts which require a linear mode of travel. For this reason the driving magnets are often designed to have linear motion. The stationary portion of the magnet is generally either "C" shaped or "E" shaped. The moving portion can then be either "I" shaped or a matching "C" or "E". The "E" magnet has a coil about the center leg. The "C" magnet with linear motion usually requires a dual coil, one on each leg, in order to give balanced operation.

The pull curve of a magnet is a characteristic of the force it exerts plotted against the travel of the magnet from open to closed positions. In the fully open position the air gap is large, the coil current is large, and the pull is generally at a minimum value. In the closed or sealed position the coil current is minimum and the pull is at a maximum value. The shape of the pull curve between these extreme positions can be shaped by the magnet design. Generally, if the moving portion of the magnet extends into the coil, the pull curve will be flatter than if this portion of the magnet remains entirely outside of the coil.

Contactors and relays must usually meet a predetermined degree of shock resistance that is related to a spring or gravity force which holds the magnet in the open position. This establishes an initial force requirement for the magnet to operate the device. A second critical point is when the contacts just touch, since all contact pressure must be overcome by the magnet force at this point in order to pull through to the sealed position of the magnet. Finally there must be sufficient force to meet a required speed of response, but not so much that the resulting slam and noise reduce mechanical life or are otherwise objectionable. Therefore all these factors enter into a good magnet design.

Shading Coils. With alternating current the pull of the contactor closing coil reverses in direction at each cycle, and each time that this reversal occurs the flux passes through zero and the contactor tends to open. Some means must be provided to overcome this tendency, or else the contactor will be extremely noisy and probably will soon hammer itself to pieces as a result of the vibration.

The tendency to open with each reversal of flux is overcome by mounting a small auxiliary coil in a slot in the contactor face. This is called a shading coil. It may be a single loop of wire or strap, or it may be several turns of wire. At any rate, it is short-circuited and depends for its action upon the flux induced in it by the main flux. With reference to Fig. 14.1, 1 represents the moving armature of the contactor and 2 the stationary frame; 3 is the main operating coil of the contactor. The shading coil 7, embedded in the slot 4, is shown as a single-short-cir-

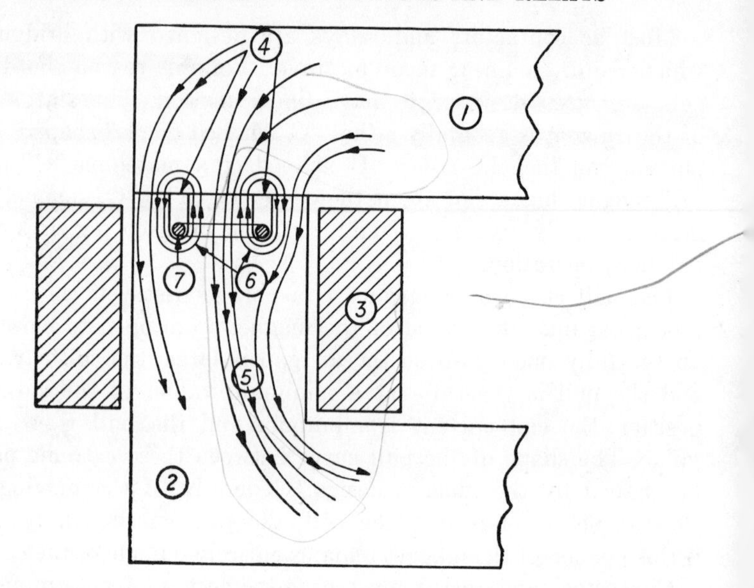

FIG. 14.1 Operation of a Shading Coil

cuited loop of wire. The main flux 5 passes partly through the shading coil and partly outside of it. This induces alternating current in the shading coil. The resistance and reactance of the shading coil itself are in such proportions that the induced current is out of phase with the main flux by approximately 120 degrees. Therefore, whenever the main flux is approaching or passing through the zero point there is an auxiliary flux from the shading coil which is holding the contactor tightly closed, and when the auxiliary flux is at the zero point the main flux has again built up to a safe value.

With well-designed shading coils, ac contactors can be made to operate very quietly. A broken shading coil will make its presence known—the contactor will immediately become extremely noisy. Such a condition should be remedied at once, as the contactor will be subject to overheating and will soon cease to operate properly.

If the surface of the armature and the magnet frame becomes dirty or rusty, the contactor will not seal properly and will become noisy. These surfaces are usually fitted very accurately by the manufacturer and are then greased to prevent rust until such time as the contactor is put into operation. The layer of grease or Vaseline should be removed when the contactor is about to be used, as the operation of the contactor will tend to keep the magnet surface clean.

Alternating-Current Coils. Direct-current contactor coils have a large number of turns and a high ohmic resistance. The current through them is limited by the resistance. In ac coils the current through the coil is limited by the impendance of the circuit, and the reactance has a greater effect than the resistance. Consequently, the resistance of an ac contactor coil is low, and the number of turns is relatively small. It is possible, therefore, to use heavier wire for these coils, which is fortunate for several reasons. With a relatively small number of turns there is a higher voltage drop per turn. Also, the ac coils are subject to continuous vibration caused by the reversal of the flux. A short-circuited turn in a dc coil does not do any particular damage, merely reducing the resistance of the coil a very slight amount. An ac coil, however, will be ruined by a short-circuited turn. The short-circuited turn acts as the secondary winding of a transformer, the primary of which is the winding of the rest of the coil. A heavy current is induced in the short-circuited turn, which will overheat and burn out.

For the above reasons, insulation is of greater importance in ac than in dc coils. It is customary to use a double-insulated wire and to impregnate the coils thoroughly with a binding compound.

Inrush Currents. With a dc magnet, the current in the coil is the same whether the contactor is opened or closed. With an ac magnet, however, the current in the coil is largely determined by the reactance of the circuit, which is lower when the contactor is open, because of the air gap in the magnetic circuit. Therefore, there will be a high inrush of current through the coil when the contactor is first connected to the supply line. The inrush may be five to twenty times as high as the current which will flow through the coil when the contactor has closed. This fact must be taken into account when ac contactors and relays are used, and care must be taken that the pilot device which handles the coil circuit has ample capacity to pass the inrush current. Since the current through the coil is automatically reduced as soon as the contactor has closed, it becomes unnecessary to insert protecting resistance as is done with some dc coils.

The fact that the inrush current is so much higher than the sealed current makes it undesirable to tie ac contactors together mechanically, since if anything happens to prevent one contactor from closing, the coil of the other will be burned out. Furthermore, when two ac contactors are tied together mechanically, there is a good possibility of their chattering and being noisy unless the mechanical link is very carefully designed and applied.

Remote Control. When controlling an ac contactor from any considerable distance, the inrush current to the coil must be taken into account, since this current will cause a relatively high voltage drop in the control line between the contactor and the pushbottom station. If the voltage drop is too high, the contactor coil will not have sufficient voltage to enable the contactor to close. In such an installation the maximum allowable resistance of the control circuit should be calculated, and the size of wire for the control circuit should be so selected that the resistance will be below that value.

The maximum allowable resistance of the control wire for a given installation may be calculated as follows.

It is first necessary to measure the ohmic resistance of the coil of the contactor and the inrush current which occurs when the contactor is open. This may be done by holding the contactor open by hand until the value of the current is read. With reference to Fig. 14.2, the line *OA* is laid out so that its length represents the resistance drop of the coil on inrush current; that is, it is a voltage equal to the ohmic resistance of the coil, multiplied by the inrush current. The line *OC* is drawn at a right angle to *OA*. From *A* the line *AB* is drawn to intersect the line *OC*. The length of *AB* should be such that it represents 90% of line voltage. This is on the assumption that the minimum voltage across the coil on inrush should be 90% of normal line voltage, in order to secure proper operation. Line *AB* then represents the voltage across the coil, and *OB* represents the reactance drop in the coil. From the point *B* the line *BD* is drawn to intersect the base line. The length of the line *BD* is selected to represent full line voltage, or in other words, this length is 1.111 times the length of *AB*. The line *AD* then represents the additional resistance drop which may be included in the line in order to obtain 90% of normal voltage across the coil on inrush. Dividing the value of *AD* by the value of the inrush current will then give the ohms allowable in the line, and from a wire table the proper size of wire can be selected.

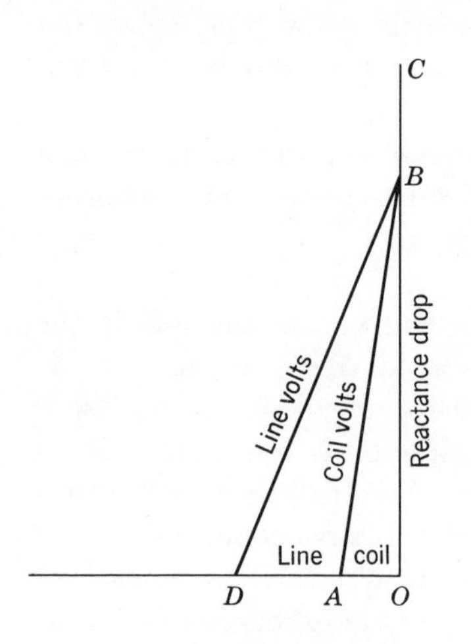

FIG. 14.2 Calculation of Line Resistance

If the line has already been constructed and is of too high a resistance, or if it is necessary to use a size of wire which will give a line of too high a resistance, the solution of the problem is to choose a small relay which will have a lower value of inrush current. The relay in turn will control the coil of the contactor which is to be operated.

Control-Voltage Transformer. There is a considerable hazard in using voltages higher than 230 V for control circuits. Although push-buttons and other pilot devices are usually designed with spacings suitable for 600 V, faults, breakages, careless wiring, and so on, may subject the operator to a shock which could be serious with higher voltages. Lower control voltages also reduce the risk of insulation breakdown and grounding in the control wiring and the pilot devices. So it is common practice to use a control-voltage transformer to provide 110 V, or some other suitable low voltage, for the control circuits.

Transformers designed for general-purpose use are usually not suitable for control work, because they are designed for relatively constant loads. When a contactor is energized, the initial current may be 15 or 20 times the current required after the contactor closes. The transformer must be able to supply the required high inrush current without dropping its voltage more than 5%. Since the transformer voltage-regulation curve will vary with the power factor of the load, the selection of the proper transformer involves an analysis of the entire control circuit.

To determine the size of transformer which is necessary for any given installation the maximum possible inrush current obtained at any one time should be calculated. The current and power factor of each contactor coil, both when first energized and when closed, are determined. Then the various combinations of closed contactors, and contactors just energized are compiled, and the total current and its power factor under the worst condition are calculated. With this information, and a set of curves for the available transformers, a suitable transformer can be selected, if one keeps in mind that the voltage must not drop below 95% of normal rated voltage, with rated controller voltage on the primary winding.

The transformer will also be required to carry the sealed current of the maximum number of contactors which are closed when the equipment is running; consequently it should have a kilovolt-ampere capacity equal to this continous current multiplied by the line voltage. For some installations this continuous capacity will be the determining factor, but for most installations it will be the inrush condition which determines the size of transformer required. The transformer must meet both conditions.

Consideration must also be given to the open-circuit secondary voltage, which should not be much higher than the secondary voltage at rated load. The reason for this requirement is that at times only one small relay or contactor of a controller may be energized, and if the voltage is too high a damaged coil may result.

Control Relays. Control relays are designed to be used as pilot circuit devices in the control circuits of other relays, contactors, solenoids, and so on. Design requirements include a relatively low operating power and the flexibility to add or change contacts in the field to meet a multitude of control circuit combinations. Small size and ease of wiring and inspection are also important considerations. Contact ratings are standardized by NEMA, as shown in Table 4.1, Chapter 4.

Control relays are available with a wide variety of physical and electrical properties. Some small devices are of the plug-in variety. The power to operate these smaller relays may be only 2 or 3 voltamperes. The larger relays provide higher electrical ratings on the contacts and require coil power in the range of 60 to 170 Va inrush and 20 to 60 Va sealed. Some of these relays have up to 12 poles which may be individually replaceable and convertible from normally open to normally closed.

Since a major requirement for reliable control circuitry has grown out of the machine tool industry, a typical control relay has been referred to as a machine tool relay. This relay has maximum voltage ratings of 300 V or 600 V and is typified by its very long operational life and flexibility of normally open and normally closed contact arrangements. Such a relay is shown in Fig. 14.3.

Contact-making elements which employ magnetic reeds in sealed tubes, commonly referred to as reed switches, are mentioned in Chapter 4 and one is shown in Fig. 4.1. A control relay which employs reed switches for its contacts is shown in Fig. 14.4. The reed contact elements are encapsulated in molded cases containing the terminals. These units are assembled into pockets provided in a base structure which contains the operating magnet and coil. Biasing magnets are provided in the normally closed elements. Extremely long electrical and mechanical life and environment-proof contacts are features of the reed relay. These relays are particularly applicable to the environments found in foundries, cement plants, textile factories, and the like.

Phase Failure Relays. Phase failure relays are used to prevent the starting of a three-phase ac motor if one of the phases has been reversed or to prevent running if one phase has been opened by the blowing of a fuse or by some other means. Depending upon the sensitivity of the relay, it may also cause a motor to be disconnected when a certain degree

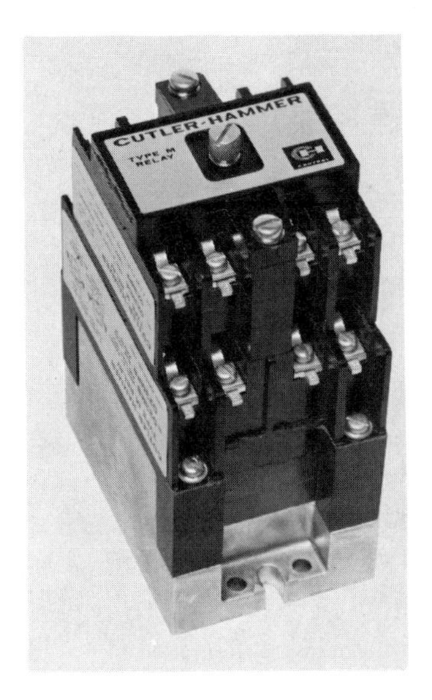

FIG. 14.3 8-Pole Control Relay

FIG. 14.4 Control Relay with Cutler Hammer "Powereed" Contact Elements

of voltage unbalance is present on the lines. Such voltage unbalance may cause improper motor operation and overloading. Accidental phase reversal will cause motor reversal, which could have serious consequences.

There are two basic types of phase failure relays. One works on the induction motor principle, and the other senses phase irregularity by electronic means. In the induction relay, a disk is acted upon by coils in two of the phases. Normally these coils produce currents and torque in the disk to cause it to rotate, overcoming a spring, and thereby causing the relay contacts to close. If one phase is reversed or opened, the contacts of the relay will open. Relays of this type have generally used series coils if they must be sensitive to a loss of one of the phases. This is because a three-phase motor will continue to run on single phase if it is not heavily loaded. When running in this manner, it will feed back the third phase on the opened line and would keep a shunt coil energized. On the other hand, there are some applications, such as elevators or hoists, where it is desirable to have a motor continue to run on single phase and only prevent an attempt to restart.

A newer type of phase failure relay uses solid-state devices in connection with a conventional relay. The relay is picked up when voltage conditions are proper. With a phase reversal, a loss of one phase, or

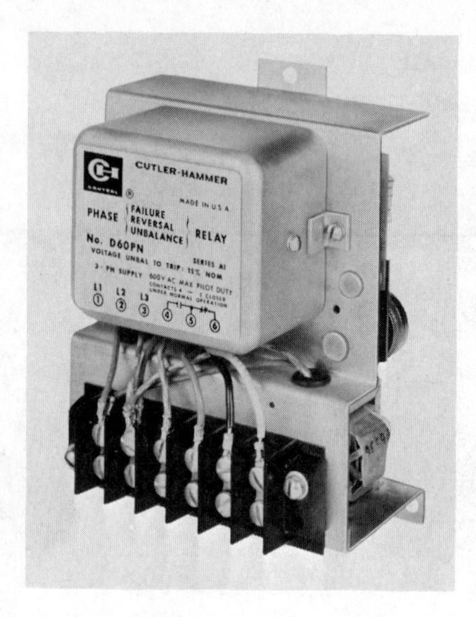

Fig. 14.5 Phase Protection Relay with Enclosure Removed

an unbalance between phases of 15% or greater, the relay will drop out. This relay is therefore sensitive enough to respond to any undesired voltage condition. Circuitry is designed to be fail-safe, and a two-second time delay is introduced to prevent nuisance tripping from voltage transients. Figure 14.5 shows this type of relay without enclosure.

Magnetic Overload Relays. Magnetic overload relays are not as generally employed as are thermal overload relays for ac motor protection; however, they are available for meeting certain special requirements. The advantage of the magnetic relay lies in being able to adjust both the ultimate trip and short-time trip values accurately. This provides field adjustment of the tripping characteristic curve, which is not possible on a given thermal overload relay.

The operating coil either carries the motor current directly, or carries a proportion of it in case a current transformer is used. The turns in this coil provide the rough calibration only. Fine calibration is determined by adjusting the position of a plunger and the orifice in an oil-filled dashpot.

This type of relay has a particular advantage for protecting certain hermetic refrigeration motors where fast tripping time is required in the case of a stalled motor. It can also be employed to allow extra long starting times for high-inertia loads.

Thermal Overload Relays. The thermal overload relay is a time-proven way of providing protection to a motor from running overloads or failure to start. An ideal overload relay would have a current-time curve such that it would trip on any given value of overload current just when the motor reaches its maximum design temperature rise. Motors with Class A insulation generally have maximum design temperature rises of 40°C or 50°C corresponding to service factors of 1.15 or 1.00 respectively. Motors with Class B or other insulation types may have considerably higher temperature rises. Furthermore, some motors will withstand stalled current for 30 sec or more, whereas others must be removed from the line in less than 20 sec when stalled. These different motor characteristics affect the design and selection of a thermal overload relay to protect a given motor.

Thermal overload relays are of two basic types, fusible alloy and bimetallic. The fusible-alloy type employs a tube and spindle element which is heated by a coil carrying the motor current, or a certain fraction of it if a current transformer is used. At a critical temperature the fusible alloy which is between the tube and spindle will melt, allowing a relative motion to occur which causes the relay to trip. The sharp melting point of the alloy and the wide range of heater coils available provide a device which can be accurately selected for motor currents in approximately 10% increments. In general, this type of relay must be manually reset.

The bimetallic relay is similar in application and is generally interchangeable with the fusible-alloy type; however, the design lends itself to two additional features. By means of a special compensating element this design can make allowance for varying ambient temperatures. Also, this design can be made to reset itself automatically. Both of these features can be optional in the bimetallic type of thermal overload relay.

Figure 14.6 shows a bimetallic type overload relay arranged for two-coil operation. By replacing the jumper in the center position of the relay with another heater coil, as shown on the right, this relay can be converted to have three coils. A coil in each line of a three-phase motor is recommended if unbalanced voltage conditions might occur in the power supply system because such unbalance could cause a higher current in one line than in the other two. This higher current could cause excessive motor heating and be undetected if it occurs in a line which has no overload heater coil.

Figure 14.7 shows a cutaway sketch of a unique bimetallic element used in the overload relay of Fig. 14.6. Heating of the bimetal disks causes the central core to move to the right, carrying a contact bias

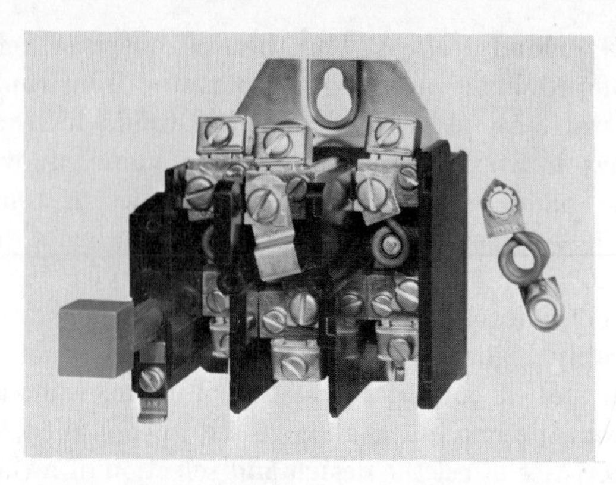

Fig. 14.6 Bimetallic Type Overload Relay and Separate Heater Coil

spring with it. At a critical temperature the contacts snap open. There are, of course, many other types of bimetallic elements.

Overload relays must be properly coordinated with the fuses or circuit breakers which provide short circuit and ground protection for the motor branch circuit. The overload relay should take care of protecting the motor against running overloads or stalled conditions. Furthermore, the overload relay must also protect other branch circuit elements, such as

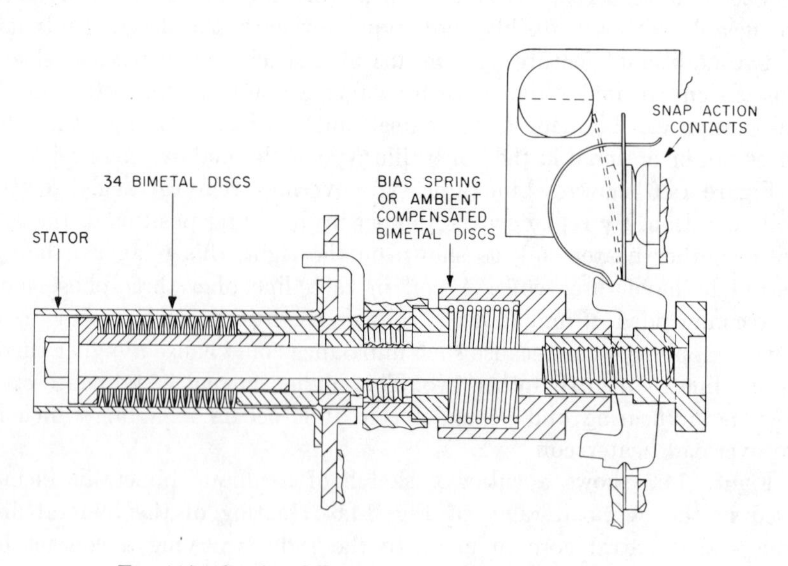

Fig. 14.7 Sectional Drawing of a Bimetallic Element

the wiring and the motor controller, under these same conditions. The short-circuit and ground protective device must take over beyond stalled motor, since neither the overload relay heater coils nor the motor contactor are designed to handle circuit faults. The fuses or circuit breaker should open a circuit fault before the overload relay coils burn out or the contactor tries to open this fault. In order to provide this proper coordination it is necessary to refer to the time-current tripping characteristics of the devices involved. In general, it can be stated that dual-element (time-delay) fuses selected on the basis of approximately 200% of the overload relay rating will adequately protect the circuit and the relay on high fault conditions and allow the relay to perform its function on motor running overloads.

Rating of Thermal Heater Coils. The ampere rating of a heater coil for a thermal overload relay is determined by testing it for the minimum current which will cause the relay to trip open, the heater being installed in an overload relay, and the relay being in a 40°C ambient temperature. This current having been determined from the average of a number of tests on duplicate heaters, the value is multiplied by 1.05 and the product becomes the heater rating. The 5% is added to make allowance for variation in the wire of the coil and for variations in the manufacturing process, and to insure that the heater will always trip at its rating. In designing a line of heater coils, a ratio between sizes of 10 to 15% is used. Assuming that the coils are 10% apart, the motor rating of a heater may be determined by dividing the heater rating by 1.25 to get the minimum motor rating, and then multiplying the minimum rating by 1.10 to get the maximum motor rating. For example, assume a heater rated 0.238 amp. The minimum motor current for which this heater should be used is 0.238 divided by 1.25, or 0.190 amp. The maximum motor current is 0.190 multiplied by 1.10, or 0.209 amp. Since the heater will trip on any current at or above its rating, a motor having a normal current of 0.209 amp will be protected on any overload of 114% or higher, and a motor having a full-load current of 0.190 amp will be protected on any overload of 125% or higher.

The above applies only to a motor having a 1.15 service factor, which means that it can operate at 115% of full load without injury. Many motors are now built with a 1.00 service factor, meaning that they have no overload capacity. In the latter case the overload relay should trip at no greater current than tolerances allow in order not to trip at 100% motor full-load current. This would generally give a maximum trip value around 110% full-load current and would be obtained by using one-size-smaller heater coil than that for the 1.15 service factor motor.

Effect of Ambient Temperature. It is customary to test and rate thermal overload heaters on the basis of 40°C ambient temperature, because that is the basis of rating the motor and represents about the worst service conditions. However, it is the usual practice to publish time-current characteristic curves on the basis of 25°C ambient temperature, because anyone making a test in the field is likely to be making it in about that temperature. Any curve may be changed from one ambient basis to another by the following equation:

$$\frac{I_a}{I_b} = \sqrt{\frac{T - a}{T - b}}$$

where I_a = amperes to trip in a given time at the lower ambient
I_b = amperes to trip in the same time at the higher ambient
a = the lower ambient in degrees Centigrade
b = the higher ambient in degrees Centigrade
T = the operating temperature of the thermal element in degrees Centigrade, that is, the melting point of a fusible alloy or the tripping temperature of a bimetallic operating element

It becomes apparent that the higher the value of T, the less effect a change in ambient will have on the relay operation. The operating temperature of the element also has a great deal to do with the time-current tripping characteristic of the overload relay and is one of the critical design parameters.

Derating Curves. When thermal relays are used as a part of an enclosed controller, it may be necessary to derate the heaters, because free ventilation is not available and the ambient temperature inside the enclosure may be relatively high. This is particularly true of small enclosures and accounts for the fact that the same heater may have several different ratings according to its application to a particluar controller. The amount of derating necessary can be determined only by test.

Standard Ratings for AC Contactors. NEMA has established standard ratings for ac contactors as given in Chapter 16. Most ac contactors are built with totally enclosed bridging contacts up through size 5. This design is illustrated in Fig. 14.8. Control circuit interlocks, either normally open or normally closed, can be added to the contactor as shown in this figure. On larger sizes of contactors it is generally necessary to provide arc chutes and more elaborate arc-quenching means to dissipate the greater amount of power when interrupting maximum rated loads. Contactors must be capable of interrupting a stalled motor at

FIG. 14.8 Size 1 3-Pole AC Contactor

rated horsepower and voltage. Design tests are usually conducted at ten times the rated full load motor current, the load being simulated by resistors in series with air core reactors.

Interrupting and Short-Time Ratings. In addition to its regular duties of starting and stopping a motor, and occasionally opening the circuit under overload, a contactor may be called upon to withstand a short circuit long enough to permit a circuit breaker to operate or a fuse to blow and relieve the condition. The magnitude of a short circuit is determined by a number of factors, among them the capacity of the generating system feeding the lines, the impedance of the lines, and the point where the short circuit occurs. A contactor is designed to handle currents within its capacity, and to do this many thousands, even millions, of times. A circuit breaker is designed to open very high currents, far beyond the capacity of a contactor, but normally it has to do this only a few times in its life. Mechanically, then, a circuit breaker cannot approach the operating life of a contactor. The usual practice is to install a circuit breaker or fuses for short-circuit protection, a contactor for operation of the motor, and an overload relay for motor protection. If a short circuit occurs, the contactor should have enough thermal capacity to carry the short circuit for the short interval before the breaker opens.

Standard ac air-break and oil-immersed contactors have the capacity to interrupt 10 times rated motor current. This is based upon the horsepower rating of the contactor. They have 1-sec thermal capacity of 15 times the current corresponding to their horsepower rating.

High Voltage Contactors. With the increase in power utilization it is becoming common to apply large motors to voltages between 2200 and 5000 V. This requires very special controllers incorporating added safeguards to personnel. The contactors are generally of the "roll-out" type for added safety and ease of maintenance. A controller using this type of contactor is shown in Fig. 14.9. The high-voltage contactor compartment is in the bottom and is separated from the low-voltage control compartment located above.

These controllers are built in two types. In one type the contacts of the controller are used both for starting the motor and for interrupting a short circuit. The other type of high-interrupting capacity controller uses the contacts of the controller for starting the motor, and fuses for interrupting a short circuit. In this type of controller the contactor need interrupt only operating overloads, and be able to carry the short-circuit

FIG. 14.9 High Voltage Controller with "Roll-Out" Contactor

current long enough to permit the fuse to open it. The fuses used are of the current-limiting type, having a temperature-resistance characteristic which to some degree limits the amount of current that can pass through them. High-interrupting-capacity controllers without fuses are built to interrupt short circuits of 25,000 and 50,000 kva. High-interrupting-capacity controllers using current-limiting fuses are built to interrupt short circuits of 150,000 and 250,000 kva.

Problems

1. An ac contactor which is used on a 220-V circuit has a coil with a resistance of 1.35 ohms. The impedance of the coil, with the contactor open, is 10.6 ohms. It is desired that the line voltage drop be no more than 10% of rated voltage. If the wire between the contactor and the pushbutton is no. 14, having a resistance of 2.525 ohms/1000 ft, how far may the pushbutton be located from the contactor?

2. When the contactor of problem 1 is operated in 440 V, the coil resistance is 5.3 ohms and its impedance with the contactor open is 40.5 ohms. Calculate the maximum allowable resistance of the line between the contactor and the pushbutton, when the line voltage drop is limited to 10% of rated voltage.

3. If the impedance of the 440-V contactor coil when the contactor is closed is 655 ohms, what is the current in the coil under the conditions of problem 2?

4. A contactor used on a 110-V circuit has a coil with a resistance of 0.35 ohm and an impedance of 2.9 ohms when the contactor is open. The coil is energized from a 550-V circuit, through 550/110-V transformer having an effective impedance of 0.13 ohm. What is the voltage on the coil?

5. The melting point of the alloy in a thermal overload relay is 98°C. If the relay trips at 52 amp in an ambient temperature of 20°C, at what current will it trip in an ambient temperature of 60°C?

6. The melting point of the alloy of another thermal overload relay is 231°C. If this relay trips at 52 amp in an ambient temperature of 20°C, at what current will it trip in an ambient temperature of 60°C?

7. If a thermal overload relay is found to trip at 50 amp in an ambient temperature of 25°C, and at 40 amp in an ambient temperature of 50°C, at what current will it trip in an ambient temperature of 10°C?

8. A dc contactor and an ac contactor are mounted side by side, and their coils are energized continuously at rated voltage for 1 hour. The temperature rise of both coils is the same. The next day a second test is made, the coils being energized intermittently. At the end of an hour the temperature rise of both coils is again found to be the same. Have the contactors been operated the same number of times? If not, which has operated the greater number of times? Explain the answer.

9. A 110-V 60-Hz coil for an ac contactor has 230 turns of wire. Approximately how many turns will be required in a coil to operate on 440 V, 25 Hz?

10. A number of tests of a certain thermal overload heater-coil design show that the average tripping current is 15.2 amp. What is the heater rating? If it is desired to allow a motor to run at 15% overload, but to protect it on any overload of more than 25%, what is the range of motor full-load currents for which this heater is suitable?

15

STATIC LOGIC

Basic Control Elements. Electrical control may be thought of as being divided into two general groupings:

1. Analog control
2. Digital control

Nearly all electrical equipment embraces some forms of both of these. As an example, a dc shunt-wound motor may be brought up to speed on a constant potential dc line, by commutating resistor in series with the motor armature. This is done by means of contactors which are *digital* devices. However, as the motor accelerates, the speed changes in a gradual (analog) manner and the armature current changes in a fashion controlled by the contactors (digital-step change inrush) and by the changing motor speed (analog).

The regulating systems described in Chapter 10 on Automatic Regulating Systems, are other examples of analog control. However, the motors are still ordinarily connected to the source of power by means of contactors, which are digital devices.

A great deal of sequencing or interlocking control is furnished to industry which involves an on-off, or digital, device. Many of the pneumatic and hydraulic solenoid valves are examples of this, and large hydraulically powered machines are sequenced through their cycle by means of appropriate energization of solenoid valves. Typical of these are machine tools, foundry machines, extrusion presses, sheet-metal punch presses, and in fact almost any industrial machine.

Among the most powerful tools used in digital switching or logic control are electromechanical relays. Electrical interlocking circuits and memory systems, incorporating combinations of relays, latching relays, stepping switches, and so on, are typical of electromechanical control.

In this chapter another tool broadly described as static logic will be introduced. This is another method of obtaining sequencing, programming, and many sophisticated logic techniques. Static logic is the basis of the modern digital computer. The term "static" generally refers to the fact that there are no moving parts in the switching circuitry. Whereas a relay has a moving armature, the transistor, which is usually the basis of static logic, has no moving parts whatever. Accordingly, it is most desirable for use in those systems whose operation requires that many millions of cycles be done in a relatively short period of time. This would naturally wear out any moving parts, but the transistor works on and on without fatigue.

Digital control systems resolve themselves into three basic groups:

1. Pilot devices such as pushbuttons, limit switches, pressure switches, and thermostats.
2. Decision-making devices such as relays and static logic control elements.
3. Output devices such as solenoids, contactors, and starters.

Definition of Terms. The meaning of the following terms will be most helpful in understanding the dicussions that follow, especially those pertaining to circuit operation.

Element: A given electrical circuit capable of performing a function.

Module: Hardware package containing one or more logic elements.

Conducting, or On: In reference to a transistor, indicates that the transistor is conducting in its power circuit; that is, emitter to collector.

Nonconducting, Open, or Off: In reference to a transistor, indicates that the transistor is not conducting in its power circuit, that is, emitter to collector.

Intelligence: The information carried by the electrical circuits, described in words, such as "Limit of travel has been reached," "Operator has pushed button," or (as a command) "Start the motor."

Signal: The voltage that carries the intelligence.

Input: Input signal.

Output: Output signal.

Fan Out: The capability of using the output signal of one device as an input to a plurality of other devices

The Transistor as a Switch. Since the transistor is the mechanism for static switching, a discussion of its operation is in order. The transistor (Fig. 15.1) is a semiconductor composed of two nonisolated circuits: (1) the base circuit, emitter to base, and (2) the power circuit, emitter

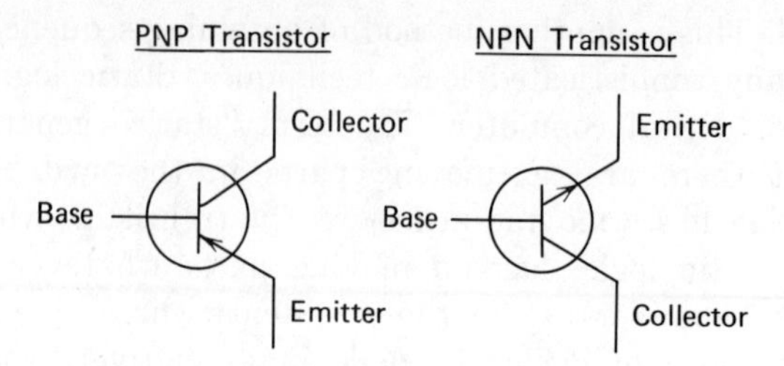

FIG. 15.1 Two Basic Types of Transistors

to collector. The base circuit controls the transistor; that is, when base current is present, the power circuit conducts. The diagram of Fig. 15.2 illustrates a single-transistor switching circuit. With the contact closed, a circuit is established from emitter to base, causing conduction in the transistor's power circuit, emitter to collector. In this condition the transistor is essentially a closed switch; that is, E and C can be thought of as connected or shorted together. Therefore, point C, the output, is "switched" on, or is brought to the +10 V potential.

Intelligence Added. In Fig. 15.3, note the addition of a pushbutton. This pilot device provides information or intelligence. Note, too, that the circuit has been changed so that the intelligence is brought to point A. This arrangement is common to most static switching. Two conditions will be considered: (1) pushbutton open, and (2) pushbutton closed.

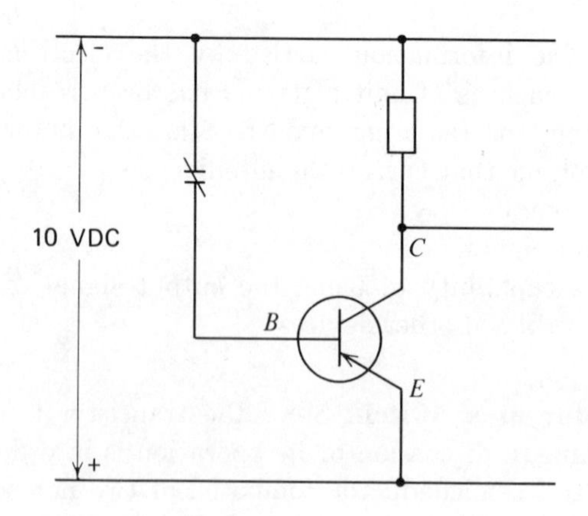

FIG. 15.2 Simplified Static Switch

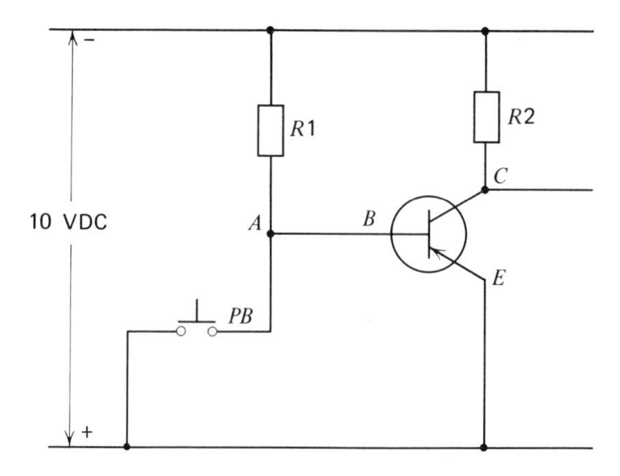

FIG. 15.3 Inverted Logic Element. Output C Is On When Pushbutton PB Is Open

With the pushbutton open, it can be seen that base current can flow from emitter E to base B through point A to the negative side of the line, thus triggering the transistor to conduct from emitter to collector (power circuit). Now, since the conducting transistor is analogous to a closed switch, a 10 V signal will appear across the output load resistor $R2$. Recall that no initiating signal was present (pushbutton open); yet a +10 V output is present. The circuit shown is usually called *inverted logic;* that is, no input provides an output.

With the pushbutton closed, intelligence is present. Now with point A electrically connected to the +10 V bus, point A acquires the same +10 V potential as the emitter of the transistor. And since base B is connected to A, there can be no base current; therefore there can be no transistor conduction. The transistor behaves like an open switch, and the drop across the load $R2$ is zero volts. In this case of *inverted logic,* it is seen that the presence of an input results in no output.

Static control systems can be made up of inverting or noninverting logic elements, or combinations of these. Inverting logic has the advantage, especially in simple systems, of requiring fewer components to accomplish the result. However, it includes the disadvantage of being slightly more difficult to understand, particularly if the individual is already schooled in normal relay switching circuitry. Each manufacturer uses one or the other of these approaches. If the student understands one form of static logic, he can readily translate this knowledge into useful circuits using other hardware. In this chapter, primarily *noninverting logic* will be described.

How Noninverting Logic Works. In Fig. 15.4, note that the circuitry to the left of the dotted line is identical to the previous circuit. The change is the addition of another transistor. One of the purposes of this transistor is to reinvert the output signal from the first transistor, thereby restoring similarity with the input signal. Since the output signal "looks" like the input signal, this circuitry is called *noninverting logic*.

Let's trace the circuit with the pushbutton *PB* open. Point *A* is virtually at the negative potential, permitting current to flow from emitter to base of transistor *T*1. This base current has the effect of turning on the power circuit of transistor *T*1. Therefore, current flows from *E* to *C*. Since the transistor *T*1 is conducting, *C* is essentially at the same potential as point *D* of the second transistor *T*2. Consequently, no current can flow from *D* to *C*. And with no base current, transistor *T*2 is nonconducting or open.

The output is off, or in other words, the point *O* is at the negative 10 V potential, except for the small drop across the resistor caused by the flow of signal current to the next logic element. This is typical of static logic circuitry. There are two values: "+" or "−," "1" or "0," and so on. An open logic circuit is not recognized. With the pushbutton open, an absence of an input (*A* negative) results in an absence of an output (*O* negative).

In Fig. 15.5, with the pushbutton closed, point *A* is electrically connected to the +10 V bus. With both base and emitter potentials at

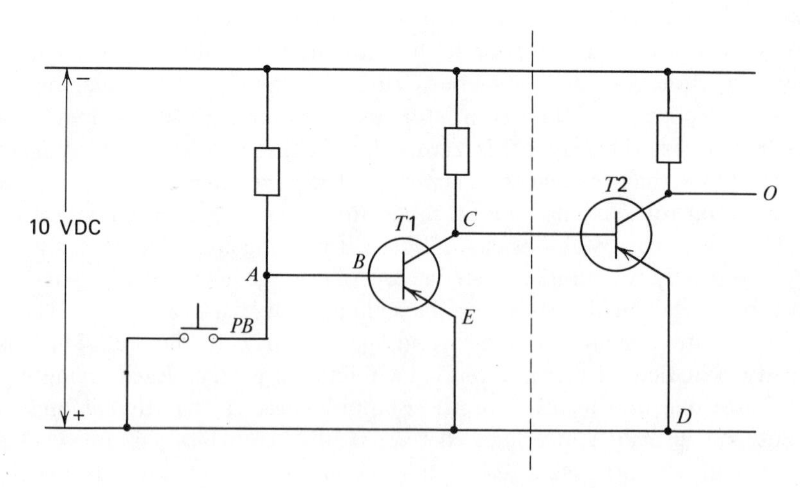

FIG. 15.4 Noninverting Logic. Output *O* Is On When Pushbutton *PB* Is Closed

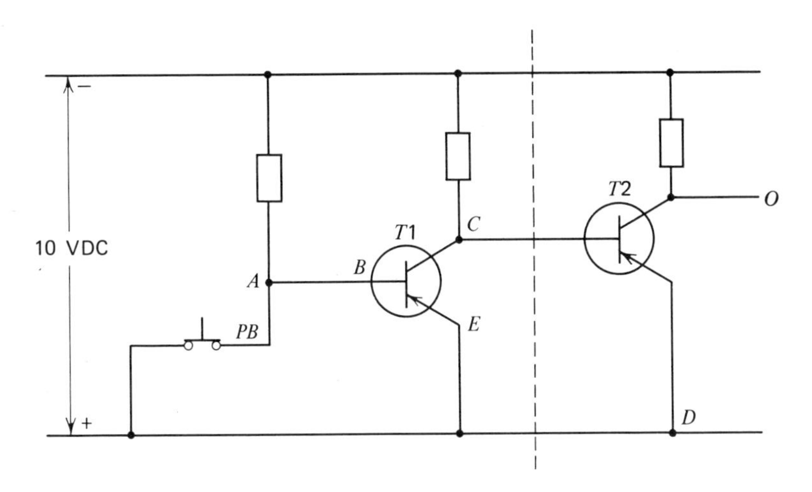

FIG. 15.5 Noninverting Logic with Pushbutton Closed

+10 V, there can be no base current flow; therefore no conduction in the power circuit of transistor *T*1. Since transistor *T*1 is nonconducting, point *C* is essentially at the negative potential, permitting base current flow through transistor *T*2. Recall that base current flow causes conduction in the power circuit of the transistor. Therefore, *T*2 conducts and can be likened to a closed switch. The result is a +10 V signal across the output load resistor—a signal in, a signal out.

Note, that any input condition that brings a signal, that is, +10 V to point *A*, shuts off the first transistor, and this causes the second transistor to conduct, thereby producing a +10 V output signal. It can also be seen that in this two-transistor switch, one transistor always conducts.

A Look at Complete Static Logic Control. An important part of static control requires methods and hardware to

(a) Provide appropriate power supplies;
(b) Get signals from pilot devices into the transistor switch;
(c) Get the output of the logic elements to suitable power amplifiers to operate machine components.

In Fig. 15.6 the relationship of the logic diagram and the circuit or elementary diagram is shown.

Power Supply. A power supply is necessary to supply the special dc voltages required for the static control circuitry from standard alternating current.

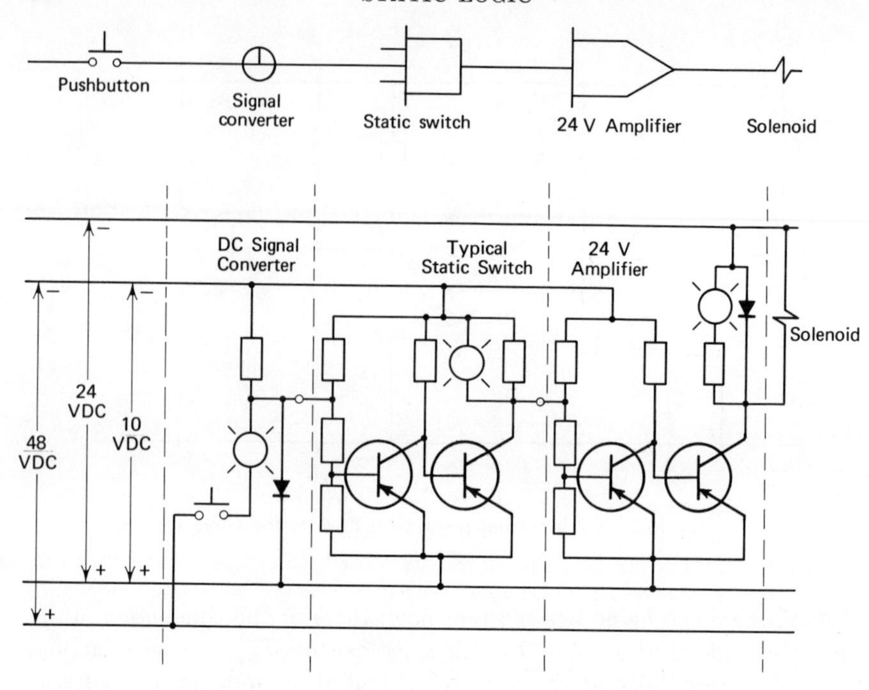

Fig. 15.6 Relationship of Logic Diagram and Elementary Diagram

Signal Converter. Pilot devices, such as pushbuttons and limit switches, do not operate satisfactorily with the low energies used in static switching. Higher voltages and higher currents must be employed to break through the contact resistance of the pilot devices. But this higher-energy signal cannot be used by the static switch, simply because its power-dissipating capabilities are very small. Therefore the purpose of the signal converter is to reduce the higher signal voltage and current used by the pilot device to a lower signal voltage and current compatible with static logic circuitry. (See Fig. 15.7.)

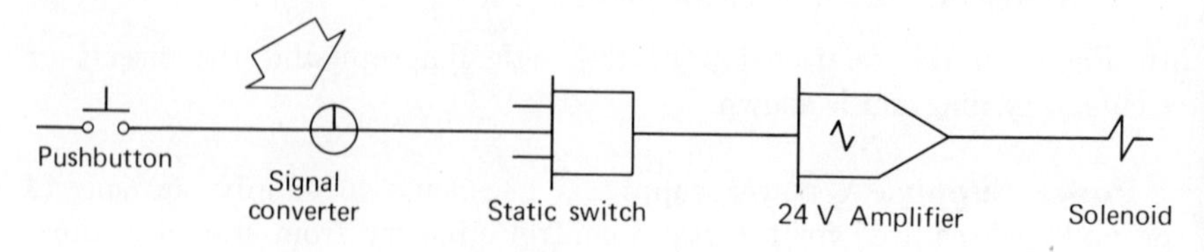

Fig. 15.7 A Look at the Signal Converter

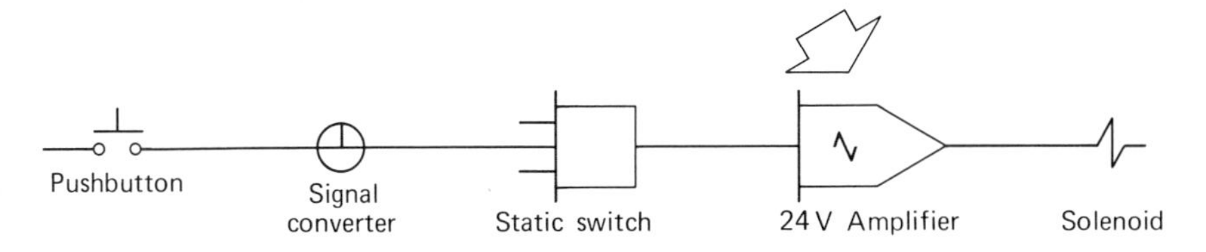

Fig. 15.8 A Typical Static Switch

Typical Static Switch. The next item is the static switch itself (see Fig. 15.8). This is the familiar two-transistor static switch of noninverting logic, already described. Only one static switch is symbolized in the diagram. Yet static switches are capable of driving other static switches, singly or as multiple units. Therefore the number of static switches used would be determined by the extent of control required.

Amplifiers. The purpose of the amplifier is to raise the output power level and drive output devices, such as relays, solenoids, and contactors. Typical is the 24 V amplifier, shown at the last static element in Fig. 15.9.

Logic Symbols and Diagrams. Introduced here are additional symbols commonly referred to as logic symbols. These symbols, the tool of the circuit designer, are nothing more than a shorthand method of describing a functional grouping of electrical components. Many of these logic symbols will be described; the detail electrical elementary (schematic) diagrams are shown on the following pages.

There is one major difference between the use of logic symbols and the use of other elementary symbols. In most elementary diagram design work, the complete electrical circuit is shown from one side of the power line to the other. This is typical of nearly all the other diagrams in this book. In logic design, the source of power is frequently not shown on the logic circuit. It is assumed to be connected to certain common inputs of each and every logic element, but it is not shown on the dia-

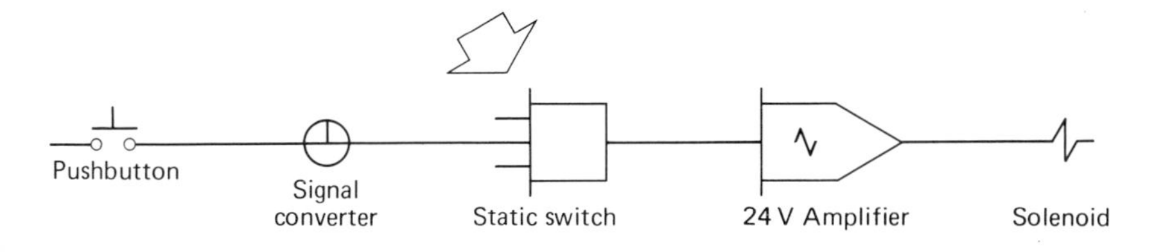

Fig. 15.9 A Look at a Power Amplifier

gram, in the interest of simplicity and ease of understanding the logic circuit. *The logic diagram shows the signal or information circuits but not necessarily the power sources and connection.*

Several different but equivalent logic symbols are in current use. One such group is discussed here. Different manufacturers use different logic symbols for their particular line of equipment. However, all of these are somewhat similiar, and the student can readily go from one to the other. NEMA has also adopted a line of static logic symbols which are tabulated in NEMA Standards for Industrial Control.

Analysis of Components—Signal Converters. The function of the signal converter is to convert the higher voltage pilot signal to a reduced dc voltage suitable for the input of static logic circuitry (see Fig. 15.10). Two basic types of signal converters are commonly used to perform this function, thus permitting a choice of the type of voltage to be applied to the pilot device.

1. *Direct-Current Resistor Type.* This signal converter operates from 48 volts D-C voltage supplied by the power supply. It is suitable for most pilot devices, provided that the contacts have good "wipe" and are subject to frequent use.

2. *Alternating-Current Transformer Type.* This type of signal converter utilizes 115 V ac across the monitored contacts, and then uses this voltage when present, to produce a 10 V dc output to the logic. It is recommended, because of its energy level, where pilot devices are located in contaminated atmospheres or subject to infrequent use.

48 V DC Signal Converter. The dc signal converter shown in the complete circuit diagram of a typical static logic control system was of the resistor type. The signal converter portion of the diagram, with pushbutton pilot device and typical static switch dotted in, is illustrated in Fig. 15.11.

When pushbutton *PB* is closed, current flows through resistors *R*1 and *R*2, which serve as a voltage divider. Note that the lamp functions

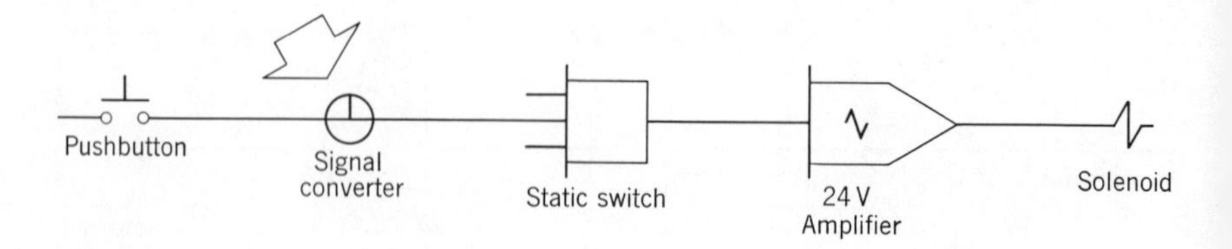

Fig. 15.10 Details of the Signal Converter

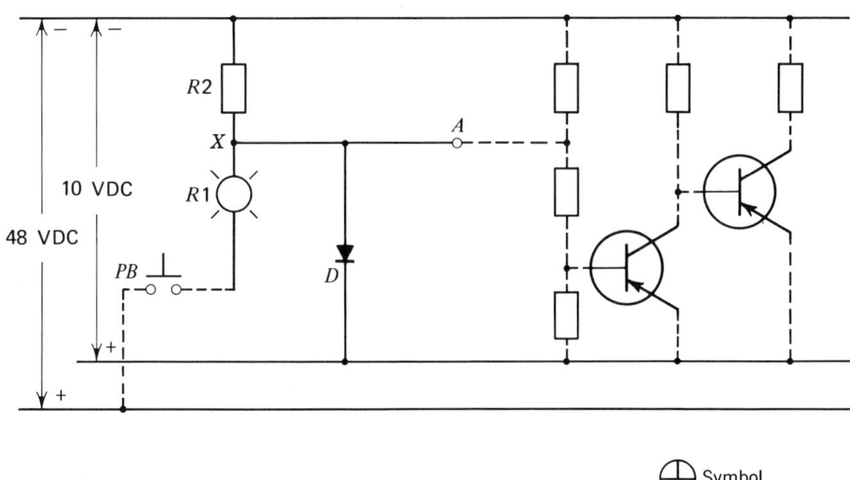

FIG. 15.11 The DC Signal Converter

both as $R1$ and as a state light. The purpose of the resistors is to reduce the 48 V, in this case, to voltage that is compatible with static logic circuitry. This voltage comes from the drop across $R2$ and appears at point A. It is, therefore, the input signal to the following logic element. Incidentally, the output of any signal converter can be regarded as a voltage at a given point (A in the illustration) which can be sensed by a static switch. This sensing is similar to the function of a control grid in a vacuum tube.

With the pushbutton closed, the voltage that appears across $R2$ must be the same as that impressed across the 10 V lines. Were this not the case, lost signals could result. Yet the voltage dividing circuit is designed to present a slightly greater voltage than 10 V across resistor $R2$. This apparent inconsistency is explained as follows. Should the bus voltages drift and cause the signal converter to present a signal slightly higher than 10 V, the voltage can always be "pulled down" to the desired value.

But if the signal voltage should go below 10 V, then voltage cannot be added to provide the necessary amount. Resistor $R2$ and lamp resistance $R1$ are intentionally chosen so that, without the diode connection the voltage at point X is substantially above 10 V. This insures a good signal from the pilot device to be impressed to the logic circuit at point A. Now to prevent this voltage from getting too high, the diode D "clamps" this potential to the 10 V maximum. Therefore the potential of point X can never be more than about ½V higher than +10 V. The final result is a signal converter that provides protection from voltage drift of the 48 V supply, the long lines which may run out to the pushbutton or other pilot device, some variation that may exist in contact resistance, and so on.

115 V AC Signal Converter. The ac transformer type signal converter both reduces and rectifies the 115 V ac signal voltage to the low dc voltage required by the logic circuitry (see Fig. 15.12). Note that the pilot device, pushbutton, PB, is in series with the 115 V primary of the transformer. The latter steps down the voltage. The rectifiers $D1$ and $D2$ plus the transformer produce a full-wave dc which is filtered by capacitor C. This filtered dc voltage output is applied to the voltage dividing resistors $R1$ and $R2$. Diode $D3$, which is connected to the +10 V bus, passes current whenever the junction of $R1$ and $R2$ attempts to go above the power supply level. This clamping action keeps the signal level at the same voltage as the power supply fed to the static devices, that is, 10 V. It should be noted that when current bleeds off through $D3$, the return path is through the connection at point B. The lamp labeled $R1$ again functions both as a resistor and as a state light.

Analysis of Components—Static Logic Elements. This section contains the basic building blocks of a control system—the static switches (Fig. 15.13). The elements will be covered one by one in detail, indicating how and why they work the way they do.

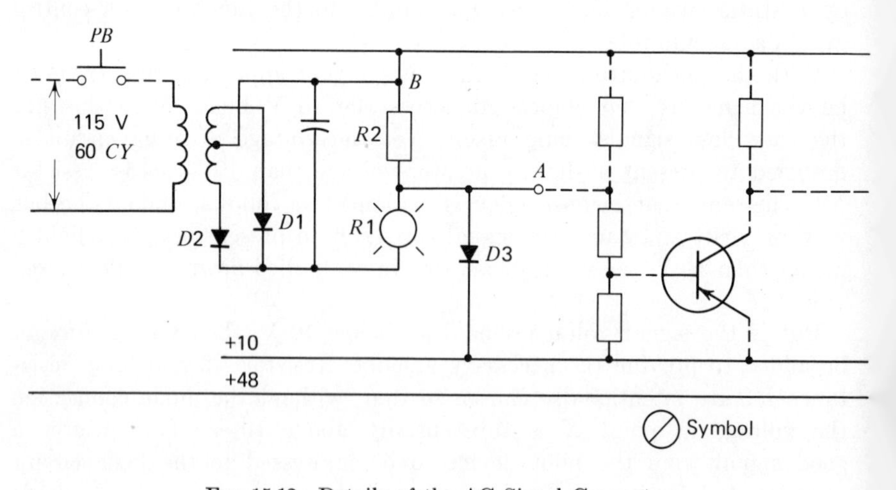

Fig. 15.12 Details of the AC Signal Converter

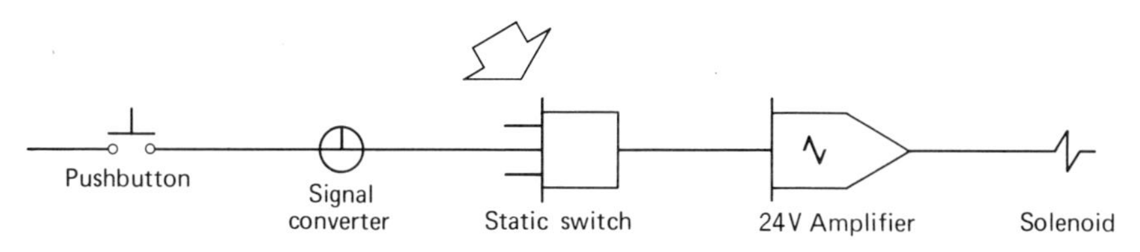

FIG. 15.13 The Static Switch Emphasized

And, Or, Not. The AND switch is a device which produces an output only when every input is energized. Referring to the AND symbol (Fig. 15.14), when all three inputs on the left are energized with a +10 V signal, the output will be +10 V. The relay equivalent (Fig. 15.15) of an AND switch consists of three normally open contacts in series, all three of which must be closed before the relay coil is energized.

Consider the circuit with all pushbuttons open (Fig. 15.16; *PB*1 only shown). No input signal can then appear at any one of the input points *A*1, *A*2, and *A*3. This no-input condition permits current to flow through the entire series-parallel voltage dividing network, consisting of resistors *R*1 through *R*7. Of special significance is resistor *R*7, since the voltage drop across it also appears across the emitter-base circuit of transistor *T*1. Its base, therefore, is held sufficiently negative to cause base current to flow from *E* to *B* in *T*1 and through resistors *R*1 through *R*6. As we have seen earlier, base current causes conduction in the power circuit. Therefore, *T*1 conducts, bringing a +10 V potential to point *C*. With point *C* at +10 V, the base path of transistor *T*2 is blocked, causing this transistor to assume its nonconducting or open state. Final result: Point *D*, the output terminal is connected only to the 0-V bus, and exhibits its no-output condition. No input signal, no output signal.

Pushbutton *PB*1 *Closed*, *PB*2 and *PB*3 *Open*. With pushbutton *PB*1 closed, one signal input is present, that is, the signal converter presents a +10 V to point *A*1. This condition may be regarded as blocking the right-hand leg of the voltage dividing network. But current still flows

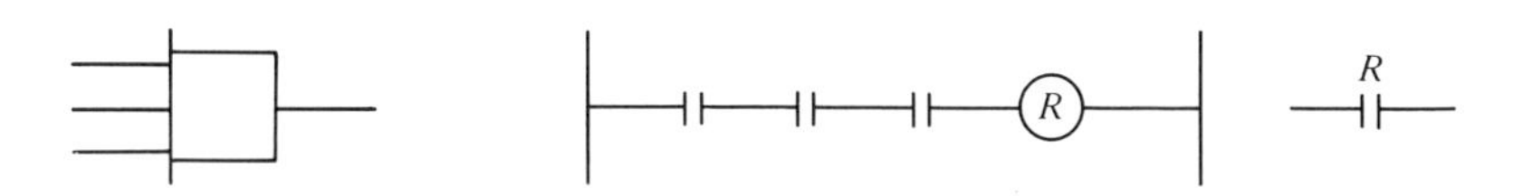

FIG. 15.14 The AND Symbol

FIG. 15.15 Relay Equivalent of the AND Symbol

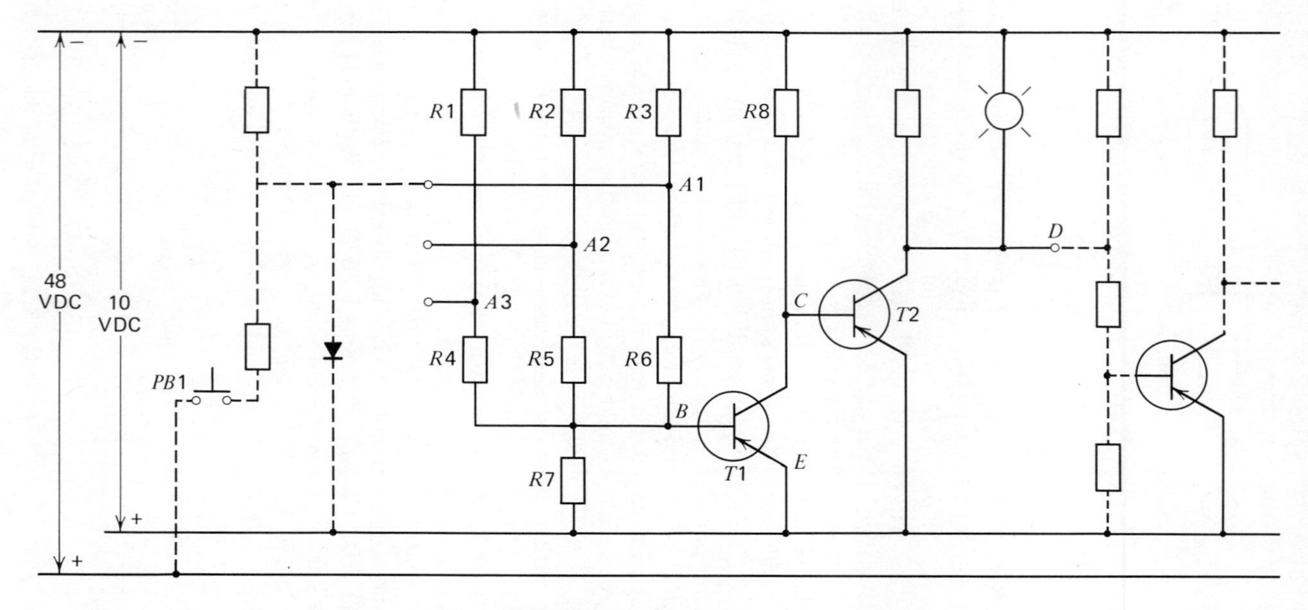

FIG. 15.16 Elementary Diagram for the AND Logic Element

through resistor $R7$, which still holds point B sufficiently negative to permit base current to flow through transistor $T1$ and the remaining two resistor legs. Therefore, transistor $T1$ maintains its conducting or ON state, while transistor $T2$ maintains its nonconducting or OFF state. The all-or-nothing concept of the AND switch is demonstrated; that is, one of a given number of inputs to an AND switch cannot, when energized, give an output.

Pushbuttons PB1 and PB2 Closed, PB3 Open. With the closing of pushbutton $PB2$ (not shown), another input is energized bringing point $A2$ to +10 V potential. Now current in the second or middle leg of the voltage dividing network is blocked. But the voltage drop across $R7$ still holds the base of transistor $T1$ negative. Therefore, base current still flows through $T1$, this time through the remaining leg of the voltage dividing network. Transistor $T1$ still stays on, transistor $T2$ still stays off. The all-or-nothing requirement of the AND switch remains intact. Two out of three energized inputs are not enough to "turn on" the AND switch.

Pushbuttons PB1, PB2, and PB3—All Closed. The closing of pushbutton $PB3$ brings point $A3$ to +10 V. Now with points $A1$, $A2$ and $A3$ all at +10 V, point B, too, is at +10 V. Therefore current can no longer flow from emitter to base of transistor $T1$. The absence of base current is equivalent to nonconduction in the power circuit. As a result, point C goes negative, permitting $T2$ base current through $T2$ and $R8$. Now transistor $T2$ conducts, turning on the state-indication light and producing a +10-V output signal at point D. The all-or- nothing concept of the AND switch is realized. With all inputs energized, an output appears.

If the application does not require the full input capabilities of the AND switch, the unused input terminals must be either connected together to the inputs used, or to the +10 V bus.

The AND switch is one of the so-called English logic elements. In following paragraphs, other English logic elements will be described and a logic system based upon AND, OR, NOT, and similar elements will be developed. A photograph of the AND board is shown in Fig. 15.17.

The OR switch can be likened to relay contacts connected in parallel (see relay equivalent diagram, Fig. 15.18). In other words, if input A, OR B, OR C is present, the output will appear. Note that the OR switch diagram is similiar to the AND switch, with this exception: the resistors have been replaced with rectifiers.

With pushbutton $PB1$ or with either of the other two pushbuttons closed, the appropriate point, $A1$, $A2$, and $A3$, will receive a +10 V signal. Therefore, base current cannot flow through transistor $T1$. As a

FIG. 15.17 The AND Logic Board

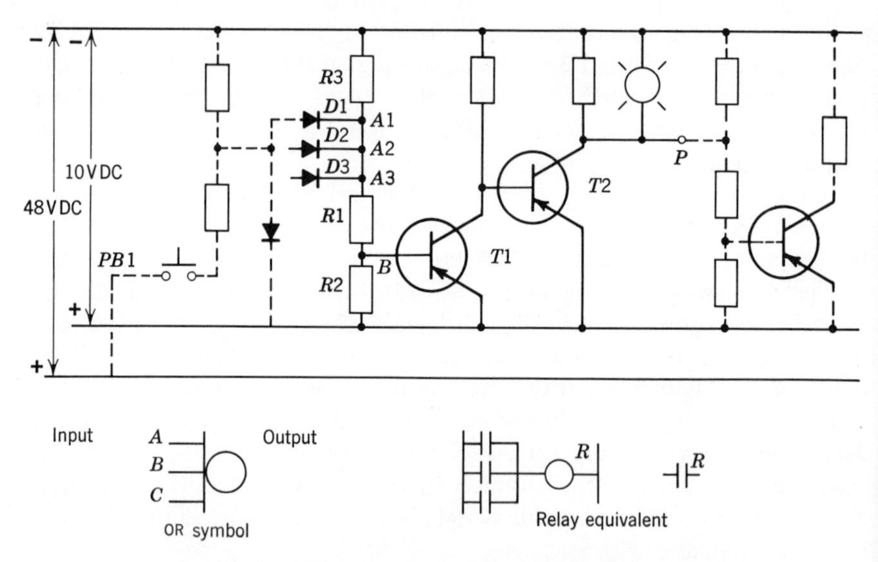

FIG. 15.18 The OR Logic Element, Its Relay Equivalent, and Its Elementary Diagram

consequence, transistor $T1$ is shut off, and, as we have seen before, transistor $T2$ is turned on. An output appears at point P.

The diodes $D1$, $D2$, and $D3$ prevent unwanted feedback. That is, they prevent a feedback circuit from one input to the state-indication light of preceding logic units that may be driving the OR switch. Such feedback could cause false state indication.

The NOT switch is a device that produces an output only when the input is not energized. But when an input does exist, then no output appears. The relay equivalent (Fig. 15.19) of a NOT switch is a single pole relay with a normally-closed contact and one input signal to the coil. The purpose of the NOT switch is to provide signal inversion where that function is not only desirable but necessary.

With the pushbutton PB closed (input present), a +10 V input signal appears at point I. Current will therefore flow from the base to emitter of transistor $T1$, and through $R2$ to the negative side of the line, causing $T1$ to conduct. Note that transistor $T1$ is an *NPN* rather than the *PNP* type. Note that the arrow points away from the base, indicating that current flows from base to emitter. Now with transistor $T1$ conduct-

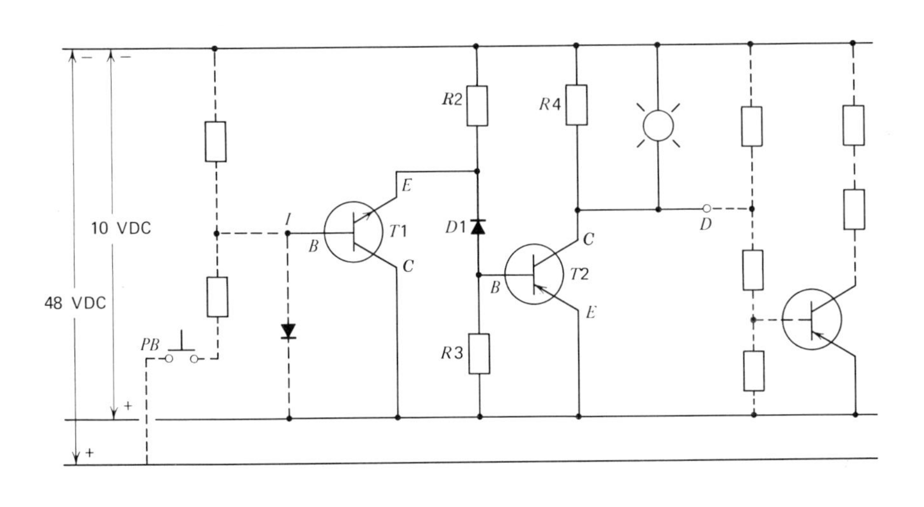

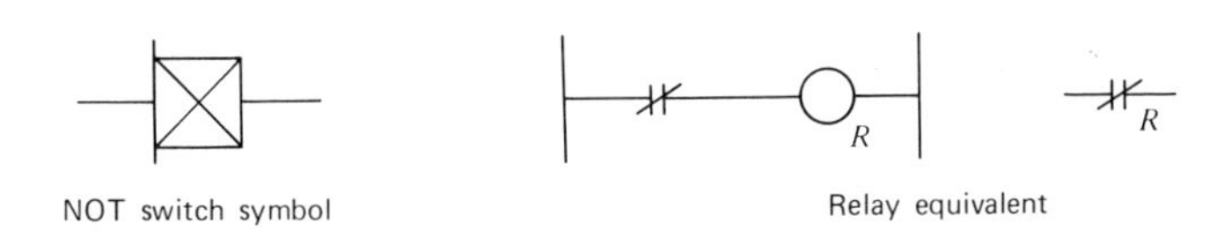

FIG. 15.19 The NOT Element, Its Relay Equivalent, and Its Elementary Diagram

ing, the junction of diode $D1$ and $R2$ is at $+10$ V, $T2$ cannot conduct, and there can be no output signal. An input results in no output.

With pushbutton PB open (input absent), no $+10$ V input signal appears at point I. Therefore the emitter of transistor $T1$ will not be at $+10$ V. Now current can flow from the emitter of transistor $T2$ to its base, through $D1$ and $R2$. Transistor $T2$ can now conduct and supply a $+10$ V output. No input results in an output.

Retentive Memory. The primary purpose of the retentive memory is to provide power loss memory. This is accomplished by a two-coil magnetically latching reed relay whose operation is similar to the conventional latched relay. In the retentive memory, however, each reed coil is driven by its own logic performing transistors (Fig. 15.20).

An input signal at either A or B will block one leg of the AND function associated with transistor $T1$. Plus 10 V applied to the inhibit terminal I will block the other leg. Thus, $T1$, denied a base current path, shuts, off, turning on $T2$. Then $T2$ energizes the reed relay coil $C1$, closing the reed contact. Terminal E is the output terminal.

Inputs C and D, in combination with the same inhibit terminal, function similarly through $T3$ and $T4$ except that they cause the reed contact to open.

Discharge diodes are provided on both coil circuits.

The two unmarked inputs associated, one with A and B and one with C and D, represent points on the front of the boards. These allow for manual setting of the reed position with a probe connected to the positive side of the 10 V supply. Only a momentary signal is necessary either to set or to reset the unit.

The inhibit terminal is a master terminal, and as such this one terminal can inhibit both the set and reset functions of both switches on the board. If not used for logic purposes, it should be tied to either the $+10$ V bus or a special reset gate.

Set-Reset Memory. The set-reset memory performs a simple memory function (Fig. 15.21). A momentary input signal, $+10$ V, applied to the set terminals produces an output signal; applied to the reset terminal it shuts the switch off. The master reset terminal, which serves all switches on the board, should be connected to a reset gate circuit. This insures that the set reset memory will assume the no-output condition when power comes on.

Assuming the no-output state, transistor $T2$ will be conducting. Its base current path uses diode $D1$ and resistor $R8$. The conducting state of $T2$ will block $T3$ base current, holding $T3$ off; $T1$ will also be conduct-

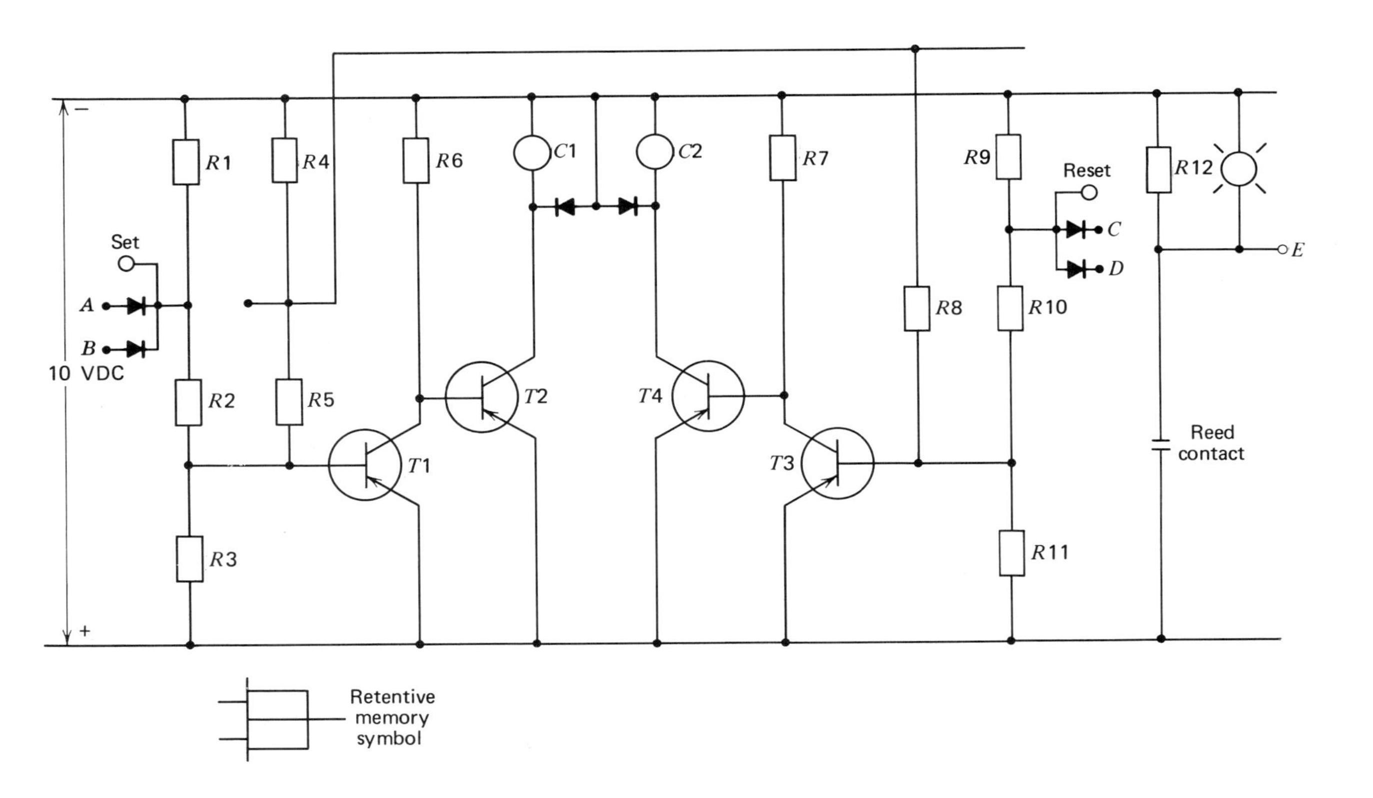

Fig. 15.20 Retentive Memory

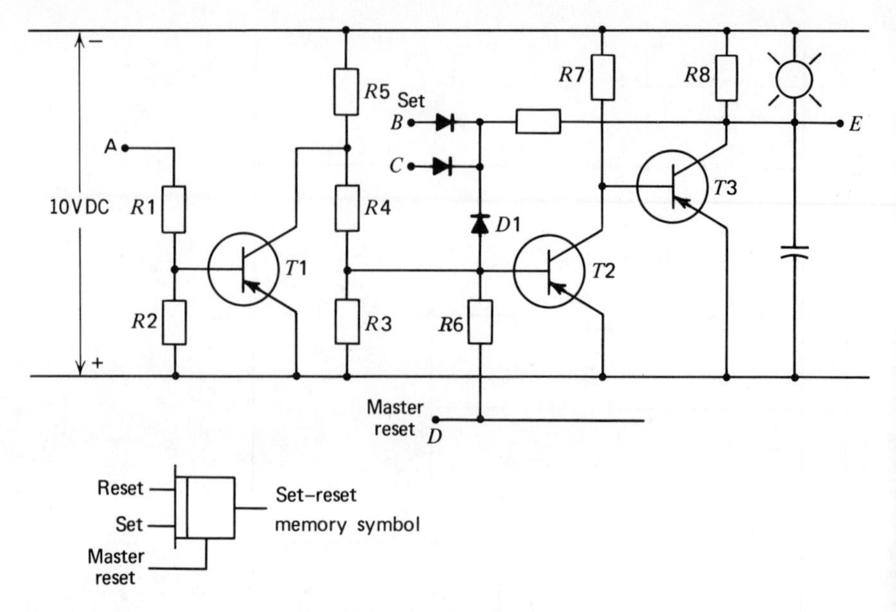

FIG. 15.21 Set-Reset Memory

ing with its base current path through $R1$ and some other logic units' load resistor. This is essential as it blocks the $R4$–R5 base current path of $T2$ so that terminals B or C can set the unit.

A setting input calls for either B or C to be driven to +10 V by the output signal of some other logic unit. This shuts off $T2$ base current; $T2$ turns off; $T3$ turns on, the output signal appears, and the $T2$ base current path through $R8$ is blocked. The set signal may now be removed.

To reset the unit, terminal A is driven to +10 V; $T1$ shuts off; $T2$ base current appears in $R4$ and $R5$, turning $T2$ on and $T3$ off. The $T2$ base current returns to the path $D1$–$R8$. The reset signal may now be removed.

When the 10 V power first appears, all the transistors start to turn on. The master reset terminal, held negative by the reset gate, insures that $T2$ has a base current path and ends up conducting.

In the interest of simplification the detailed circuitry will not be shown for the more complex logic elements to follow.

Timer. *Time delay* is another function which is frequently used in logic design. In relay logic a pneumatic timer or synchronous electrical motor timer customarily serves this function. Time delay can also be accomplished by a variety of rc networks, together with transistors for amplification and switching. The timer customarily appears in two forms:

1. ON-delay or type-E timer (timing begins after energization of the input).
2. OFF delay or type-D timer (timing begins after deenergization of the input).

The timer function is represented in Fig. 15.22.

It is noted in the timing diagrams below that there is short reset-time delay in both the ON-delay and OFF-delay timers. This is usually inherent in a timing circuit and is the minimum time that the energy can be dissipated from storage elements such as capacitors. In fact, an ON-delay

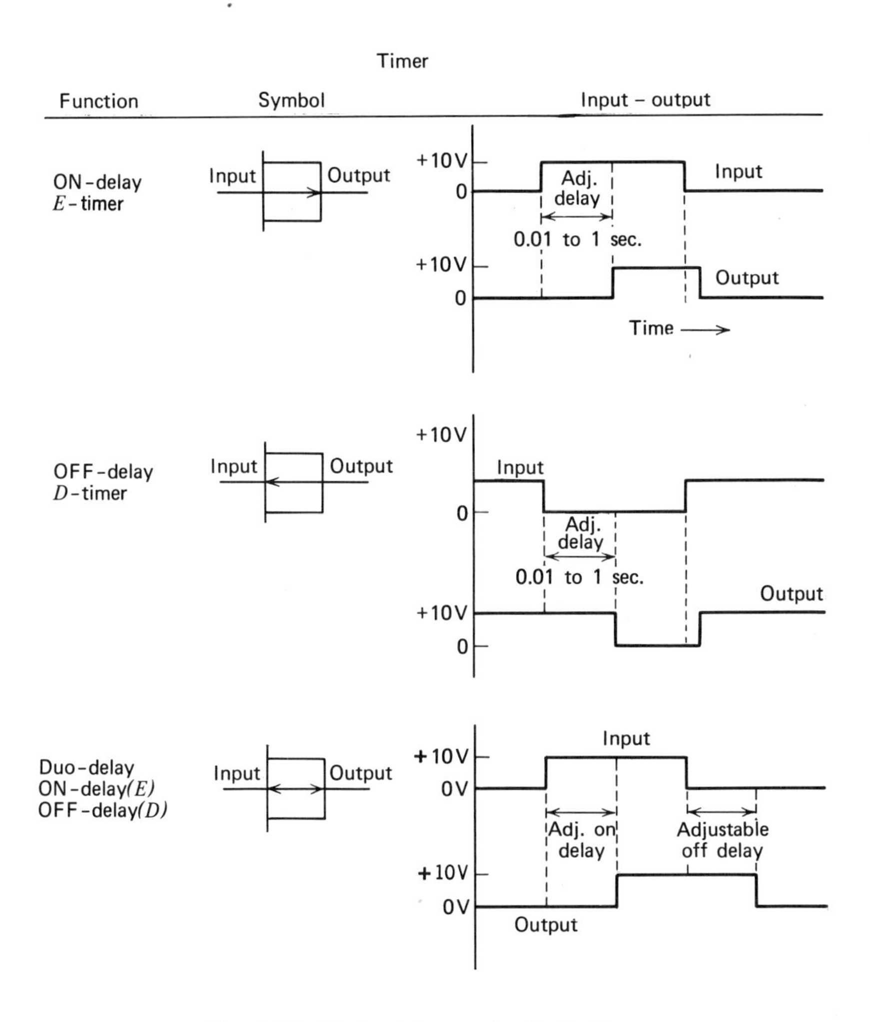

FIG. 15.22 Timing Diagram for Static Timer

timer might be looked upon as a duo-delay timer with the OFF-delay set at the minimum possible value.

Shift Register. The shift register is a memory element used for the purpose of storing binary bits or signals and shifting or sequencing them in accordance with a particular desired control program.

The shift register can be thought of as a device for handling information somewhat as a conveyor might handle boxes. This particular register represents a single-channel shift register, five positions long, possessing the capabilities of series or parallel operation for both set-in and readout. A flip-flop provides memory for each of the five positions.

Referring to the timing diagram (Fig. 15.24), the *shift pulse,* the *serial input,* a *typical parallel input* for No. 3, and the *five outputs* are shown. Starting with no bits in the register, a bit is inserted in position 1 by means of the serial input. Note that an output appears at output 1.

When the shift pulse is applied, this bit is removed from output 1 and appears on output 2. Note that the shift pulses may be inserted at random on any desired timing basis, but the pulse itself is usually required to be of a specified pulse width. However, they may be a few microseconds apart, or they may be many seconds or hours or even days apart.

With the next application of the shift pulse, the output is shifted to No. 3, and is removed from No. 2. In like fashion, the "bit" may be shifted all the way to output 5 and finally shifted out of the register entirely. The sketch at the bottom of the timing chart shows the way the state lights appear on the edge of the shift register board under each situation. See Fig. 15.23.

It is also possible to insert a bit at any one of the five positions. Only one of these is shown for simplicity, the one for output 3. If a bit is put into the parallel input 3, it appears immediately at output 3 without having to be shifted through outputs 1 and 2. Now when a shift pulse is applied, this output moves to No. 4. If a serial input is now placed into the shift register, an output, of course, appears on No. 1, as previously explained. There now is an output on No. 1 and No. 4 and when the next shift pulse appears, the outputs are shifted to No. 2 and No. 5, respectively.

Further operation of the shift pulse causes the action as shown in the timing diagram (Fig. 15.24).

The shift register is a very useful device for control purposes where certain pieces of machinery, packages on a conveyor, or internal control functions move along in a fixed position or pattern *relative to each other,* and cause further control actuations at various points. It might be desirable to use output 3 to start motor 1, output 5 to energize solenoid

FIG. 15.23 Five-Bit Shift Register Static Logic Board

valve 3, and output 4 might not be used at all, except for memory storage of a bit between 3 and 5.

The Stepper. The stepper is a limited form of shift register which finds its greatest usage in sequencing systems in a manner similar to an electromechanical stepping switch. This particular circuit has five positions through which a single bit may be moved. A flip-flop provides memory for each position. The operation of the stepper is very similar to the shift register except that an output can appear at one and only one position.

Referring to the stepper timing chart (Fig. 15.25), a bit is first set into the first point by means of the set terminal. Note that an output appears at output 1.

Pulsing of the step circuit steps the signal from output 1 to outputs 2, 3, 4, and 5 as shown. Again the step pulses, through required to be of a specified pulse duration, can occur at any time intervals—microseconds, hours, days.

Binary Counter. The binary counter is designed for use either as a straight binary or a binary coded decimal counter. In electromechanical relay-type counters, a solenoid is energized and a pawl and ratchet mechanism pulses a mechanical assembly until it reaches the count setting, causing an output. In the static binary counter there are four outputs in binary form available on each binary counter board. These are desig-

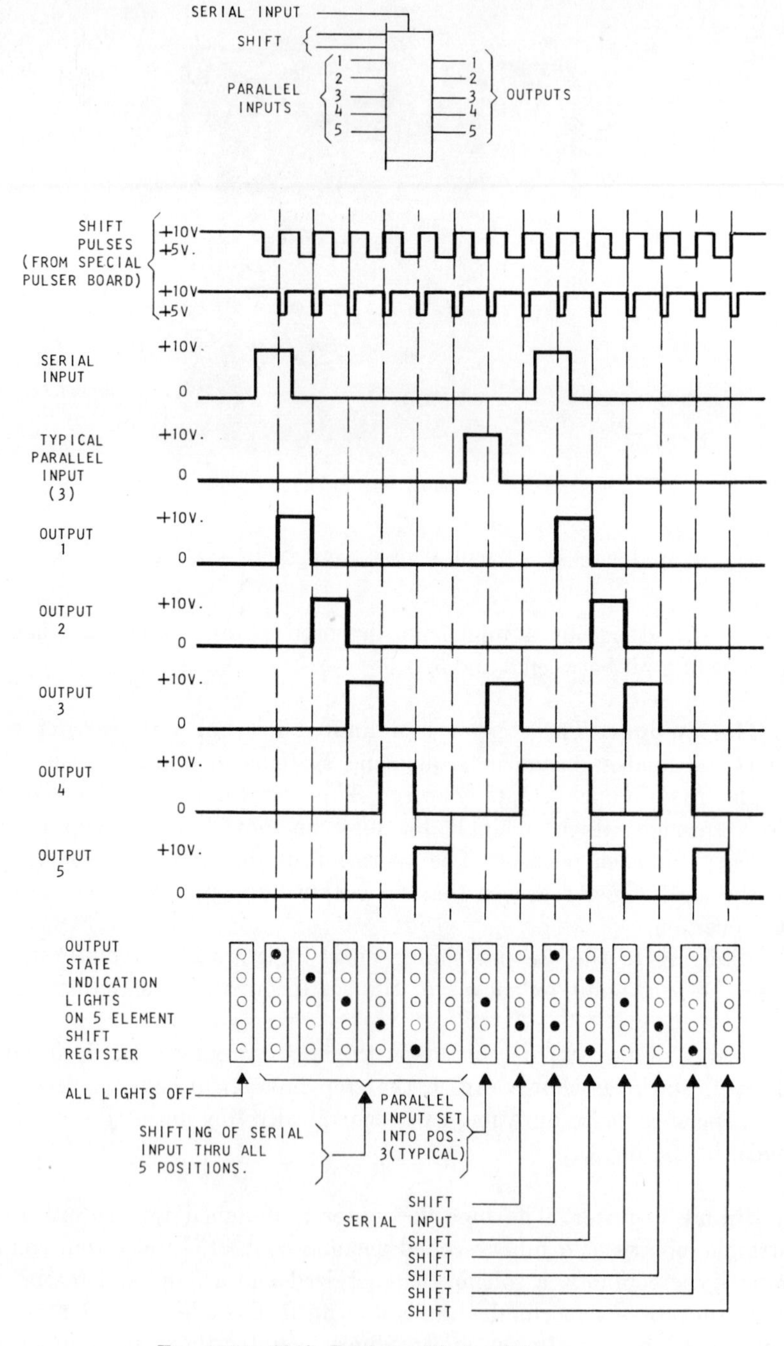

Fig. 15.24 Shift Register Timing Diagram

nated as output 1, output 2, output 4, and output 8. To get the decimal equivalent of this, this output of course must be decoded such that an output from output 2 and output 4 is recognized as the decimal 6, an output from 4 and 8 is recognized as the decimal 12, and so on.

Referring to the binary counter timing chart, it is seen that with each count pulse the state of the output is changed. Notice that output 1

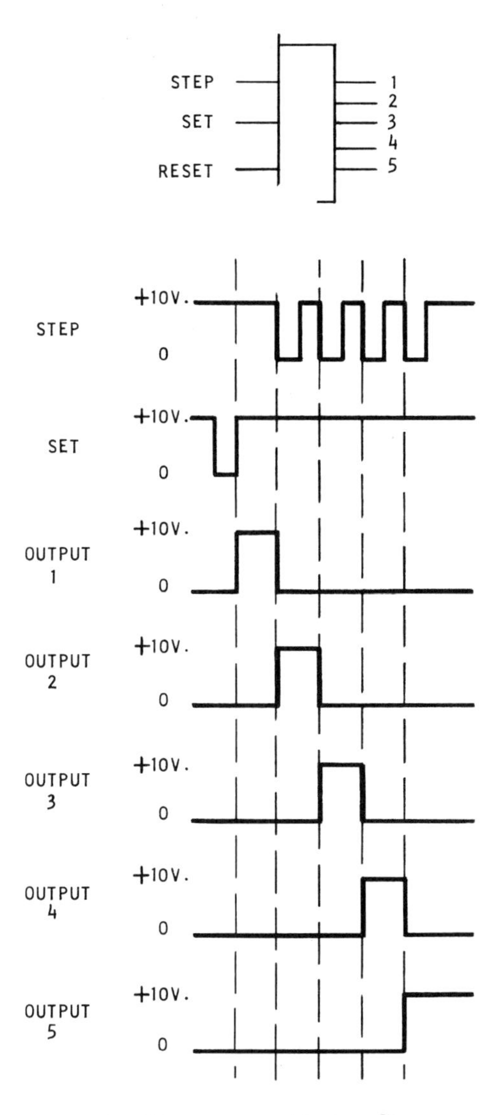

FIG. 15.25 Stepper Timing Chart

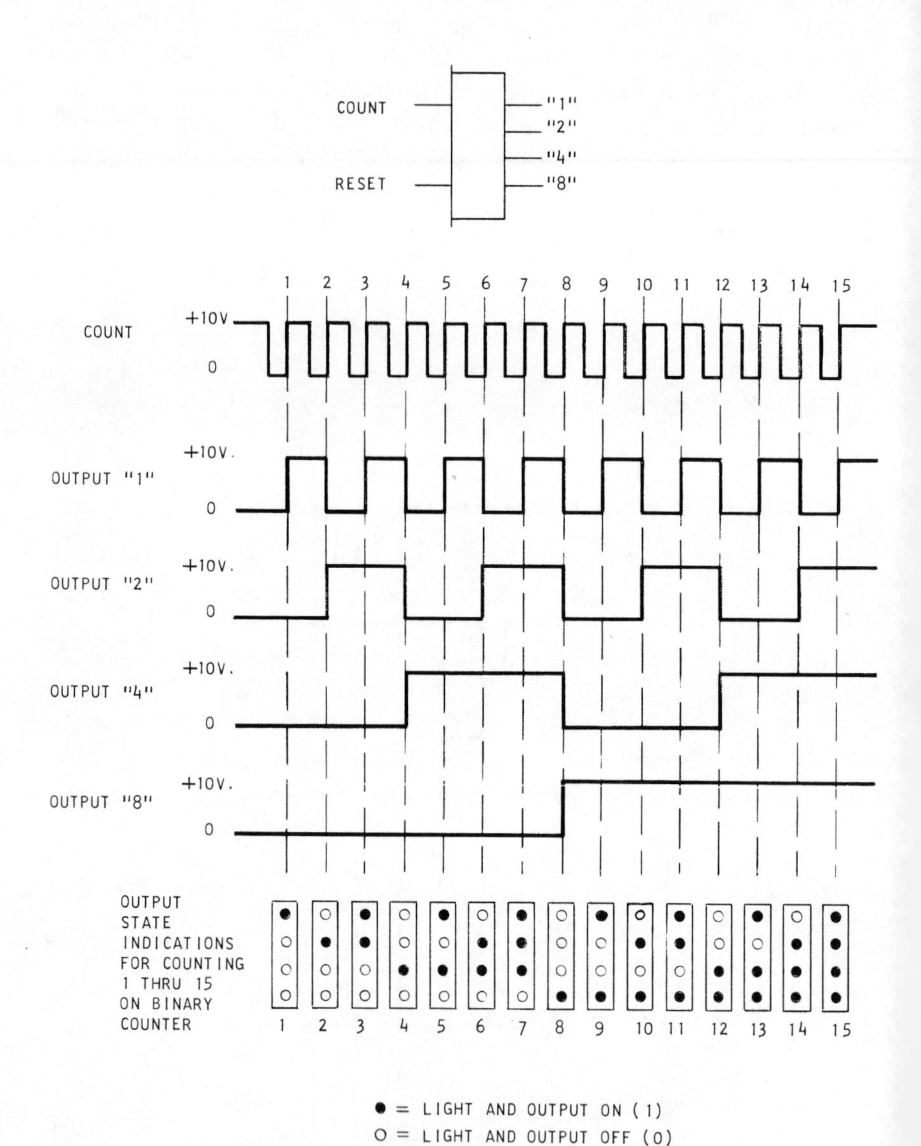

FIG. 15.26 Binary Counter Timing Chart

goes on with the first count pulse, off with the second count pulse, on with the third count pulse, and so on.

Output 2 goes on with the second count pulse, off at the fourth count pulse on at the sixth, and so forth.

Output 4 goes on at the fourth count pulse, off at the eighth, on at the twelfth. The outputs show as state indications, as well as electrical signals as indicated in the view at the bottom of the timing chart (Fig. 15.26).

If it was desired, for example, to have a certain action take place at a count of 14, outputs 2, 4, and 8 could be connected to an AND element and when all three of these were energized, which occurs only at count 14, the output of the AND switch would be energized and the desired control action would take place.

Other outputs could be similarly used as desired simply by circuit selections.

Analysis of Components—Output Amplifiers

Introduction. The output of static switches is not sufficient to provide energy to power-consuming devices, such as pilot lights, relays, and solenoids. If these devices are to be used, power amplification is required. Some of the devices available are listed in this section.

10 V Power Amplifier. This device is used to accept logic level signals from other logic elements and drive a small relay such as the output relay shown in the chart (Fig. 15.28). The relay is then used to operate ac contactor coils in another circuit. The most common arrangement is for the logic controls to have certain outputs to other control, typically motor control centers involving separate sources of power and high-power-level circuitry. It is most desirable to isolate this from the logic circuits, and the relay provides this isolation. The 10 V power amplifier is the driver for the relay.

24 V Power Amplifier. This device accepts logic level signals the same as the 10 V power amplifier, but is most commonly used to drive solenoid

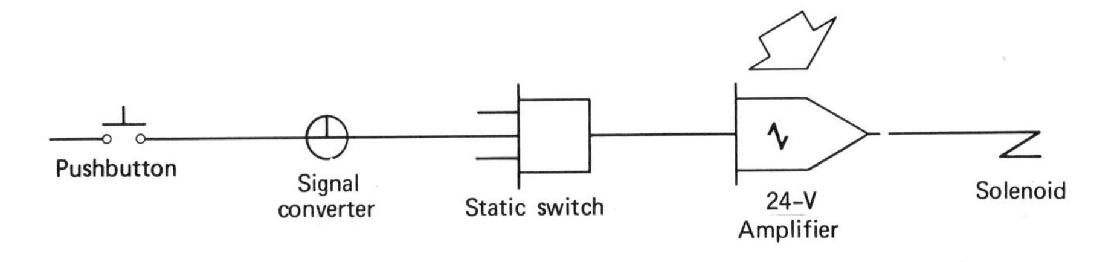

FIG. 15.27 Power Amplifier Emphasized

valves directly on 24 V dc (Fig. 15.27). Note that neither the 10 V nor the 24 V power amplifier is suitable to drive logic signals. The diode protection and isolation provided in the circuitry for these elements prevent them from being used to drive other logic elements.

SCR Amplifiers. Occasionally it is desired to have a very high-power static output. Ordinarily it is satisfactory to use the 10 V power amplifier and the relay to provide this power level signal. However, if it must be completely static, the SCR amplifier provides this output. Note that this output is 115 V ac. This unit is basically physically large as compared to the other output amplifiers, and it is somewhat more expensive.

A chart of the various amplifiers is shown in Fig. 15.28.

Useful Circuit Ideas

The Language of Logic. Indeed, it can be said that the language of logic is most logical! For example, consider the AND function, symbolized in Fig. 15.29, with its relay equivalent to the right.

By definition, an AND element will have an output only when all inputs have been energized. Relay circuitry has no word for this requirement. But note how the language of logic takes care of this control requirement.

	Symbol	Input	Output	Most Common Usage
10-V Power AND	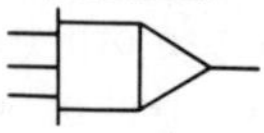	3-INPUT AND	0.5 amp 10 V dc	Relay driver
24-V Power AND	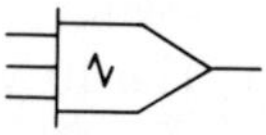	3-INPUT AND (10 V)	1 amp 24 V dc	Dc Sol. valve driver
SCR Amplifiers	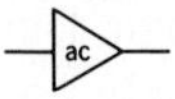 ac	1-INPUT to 10 V Reed relay coil for isolation	5 amp 115 V ac	Ac Sol. valve; Ac contactor coil; Small single-phase ac motors
Output Relay		1-INPUT from 10-V power AND (10-V coil)	1 PDT 10 amp 115–230 V	Sol. valve, ac or dc contactor coil (to obtain isolation from logic)
Output Relay		1-INPUT from 24-V power AND (24-V coil)	1 PDT 10-amp 115–230 V	Sol. valve, ac or dc contactor coil (to obtain isolation from logic)

FIG. 15.28 Output Amplifier Chart

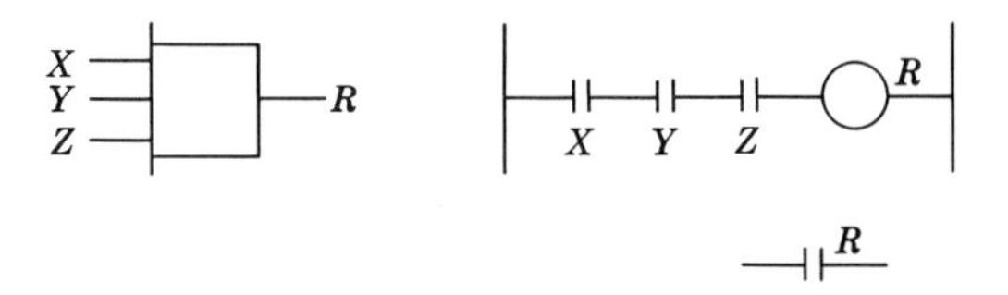

Fig. 15.29 The AND Element and Its Relay Equivalent

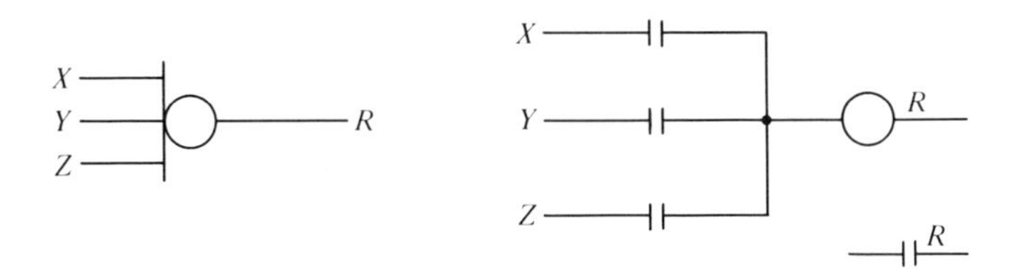

Fig. 15.30 The OR Element and Its Relay Equivalent

If inputs X AND Y AND Z are energized, an output will be present at R. Note that the word AND describes the function.

The same logical reasoning applies to the OR function. Here again, the word OR means just what it says. That is, referring to the diagram (Fig. 15.30), if input X OR Y OR Z OR any combination of these inputs is energized, an output will appear at R.

The NOT function is equally descriptive. An output is produced only when the input is NOT energized. Therefore, in Fig. 15.31, if the input X is NOT energized, an output R will appear.

The simple, straightforward English of direct static logic is a boon when designing circuits. It is possible actually to "talk out" a circuit! For example, "Limit switch $LS1$ must close AND limit switch $LS2$ must close AND pressure switch PS must close before the motor will start." Obviously, this calls for an AND switch.

Common Combinations. Suppose a control system required seven inputs, all to be energized before an output could occur. The relay circuit for such a requirement is shown in the sketch Fig. 15.32. To logically perform this function, simply use three 3-INPUT AND switches.

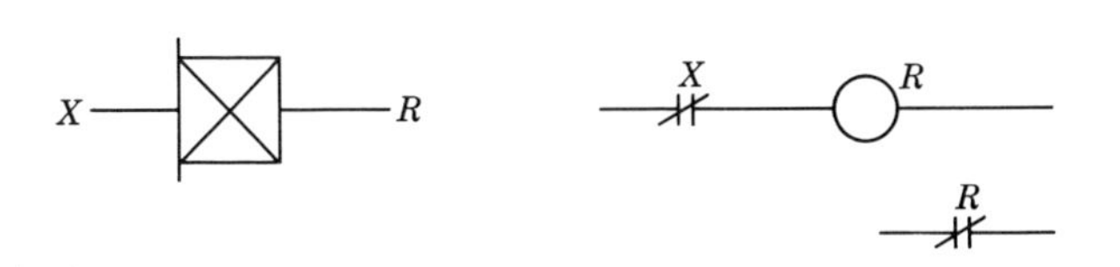

Fig. 15.31 The NOT Element and Its Relay Equivalent

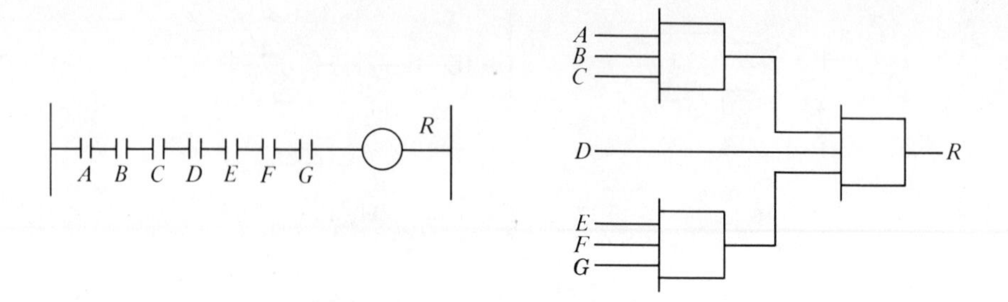

Fig. 15.32 The "7 INPUT AND"

Now, suppose that seven inputs are required to provide an AND function with a power output. The logic diagram for such an arrangement is shown in Fig. 15.33. OR switches can be given the same treatment. For example, the five pairs of contacts in parallel in the relay circuit can be easily converted to static logic by using two 3-INPUT OR switches (Fig. 15.34*a*). Figure 15.34*b* shows how three normally closed contacts in series could be handled in static logic.

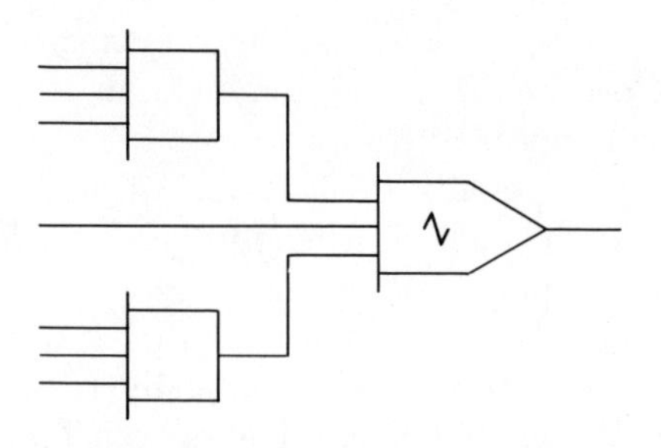

Fig. 15.33 The Logic AND of Seven Things to Control a Solenoid Driver

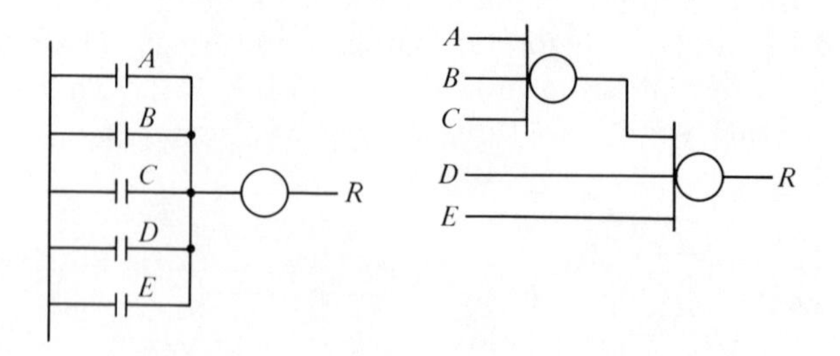

Fig. 15.34(a) The Logic OR for Five Inputs

FIG. 15.34(b) Logic Equivalent of Normally Closed Contacts in Series

Simplified Examples. Electrical control people are an imaginative lot. Given a control problem, they can frequently improve over existing control schemes. Call it a trick of the trade, a clever circuit, or a short cut, it amounts to the same thing—creative design circuits. Following are simplified examples of such creative design circuits.

The AND gate provides excellent control for simultaneous starting or stopping—with motors, for example.

The relay diagram (Fig. 15.35*a*) illustrates the function desired. Here closing contact A simultaneously energizes coils $R1$, $R2$, and $R3$, provided that contacts B, C, and D are closed. Contact A, therefore, provides the "gating signal."

Now, note how the same function is efficiently accomplished by three AND switches (Fig.15.35*b*). An input to each AND switch is "tied" to the gating bus. If the gate is at 0 V and then goes to +10 V, all the AND switches that have their other inputs satisfied will be turned on simultaneously.

The AND SEQUENCER might be useful in conveyor control. Here's what it can do.

1. Start all conveyors in sequence at the closing of a single pushbutton.
2. Stop all conveyors simultaneously at the opening of a single pushbutton.

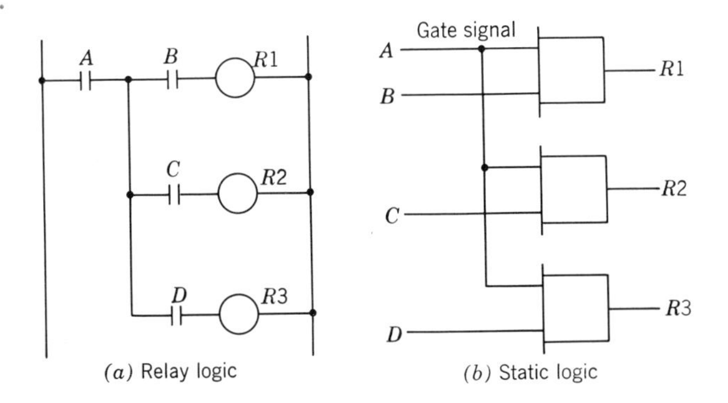

FIG. 15.35 Combinations of AND Arrangements

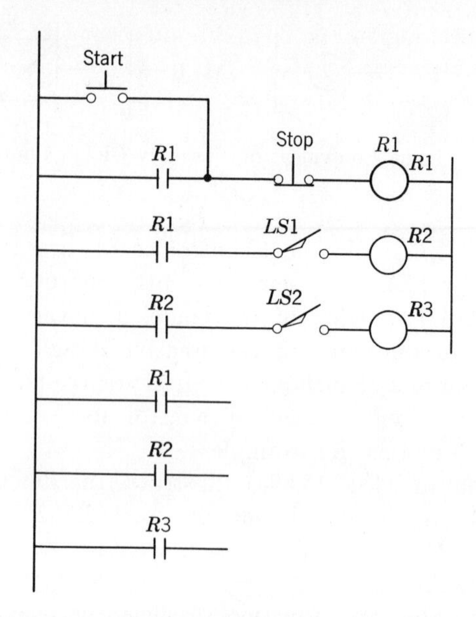

FIG. 15.36 Partial Relay Circuit for Conveyor Control

3. Continously monitor all conveyors so that all conveyors ahead of a trouble point are stopped the instant malfunction is sensed by the zero speed switch of a conveyor.

Figure 15.36 is a diagram of an AND SEQUENCER circuit, using relays.

Figure 15.37 is the static logic diagram of the AND SEQUENCER. Pushing the START button causes pickup of the first AND switch, providing an output $R1$ which can be used with appropriate amplification to control a motor. As can be seen, the output $R1$ also provides one of the input

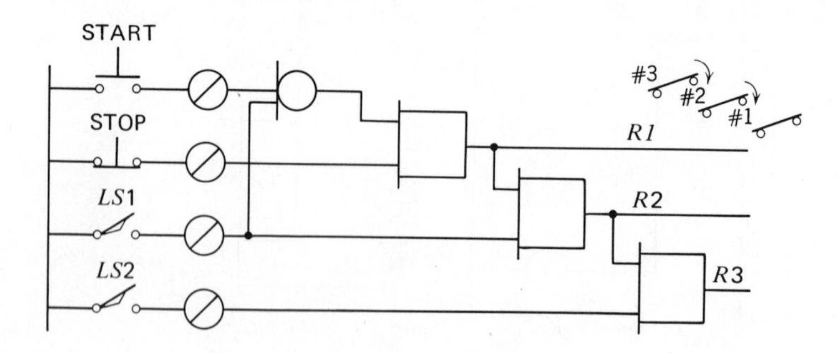

FIG. 15.37 Logic Equivalent of Conveyor Relay Circuit

signals to the next logic AND. Therefore an output at $R2$ occurs whenever an output at $R1$ is present and when $LS1$ is closed. Further circuit examination reveals that an output at $R3$ will exist when an output at $R2$ is present and when $LS2$ is closed. Symbols $LS1$ and $LS2$ are zero speed switches; $LS1$ should monitor conveyor number 1.

Fan-Out Limitations. The fan out capability of a given switch is its ability to serve as an input signal to several other static switches. The term comes from the fact that the signal wires fan out from the output terminal of the given switch to a plurality of input terminals.

Since the capacity of a transistor is limited, it is only reasonable to assume that the ability of a transistor switch to drive other switches is limited. In fact, two limits exist. One is the ability of the driving switch to provide current. The second relates to controlling the voltage at the input terminal when no output signal exists. A typical fan-out situation is illustrated in Fig. 15.38, where the AND element is used to drive a set-reset memory, another OR element, and a NOT element. Similarly, additional elements might be connected to this output of the AND. However, there is a limit to the number that can be handled strictly from a power capability standpoint. The fan-out limitation varies from company to company, but is specified by the manufacturer for his line of logic. A limitation of 7 to 10 logic signals connected to a single logic output will usually not limit the designer unnecessarily.

Mounting and Wiring. The circuitry of static logic must be assembled into some standardized hardware for interconnection to provide the complete control system. Most static switching logic elements are mounted on boards, such as shown in Fig. 15.39. Each board contains several switching elements, each element being an independent switch. Boards are usually of the plug-in type, and since all boards have the same

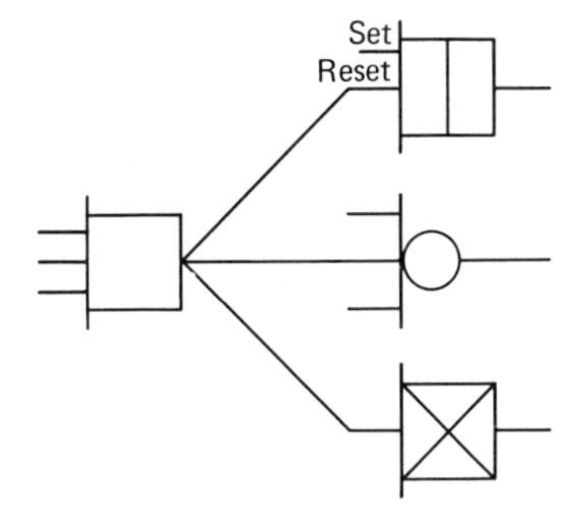

FIG. 15.38 Fan Out. Control of Several Logic Elements from a Single Logic Element

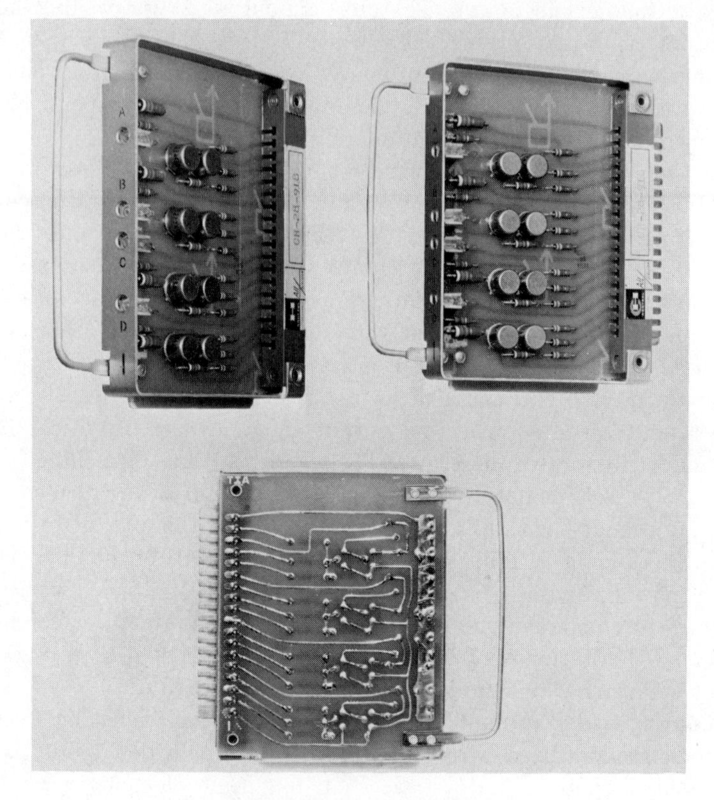

FIG. 15.39 The Static Logic Board

dimensions and conform physically, they can be arranged in buckets to suit circuit convenience. Each circuit point on the receptacle provides for two taper pins or other insertable contacts connected in parallel, to facilitate wiring.

Buckets. The bucket is a sort of "electronic bookcase" for the boards. The one illustrated in Fig. 15.40 has a capacity of 20 boards, one of which is plugged in. Basically, it consists of a steel framework, 20 receptacles, and the necessary bus work at the rear. This basic configuration is designed to be mounted in a cutout of a steel panel, or in a conventional 20-in. wide relay rack. Because of the separate elements per board, the bucket contains an equivalent of approximately 80 relay circuits. If desired, buckets can be surface-mounted to panels by means of a family of hinged stand-offs that permit rear access.

Rear of Bucket. The standard bucket includes receptacles and two bus bars. In wiring the buckets, it is first necessary to make connection between each plug and its respective "A" or "V" bus bar. A separate wire for each plug is recommended to assure a "solid" voltage supply.

The "A" bus bar also serves as the positive side of the line for the 24-V circuit. The negative side of the 24-V supply should be brought in by wire to the affected boards. The bus bars on the rear of the bucket should be connected to the voltage source with No. 12 wire.

The recommended wire size for logic wiring is No. 22. These wires should be terminated with taper pins for plug insertion, and spade terminals for connection to bus bar.

This configuration of boards and bucket is only one of many mechanical arrangements which may be used. Nearly all manufacturers employ some form of printed circuit cards plugged into sockets on a metal frame.

Coding and Diagram System. Symbology and identification coding varies in the electrical control industry. The system shown has proved very successful, and its use is suggested since it is built around the marking already present on the boards.

The following pages are devoted to the interpretation of the symbols that appear in diagrams, and the marks that are printed in the logic boards.

The front of a logic board shows several markings (Fig. 15.41). At the top appears a symbol that identifies the function of the board—an AND function in this case. At the bottom, appears a function number which identifies a specific circuit. The familiar convention of "from left to right" is used on diagrams. That is, inputs are on the left; outputs, on the right.

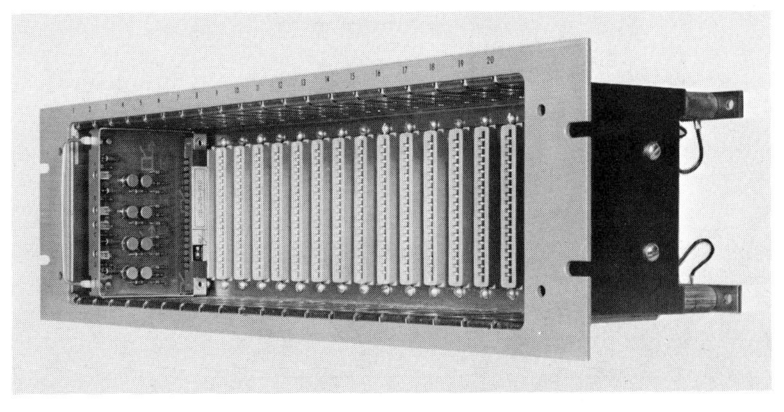

FIG. 15.40 Twenty-Board Logic Bucket, Showing One Board Mounted

Terminal Markings. Because more than one element is on a board, identification of terminals is necessary. The letters "A" to "V," with omissions, identify the specific terminals on the plug of the board. The "A" and "V" terminals of all boards are used for the 10 V power supply.

The symbol for one element of an AND board is shown in Fig. 15.43, just as it would appear in a diagram. The letters inside the small circles identify the plug markings. For example, the circled *T* in the Figure indicates that this termination goes to the *T* terminal on the plug. The receptacle on the bucket has the identical markings that appear on the rear of the plug.

Speedy indentification of state-indication lights on logic boards makes for fast troubleshooting. As can be seen in Fig. 15.44, lights for each element are identified by letters *A, B, C,* and *D.*

Bucket Information. The front of all buckets have slots numbered, left to right, at the top. Slots have corresponding numbers on the rear of buckets, at the top. Note, too, the numbers at the bottom of the panel (Fig. 15.45). If the bucket is marked in this manner, when the control is wired, a speedy eye check instantly reveals any board that is not in its proper slot.

Bucket Identification. Although the buckets are not physically identified, they are nevertheless always arranged in an alphabetical sequence from left to right—*A, B, C, D.* If desired, buckets can be identified in this manner with small squares of pressure-sensitive tape, as shown in Fig. 15.46.

The Final Symbol. Figure 15.47 shows a typical AND symbol that might appear on a logic diagram. A simple "reading" of the code is all that is required to trace circuitry systematically and determine the

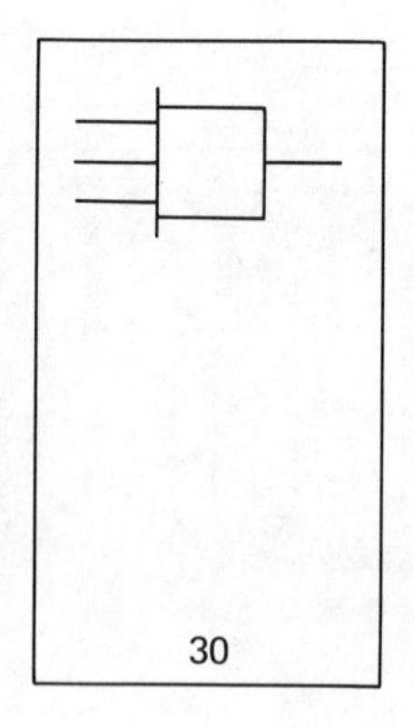

FIG. 15.41 Front Appearance of Logic AND

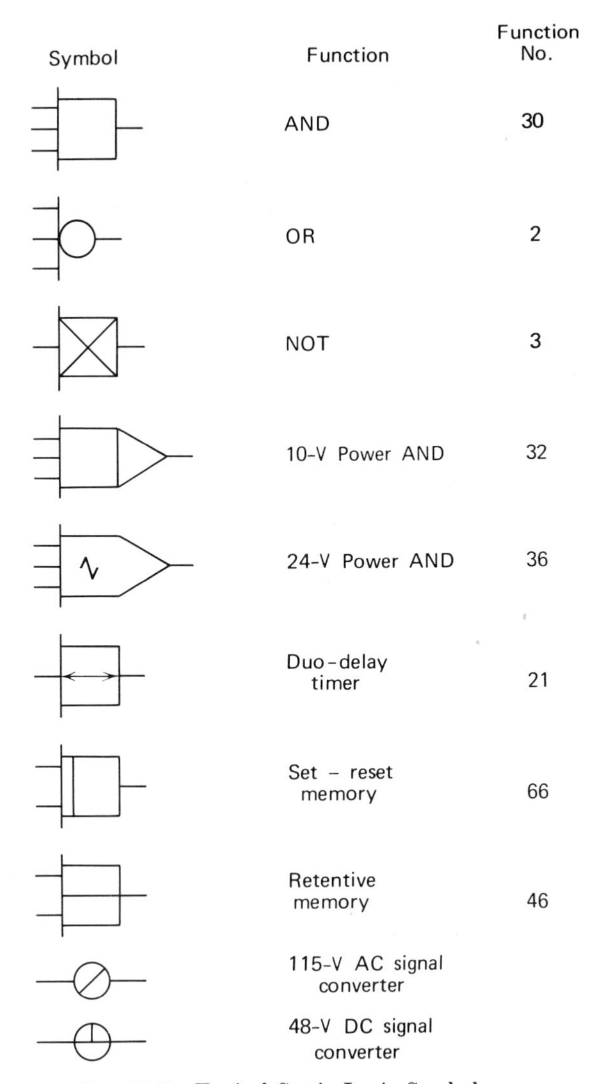

FIG. 15.42 Typical Static Logic Symbols

physical location of the portion of control under test. Here, for example, is what the code illustrated in Fig. 15.47 reveals: The AND element can be found in panel No. 3, bucket *C*, board No. 5, and its state-indication light is identified by letter *D*.

With such complete information, the logic diagram is the only "tool" required to effect fast testing for speedy repair or replacement.

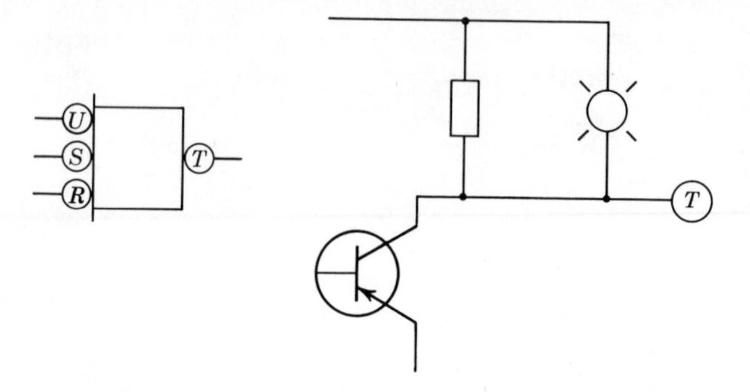

FIG. 15.43 Logic AND Element with Terminal Designations

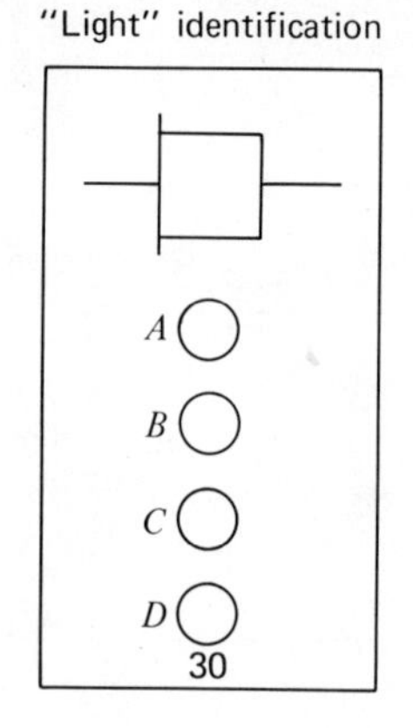

FIG. 15.44 Logic AND Board Appearance Showing Indicating Lights

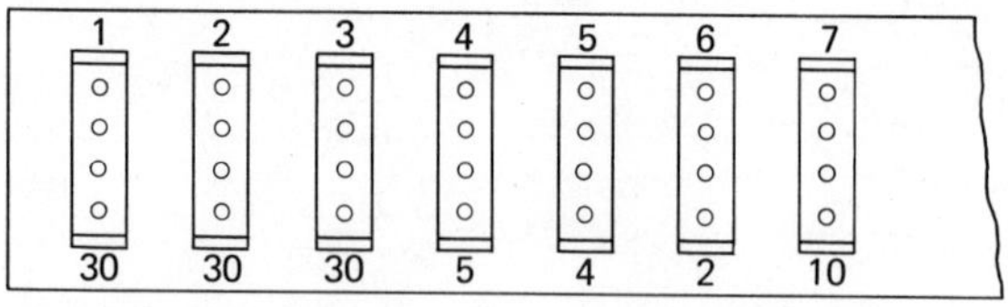

FIG. 15.45 Arrangement of Boards in the Bucket

Microelectronic Integrated Circuits (IC). The integrated circuit is coming to be more and more important in the design of industrial control digital circuits. Simply stated, the integrated circuit is the electrical equivalent of a large number of discrete components cleverly fabricated in a mechanical configuration of very small size and high reliabil-

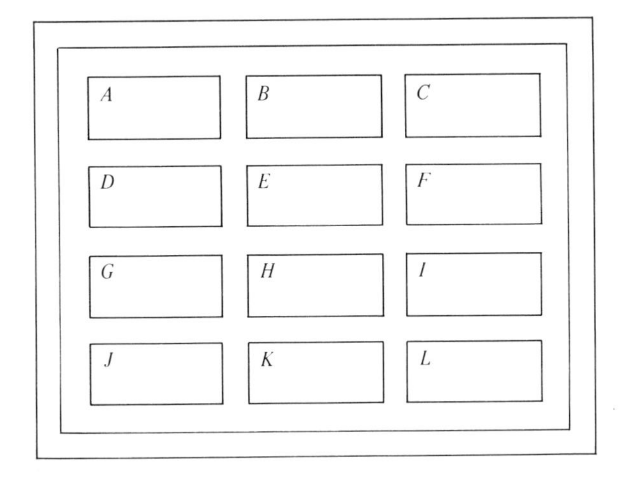

FIG. 15.46 Arrangement of Buckets on the Control Panel

ity. The entire circuit consisting of transistors, diodes, resistors, and so on, is made as one entity, somewhat as a single transistor is made. Integrated circuits are divided into essentially three basic categories.

RTL-Integrated Circuits. These were the first type to be commercially available, since they represented the most direct transition from the discrete logic component circuitry then being used. A typical circuit is shown in Fig. 15.50. Note that this is a form of inverting logic whereby, when A OR B OR C is positive, Z is not positive (is negative). When A AND B AND C are not positive (are negative), Z is positive.

When any one of the inputs is present (positive), the transistors conduct and the output Z is connected to ground (negative).

DTL-Integrated Circuits. In this version (Fig. 15.51) the resistors are primarily replaced by diodes and hence the name *diode-transistor logic.* This comes about because it is easier to fabricate diodes on the microcircuit chip than it is to fabricate resistors. The diodes, in addition to consuming less power, also provide some noise immunity by their inherent threshold characteristic. The particular version shown in the circuit is really the 3-INPUT AND configuration. Note that an output, although inverted, appears at Z when and only when A AND B AND C are connected to the positive V bus. When any one of them is negative, the output Z is connected to the positive bus through the resistor.

TTL-Integrated Circuits. The transistor-transistor logic is a more recent development in IC's. Since transistors, as well as diodes, are easier to fabricate by this process than are resistors, it is apparent that many

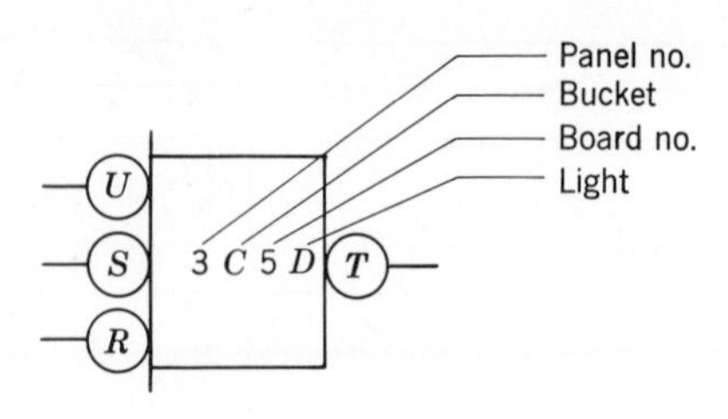

FIG. 15.47 Logic Element with Diagram Designation

FIG. 15.48 Logic Panel "Five-Pack" Showing Board Marking and State Indication Lights

FIG. 15.49 A Large-System Logic Panel for a Blast-Furnace Charging Control

sophisticated forms of electronic circuitry can be designed into a single integrated circuit package. One such circuit is shown in Fig. 15.52. The TTL circuit is ideally suited for use in some forms of digital logic where high speed is extremely important. Such an example is a modern digital computer.

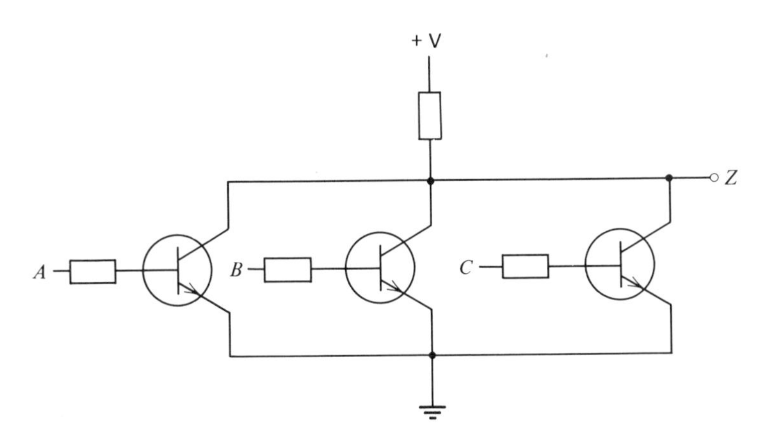

FIG. 15.50 Circuit for Three-Input RTL-Integrated Circuit

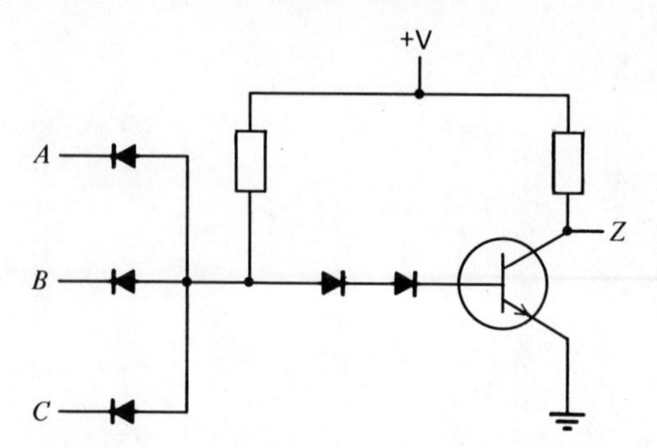

FIG. 15.51 Circuits for DTL-Integrated Circuit

Integrated circuits are available from a wide variety of manufacturers. Since the integrated circuits are so small, one of the immediate problems is the terminals which must be provided to get wires and signals in and out. The package simply does not have room for terminals. Because of this problem, a number of packaging techniques have been used, nearly all of them associated with the mounting of these chips on printed circuit

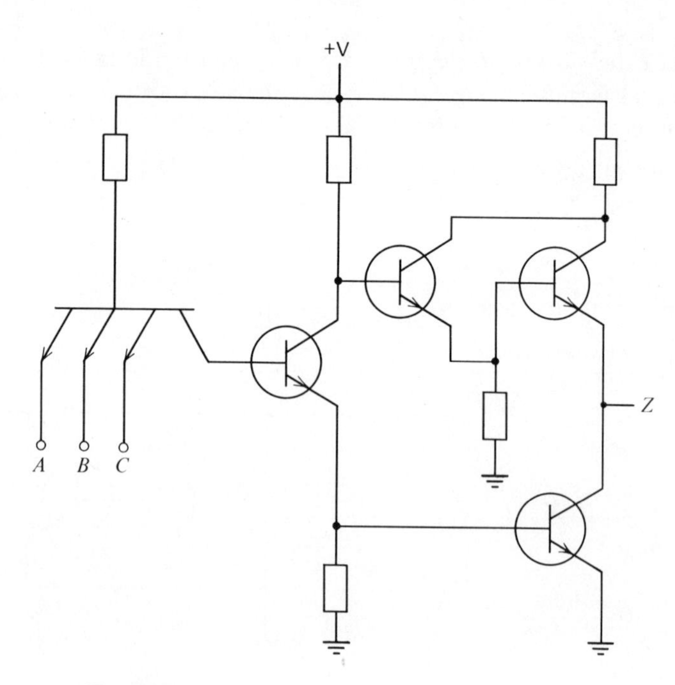

FIG. 15.52 Circuit for TTL-Integrated Circuit

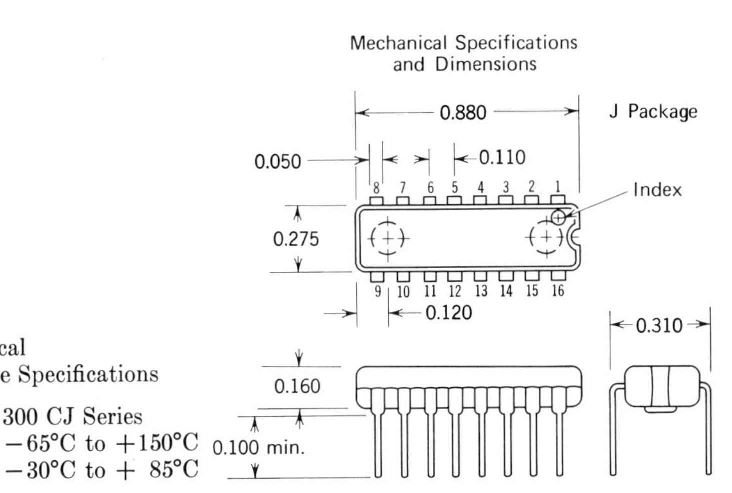

Typical
Temperature Specifications

	300 CJ Series
Storage	−65°C to +150°C
Operating	−30°C to + 85°C

ELECTRICAL SPECIFICATIONS at 25°C and V_{CC} = 12V (Unless Otherwise Specified)

Parameter	Symbol	Limits Min	Limits Typ	Limits Max	Units	Test Conditions
Output voltage "one" state	V_{OH}	10.5	11.3		V	V_{IN} = 5.0V
Output voltage "zero" state	V_{OL}		1.2	1.5	V	V_{IN} = 6.5V, I_{SINK} = 8.5 mA (60 mA for 301)
Output voltage	V_{OL}			500	mV	V_{IN} = 6.5V
301				1.8	V	I_{SINK} = 60 mA
302				0.500	V	I_{SINK} = 30 mA
302				1.2	V	I_{SINK} = 60 mA
323				0.500	V	I_{SINK} = 10 mA
370				0.500	V	I_{SINK} = 6.8 mA
Input current (1 load)	I_{IN}		1.2	1.7	mA	V_{IN} = 1.5V
Noise immunity	NI					
("zero" state)		3.5	4.5		V	
("one" state)		4.0	5.5		V	
Fan out	FO					
gate		5	9			
buffer		36	50			
Leakage current	I_L		0.1	1.0	μA	V_{IN} = 12V, All other inputs grounded
Output voltage "one" state loaded	V_{OHL}	7	8		V	V_{IN} = 5.0V, I_O = −5.0 mA (Except: 302, 323, 331, 361, and 362)
Propagation delay (gates)			60		nsec	
Clock rate (F/F)						
311			4		MHz	
312			8		MHz	

FIG. 15.53 Typical Configuration and Ratings of a Line of Integrated Circuit Logic (Data Courtesy of Amelco Semiconductor Division of Teledyne, Inc.)

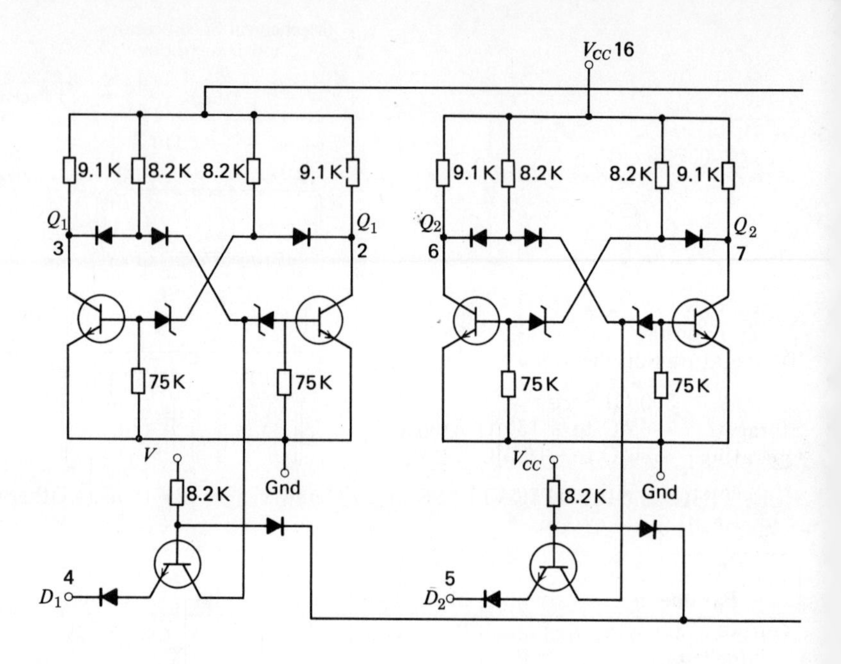

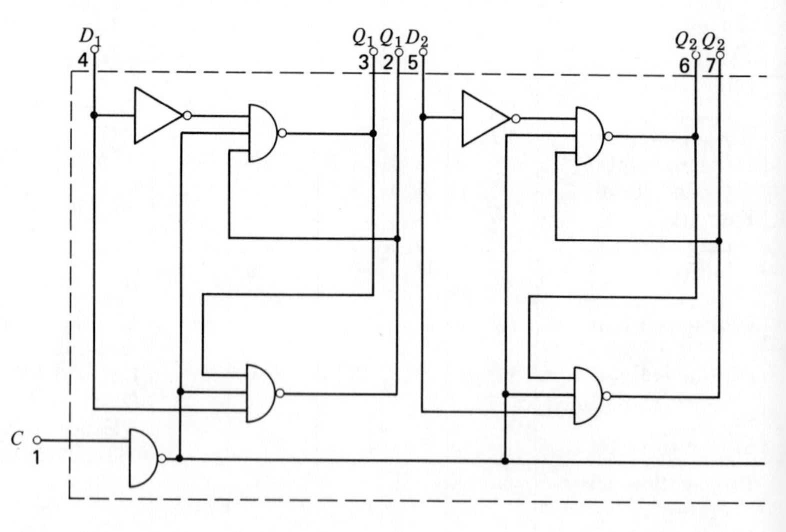

Fig. 15.54 Elementary, Logic Diagram, and Pin Connections for IC Flip-Flop (Courtesy of Amelco)

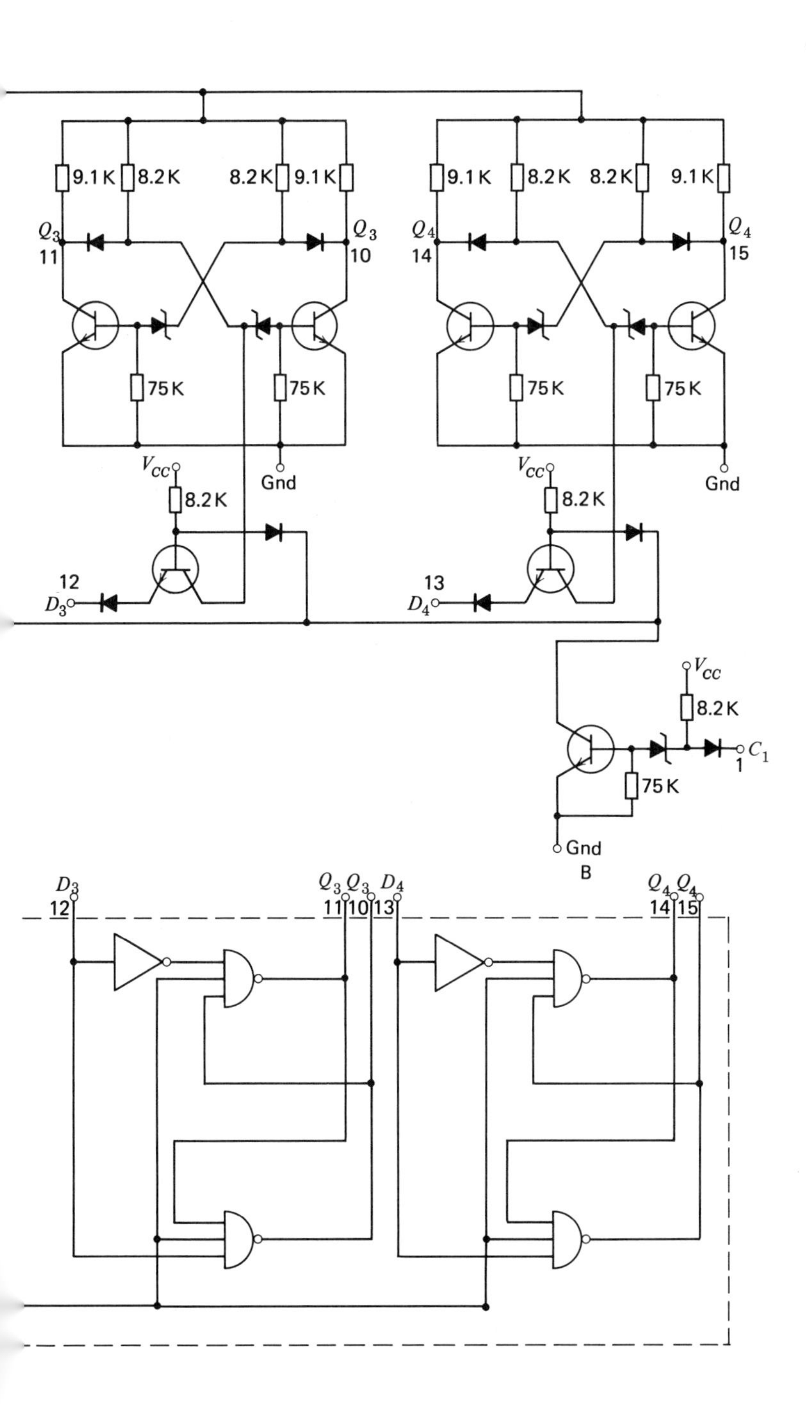

9.1 K
8.2 K
8.2 K
9.1 K
Q_3
11
Q_3
10
75 K
75 K
V_{CC}
Gnd
8.2 K
12
D_3
9.1 K
8.2 K
8.2 K
9.1 K
Q_4
14
Q_4
15
75 K
75 K
V_{CC}
Gnd
8.2 K
13
D_4
V_{CC}
8.2 K
C_1
1
75 K
Gnd
B
D_3
12
Q_3 Q_3 D_4
11 10 13
Q_4 Q_4
14 15

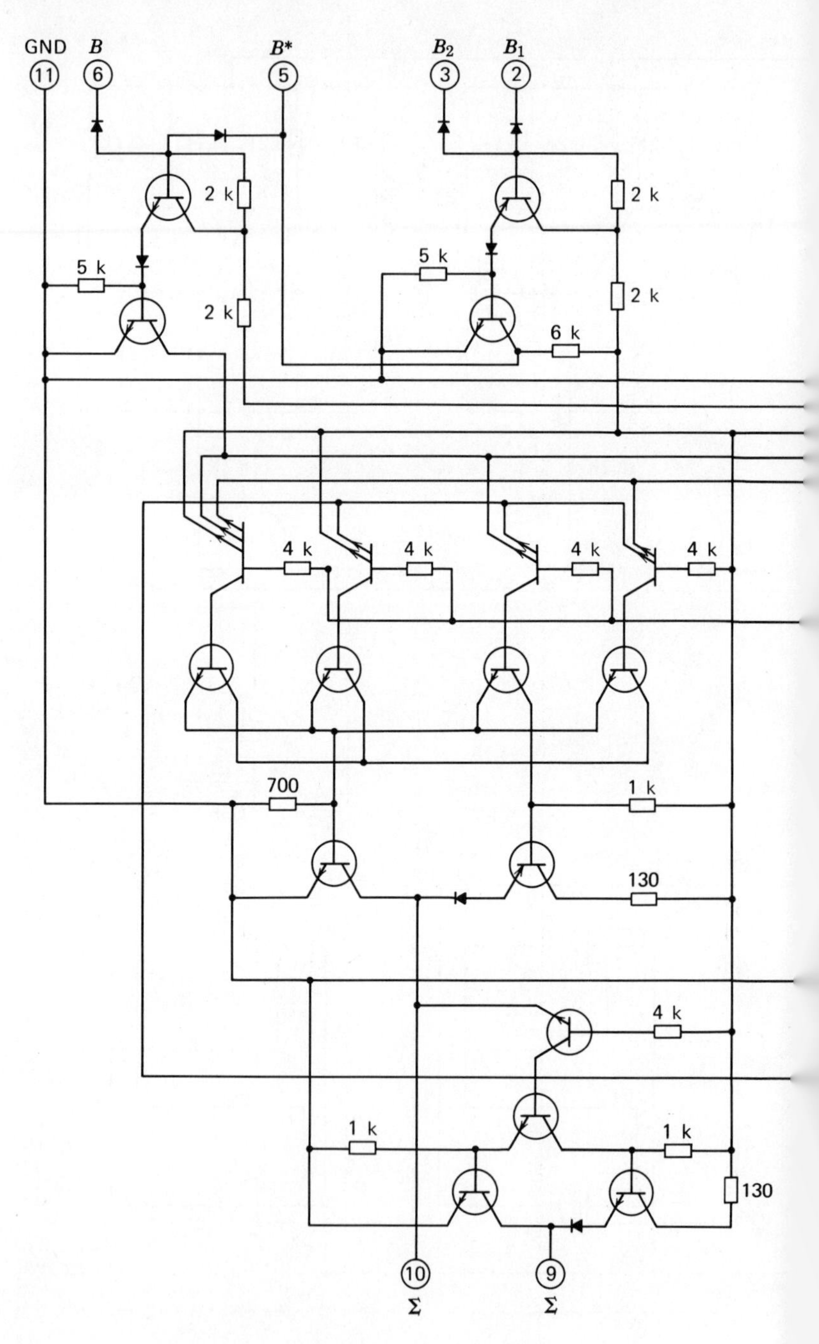

FIG. 15.55 Elementary Diagram for a Gated Full Adder
(Courtesy of Texas Instrument Company)

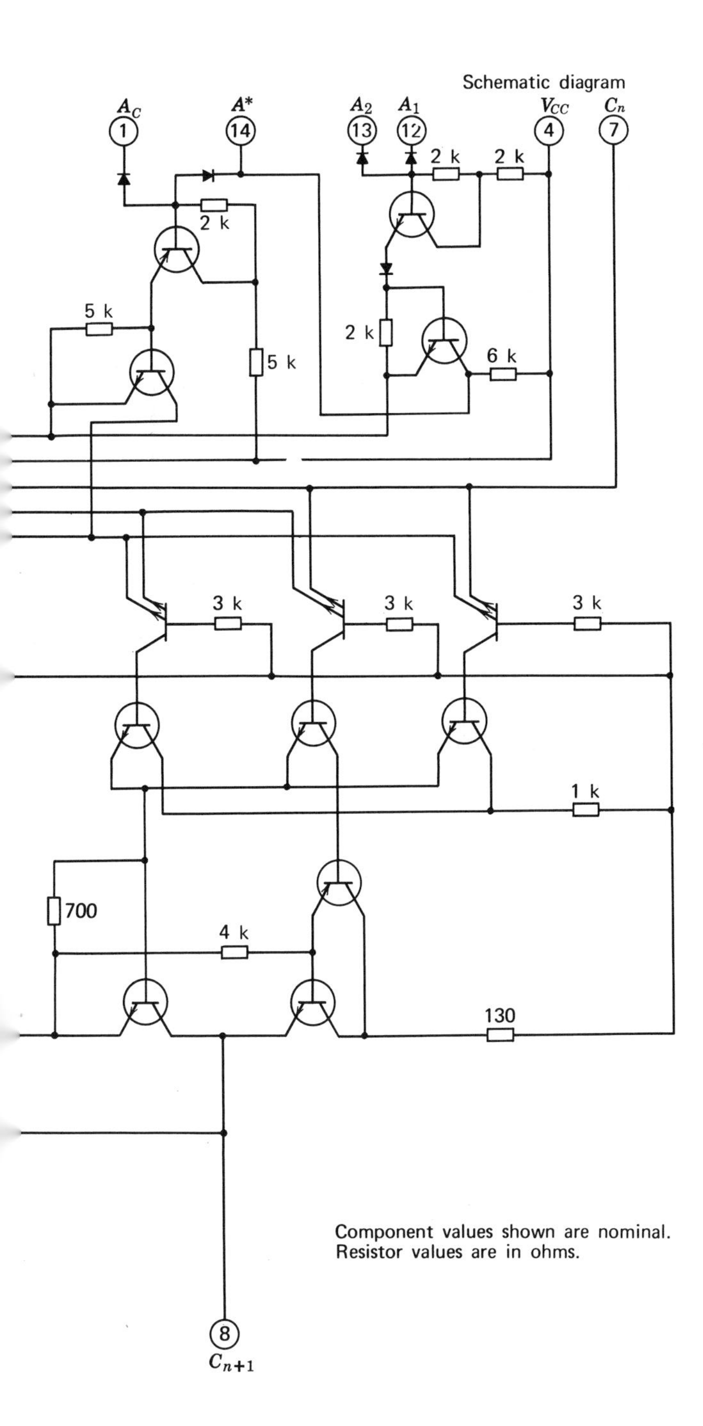

Component values shown are nominal.
Resistor values are in ohms.

cards, or into receptacles which are soldered into the PC cards. However, this only amplifies the problem since now all of these terminals must somehow be brought out from the board to a plug and receptacle, a set of wire leads, or some other suitable termination.

Because of this difficulty, and because the IC's are basically small in the first place, one technique is to arrange each circuit board to contain a complicated system function. In this way the "interwiring" is done as a part of the printed circuit board and only the minimum number of inputs and outputs are brought out. Where it is possible to group these functions, a great saving in wiring cost as well as space may be achieved.

It is beyond the scope of this chapter to describe in detail the vast quantity of logic circuits that are available with integrated circuits. Figure 15.53 is typical of one such line of integrated circuit logic. Note that this is available in two mechanical configurations, the round *G* package and the in-line "flat pack" *J* package. Note that each of these is a 16-pin configuration. Another standard package is similar to the *J* package, but with only 14 pins. These pins are arranged on specified centers for ease in mounting in printed circuit boards accurately drilled to these center distances.

Figure 15.54 is an example of a circuit arranged in the package shown in Fig. 15.53.

Figure 15.55 is an example of a more complex electrical function embodied entirely on a single IC chip.

MSI and LSI. Medium-scale integration (MSI) and large-scale integration (LSI) are further extensions of the electronic integrated circuit manufacturing technique. As a means of comparison of the relative sizes of this equipment, let us compare this with the DSL logic board. This line of logic, as previously described, is mounted on boards about 4-in. square. As can be seen from the photographs, the board is quite crowded with discrete components for the shift register, which represents about the maximum density that can be accomplished with discrete components. One of these boards can be essentially replaced by a single integrated circuit chip in a 14- or 16-pin "flat pack." This occupies about $\frac{1}{64}$ of the area, or approximately 20 of these "flat packs" could be mounted on a single board the size of the DSL board. In other words, a single *board* of IC's might be the equivalent of the whole *bucket* full of discrete component boards.

In MSI a single package might be the equivalent of a maximum of 25 or 50 transistors and large-scale integration (LSI) would have the equivalent of 100 flip-flops (300–400 transistors) in a single package.

16

THE POLYPHASE SQUIRREL-CAGE MOTOR

Because of its mechanical simplicity, the squirrel-cage motor is ideal for constant-speed applications. It can safely be installed in out-of-the-way places or in places where gas, dirt, or moisture-laden atmospheric conditions prevail, and under these conditions it will perform satisfactorily with little attention. Since the motor has no commutator and no brushes, and is strongly constructed, it is able to stand high inrush currents without injury, and it is also easy to service and maintain.

Since both the stator and the rotor of a squirrel-cage motor carry alternating current, the motor is nonsynchronous, and its speed is not strictly constant. The speed will vary with the load on the motor and with the frequency of the power supply. However, the speed is essentially constant, and the motor does not lend itself readily to speed variation or regulation.

For further information on the speed control of a polyphase squirrel-cage motor, see Chapter 18.

Construction. Figure 16.1 shows the construction of a squirrel-cage motor. The stator, or stationary member, is a laminated framework into which are wound wire coils for connection to the power supply. The laminations are made of high-grade annealed sheet iron, and the outer framework usually of fabricated steel plate. The rotor also is a laminated structure, having slots into which copper or aluminum bars are fitted. All the bars are connected together at each end by metal end rings. Blowers are used on each end of the rotor to supply cooling air, and air passages are provided in the larger rotors to permit circulation of the air through the rotor iron. The slots in the rotor are usually slightly skewed to reduce magnetic vibration and to insure a uniform torque for all rotor positions. The end plates of the motor, which house the bearings for the shaft, are usually cast of iron or steel. The motor takes its name from the similarity of the rotor construction to that of a squirrel

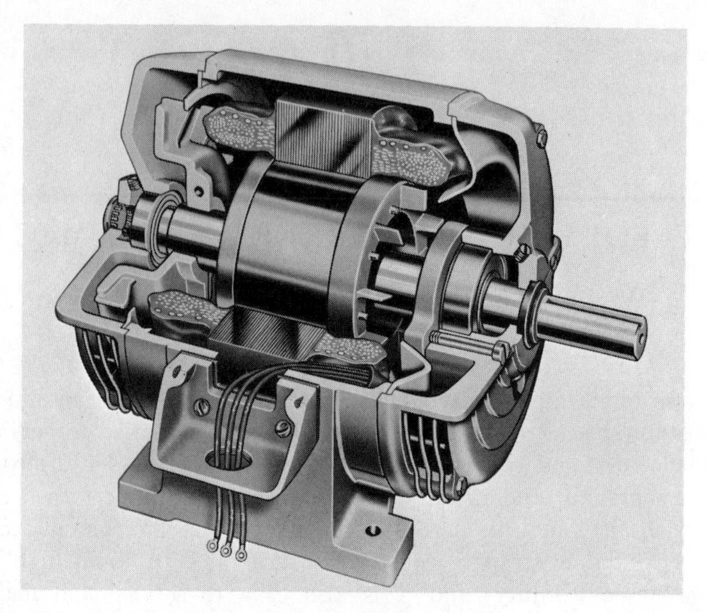

FIG. 16.1 Disassembled View of Wagner Type-CP Totally Enclosed Fan-Cooled Motor (Courtesy Wagner Electric Corporation)

cage. Since there are no brushes, no commutator, and no wire windings, it follows that no connections can be made to the rotor.

Rotating Field. When polyphase alternating current is applied to a stator winding, the resultant magnetomotive force is the vector sum of those phases. The resultant field rotates around the stator at a speed depending on the frequency of the supply voltage. This is shown in Fig. 16.2, in which the top sketch represents a three-phase two-pole stator winding, the center sketch shows the relation of the phase currents, and the bottom sketch shows the vector relationship.

Referring to the top sketch, winding 1 will create a magnetic field in a north-south direction, varying from a maximum north pole directly north to a maximum south pole directly north. Winding 2 will create a field 120 deg out of phase with winding 1, having a maximum north pole 30 deg south of west. Winding 3 will create a field 240 deg out of phase with winding 1, having a maximum north pole 30 deg south of east. Referring now to the center sketch, and calling the north direction the positive or north, pole, it is evident that at point A in the cycle, phase 1 is at its maximum positive value, and phase 2 and phase 3 are each at half of their maximum negative value. Adding these vectorially shows that the resultant north pole is at a point in the winding directly north.

At point B the conditions have changed so that the resultant north pole is 60 deg west of north. At point C in the cycle, the north pole is at 30 deg south of west, and at point D it is directly south. In the remaining half of the total cycle, the north pole continues to move around the stator winding, until it has returned to its starting point, directly

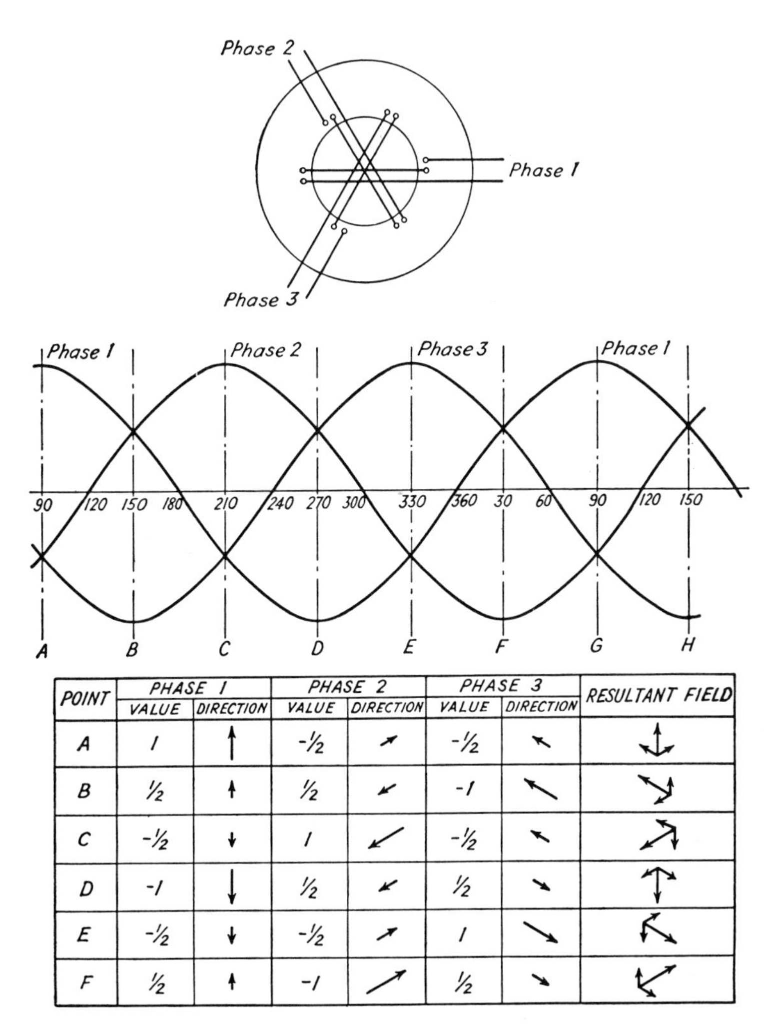

POINT	PHASE 1		PHASE 2		PHASE 3		RESULTANT FIELD
	VALUE	DIRECTION	VALUE	DIRECTION	VALUE	DIRECTION	
A	1	↑	-½	↗	-½	↖	
B	½	↑	½	↙	-1	↖	
C	-½	↓	1	↙	-½	↖	
D	-1	↓	½	↙	½	↘	
E	-½	↓	-½	↗	1	↘	
F	½	↑	-1	↗	½	↘	

FIG. 16.2 Rotation of a Three-Phase Motor Field

north. Points A, B, C, and so on, have been selected for easy calculation. Measurements made from the curves at any point will show the position of the north pole at that point. The demonstration therefore shows that for each complete cycle of the alternating current the stator field rotates 360 electrical degrees, which in the two-pole machine is also 360 mechanical degrees around the winding. If the power is supplied at 60 Hz, the speed of field rotation will be 3600 rpm. The three individual phase windings may be located in only a portion of the stator frame, with one or more additional sets of windings located in the rest of the frame, which is just another way of saying that the stator may be wound with two, four, six, or more poles. Since the magnetic field rotates 360 electrical degrees for each cycle and passes through one pair of poles in that period, it is evident that with a four-pole machine the field will rotate completely around the stator in two cycles. The general equation for the speed of rotation is

$$\text{Speed in rpm} = \frac{60 \times \text{Frequency}}{\text{Pairs of poles}}$$

Motor Speed. When power is applied to the stator of the motor, and a rotating field is set up, the squirrel-cage structure of the rotor becomes essentially the secondary of a transformer. A voltage is induced in the rotor bars, its values being determined by the ratio of the stator and rotor turns to the rate of change of the field flux. Current will flow in the rotor bars, in an amount limited by the rotor impedance, and the rotor will be magnetized. As the stator field revolves, the rotor field will be impelled to follow it, and the rotor will start to turn. Considering the two-pole 60-Hz machine, with the rotor stationary, the rate of change of the field cut by any given rotor bar will be 3600 times per minute. As the rotor speeds up, the rate at which the field is cut by a given bar becomes less and less, with consequent reduction in the induced voltage and the generated current. The maximum theoretical speed, called the synchronous speed, is reached when the rotor is revolving at the same speed as the stator field, or 3600 rpm. At that speed there would be no field flux cut by the rotor bars, no induced voltage, and no generated current. There would also be no torque, as the rotor would not be magnetic without current flowing in it, and for this reason synchronous speed can never quite be attained. Some torque is required, even without load on the motor, to keep the rotor turning against friction and windage. The difference between the actual speed and the synchronous speed will be just enough to permit the rotor conductors to cut enough field flux to induce the voltage and current required to produce the necessary torque. This speed difference is called slip.

Standard polyphase squirrel-cage motors are built for 115, 230, 460, 575, 2300, and 4600 V, 60 Hz, and 380 V, 50 Hz. The synchronous speeds for 60 Hz are given in Table 16.1.

TABLE 16.1

POLYPHASE SQUIRREL-CAGE MOTORS

Number of Poles	Synchronous Speed (60 Hz)
2	3600
4	1800
6	1200
8	900
10	720
12	600
16	450

Motor Torque. The motor torque is determined by the design of the machine and in particular by the resistance of the rotor conductors. The standard motor is built with a relatively low-resistance rotor and has a relatively low starting torque and low running slip. A motor built with a relatively high-resistance rotor will have a higher starting torque but will also have a higher slip. Such a motor is called a high-torque motor.

Figure 16.3 shows typical characteristic curves for a standard squirrel-cage motor. The starting torque is shown as about 145% of the normal full-load, full-speed value. It may vary, in different motors, from 110 to 175%. The maximum torque available, called the pullout torque, is shown as 280%. This value may vary 200 to 300% of normal. The slip at 100% torque is shown as about 3% of synchronous speed. The inrush current obtained when starting the motor on full voltage, by direct connection to the power supply, is usually from six to ten times full-load running current but may be higher if an exceptionally low-resistance rotor is used.

Figure 16.4 shows typical characteristic curves of a high-torque squirrel-cage motor. The starting torque is shown as about 200% of the normal full-load, full-speed value. The pullout torque is shown as

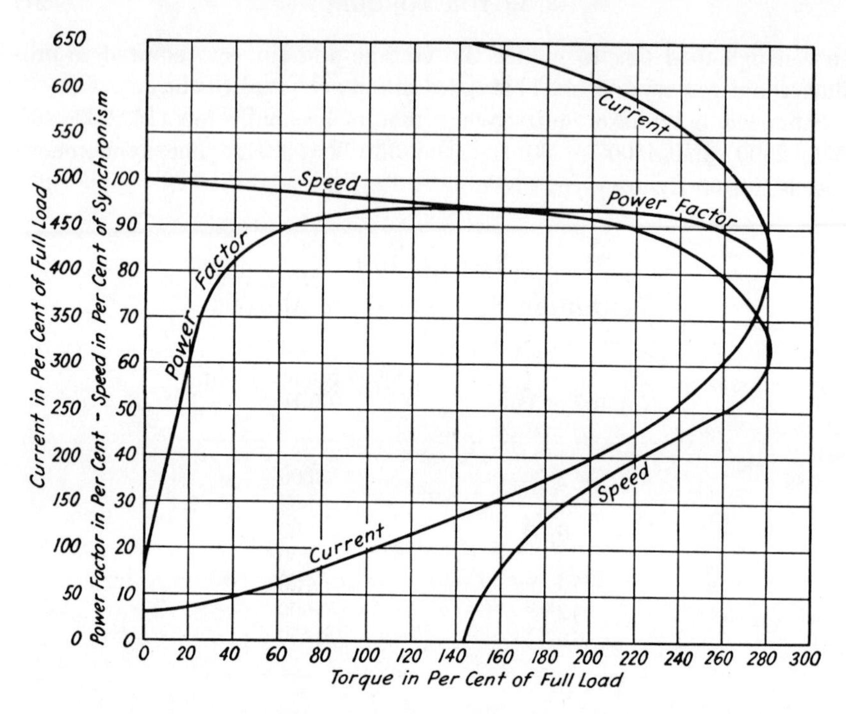

Fig. 16.3 Characteristic Curves of a Standard Squirrel-Cage Motor

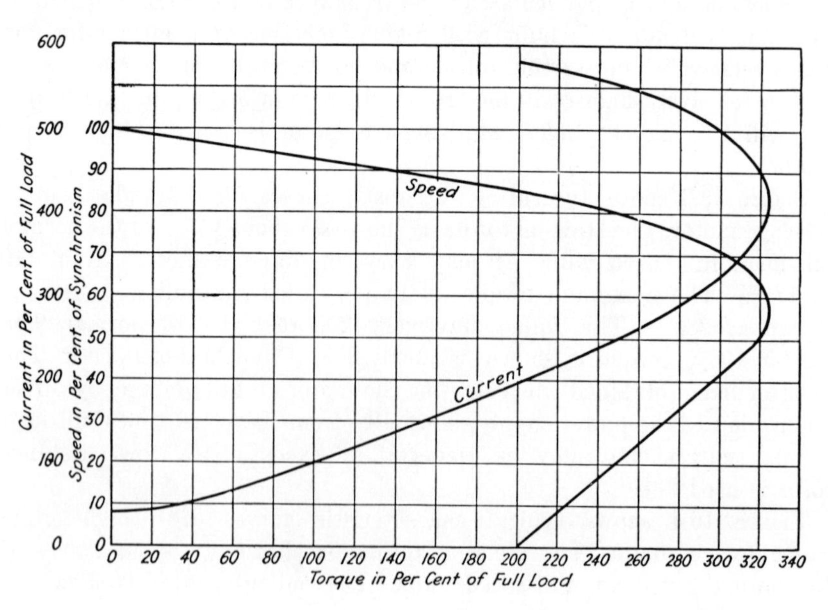

Fig. 16.4 Characteristic Curves of a High-Torque Squirrel-Cage Motor

about 325%, and the slip at full load is about 7% of synchronous speed. Because of the higher rotor resistance, the starting inrush current is relatively low, usually being from 400 to 500% of the normal full-load value. The higher-resistance rotor, therefore, has the advantage of giving a greater starting torque with a lower starting current. The higher running slip, however, results in higher heat losses in the rotor and also causes the motor speed to vary more with changes in the load.

Double-Cage Motor. In order to obtain the advantages of both the low-resistance rotor and the high-resistance rotor, motors are built with two squirrel-cage windings embedded in the same rotor core. Such a motor is called a double-cage motor. The rotor bars are placed in the same slots, one layer above the other. The inner squirrel cage is designed to have low resistance and high reactance, and the outer one has high resistance and low reactance. At standstill the rotor current has line frequency, and the larger part of it flows through the low-reactance outer winding. As the motor accelerates, the frequency of the rotor current decreases and the reactance of the inner cage becomes less effective, so that a constantly increasing proportion of the current flows through

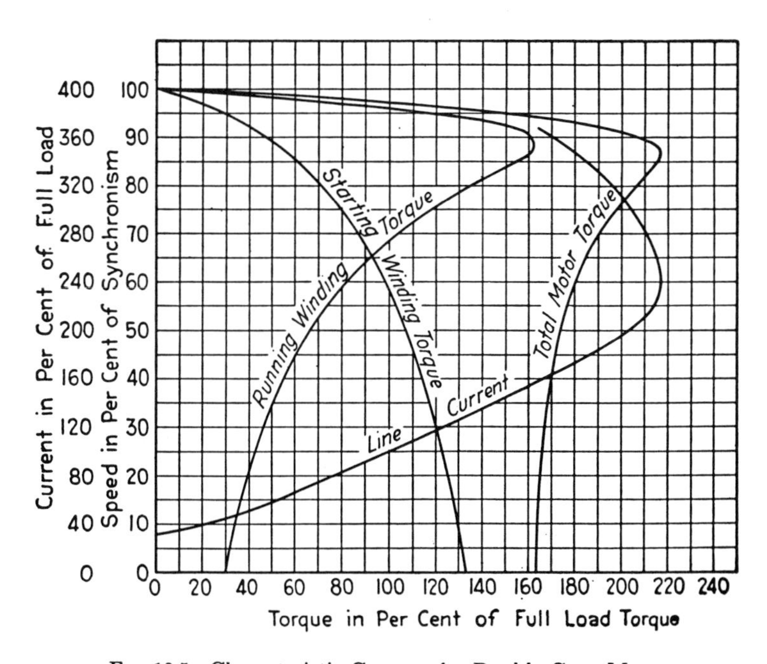

FIG. 16.5 Characteristic Curves of a Double-Cage Motor

the inner cage. The total torque is the sum of the torques of the two windings.

Figure 16.5 illustrates the characteristics of the motor. The starting torque will be from 150 to 250% of the normal full-load running torque, and the maximum pullout torque will be from 175 to 225% of normal torque. The slip will be from 3 to 5% of synchronous speed. The starting currents of a motor of this type are low enough to permit the motor to be started directly from the line in sizes up to 40 hp without exceeding the limitations set by power companies. Above this size, the inrush, when starting directly from the line, will not exceed that of a standard motor started on 80% voltage.

Starting Methods. Theoretically, there is no reason why any squirrel-cage motor could not be started by connecting it directly to the power lines. If this were done, the inrush current would be from four to ten times the normal running current of the motor. This inrush would not damage the motor, but it might cause too great a disturbance on the power line. It might also impose too great a shock on the machine being started. It is often necessary, therefore, to use reduced voltage starting.

There are five common methods of starting squirrel-cage motors.

1. Across-the-line starting, which connects the motor direct to full line voltage by means of a manually operated switch or a magnetic contactor.
2. Autotransformer starting, which starts the motor at reduced voltage from an autotransformer and then connects it to line voltage after the motor has accelerated.
3. Primary resistor starting, which introduces a fixed or variable resistor in the primary of the motor during the accelerating period and then short-circuits this resistor to apply full voltage to the motor after it has accelerated.
4. Star-delta starting, which necessitates special motor connections. This method gives approximately 58% of line voltage at the motor terminals, 58% full-load current, and 35% torque.
5. Part-winding starting, with which a part of the motor winding is connected to the lines as a first starting step, and, after a time delay, the remainder of the motor winding is connected to the lines.

In addition to these five methods, starting through a primary reactor instead of a primary resistor is sometimes done.

Across-the-Line Starting. The advantages of across-the-line starters are evident in that they are simple, easy to install and maintain, and inexpensive. Their disadvantage is that the motor draws an inrush current of four to ten times full-load current. With a small motor this does not make any difference, nor does it with a large motor if the power supply is adequate. However, most power companies object to the line disturbance caused by connecting large induction motors directly to the line. The power-generating companies have developed rules and values as suggested practice covering the installation of squirrel-cage motors on central-station distributing systems.

Manual Starters. Ratings for across-the-line manually operated starters (including reversing), with or without overload relay or other auxiliary devices, in any enclosure, for use with any type of induction motor, are given in Table 16.2.

TABLE 16.2

RATINGS OF MANUAL LINE STARTERS

Size	Horsepower at 115 V, Three-Phase	Horsepower at 230 V, Three-Phase	Horsepower at 460–575 V, Three-Phase
M-0	2	3	5
M-1	3	7½	10

Several kinds of devices are used for manual motor starters. Where a small compact starter without overload relay is desired, a suitably constructed three-pole snap switch in sizes 00 and 0 is satisfactory.

In sizes 0 and 1, starters are available which are essentially like a magnetic contactor but without the operating magnet and coil. The contactor mechanism is arranged to be closed and opened manually by pushbuttons mounted in the cover of the starter and operating through a positive make-and-break toggle mechanism. These starters are provided with overload relays and are free-tripping, so that the mechanism cannot be held closed on overload.

Controllers of the drum type are widely used, especially in the smaller sizes. They are available for separate mounting on any suitable surface, and also without cover and with head plates suitable for mounting in a cavity in a machine. This type has a pistol-grip handle, which turns an insulated cylinder on which contact segments are mounted. The segments engage stationary contact fingers mounted in the drum frame.

They can be arranged with a self-centering spring if it is desired to make contact only as long as the lever is held in the on position. They are also used with a rope drive instead of a handle for such applications as floor-operated cranes and hoists.

Magnetic Starters. The NEMA standard ratings for magnetic across-the-line starters are given in Table 16.3. The horsepower ratings apply also to two-phase three-wire starters, but the current ratings on this type of power supply are reduced to 90% of the three-phase ratings. The ratings apply to all starting and reversing controllers for nonplugging and nonjogging duty requiring five or less openings per minute.

When the controller is subject to rapid jogging service or frequent plug-stop service, and so must repeatedly open the stalled-motor current, it is recommended that the starter ratings be reduced in accordance with Table 16.4. Rapid service is defined as being in excess of five operations a minute.

In their simplest and most widely used form, magnetic line starters consist of a three-pole magnetic contactor and a thermal overload relay (see Fig. 16.6). These devices are mounted in a suitable enclosing case, which may be of the general-purpose sheet-metal construction or may be dust-tight, water-tight, explosion-resisting, or whatever may be re-

TABLE 16.3

RATINGS OF MAGNETIC ACROSS-THE-LINE STARTERS

Size Numbers	8-hour Rating of Contactor (amp)	Horsepower at 110 V, Three-Phase	Horsepower at 220 V, Three-Phase	Horsepower at 440–550 V, Three-Phase
00	9	3/4	1½	2
0	18	2	3	5
1	27	3	7½	10
2	45	—	15	25
3	90	—	30	50
4	135	—	50	100
5	270	—	100	200
6	540	—	200	400
7	810	—	300	600
8	1215	—	450	900
9	2250	—	800	1600

TABLE 16.4

Ratings for Jogging Service

Size Numbers	8-hour Rating of Contactor, amperes	Horsepower at 115 V, Three-Phase	Horsepower at 230 V, Three-Phase	Horsepower at 460–575 V, Three-Phase
0	18	1	1½	2
1	27	2	3	5
2	45	—	10	15
3	90	—	20	30
4	135	—	30	60
5	270	—	75	150
6	540	—	150	300

quired by the installation conditions. Start and stop pushbuttons may be mounted in the cover of the case; then the stop button is usually combined with the overload resetting device. A separately mounted start-stop pushbutton may also be used, and the case mount only the reset button. When both local and remote control are desired a small

Fig. 16.6 NEMA Size-1 Starter with Cover Removed

three-position switch is mounted in the starter cover to give local control, off, and remote control. The starters are also furnished as open-type components for mounting in a cavity in a machine or for multimotor control panels.

The control circuit is very simple, since it involves only energizing the contactor coil when the start button is pressed and deenergizing it when the stop button is pressed or when the overload relay trips. With a three-wire momentary-contact type of pushbutton, low-voltage protection is obtained. With a two-wire snap-switch type of button, low-voltage release is obtained.

Line starters are also built with self-contained outside-operated disconnect switches. The disconnect switch may be of the fusible or nonfusible type. Starters are also built with self-contained outside-operated circuit breakers, which act as a disconnect device and a protection against short circuit.

All the above types of line starter are also built for reversing service. The single magnetic contactor is then replaced with a pair of mechanically interlocked contactors. As a rule, only the reset button is mounted in the cover of a reversing starter, the other buttons being separately mounted.

Line starters are made by many manufacturers. Their design is not so simple as it might appear, because the designer is always pulled in opposite directions by the desire to make the device as small and compact as possible and by the desire to make it readily accessible for mounting, wiring, and maintenance. However, the starters are simple, easy to service, inexpensive, and relatively trouble-free, which accounts for the fact that they are probably the most popular item of any control manufacturer's line of apparatus.

Reduced-Voltage Starting. The principle of both the autotransformer starter and the primary resistance starter is to reduce the voltage across the motor terminals at starting. Since the current inrush, or starting current, varies almost directly with the applied voltage, the current can be reduced by applying less than full line voltage to start. However, whereas the starting current varies almost directly with the applied voltage, the starting torque varies as the square of the applied voltage. If the applied voltage is reduced 50%, the starting current will be reduced to 50% but the starting torque will be reduced to 25% of the full voltage value. Any starter, therefore, must be so adjusted as to give the proper compromise between the torque which is required and the current which is taken from the line. An examination of the curves in Fig. 16.7 will show the relation between starting current and torque. The solid curves,

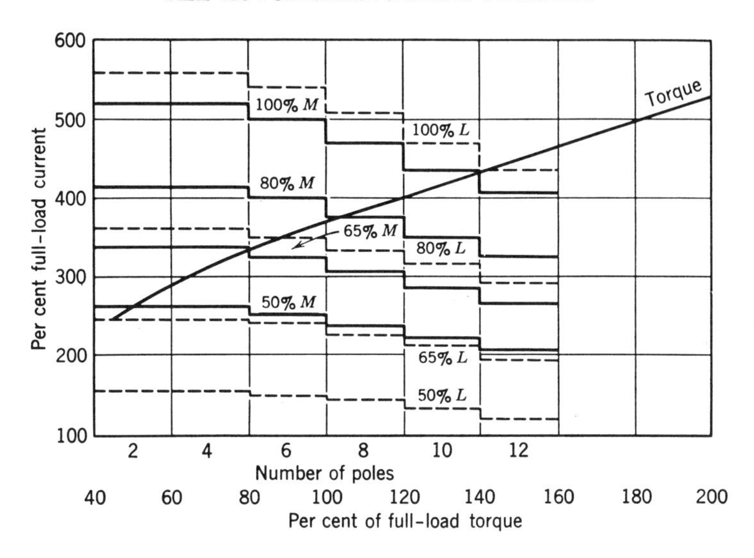

FIG. 16.7 Characteristics of a Squirrel-Cage Motor

marked *M*, give the starting inrush to the motor in percentage of full-load current for average commercial motors of 2 to 12 poles. The dotted curves, marked *L*, give the corresponding line currents with an autotransformer starter. The torque curve gives the starting torque, in percentage of full-load torque, for different values of motor-starting current.

For example, consider a 25-hp 230-V three-phase 60-Hz 1200-rpm motor driving an exhaust fan. If it were connected directly to the line, the starting current would be about 500% of the full-load current. The corresponding starting torque would be about 180% of full-load torque. Not nearly that much starting torque is needed; furthermore, we may assume not more than 200% of full-load current to accelerate. By using the 50% tap on an autotransformer starter, the starting current drawn from the line can be kept down to 150% of full-load current. The corresponding inrush to the motor is 250% of full-load current, giving a starting torque of about 45%, which is sufficient in this case. Therefore, by using an autotransformer starter, the current drawn from the line has been reduced from 500% to 150% of the full-load value, keeping within the assumed limitation and still having sufficient torque to start the load.

The Autotransformer Starter. The autotransformer starter consists of two autotransformers connected in open delta, and the motor is connected as shown in Fig. 16.8. Some manufacturers use three autotrans-

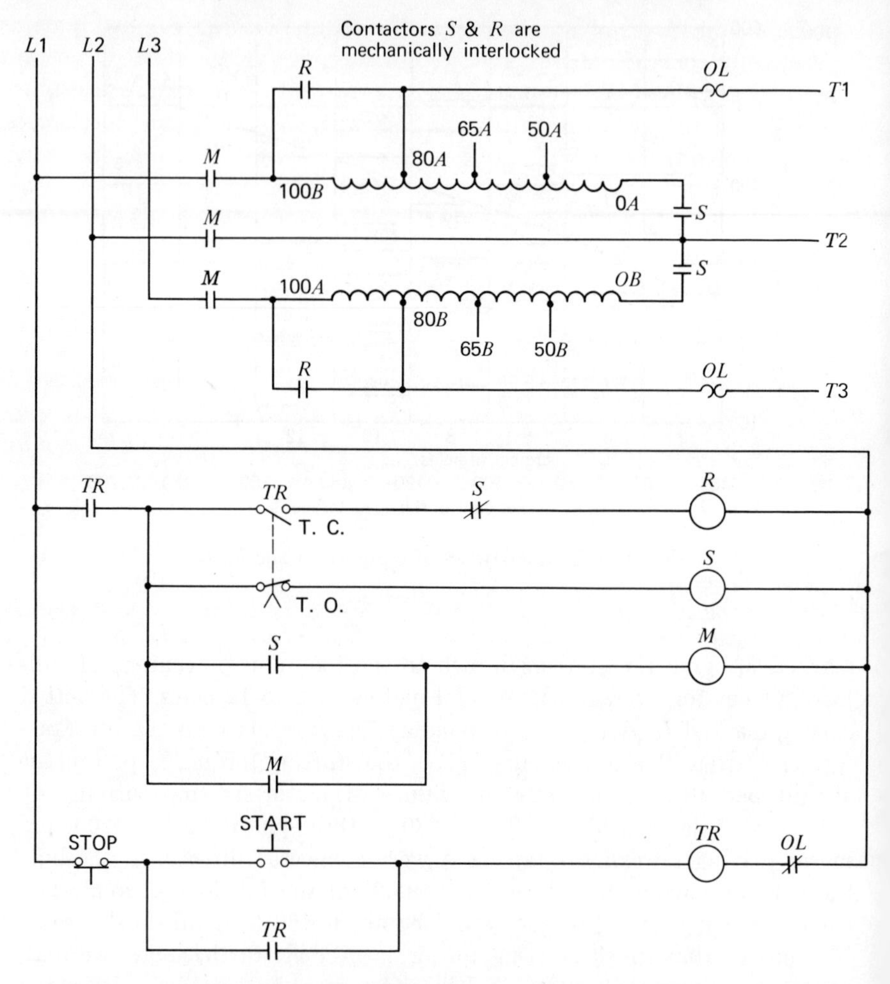

FIG. 16.8 Connections of an Autotransformer Starter (Korndorfer Method)

formers, but two are satisfactory because with two the current in the third phase is only about 15% greater than in the other two phases, and this unbalance is permissable.

Three taps are usually provided, giving 50, 65, and 80% of full line voltage. The motor current varies directly as the impressed voltage; the line current varies as the square of the impressed voltage. The starting torque consequently varies directly as the line current, neglecting transformer losses. The chief characteristics of the starter as compared to other types are low line current, low power from the line, and a low power factor. A disadvantage is that the torque which is applied remains practically constant for the first step of starting and practically constant at another value for the second step, whereas with the primary resistance starter the torque varies, increasing steadily as the motor accelerates. Another disadvantage is that, in transferring from the tap on the transformer to the line voltage direct, the motor is momentarily disconnected from the line. For the above two reasons acceleration is not so smooth with the autotransformer starter as it is with the resistor type.

Figure 16.9 shows a manually operated autotransformer starter. The transformer is mounted at the top, and the contacts at the bottom. An overload relay is included, and also a low-voltage release magnet.

An automatic transformer starter, as shown in Fig. 16.8, consists of a three-pole main contactor, a two-pole start contactor, and a two-pole run contactor. The latter two are mechanically interlocked. A timing relay makes the transfer from transformer to line by dropping out S and picking up R. This method is known as the Korndorfer connection, since the motor is not disconnected from the line during the transition from reduced voltage to full voltage.

High-voltage transformer starters function in the same manner as those built for lower voltages. A potential transformer is included, to

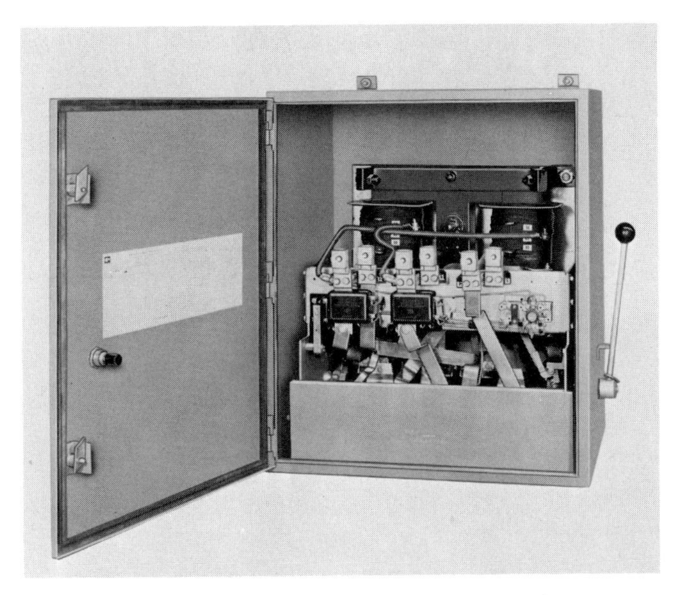

Fig. 16.9 Manually Operated Autotransformer Starter

apply low voltage for the control circuits. If overload relays are included, as they generally are, they are energized through current transformers.

The Primary-Resistor Starter. With the primary-resistor starter the motor is connected to the line through a primary resistor, and the reduced voltage at the motor terminals is obtained by means of the voltage drop across the resistor. Accordingly, the line current is the same as the motor current, and this current, and power taken from the line are much higher than in the autotransformer starter. It follows that the efficiency of this starter is less than that of the autotransformer starter. Its advantages are smoothness of acceleration, high power factor, and low cost in the smaller sizes. The smoothness of acceleration results be-

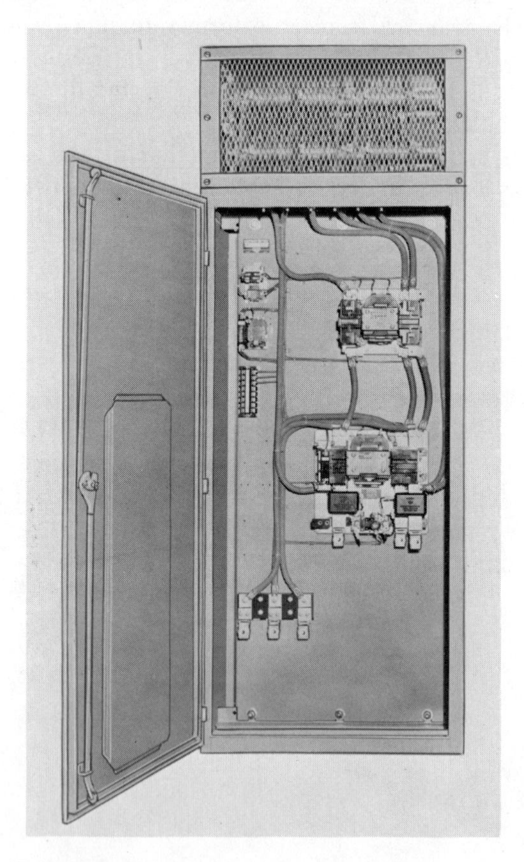

Fig. 16.10 Size-5 Primary-Resistance-Type Starter

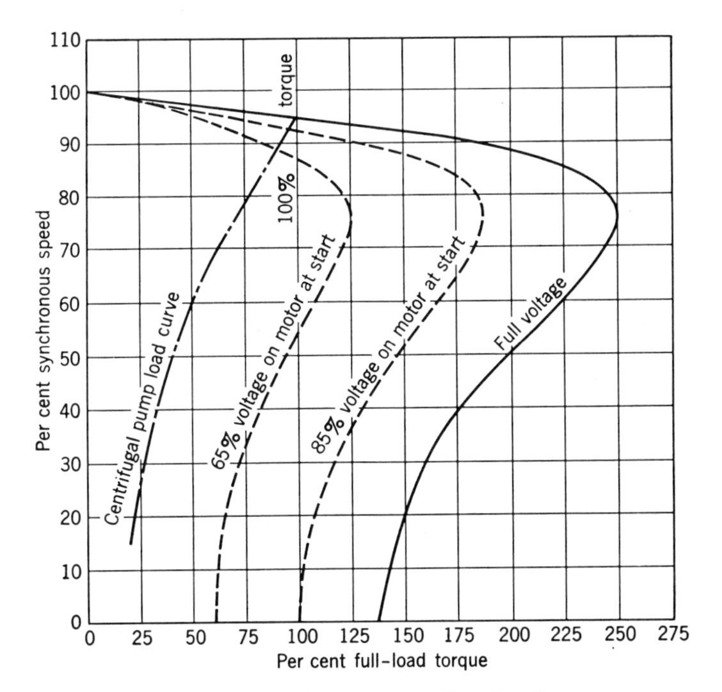

FIG. 16.11 Primary-Resistance Starting Curves

cause, as the motor accelerates, the current taken becomes lower, and consequently the voltage drop across the resistor becomes lower, and the voltage at the motor terminals rises. The torque delivered by the motor is constantly increased as the motor speeds up. The motor will accelerate faster with a given initial torque than when started by an autotransformer. Furthermore, the motor is not disconnected from the line at the transfer period, but the resistor is simply short-circuited without being disconnected. Consequently, the motor does not lose speed during the transfer period, and the accleration is smoother. So far as cost is concerned the primary-resistor starter is cheaper in the small sizes.

The Size-5 primary resistor starter shown in Fig. 16.10 has time-limit acceleration. It consists of a Size-4 start magnetic contactor and a Size-5 run contactor. The start contactor connects the motor to the line through a step of resistor in each phase. After a timed period, provided by a timing relay, the run contactor closes to bypass both the resistor and the contact of the start contactor. The start contactor is then dropped out with an interlock on the run contactor. The overload relay and current transformers are located on the Size-5 contactor where the motor connections are made. Main line terminals are located at the bottom.

The curves of Fig. 16-11 are drawn for different values of primary resistance and show the acceleration of the motor in each case. It is evident that with a light load to start, as for instance a centrifugal pump, the motor will reach nearly full speed with the resistance in circuit. For this reason it is common practice to have only one step of resistance. Two or more steps are sometimes used where it is necessary to limit the increments of current which may be drawn from the line.

A multiplicity of starting steps is obtained in the Allen-Bradley carbon-pile starter (Fig. 16.12). The resistor of this device consists of a number of graphite disks assembled in a steel tube. When the disks are loose, the contact resistance is very high. The application of pressure lowers the resistance. A solenoid, operating against an oil dashpot, compresses the disks in a set time. When the solenoid has completed its travel and the disks are fully compressed, an interlock on the solenoid

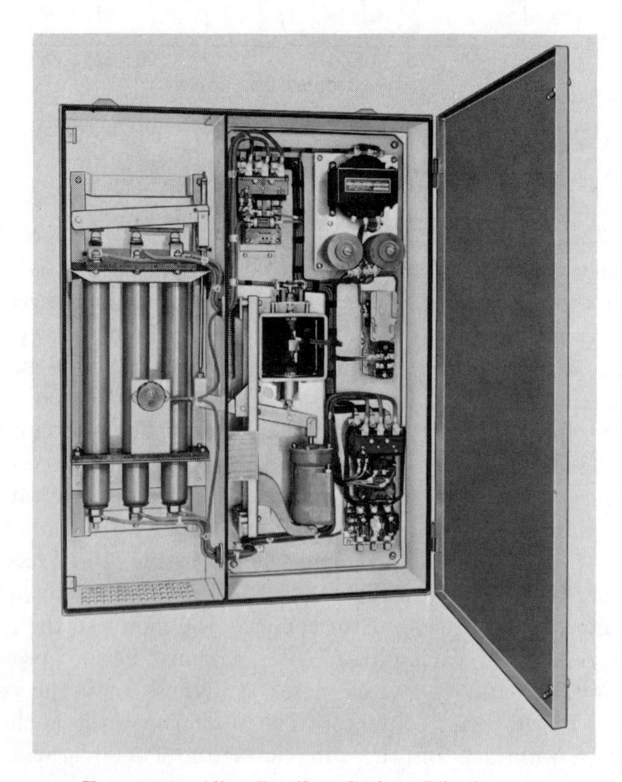

Fig. 16.12 Allen-Bradley Carbon-Pile Starter

TABLE 16.5

Method of Starting	Starting Current Drawn from the Line as a Percentage of Full-Load Current	Starting Torque as a Percentage of Full-Load Torque
Connecting motor directly to the line full potential	470	160
Autotransformer 80% tap	335	105
Resistor starter to give 80% applied voltage	375	105
Autotransformer 65% tap	225	67
Resistor starter to give 65% applied voltage	305	67
Star-delta starter	158	54
Resistance starter to give 58% applied voltage	273	54
Autotransformer 50% tap	140	43
Resistor starter to give 50% applied voltage	233	43

energizes a contactor to short-circuit the resistance. Since direct current is more suitable for operating the solenoid, a rectifier is generally included to supply it.

Table 16.5 presents a comparison of three types of starters used with a 60-hp 440-V three-phase 60 Hz 900-rpm eight-pole motor.

TABLE 16.6

Line current	Considerably greater with primary resistor for same voltage at motor terminals.
Power factor	Higher with primary resistor.
Power from line	Considerably greater with primary resistor for same voltage at motor terminals.
Torque	With the autotransformer the torque does not change much as motor accelerates. It increases as the motor accelerates with primary resistor.
Smoothness of acceleration	Primary resistor much better.
Cost	Primary resistor is generally cheaper.
Maintenance	Not much difference.
Reliability	Not much difference.
Efficiency	Autotransformer higher, particularly on low-voltage taps.
Line disturbance	Initial first cycle transient higher with autotransformer because of magnetizing current.

Table 16.6 gives the relative characteristics of the autotransformer starter and the resistor starter.

Line Current. The current taken from the line in starting is less with the autotransformer starter than with the primary-resistor starter particularly on the lower taps; that is, the difference is great on the 50% tap and on the 65% tap, but on the 80% tap the difference is not so great. With a primary-resistor starter the motor current is the line current. With the autotransformer starter, the current is nearly, but not quite, in proportion to the ratio of transformation, the difference being due to the magnetizing current of the transformer. This does not necessarily mean that the line voltage will drop lower when the resistance starter is used or that the line disturbance will be greater.

Power Factor. The power factor of the line at the moment of starting is materially higher with the primary-resistor starter than with the autotransformer. When adjusted to give 65% of line voltage at the motor terminals for starting, the power factor with the primary-resistor starter varies from 80 to 90% depending on the motor size, whereas with an autotransformer under the same conditions the power factor of the line varies from 30 to 60%, depending on the size of the motor.

Power from the Line. The power taken from the line is greater with the primary-resistor starter than with the autotransformer. The autotransformer is a voltage-changing device; the primary resistor is an energy-consuming device. Also, the fact that the power factor is high for the primary-resistor starter causes a greater power loss. When line disturbance is not taken into account, and with 65% voltage applied to the motor, the power taken from the line at the moment of starting with an autotransformer is about 50% of that which would be taken with a primary resistor. Similarly, when connected to the 80% tap, the power taken from the line with an autotransformer is about 60% of that taken with a primary resistor. However, if the time taken to accelerate to full speed is considered, the difference in energy taken during the starting period is not so great, because the primary-resistor starter will accelerate the motor in a shorter time than the autotransformer starter will. This is because of the increasing torque as the motor accelerates. As a matter of fact, the power lost by the resistor starter in most installations does not amount to very much in actual cost. Calculations based on a 50-hp 230-V three-phase motor, started by a primary-resistor starter, assuming 5 sec to start and estimating power at 2 cents per klowatt-hour, show that the power wasted in the resistor would cost approximately 0.2 cent. It is evident that this motor could be started a great many times each day before the power loss would become much of an item.

Torque. The torque developed by a squirrel-cage motor is independent of the method used to reduce the voltage at its terminals and depends only on the actual voltage impressed on the motor, varying as the square of that impressed voltage. For any application a certain torque is required for starting. This torque will, of course, be obtained with either starter. Therefore, the line current will vary, being greater for the resistor starter. For a definite given line current the torque that will be developed at the motor is materially higher when an autotransformer is used because higher voltage can be applied at the motor terminals with the same line current.

Smoothness of Acceleration. This is the principal advantage of the primary-resistor starter. When the motor is connected to the line through the resistor, a certain inrush takes place. As the motor accelerates up to speed, the current required decreases, automatically increasing the voltage at the motor terminals. Thus the torque at the motor builds up as the motor accelerates. With the autotransformer starter the applied voltage is constant as long as the motor is connected to the transformer tap. Accordingly, the acceleration of the motor is smoother and the motor accelerates faster with the primary-resistor starter.

Cost. For motors of moderate sizes the size and cost of the primary-resistor starter are less than for the corresponding autotransformer starter. This difference varies somewhat with the voltage reduction supplied, since more resistance material is required to reduce the motor voltage to 65% than to 80%.

Maintenance and Reliability. In maintenance and reliability there is not much difference between the two types of starters, provided that they are both of the air-break type. An autotransformer starter of the oil-break type is a little more difficult to maintain.

Efficiency. Because the primary-resistor starter takes more current from the line, and more power from the line in starting, its efficiency of starting is lower than that of the autotransformer starter.

Line Disturbances. The amount of line disturbance caused by starting depends not only on the amount of current taken from the line but also on the conditions of the line itself: that is, whether its capacity is large or small in relation to the motor being started, the amount of load already on the line, the power factor of the original load, and the starting power factor of the new load. Because of the low power factor of the autotransformer starter, more line disturbance may be caused by its use than by the use of a primary resistor, even though the current drawn from the line is lower.

Furthermore, since there is a high magnetizing current transient during the first half-cycle of power application to the autotransformer, it is important to recognize the effect this might have on instantaneous trip breakers or on induced electrical interference with sensitive nearby circuits. This transient is of extremely short duration and exists whenever any magnetic device is energized, so it is also true of the motor itself. A series resistor will have a limiting effect on such transients as well as on the starting current of the motor, whereas the autotransformer adds a transient of its own.

Summary. When deciding which type of starter to choose for a given installation, consideration should be given to the various characteristics discussed, and the decision should be based on which of these characteristics are the most important for the installation in question. The questions of power loss during starting, reliability, ease of maintenance, ease of operation, safety, and efficiency will ordinarily not influence the decision. If low line current during starting is the vital requirement, the autotransformer starter should be chosen. If smooth starting, high power factor, or high torque is a vital requirement, then the resistance starter is better. If line disturbance is the criterion, the whole installation and the existing line conditions should be carefully considered. If none of the above factors is of vital interest, the decision will probably be made on a basis of cost.

Part-Winding Starters. In order to use the part-winding starting method it is necessary that the motor winding be in two parts, and that at least six terminal leads be provided on the motor. The method is therefore applicable to those motors which are designed for use on either of two voltages, the windings being in parallel on the lower voltage and in series on the higher voltage. For example, a 230/460-V motor could be used on 230 V with a part-winding controller. The controller would then be arranged to connect one section of the winding to the supply lines as soon as the starting button was pressed. Then, after a time delay provided by a timing relay, a second contactor would connect the other section of the motor winding to the supply lines, in parallel with the first section.

In this way the starting current is reduced to approximately one-half of what would be required if both winding sections were connected at the same time, as they would be with a standard three-lead motor. The starting torque when the first winding section is connected will be less than half of the torque that would be obtained if both sections were connected at the same time.

Controllers are also built with a step of resistance connected in the circuit of the first winding section. Three starting steps are then available.

Contactors used for part-winding starters need capacity to handle only the circuit which they control, and so may be rated at one-half of the rating that would be required to handle the whole motor. Overload relays are provided for each section of the winding.

Figure 16.13 shows the part-winding starter connections for two types of motors. The first is wye-connected and could be a dual-voltage motor if used on the lower voltage with this type of starter. The other motor is a dual-winding, delta-connected motor. In all cases where part-winding starting is considered it is important to ascertain that the motor is suitable for this type of starting. Furthermore, the motor torque characteristics on one winding often fall to a low value at about half synchronous speed, so it is general practice to bring in the second winding after only a few-seconds time interval. This is sufficient to hold down the initial power inrush, and the starter is often referred to as an increment-type starter.

Reversing. The rotation of a three-phase squirrel-cage motor may be reversed by reversing the line connections to any two of the stator terminals. Similarly, a two-phase machine may be reversed by reversing the line connections to one of the phases. The motor may be plugged for a rapid reversal by connecting it for the reverse direction while it is still running in the forward direction. When this is done, the current inrush obtained is only slightly higher than that obtained when starting from rest.

Reversing controllers for squirrel-cage motors are usually of either the drum type or the magnetic type. The magnetic controller consists of a pair of magnetic contactors, mechanically interlocked so that they cannot both close at the same time, and an overload relay. The controlling pushbutton station has three buttons, for forward, reverse, and stop.

For reversing service, magnetic across-the-line controllers and mag-

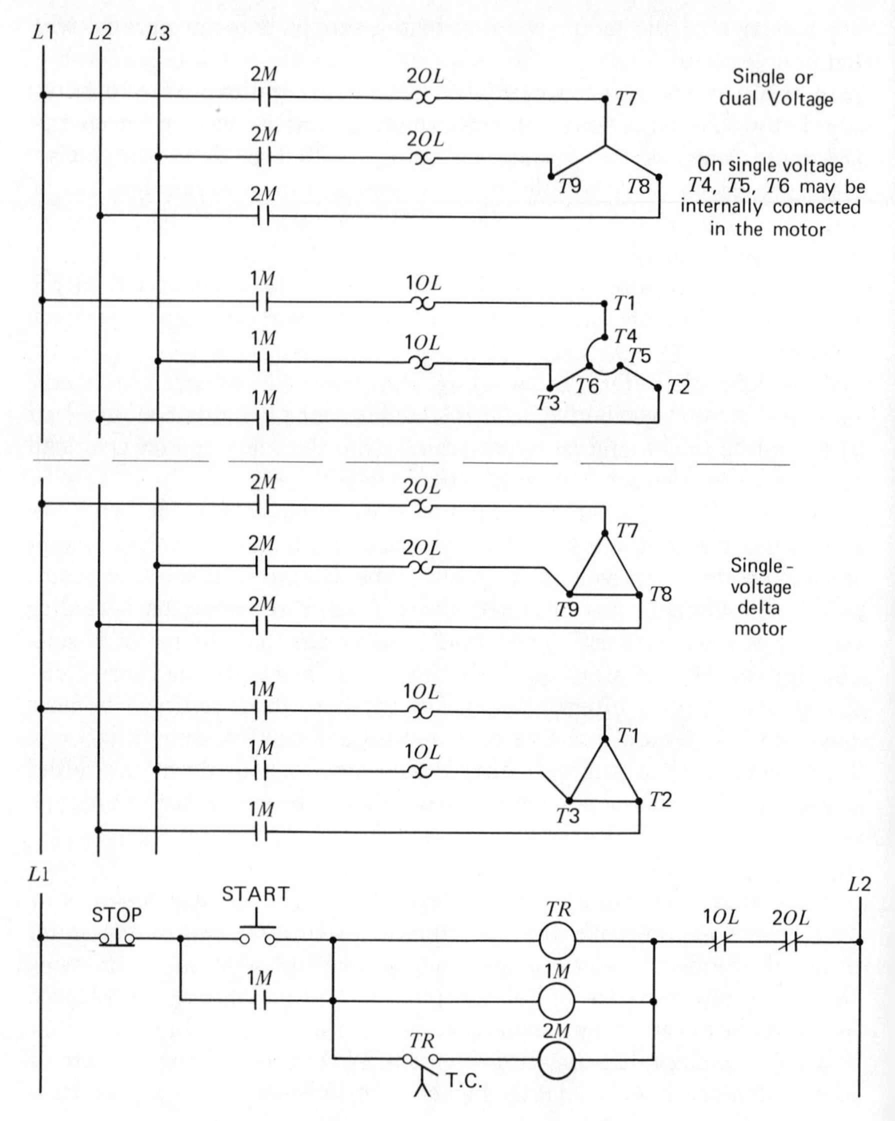

FIG. 16.13 Connections for a Part-Winding Motor Starter

netic reduced-voltage controllers are rated the same as for nonreversing service.

Stopping. The most frequently used method of stopping is simply to disconnect the motor from the supply lines and let it drift to rest. When a quick stop is desired, a magnetic brake may be used. A quick stop may also be obtained by plugging the motor and then opening the reverse contactor just as the motor has stopped and before it starts to reverse. To do this, some sort of switch is required which will close its contacts only in one direction of motor rotation. The switch is coupled to the motor and driven by it. A friction switch of this type is shown in Fig. 4.11. It is desirable to arrange the control circuits so that a movement of the driven machine by hand will not start the motor.

Figure 16.14 shows the connections for a plug-stop control. In the OFF position all contactors and relays are open, and the friction switch *FS* is also open. When the run button is pressed, the undervoltage relay *UV* and the forward direction contactor *F* close. Contact *UV*1 maintains them both closed. A normally closed contact on the run button opens the circuit to the reverse-direction contactor *R* and to relay *CR*. Contact *UV*2 also opens the circuit to *R*. The motor starts to run forward, and *FS* closes its contacts. When the run button is released, the relay *CR* gets a circuit through *UV*1 and the interlock contact *F*1 of contactor *F*. Relay *Cr* closes, maintaining itself through its contact *CR*2, and opening the circuit to the run button by contact *CR*1. The circuit is now set up for stopping, and when the stop button is pressed *F* and *UV* are deenergized and drop open, disconnecting the motor from the lines. The reverse-direction contactor *R* now gets a circuit through *FS*, *CR*2, and *UV*2, and closes, plugging the motor. The motor slows down and stops, and, as soon as it moves the least amount in the reverse direction, the friction switch *FS* opens its contacts, dropping out *R* and *CR*, and finally disconnecting the motor. During the plugging period, operation of the run button cannot do any harm, because *CR*1 is open. With the motor shut down, accidential closing of *FS*, as by hand rotation of the motor, cannot cause any operation, because *CR*2 is open.

Another method sometimes used for a quick stop is the application

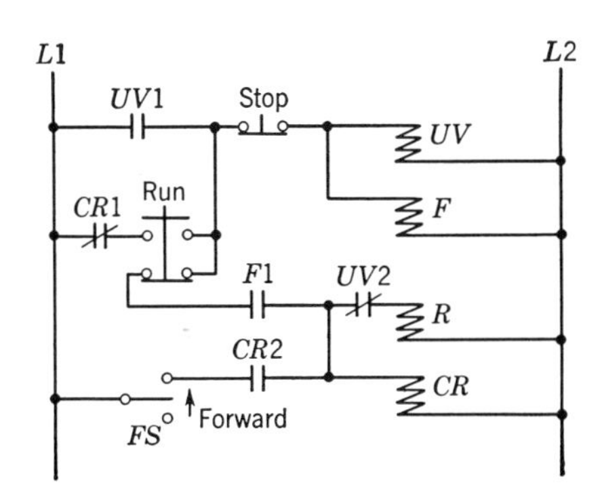

FIG. 16.14 Connections for Plug-Stop Control

of direct current to one phase of the stator winding. The squirrel-cage rotor, turning in a dc field, is brought to rest by dynamic braking.

Multispeed Squirrel-Cage Motors. Although speed regulation of the squirrel-cage motor is not practical, it is possible to get two, three, or four different constant speeds by special arrangement of the stator windings. Two separate and independent windings may be used, each being wound for any desired number of poles. With this arrangement any desired combination of the possible motor speeds may be obtained. Two speeds may also be obtained by regrouping a single stator winding to give a different number of poles. With such an arrangement the ratio of the two speeds is always 2 to 1, as, for example, 1800 rpm and 900 rpm. Three-speed motors have one winding reconnected to give two speeds, and a separate winding for the third speed. Four-speed motors have both windings reconnected.

Multispeed motors may be wound for constant horsepower, constant torque, or varible horsepower and torque.

$$\text{Horsepower} = \frac{\text{Torque (ft-lb)} \times \text{Speed (rpm)}}{5252}$$

With the constant-horsepower design, the torque is inversely proportional to the speed, and the horsepower is the same at each speed. With the constant-torque design, the horsepower varies directly with the speed, and the torque is the same at each speed. With the variable-torque and variable-horsepower design, both horsepower and torque decrease with a reduction in speed, the torque varying directly with the speed and the horsepower with the square of the speed. Any of these characteristics can be obtained either by regrouping poles or by separate windings. Consequent-pole motors are built for three-phase service only; where only two-phase power is available, it is necessary to transform to three-phase. This may be done by using Scott-connected transformers. Separate winding motors are built for either two- or three-phase service.

For constant torque the stator windings are connected in series-delta on the low speed and parallel-star on the high speed.

For constant horsepower the windings are connected in parallel-star on the low speed and series-delta on the high speed.

For variable torque and horsepower, the connections are in series-star on the low speed and parallel-star on the high speed.

Figures 16.15, 16.16, 16.17, and 16.18 show typical arrangements for multispeed motors.

Manually Operated Controllers. Because of the necessity of switching a number of circuits to obtain the different speeds, a drum-type controller

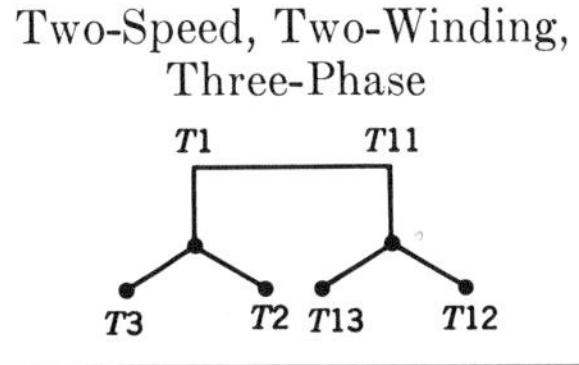

Speed	Connect Lines L1	L2	L3
	To Motor Terminals		
1	$T1$	$T2$	$T3$
2	$T11$	$T12$	$T13$

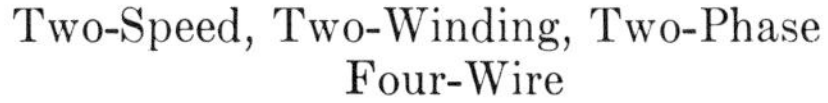

Speed	Connect Lines L1	L2	L3	L4
	To Motor Terminals			
1	$T1$	$T2$	$T3$	$T4$
2	$T11$	$T12$	$T13$	$T14$

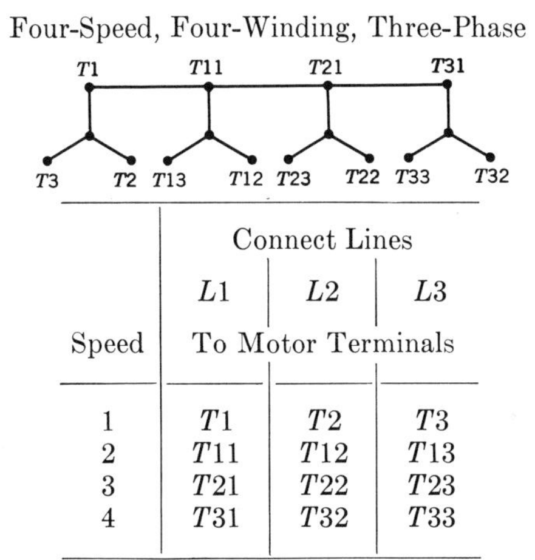

Speed	Connect Lines L1	L2	L3
	To Motor Terminals		
1	$T1$	$T2$	$T3$
2	$T11$	$T12$	$T13$
3	$T21$	$T22$	$T23$
4	$T31$	$T32$	$T33$

FIG. 16.15 Multispeed Motor Connections. Separate-Winding Types

is the most suitable manually operated device. It is not difficult to arrange the contact cylinder and fingers to give any desired switching sequence. A drum, being a compact device, is particularly suited to mounting in or on machine tools, where space is limited. Although the drum can readily provide the switching, auxiliary apparatus is necessary to obtain low-voltage protection, reduced-voltage starting, and overload protection. The larger drums also require a contactor to interrupt the circuit during switching.

Low-voltage protection is readily secured by means of a magnetic contactor, the coil of which is energized through interlocks in the drum as it is moved from the OFF position, and maintained through an interlock

Two-Speed, Single-Winding

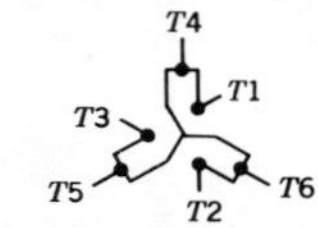

Speed	Connect Lines L1	L2	L3 To Motor Terminals	Connect Together
1	T1	T2	T3	—
2	T6	T4	T5	T1, T2, T3

Typical speeds—600, 1200

Three-Speed, Two-Winding

Alternate (1-2-1) Arrangement. Typical Speeds 600, 900, 1200

Speed	L1	L2	L3	Connect Together
1	T1	T2	T3	—
2	T11	T12	T13, T17	—
3	T6	T4	T5	T1, T2, T3

Three-Speed, Two-Winding

Tandem (1-1-2) Arrangement. Typical Speeds 600, 1200, 1800

Speed	L1	L2	L3	Connect Together
1	T1	T2	T3	—
2	T6	T4	T5	T1, T2, T3
3	T11	T12	T13, T17	—

Three-Speed, Two-Winding

Tandem (1-2-2) Arrangement. Typical Speeds 600, 900, 1800

Speed	Connect Lines L1	L2	L3 To Motor Terminals	Connect Together
1	T1	T2	T3, T7	—
2	T11	T12	T13	—
3	T16	T14	T15	T11, T12, T13

Four-Speed, Two-Winding

Alternate (1-2-1-2) Arrangement. Typical Speeds 600, 900, 1200, 1800

Speed	L1	L2	L3	Connect Together
1	T1	T2	T3	—
2	T11	T12	T13	—
3	T6	T4	T5	T1, T2, T3
4	T16	T14	T15	T11, T12, T13

Four-Speed, Two-Winding

Tandem (1-1-2-2) Arrangement. Typical Speeds 600, 1200, 1800, 3600

Speed	L1	L2	L3	Connect Together
1	T1	T2	T3	—
2	T6	T4	T5	T1, T2, T3
3	T11	T12	T13	—
4	T16	T14	T15	T11, T12, T13

Fig. 16.16 Multispeed Motor Connections. Variable-Torque Reconnected-Winding Types

Speed	Connect Lines L1 To Motor Terminals	L2	L3	Connect Together
1	T1	T2	T3	—
2	T6	T5	T5	T1, T2, T3

Typical speeds—600, 1200

Three-Speed, Two-Winding

Alternate (1-2-1) Arrangement. Typical Speeds 600, 900, 1200

Speed	L1	L2	L3	Connect Together
1	T1	T2	T3, T7	—
2	T11	T12	T13, T17	—
3	T6	T4	T5	T1, T2, T3, T7

Three Speed, Two-Winding

Tandem (1-1-2) Arrangement. Typical Speeds 600, 1200, 1800

Speed	L1	L2	L3	Connect Together
1	T1	T2	T3, T7	—
2	T6	T4	T5	T1, T2, T3, T7
3	T11	T12	T13, T17	—

Three-Speed, Two-Winding

Tandem (1-2-2) Arrangement. Typical Speeds 600, 900, 1800

Speed	Connect Lines L1 To Motor Terminals	L2	L3	Connect Together
1	T1	T2	T3, T7	—
2	T11	T12	T13, T17	—
3	T16	T14	T15	T11, T12, T13, T17

Four-Speed, Two-Winding

Alternate (1-2-1-2) Arrangement. Typical Speeds 600, 900, 1200, 1800

Speed	L1	L2	L3	Connect Together
1	T1	T2	T3, T7	—
2	T11	T12	T13, T17	—
3	T6	T4	T5	T1, T2, T3, T7
4	T16	T14	T15	T11, T12, T13, T17

Four-Speed, Two-Winding

Tandem (1-1-2-2) Arrangement. Typical Speeds 600, 1200, 1800, 3600

Speed	L1	L2	L3	Connect Together
1	T1	T2	T3, T7	—
2	T6	T4	T5	T1, T2, T3, T7
3	T11	T12	T13, T17	—
4	T16	T14	T15	T11, T12, T13, T17

FIG. 16.17 Multispeed Motor Connections. Constant-Torque Reconnected-Winding Types

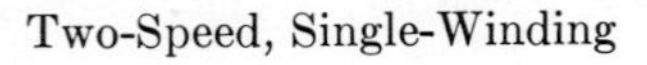

Two-Speed, Single-Winding

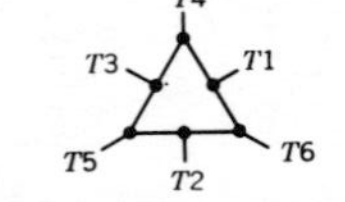

Speed	Connect Lines To Motor Terminals L1	L2	L3	Connect Together
1	T1	T2	T3	T6, T4, T5
2	T6	T4	T5	—

Typical Speeds—600, 1200

Three-Speed, Two-Winding

Alternate (1-2-1) Arrangement. Typical Speeds 600, 900, 1200

Speed	L1	L2	L3	Connect Together
1	T1	T2	T3	T4, T5, T6, T7
2	T11	T12	T13, T17	—
3	T6	T4	T5, T7	—

Three-Speed, Two-Winding

Tandem (1-1-2) Arrangement. Typical Speeds 600, 1200 1800

Speed	L1	L2	L3	Connect Together
1	T1	T2	T3	T4, T5, T6, T7
2	T6	T4	T5, T7	—
3	T11	T12	T13, T17	—

Three-Speed, Two-Winding

Tandem (1-2-2) Arrangement. Typical Speeds 600, 900, 1800

Speed	Connect Lines To Motor Terminals L1	L2	L3	Connect Together
1	T1	T2	T3, T7	—
2	T11	T12	T13	T14, T15, T16, T17
3	T16	T14	T15, T17	—

Four-Speed, Two-Winding

Alternate (1-2-1-2) Arrangement. Typical Speeds 600, 900, 1200, 1800

Speed	L1	L2	L3	Connect Together
1	T1	T2	T3	T4, T5, T6, T7
2	T11	T12	T13	T14, T15, T16, T17
3	T6	T4	T5, T7	—
4	T16	T14	T15, T17	—

Four-Speed, Two-Winding

Tandem (1-1-2-2) Arrangement. Typical Speeds 600, 1200, 1800, 3600

Speed	L1	L2	L3	Connect Together
1	T1	T2	T3	T4, T5, T6, T7
2	T6	T4	T5, T7	—
3	T11	T12	T13	T14, T15, T16, T17
4	T16	T14	T15, T17	—

Fig. 16.18 Multispeed Motor Connections. Constant-Horsepower Reconnected-Winding Types

on the contactor. When voltage fails, the drum must be returned to the OFF position to restart the motor. If an overload relay is used, it will be arranged to trip out the contactor.

Reduced-voltage starting may be secured by substituting any type of reduced-voltage starter, as an autotransformer starter or a primary-resistor starter, for the low-voltage-protection contactor.

There are several methods of using contactors to open and close the motor circuit, and switching only nonarcing circuits in the drum. One method is to incorporate a switch into the drum handle, arranging it to close its circuit only when the drum handle is in one of the running positions. The contactor coil is connected through the switch and so will open whenever the drum handle is moved from one position to another. The switch is arranged mechanically so that the drum handle cannot be moved from one position to another until the switch has been operated and the contactor de-energized. Another method is to use a quick-acting relay, arranged to drop open whenever the drum is moved from one position to another, and to deenergize the contactor. When the drum is again in an operating position, and when the contactor is open, a circuit is again completed to the relay, which closes and reenergizes the contactor. There are also mechanical devices for opening the contactor between operating positions.

To obtain suitable overload protection for a motor with separate speed windings, it is necessary to have one overload relay for each winding. For reconnected variable-torque motors one overload relay per winding is sufficient. For reconnected constant-horsepower motors, and reconnected constant-torque motors, two overload relays per winding are necessary.

Magnetic Multispeed Motor Controllers. Magnetic controllers are used for those applications where the motor is located at a distance from the operator and it is desirable to install the controller near the motor. They are necessary where automatic starting, from a thermostat or other pilot device, is used, and they are desirable for applications where the service is severe. They are made for two-, three-, and four-speed motors, of either the separate-winding or the reconnected-winding type. Though across-the-line starting is generally used, controllers are also built for reduced-voltage starting. These are of primary-resistance type, a magnetic contactor being provided to short-circuit the resistance during the accelerating period. The resistance contactor is timed by a timing relay. The arrangement is the same as that for a single-winding motor.

The three common forms of control are known as selective, compelling, and progressive.

Selective control permits starting the motor on any desired speed winding. To change the speed of a running motor to any higher speed, it is only necessary to press the desired speed button. To change to a lower speed, it is necessary first to press the stop button and then to press the desired speed button. The shock to machinery when changing speeds is greater when the speed is reduced than when it is increased, and so this control method allows the motor to decelerate somewhat before it is connected at the lower speed.

Compelling control provides that in accelerating the motor from rest it must always be started on the low-speed winding. To reach higher speeds, the pushbuttons must be operated in the sequence of speeds thus compelling the operator to accelerate the motor gradually. To change to a lower speed, it is necessary first to press the stop button and then to proceed as if starting from rest.

Progressive control provides automatic, timed acceleration of the motor to the selected speed by energizing the speed windings progressively from the lowest to the desired speed. To start the motor from rest, or to change the speed of a running motor to a higher speed, it is only necessary to press the desired speed button; the controller will automatically go from speed to speed until the desired one is reached. To change to lower speed, it is necessary first to press the stop button and then to proceed as if starting from rest.

All these forms of magnetic starter are provided with overload relays. Low-voltage release is obtained with a two-wire pilot device, and low-voltage protection with a three-wire pilot device. Selector switches with "automatic-off-manual" marking are available, as they are for single-winding motor starters, with the difference that on the "manual" side they have a position for each motor speed.

Figure 16.19 shows the connections for a two-speed reconnectable motor selective controller starting on full voltage. The high speed contactor could be a five-pole device, but since these are not generally available it is made up of one three-pole and one two-pole contactor.

Figure 16.20 shows the connections for a two-speed compelling controller, also for a reconnectable motor. Introducing the normally closed element of the SLOW pushbutton allows transfer from high speed to slow speed without having to operate the STOP pushbutton first. This can be undesirable because of the severe braking torque that occurs when the slow connection is made with the motor running above slow speed.

Considering the various kinds of motors, speed combinations, types of windings, and control schemes, it is evident that there are a good many possible combinations. The two examples given are typical, and other combinations can be worked out from them.

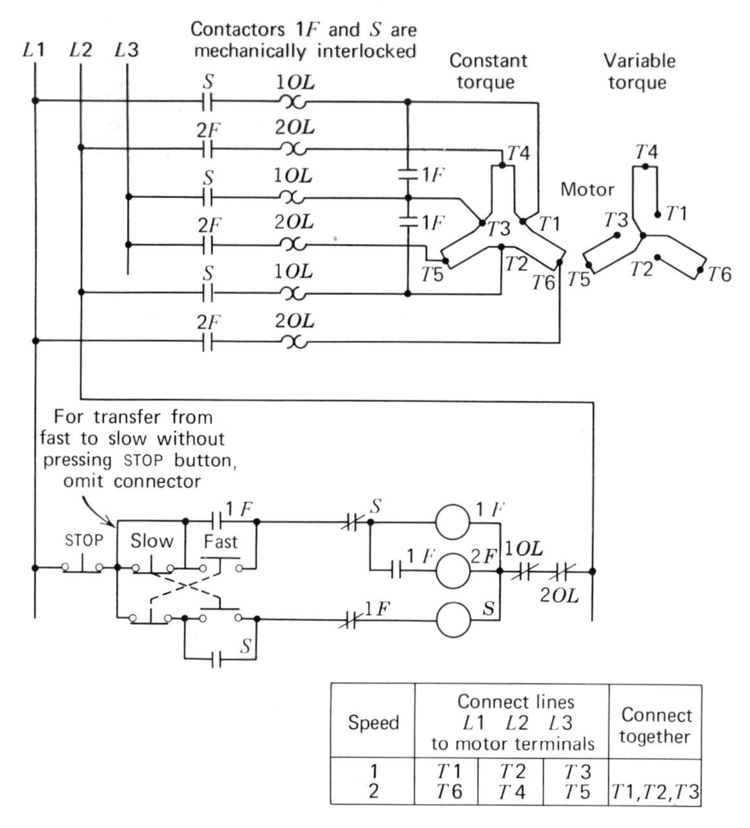

Speed	Connect lines L1	L2	L3 to motor terminals	Connect together
1	T1	T2	T3	
2	T6	T4	T5	T1,T2,T3

FIG. 16.19 Connections for Two-Speed Selective Controller

Applications of Squirrel-Cage Motors. For reasons already stated, the standard squirrel-cage motor is very popular and is used wherever the installation and service conditions permit. It is not suitable where the starting torque is high but is best applied where the starting load is light. For example, a centrifugal pump, working against a constant head of water, does not start to deliver water until it is well up to speed, and the starting load is low. Fans have similar characteristics. Many machine tools are suitable applications. Motor-generator sets are often driven by these motors.

The high-torque motor is used for slow-speed freight elevators where speed control is not required. It offers good starting torque, simplicity, and low cost. Motors for this service may have as much as 15% slip at full load. Punch presses, printing presses, and washing machines are other applications of the high-torque motor.

The double-cage motor is used where it is desirable to limit the line current when starting, and also where high torque is needed. Crushers air compressors, and conveyors starting under load are typical applications.

The constant-torque multispeed motor is used to drive printing presses, compressors, dough mixers, tumblers, constant-pressure blowers, conveyors, elevators, and stokers.

The constant-horsepower multispeed motor is used to drive lathes boring mills, other metal- and wood-working machinery, and similar machines in which a higher torque is required at the low speed.

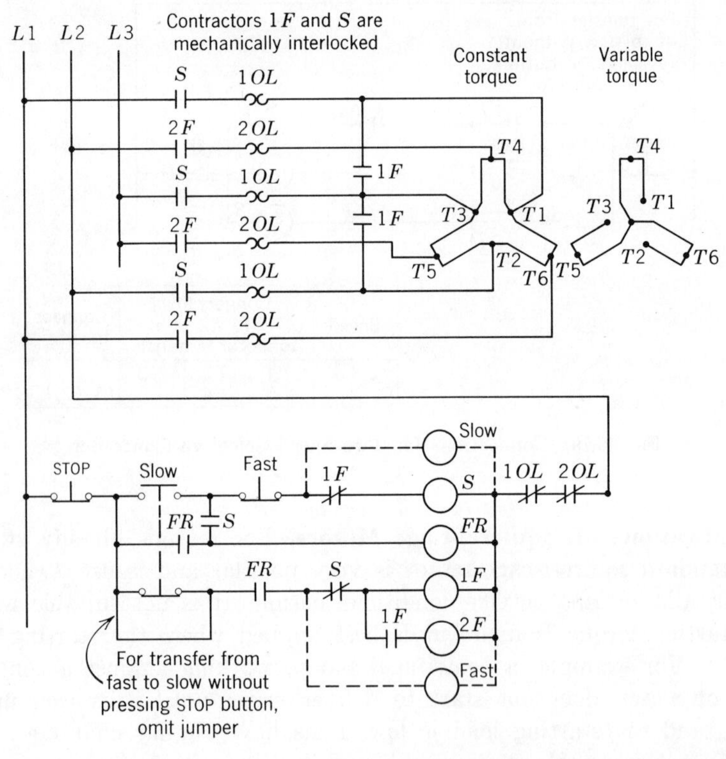

Speed	Connect lines $L1$ $L2$ $L3$ to motor terminals			Connect together
1	$T1$	$T2$	$T3$	
2	$T6$	$T4$	$T5$	$T1, T2, T3$

Fig. 16.20 Connections for Two-Speed Compelling Controller

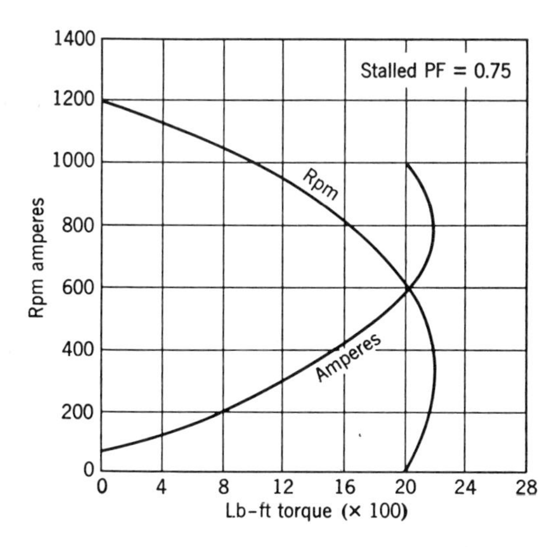

Fig. 16.21 Speed-Torque Curve of a Squirrel-Cage Motor

Variable-torque motors are used to drive machines whose load varies approximately as the square of the speed, as, for instance, fans, blowers, and centrifugal pumps.

When selecting a motor it will be found economical to choose a high-speed motor if possible, and an open motor rather than an enclosed motor if conditions will permit. For example, a 10-hp, 900-rpm motor weighs about 300 lb, whereas a 10-hp, 1800-rpm motor weighs about 135 lb. An enclosed motor has a lower horsepower rating than the same size of open motor, which means that for a given horsepower rating the enclosed motor will probably be larger than the open motor.

Design of Primary Resistor. Resistance is used in the primary circuit of a squirrel-cage motor to limit the starting torque or the starting current. It is necessary to have some data on the motor characteristics, which may be obtained from a motor curve like that of Fig. 16-21. Suppose that with this motor it is desired to limit the starting torque to 600 lb-ft.

E = line voltage = 440
PF = stalled power factor = 0.75

From the curve,

I_s = stalled current = 1000 amp
T_0 = stalled torque = 2000 lb-ft

The impedance of the motor is

$$Z_s = \frac{E}{1.73 \times I_s} = \frac{440}{1.73 \times 1000} = 0.254 \text{ ohm}$$

Calling the total impedance of motor and resistor Z, the inrush current will be

$$I = \frac{E}{1.73 \times Z} = \frac{440}{1.73Z}$$

$$Z = \frac{440}{1.73I} \tag{16.1}$$

The voltage across the motor will be

$$E_m = \frac{Z_s \times E}{Z} = \frac{0.254 \times 440}{Z}$$

$$Z = \frac{0.254 \times 440}{E_m} \tag{16.2}$$

Combining equations 16.1 and 16.2 gives

$$\frac{440}{1.73I} = \frac{0.254 \times 440}{E_m}$$

$$E_m = 0.44I \tag{16.3}$$

This gives one relation of E_m to I, and it is now necessary to get another equation between them. The torque obtained with the resistance in circuit will be proportional to the current and to the voltage.

$$T = \frac{I \times E_M \times T_0}{I_s \times E}$$

$$600 = \frac{I \times E_m \times 2000}{1000 \times 440}$$

$$E_m I = \frac{600 \times 440 \times 1000}{2000}$$

Reducing this somewhat, we have

$$E_m = \frac{132{,}000}{I} \tag{16.4}$$

Now combining equations 16.3 and 16.4 gives

$$0.44I = \frac{132{,}000}{I}$$

$$I = 548 \text{ amp} \tag{16.5}$$

and, from equation 16.3,

$$E_m = 241 \text{ V} \tag{16.6}$$

From equation 16.2,

$$Z = \frac{0.254 \times 440}{241}$$

$$Z = 0.465 \text{ ohm} \tag{16.7}$$

Referring to Fig. 16.22, the line AD is drawn at such an angle to the base line AC that the cosine of the angle is 0.75.

Choosing a convenient scale, lay off $AD = 0.254$.

Then AD is the impedance of the motor, AB is the motor resistance, and BD the motor reactance. From D draw the line DE parallel to AC, and select the point E so that the line $AE = 0.465$. Then AE is the total impedance of motor and resistor, and DE is the value of the resistor alone. Then DE will be found to be 0.24 ohm.

If the current inrush obtained in starting is the limiting factor, and the torque of secondary importance, the calculations are much simpler. With I known, the total impedance Z is immediately determined from

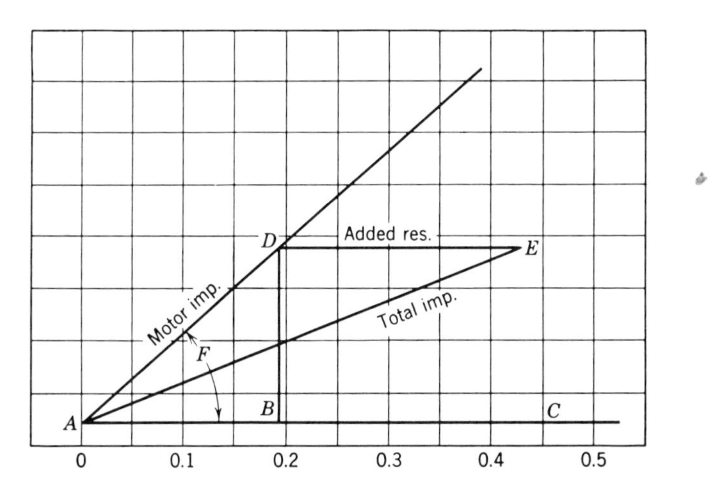

Fig. 16.22 Calculation of Primary Resistance for a Squirrel-Cage Motor Controller

equation 16.1. The controller resistance is then obtained from the vector diagram in the manner described.

Problems

1. What is the synchronous speed of an eight-pole 60-Hz squirrel-cage motor?

2. What is the synchronous speed of a 12-pole 50-Hz 380-V squirrel-cage motor?

3. At what speed will the motor of problem 2 run if it is connected to a 230-V 30-Hz supply line?

4. A squirrel-cage motor is driving a machine at 1780 rpm. If the rotor of the motor is removed, and replaced with a rotor having twice the resistance of the original rotor, at what speed will the machine be driven?

5. Referring to Fig. 16.2, draw a set of curves which will produce a field that rotates in a clockwise direction.

6. Referring to Fig. 16.21, calculate the ohms of a primary resistor which will limit the inrush current on starting to 450 amps.

7. What starting torque will be obtained with the resistor of problem 6?

8. If suitable cast-iron grids having a resistance of 0.01 ohm each are available, how many will be required for the resistor of problem 6?

9. It is desired that the motor of Fig. 16.21 be braked to stop, by the application of direct current to two of the motor leads. It is also desired that the braking current be limited to 600 amp. If the direct current is obtained from a small motor-generator set, what will be the voltage of the generator?

10. A four-speed 60-Hz multispeed motor has a two-pole winding and a six-pole winding, each of which may be reconnected. What four speeds will be obtained?

11. Draw an elementary diagram for the main and control circuits of a primary-resistor starter, as shown in Fig. 16-10, consisting of:

Three-pole line contactor
3 overload relays
Three-pole accelerating contactor
3 blocks of resistor grids
Timing relay magnetically operated
Start-stop pushbutton

12. Draw an elementary diagram of a primary-resistor increment starter, consisting of:

2 three-pole reversing contactors
3 accelerating contactors
3 overload relays
Resistors in three phases
Timing relays mechanically operated
Forward-reverse-stop pushbutton

13. A manufacturing company requests a quotation on a nonreversing increment starter for a 50-hp three-phase 230-V 122-amp squirrel-cage motor having a locked-rotor current of 10 times its full-load running current. The power company limits the starting current to 210 amp/sec. Prepare a list of the devices which are required to build the starter.

14. Calculate the ohmic value of a primary resistor for a motor and controller of the following characteristics:

Line voltage	460	
Stalled amperes	1200	
Stalled power factor	0.80	
Stalled torque	2300	lb-ft
Starting torque	500	lb-ft

15. Calculate the ohmic value of a primary resistor for a motor and controller of the following characteristics:

Line voltage	230	
Stalled amperes	1100	
Stalled power factor	0.80	
Stalled torque	1150	lb-ft
Starting amperes	500	

16. Calculate the starting torque with the motor and controller of problem 15.

17. Two squirrel-cage motors are to be used to drive a machine, and it is desired that they divide the load equally between them. What type of motor should be used, and why?

17

THE WOUND-ROTOR MOTOR

The wound-rotor induction motor is like the squirrel-cage motor, but instead of having a series of conducting bars placed in the rotor slots it has a wire winding in the rotor. If the winding is permanently short-circuited, the rotor is just another form of squirrel cage. However, if the ends of the rotor winding are brought out to three continuous slip rings, and brushes are arranged to ride on the slip rings and afford a method of connecting to them, the motor offers possibilities for application widely different from the squirrel-cage motor. Strictly speaking, a wound-rotor motor may be of the first-mentioned type, without slip rings, but in this discussion the terms "wound rotor" and "slip ring" will be used synonymously to mean a motor with slip rings.

The stator of the wound-rotor motor is of the same construction as that of the squirrel-cage motor. The brushes are mounted in the end bearing bracket, and connections from them are led to a terminal box on the motor frame.

Since the motor operates on the principle of the rotating stator field, the equation for its speed and the synchronous speeds obtainable are the same as those of the squirrel-cage motor.

In chapter 16, it was shown that when a motor is constructed with a relatively low-resistance rotor it will have a low running slip, which is desirable, but will draw a high starting current and will have a low starting torque, which are undesirable features. Conversely, a motor with a relatively high-resistance rotor will have the desirable features of higher starting torque and lower starting inrush, but will have a high slip at full load. With the squirrel-cage rotor, the choice must be made in the design, and nothing further can be done about it. The slip-ring motor, however, offers the possibility of connecting an external resistor into the rotor circuit for starting, and then cutting it out of circuit for running, and so obtaining all the desirable characteristics. The external resistance in the rotor circuit also affords a means of speed regulation.

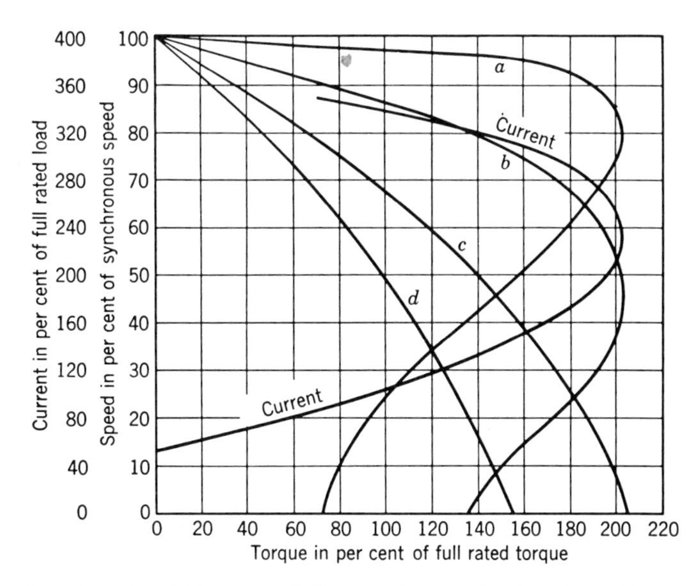

FIG. 17.1 Speed-Torque and Current Curves of a Slip-Ring Motor

Figure 17.1 shows the speed-torque curves for a slip-ring motor with various amounts of resistance connected into the rotor circuit. Curve *a* is that of the motor with the rotor short-circuited. The starting torque is about 72% and the starting current is about 350% of full-load current. Curve *b* shows the conditions when a small amount of resistance has been added in the rotor circuit. The starting torque is now about 138% and the starting current 320%. Curve *c* shows the conditions when just enough resistance has been added in the rotor circuit to give the maximum starting torque of about 204%. The inrush current is about 230%. A further increase in the resistance will reduce the starting torque as well as the starting current. Curve *d* represents a common starting condition, where enough resistance is inserted to allow a starting current of about 150% and a torque of about the same per cent of the rated value. It is, therefore, possible to start a slip-ring motor in the same way that a dc motor is started, using enough resistance to limit the starting current to the desired amount and then cutting it out in one or more steps until the motor is fully accelerated and its rotor is short-circuited.

The motor may be reversed by reversing any one phase of the stator windings.

The speed-torque curves show that speed regulation may readily be secured by means of resistance in the rotor circuit. Considering curve *d*, the motor, if fully loaded, will run at 48% of full speed. At half load it will run at 78% speed. With some resistance cut out, so that curve *c* applies, the motor will run on full load at 65% speed and on half-load at 82% speed. Similarly, any desired speed at a given load may be obtained. It will be noted that the curves become flatter as greater amounts of resistance are inserted, which means that there will be a greater speed change with changes in the load.

To calculate resistors for a slip-ring motor, it is necessary to know the characteristics of the rotor winding: that is, the voltage across the open-circuited slip rings at standstill, and the current in the rotor leads to the slip rings at full load. At standstill, the rotating stator field induces, in the rotor windings, a voltage which is determined by the ratio of stator and rotor turns. The frequency of the rotor voltage will be the same as that of the stator current, which is supply-line frequency. When the rotor starts to turn, the voltage and frequency decrease, finally approaching zero as the rotor approaches synchronous speed. Their value at full running speed depends upon the slip. If the full-load slip is 5%, the rotor voltage and frequency will be 5% of normal, and this voltage will be just enough to cause full-load current to flow against the rotor impedance. So far as calculations for resistors are concerned, it does not make any difference whether the rotor windings are connected in star or in delta; it is only necessary to know the open-circuit voltage and the load amperes.

With an overhauling load, and with the slip rings short-circuited, the motor will run slightly above synchronous speed. With external resistance in the rotor circuit, the speed will be increased.

Controllers for Slip-Ring Motors. A controller for a slip-ring motor usually consists of some form of switch or contractor, to connect the primary winding to the supply lines, and some form of resistance-commutating device for the rotor circuit. The controller may be manually operated, semimagnetic, or fully magnetic. It may be reversing or non-reversing, and it may be for plain starting or for speed regulation. Across-the-line starting is not used, as there would be no reason for the more expensive slip-ring motor if its advantages were not exploited. For the same reason, autotransformer starting and primary-resistance starting are not used.

Face-Plate Starters. Figure 17.2 is a diagram of a face-plate starter for the rotor circuit. The starter has a manually operated lever provided with contact brushes on both ends. Two sets of copper segments are mounted on the starter base. The resistor material is connected to the

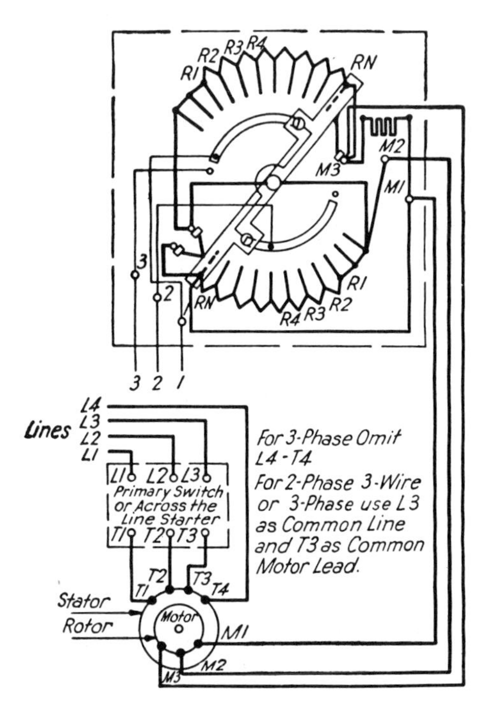

FIG. 17.2 Diagram of an AC Face-Plate Starter

segments, and the lever bridges the two sets. The resistor consists of three sections connected in delta. It is commutated in two phases only, the third section being a fixed step, as this allows a simple construction of the operating lever. The resistor is balanced in the three phases in the initial starting position and in the final running position. The intermediate starting steps are unbalanced. The unbalancing results in a lower torque than would otherwise be obtained, but the resistance is proportioned to offset this somewhat; since this type of starter is used only with small motors, the unbalancing is not a serious matter. The starting lever has a spring to return it to the all-resistance-in position when it is released, and it is provided with a low-voltage-release magnet, the coil of which is connected across two phases of the primary circuit, behind the stator contactor. This insures that the motor will not be started with the resistance short-circuited, as the lever cannot be left in the full on position with the stator circuit open. The stator contactor is a standard across-the-line magnetic starter with thermal overload relay, as used for a squirrel-cage motor.

Face-Plate Speed Regulator. A face-plate speed regulator must be arranged to commutate resistance in all three phases at the same time, because the regulator may be left on any resistance point, and all running speed points should be balanced. Running with an unbalanced resistance in the rotor circuit will cause vibration of the motor and driven machinery. It is not so difficult to build a three-arm regulator, because the spring-return feature is not required. The usual practice is not to provide any interlock between the regulator arm and the stator contactor. The resistor is star-connected and is cut out step by step in all three phases. Figure 17.3 is a diagram of the regulator.

Drum Controllers. Drum controllers for slip-ring motors are made in several types, as follows:

Nonreversing, with circuits for rotor control only.
Reversing, with circuits for stator and rotor control.
Motor driven, with circuits for rotor control only.

All these may be used with a resistor suitable for starting duty only or with a speed-regulating resistor. The manual types may have either a radial drive lever or a straight-line drive lever, or they may be provided with a handwheel drive. The general construction of an ac drum controller is the same as that of a dc drum, but the circuits are arranged to commutate resistance in the three rotor phases. The drum starter generally does not cut out resistance in all three phases simultaneously. The number of drum fingers required can be considerably reduced, with

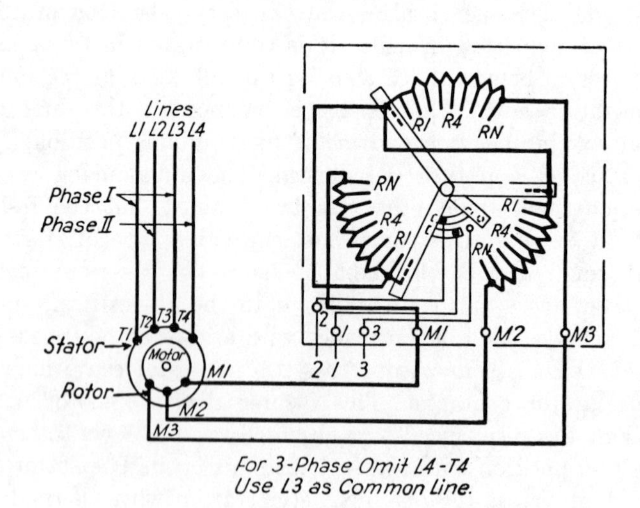

FIG. 17.3 Diagram of an AC Speed Regulator

corresponding reductions in the size and the cost of the controller, by cutting out steps alternately in the three phases. This unbalances the currents in the three phases and reduces the starting torque somewhat, but neither effect is serious when the drum is used for starting only. Drums for regulating duty have several balanced speed points in addition to some unbalanced points. Figure 17.4 shows the connections for a drum-type starter. The primary circuit of the motor is not handled in the drum but by an auxiliary magnetic contactor. This contactor may be interlocked with the drum to give either low-voltage release or low-voltage protection, as desired, and in either case the stator cannot be closed unless the drum is in the all-resistance-in position.

Reference to the diagram will show that, with the drum in the OFF position and with the primary switch closed, the resistances in the three phases are equal. When the drum is moved to the first running point, resistance step $R1$–$R2$ is short-circuited. On the second point, resistance

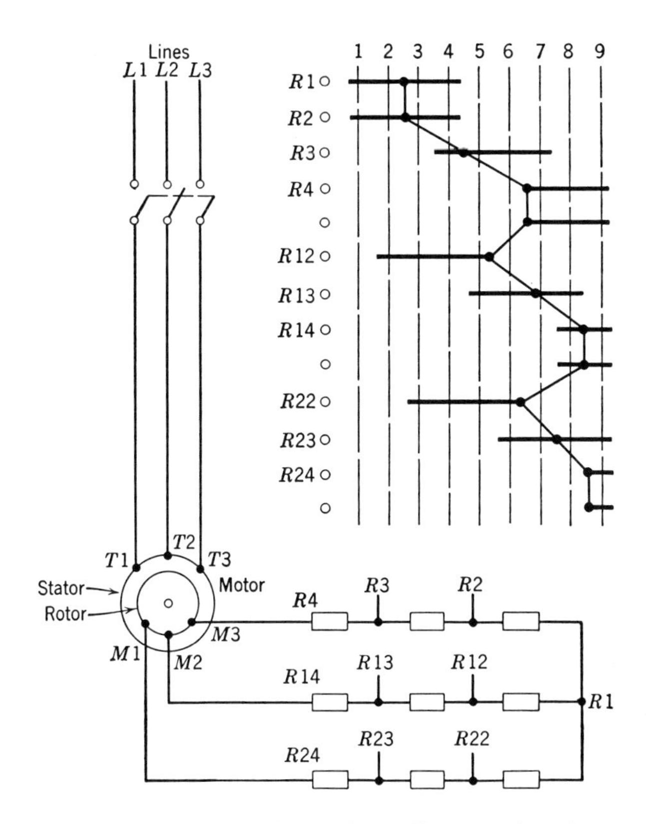

FIG. 17.4 Connections for a Drum Controller for a Slip-Ring Motor

FIG. 17.5 Drum Controller for a Slip-Ring Motor

$R1$–$R12$ is short-circuited. These two points give unbalanced conditions. On the third point, resistance $R1$–$R22$ is cut out, and the resistance is balanced again. The same procedure is followed for the remaining points, every third point being balanced. It will be noted that the resistor is star-connected. This is the most common connection.

Reversing drums for slip-ring motors have the reversing contacts included in the drum, these contacts changing the primary circuit of the motor. The rotor circuits are similar to those of nonreverse drums. A typical drum controller is shown in Fig. 17.5.

Motor-Driven Drums. In some applications, particularly large motors, close automatic speed regulation is desired; for this purpose motor-driven

drums are used. Stokers and blowers in power plants and large air-conditioning units are typical examples. The drums usually operate to vary resistance in the rotor circuit, the stator being connected to the power lines through a conta[illegible] The resistor is usually mounted separately from the drum controller.

NEMA standard sizes for motor-driven drums are given in Table 17.1. All speed points are balanced.

TABLE 17.1

MOTOR-DRIVEN DRUMS

8-hour Rating (amp)	Number of Speed-Regulating Points
300	13
300	20
600	13
600	20

Figure 17.6 shows a motor-driven drum with the covers removed from the motor drive and from the contact mechanism. The frame of the

FIG. 17.6 General Electric Motor-Operated Drum Controller with 20 Balanced Speed Points, Rated at 600 amp and 1000 V

drum is of fabricated steel with cast top and bottom plates. The contacts, of copper, are opened and closed by cams on the operating shaft. Each contact arm carries two contacts, both of which are connected to the middle leg of the resistor. They make contact simultaneously with stationary contacts, one connected to each of the outside legs of the resistor. Therefore, the closing of each contact arm short-circuits the resistor at the point to which the contacts are connected.

The motor-operating mechanisms consists of a standard fractional-horsepower pilot motor, connected to the drum shaft through suitable gearing. It also includes a positioning switch arranged to insure that the pilot motor, once energized, will run long enough to move the drum from full-on one position to full-on another position. This is sometimes called a step-by-step device, and its purpose is to prevent operation with the drum contacts only partly engaged. Limit switches are also included, to stop the pilot motor at each limit of the drum rotation, and a manually operated lever is provided so that the drum may be operated by hand if for any reason the pilot motor cannot function.

Motor-driven drums are sometimes built with sliding copper contact segments and contact fingers, similar to the manual drum construction (Fig. 7.13). With this construction it is difficult to get a large number of balanced points, since the required number of contact fingers becomes too great to get into a reasonable space. If unbalanced points are used, commutating resistance in one leg at a time, the drum must be arranged so that these are accelerating points only, and any running position must be a balanced point.

Magnetic Starters. Magnetic controllers used simply for starting a slip-ring motor include a magnetic contactor and overload relay for the stator winding and a suitable number of magnetic contactors to commutate the resistor in the rotor circuit. In starting, the stator contactor is closed when the starting button is operated. Electric interlocks on the stator contactor complete a circuit for the resistor contactors, which close in sequence under the control of some type of accelerating device. Timing relays are normally used. If reversing is required, the stator contactor is replaced by a pair of mechanically interlocked contactors, one of which closes in each direction of travel.

The horsepower ratings of these controllers are the same as those of across-the-line starters for squirrel-cage motors, and they are made in sizes from NEMA size 1 (27 amp) up. Of course, the resistor contactors must be selected to suit the rotor current, which will vary widely with different motors of the same horsepower rating. The resistor is usually connected in star, and three-pole contactors connected in delta

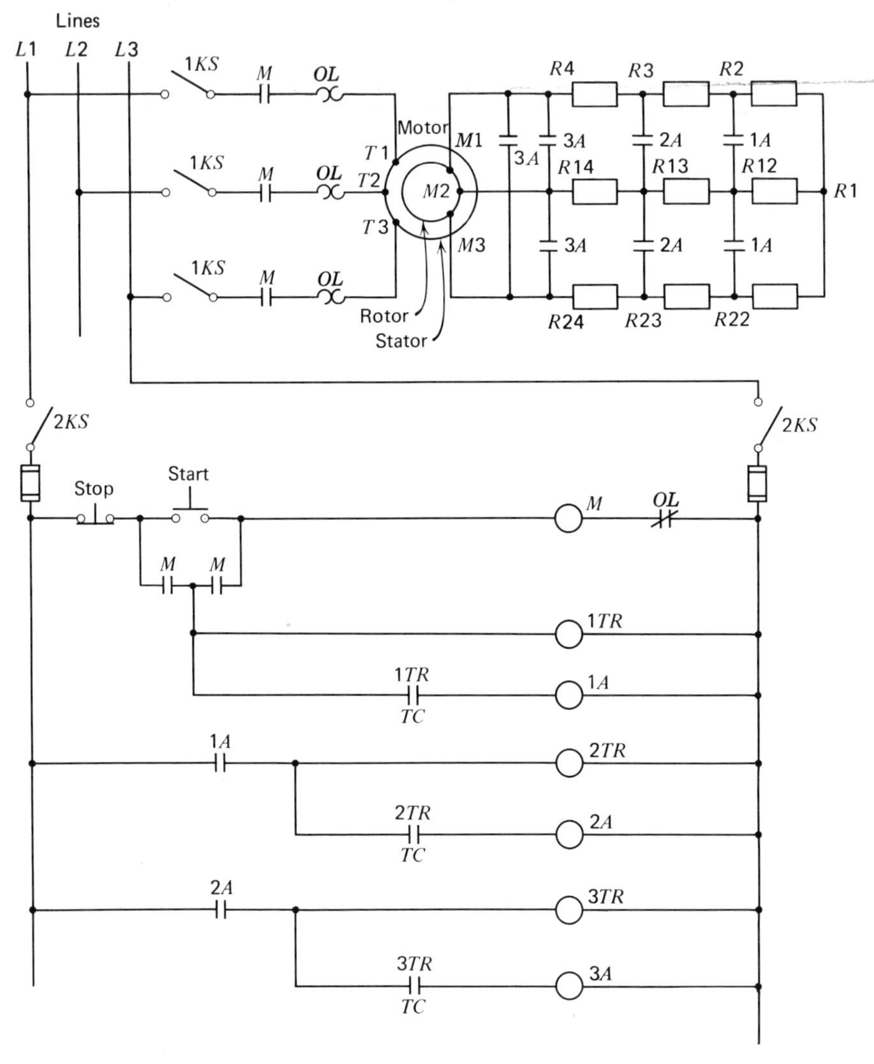

FIG. 17.7 Acceleration by Individual Timing Relays, Magnetically Operated

are used to short-circuit the sections as shown in Fig. 17.7. Interlocks on the line contactor *M* provide a circuit to energize the coil of timing relay 1*TR*, which, after a time delay, closes its contacts and provides a circuit to energize the coil of accelerating contactor 1*A*. When 1*A* closes, its electric interlock sets up the circuit for the coil of timing relay 2*TR* and, after time delay, for the coil of accelerating contactor

2*A*. The sequence may be continued for any desired number of accelerating contactors. With large motors a delta-connected resistor or parallel-star or parallel-delta connections may be economical. These connections permit the use of smaller contactors, and in the larger sizes the difference in the cost of one size of contactor over another is considerable. In the smaller sizes the difference in cost is not so great, and the saving does not warrant the extra complication of the circuit.

Speed Regulation. Speed regulation may be secured by means of the same kind of controller, except that the resistor contactors are arranged to be individually energized from a master controller or some form of multipoint pilot device. The resistor must then be designed for continuous duty on any point.

Speed Setting. Speed setting is desirable on printing presses and some other machines. This means that the speed-regulating master may be left on any desired speed, and, when the motor is stopped and restarted from the pushbutton station, it will run at the speed at which it was originally running. There is no difficulty about this if the amount of speed reduction is small, but if it is large the resistor required for the speed reduction will limit the starting current to an amount too low to start the motor. A printing press, for example, may have a point in its cycle at which a very high torque is required. When the press is running, the inertia of the machinery will carry it through the high-torque point easly; but if it happens to stop at that particular point, it will take a high torque to restart it. The difficulty may be ovecome by connecting a so-called high-torque contactor into the rotor circuit at the point which will short-circuit just enough resistance to provide the maximum motor torque. When the starting button is operated, the high-torque contactor closes at the same time that the line contactor closes, and it remains closed as long as the operator keeps his finger on the button. When the motor has accelerated, the operator releases the button and the high-torque contactor opens, inserting additional resistance as determined by the setting of the speed-selecting master.

If a relatively small number of speeds is desired, one resistor contactor is used for each speed point, but, if a large number of speeds is necessary, that arrangement would result in a large and expensive controller. To secure a large number of speeds with a small number of contactors, the resistor steps are tapered in ohmic value according to a geometric ratio, and contactors are closed in various combinations instead of a fixed sequence. For example, with four resistor steps, the ohmic values of the steps would be in the ratio 1, 2, 4, 8, and there would be four

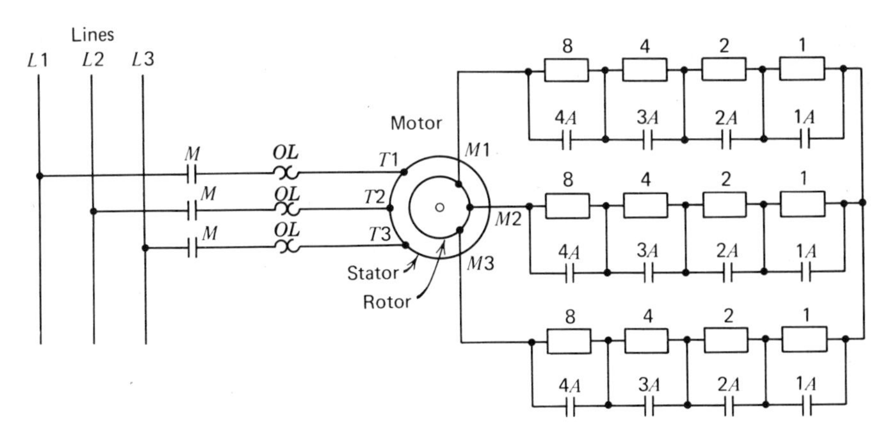

FIG. 17.8 Connections for a Speed Regulator with 16 Speed Points

contactors, one arranged to short-circuit each step. Referring to Fig. 17.8, if the total resistance were 15 ohms, closing contactor 1 only would leave 14 ohms in circuit. Closing 2 only would give 13 ohms. Closing numbers 2, 3, and 4 would give 1 ohm. It will be evident that with the four contactors it is possible to get 16 speeds. Using five contactors, it is possible to get 32 speeds.

Crane and Hoist Control. Magnetic control for a crane bridge or trolley motor is made up of a pair of magnetic reversing contactors with a disconnecting knife switch and suitable overload relays for the motor stator, and a number of magnetic contactors to commutate resistance connected in the rotor winding. A multipoint master gives about five speeds in each direction. The first point in either direction closes one of the direction contactors, and, with all resistance in circuit, about 80 per cent torque is obtained. This is a low-torque starting point and also a plugging point. Succeeding positions of the master close the rotor contactors in sequence, giving successively higher speeds, until on the last point the rotor is short-circuited. The rotor contactors may be controlled either by series current relays or by any of several methods of timing. The resistor is usually star-connected, so that two-pole contactors may be used to commutate it. The stator contactors are normally three-pole devices.

When a crane hoist is equipped with a wound-rotor motor, and the gearing is such that the load will overhaul and drive the motor when lowering, some special provision must be made to permit safe handling of the load in the lowering direction. Small cranes which do not require accurate stopping or spotting of the load may simply be equipped with a foot-operated brake, no provision being made for a slowdown before stopping. Some cranes are equipped with a device called a load brake, which remains released as along as the motor is driving the load down but automatically sets to provide a braking action when the load begins to overhaul the motor. With either of these arrangements, the control need be only simple reversing equipment, as previously described. Brakes for retarding the load are subject to a good deal of wear and to heating, and so, particularly for larger motors, methods of obtaining electric braking have been developed. Two methods are common, one known as countertorque lowering or plugging lowering, and the other as the unbalanced-stator method.

Countertorque Lowering. Figure 17.9 shows the connections for a controller with the countertorque method of lowering a load. The controller has the following devices:

H	Two-pole hoisting contactor
L	Two-pole lowering contactor
M	Two-pole main contactor
OL	2 overload relays
KS	Main and control circuit knife switches
1*A*–2*A*–3*A*	3 single-pole speed-selecting contactors
4*A*–5*A*–6*A*–7*A*	4 double-pole speed-selecting contactors
SR	A three-coil series relay
UV	Double-pole undervoltage relay
1*C*	Double-pole control relay

The resistor steps *R*1, *R*11, and *R*21 are so high in ohmic value that it is not necessary to delay or control the operation of contactors 1*A*, 2*A*, and 3*A* by any accelerating means. Contactor 4*A* is controlled by the series relay *SR*. No method of controlling 5*A* and 6*A* is shown. Any of the common accelerating means, timing relays, series relays, frequency relays, and so on, may be used for that purpose.

The undervoltage relay is closed in the off position of the master; thereafter it maintains its own circuit and sets up a circuit for one side of coils *M*, *H*, and *L*. If power fails, or if an overload occurs, *UV* opens, and the master must be returned to the off position to restart.

To hoist, the master is moved to the first position in that direction, and contactors *M*, *H*, 1*A*, 2*A*, and 3*A* close. The resistor remaining in circuit will allow a starting torque of about 80% of full-load torque. This is provided to permit taking up a slack cable. On the second point contactor 4*A* closes, increasing the torque to about 150% of full-load torque. The series relay *SR* is set to remain closed and does not delay the closing 4*A*. However, if the motor had been running in the lowering direction and was then plugged to the hoisting direction, the relay would function on the high inrush current and would delay the closing of 4*A*. Contactor 4*A* is, then, the plugging contactor, and resistor steps *R*2, *R*12, and *R*22, are the plugging resistor steps. On the third and fourth points hoisting, contactors, 5*A* and 6*A* are closed, further increasing the speed. Contactor 7*A* closes on the fifth point, and the motor runs at full speed. The operator may move the master back or ahead to any hoisting point, selecting the one that gives him the speed he desires.

On the first point in the lowering direction, contactor *M* closes, but the motor is not energized because *L* does not close. No other contactors are closed on this point or on points 2, 3, and 4. Nothing happens until point 5 is reached, and on that point the motor is energized in the lowering direction by the closing of contactor *L*. If the master is left on that point, contactors 4*A*, 5*A*, 6*A*, and 7*A* will close in sequence, under the control of their acclerating means, and the motor will be brought up to full speed. This point is used to drive down an empty hook or a light load. It may also be used to lower an overhauling load, the lowering speed will then be a little higher than synchronous speed. On the fifth point the relay 1*C* is energized by a circuit through an interlock contact of *L*.

Now, if the master is moved back to point 4, *L* will open and *H* will close, circuit for *H* being made through the contact of 1*C*. All seven rotor circuit contactors are opened and the motor is connected for hoisting, but with a high resistance in the rotor. The torque in the hoisting direction is too low to hoist the load, but, since it is opposing the pull of the overhauling load, it acts as a brake and retards the speed. Moving the master back to point 3 will further increase the braking torque by closing 1*A* and decreasing the resistance in the rotor circuit. The speed will be decreased. On point 2 contactor 2*A* closes, further increasing the braking torque, and similarly on point 1 contactor 3*A* closes, increasing it again. The master may be moved back or ahead to any of the first four points, increasing or decreasing the braking torque to secure the desired control of the load, and on these points the motor will be connected in the hoisting direction so long as the load is heavy enough to overcome the hoisting torque applied.

When the master is moved to the off position, the magnetic brake (not shown) is set to stop the motor. Contactors, *M*, *H*, 1*A*, 2*A*, and 3*A* are kept energized until the motor stops, so that the countertorque may assist the brake. This is done by the relay 1*VR*, which may be a voltage relay or a frequency relay, having its coil connected across the rotor of the motor and set to open at a voltage or a frequency corresponding to practically zero speed.

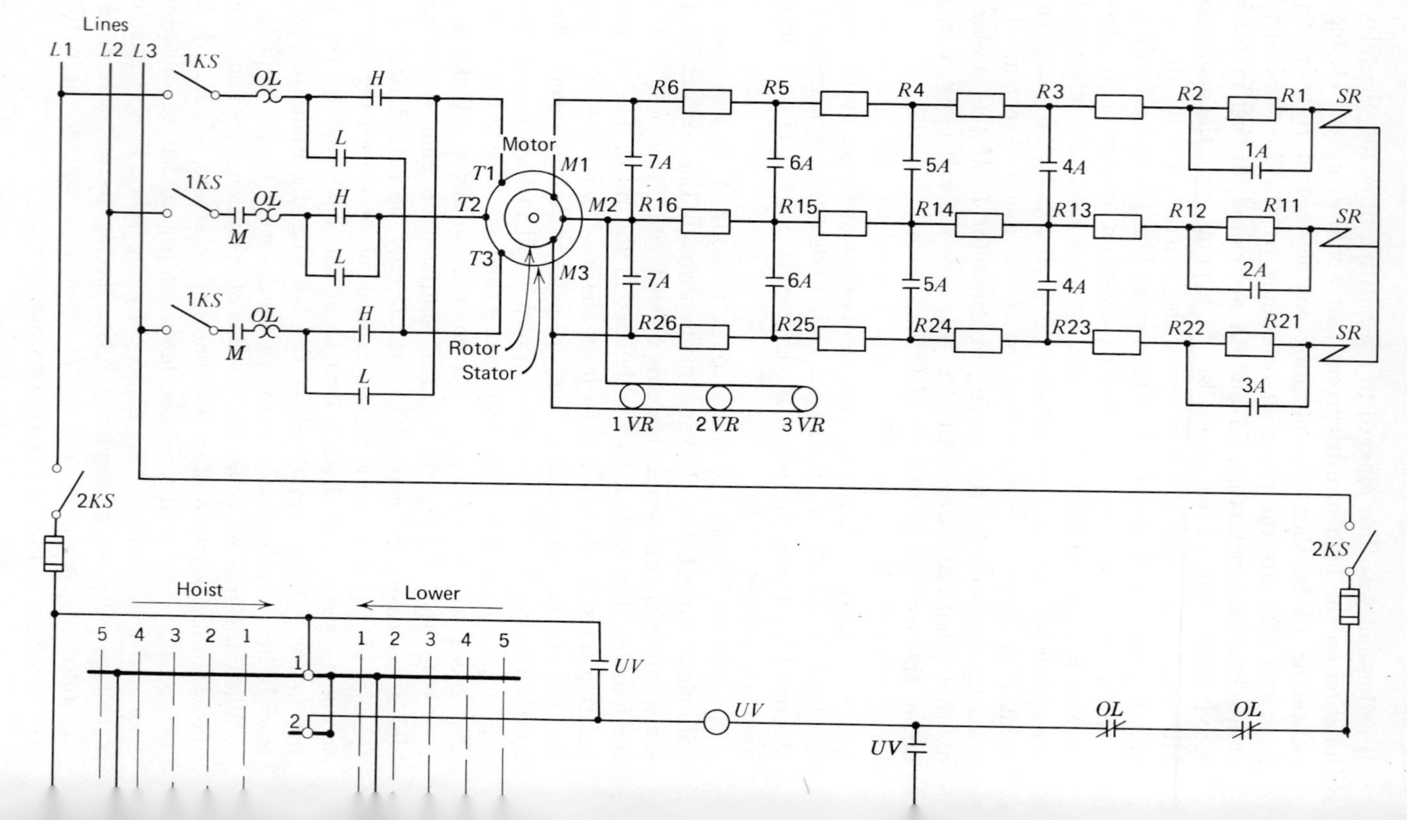
Lines
L1 L2 L3
1KS
OL
H
L
M
Motor
T1
T2
T3
M1
M2
M3
Rotor
Stator
R1
R2
R3
R4
R5
R6
R11
R12
R13
R14
R15
R16
R21
R22
R23
R24
R25
R26
SR
1A
2A
3A
4A
5A
6A
7A
1VR
2VR
3VR
2KS
Hoist
Lower
5 4 3 2 1
1 2 3 4 5
UV
OL

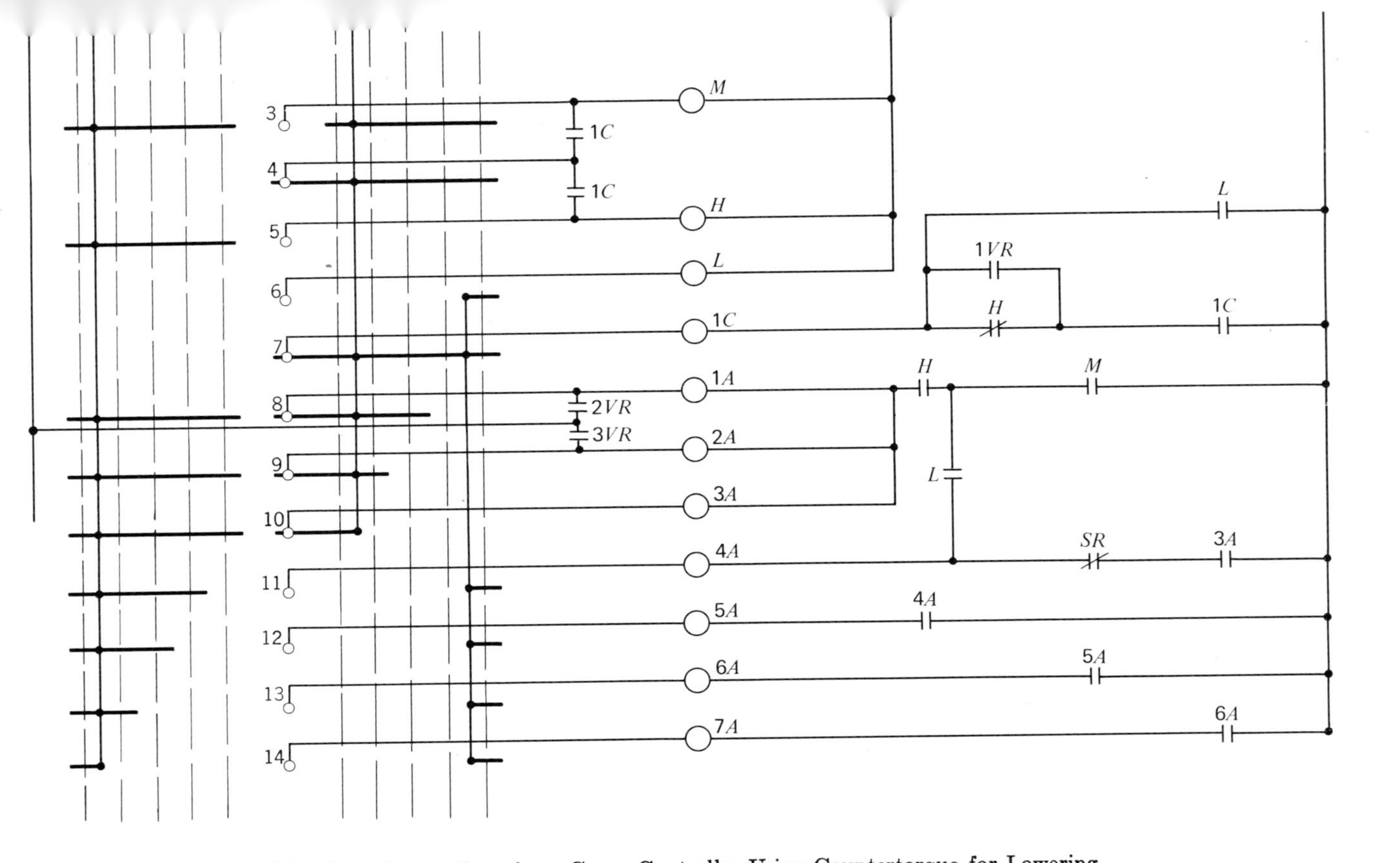

Fig. 17.9 Connections for a Crane Controller Using Countertorque for Lowering

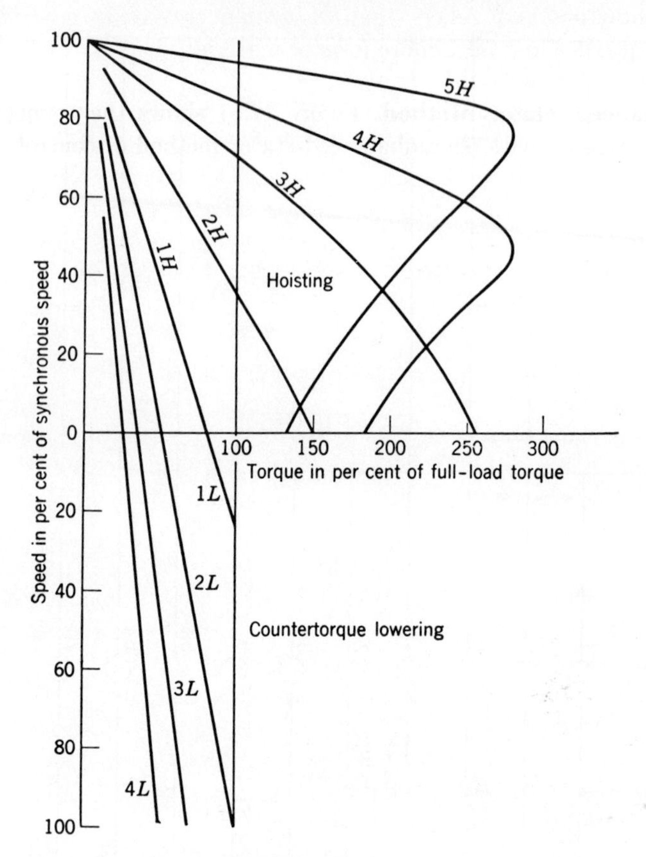

FIG. 17.10 Speed-Torque Curves for a Crane Controller Using Countertorque for Lowering

Referring to Fig. 17.10, which shows the speed-torque curves obtained with a controller of this kind, it will be evident that the curves on the countertorque points are rather flat, so that changes in the load will result in relatively great changes in speed. Also, if the operator misjudges the load and tries to lower with too little retarding torque, the motor is likely to overspeed dangerously. Relays $2VR$ and $3VR$ are used to guard against this possibility. They are voltage or frequency relays, like $1VR$, and have their coils connected across the rotor slip rings. Relay $2VR$ is set so that, if the master is on point 3 and the motor begins to overspeed, the relay will close and energize contactor $2A$. If the speed is still too high, relay $1VR$ will close and energize $3A$.

Countertorque control is used for cranes and hoists, for coal-handling bucket hoists, and for car-hauling hoists in mines. It offers a method of obtaining speed control of as overhauling load electrically. It requires more contactors and relays than a straight reversing controller, and it also requires about 50% more resistor material.

Unbalanced Stator Method. Figure 17.11 shows the connections for a hoist controller with the unbalanced-stator method of control. In hoisting, the circuit is the conventional three-phase arrangement, with resistance connected in the rotor circuit for speed regulation. In lowering, the voltages applied to the stator are unbalanced. An autotransformer is connected across two of the power lines, and a series of contactors connects the motor to one of several voltage taps. Voltages both lower and higher than normal are applied.

When a motor stator is connected to an unbalanced voltage supply, the effect is the same as if two rotating fields were produced, rotating in opposite directions. One tends to rotate the motor clockwise; the other counterclockwise. The relative strength of the two fields may be varied by varying the amount of unbalance of the stator voltages and by varying the resistor in the rotor circuit. The resultant torque on the motor shaft is equal to the algebraic sum of the two opposing torques. The system is comparable to two identical motors mechanically coupled to the same load and connected for rotation in opposite directions. Such a system could be made to run at any desired speed, and in either direction, by varying resistance in the two rotor circuits.

By properly selecting the unbalancing voltage and the rotor-circuit resistance, steep speed-torque curves, like those of a dc dynamic lowering hoist, may be obtained. Typical curves are shown in Fig. 17.12, and it will be evident that their characteristics are much more desirable than those of the reverse-torque method of control.

In hoisting, the low torque on the first point is obtained with all rotor resistor in circuit and with one phase of the rotor open. This is a slack cable take-up point. The other four curves show the torques and speeds obtained by successive closing of the rotor circuit contactors. In all these points the stator voltage is balanced.

In lowering, the first, second, and third speeds are obtained by the selection of suitable taps on the transformer and with a rotor circuit resistance of 50% of E/I. The stalled torque of 10% on the first point is selected to provide a minimum lowering speed without applying enough torque to hoist an empty hook. The fourth and fifth speed points are obtained will full voltage on the motor and rotor resistances of 50 and 100% of E/I, respectively. When transferring from the third to the fourth speed point, the total resistance is first inserted to reduce the current inrush and the motor is permitted to accelerate curve $5L$. A timing relay then operates to reduce the resistance and transfer the motor operation to curve $4L$.

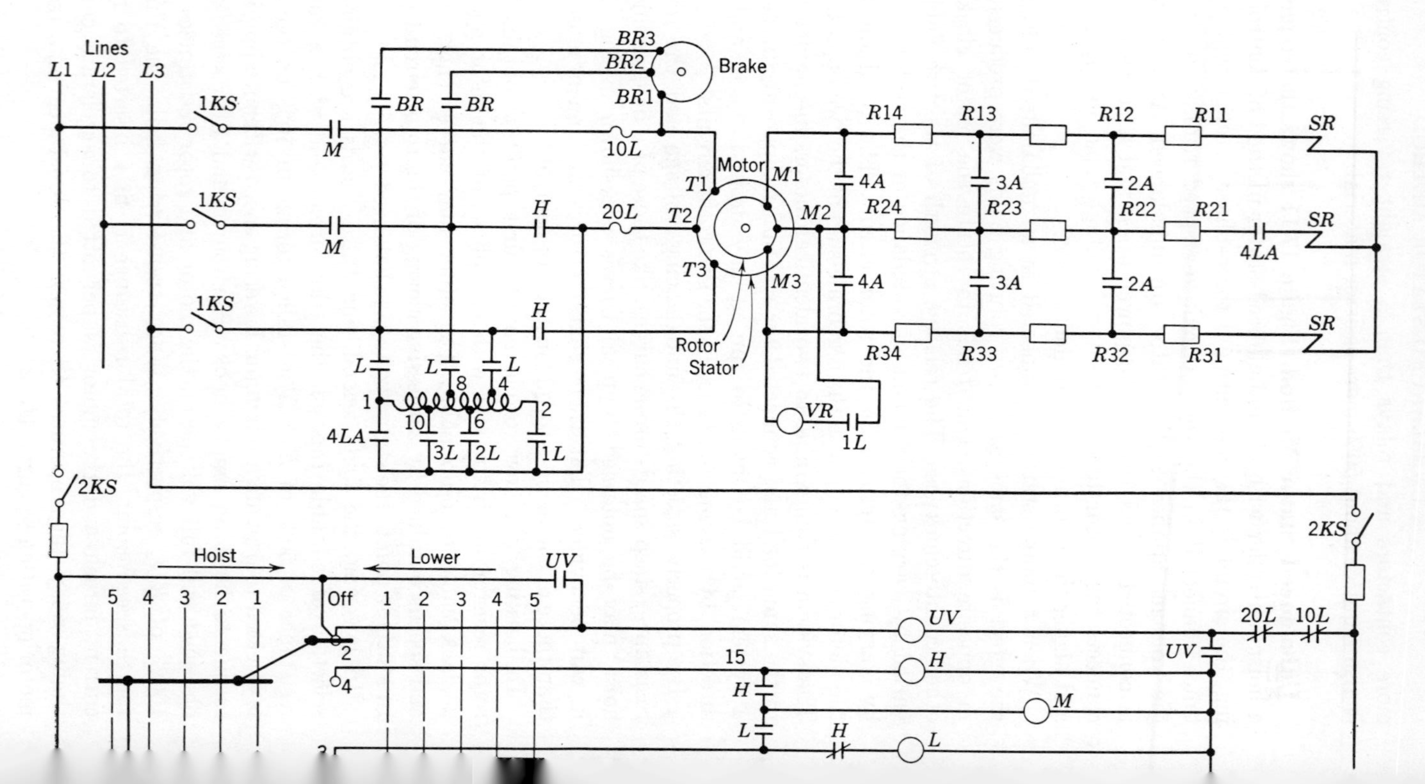
Lines
L1
L2
L3
1KS
2KS
M
BR
L
H
4LA
1L
2L
3L
10L
20L
BR1
BR2
BR3
Brake
Motor
T1
T2
T3
Rotor
Stator
M1
M2
M3
VR
R11
R12
R13
R14
R21
R22
R23
R24
R31
R32
R33
R34
2A
3A
4A
SR
UV
15
Hoist
Lower
Off
1
2
3
4
5

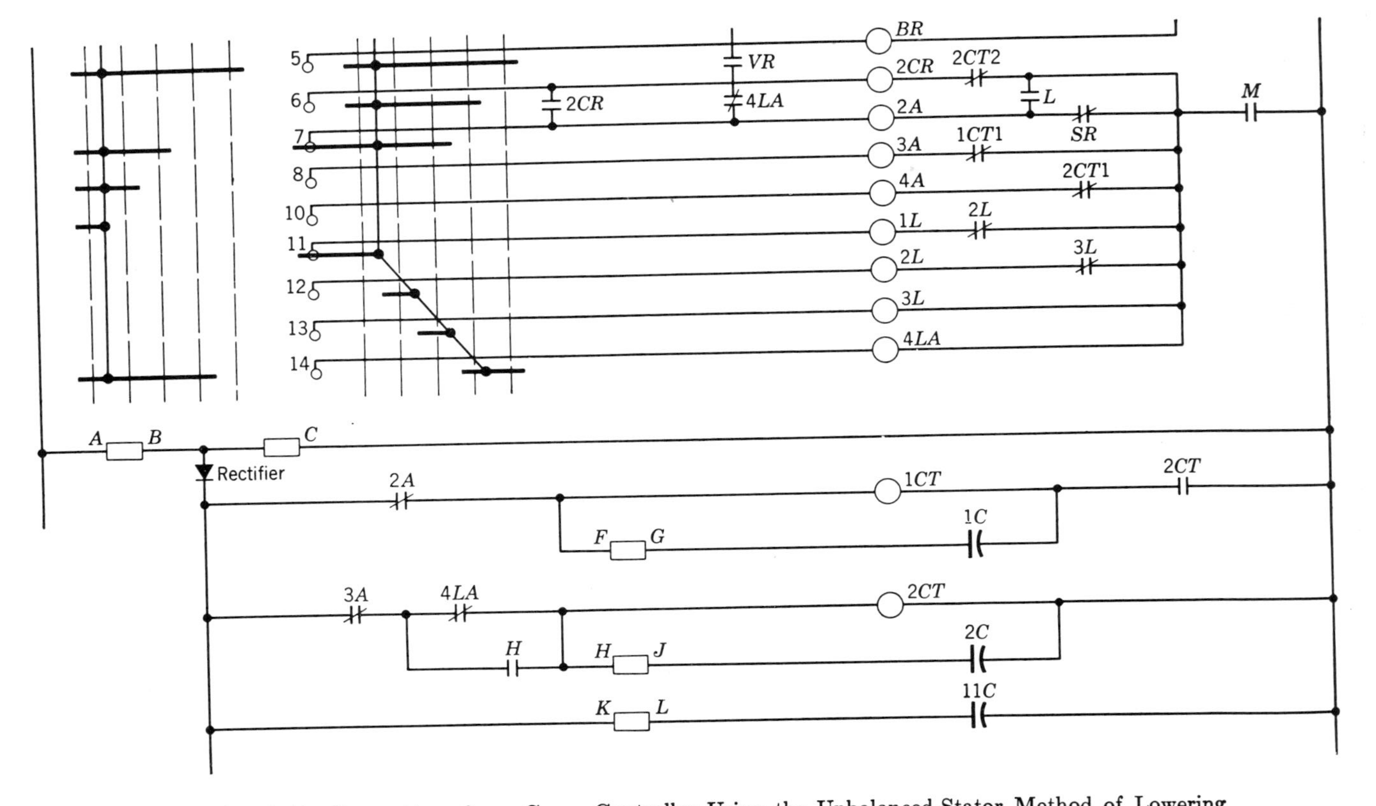

Fig. 17.11 Connections for a Crane Controller Using the Unbalanced-Stator Method of Lowering

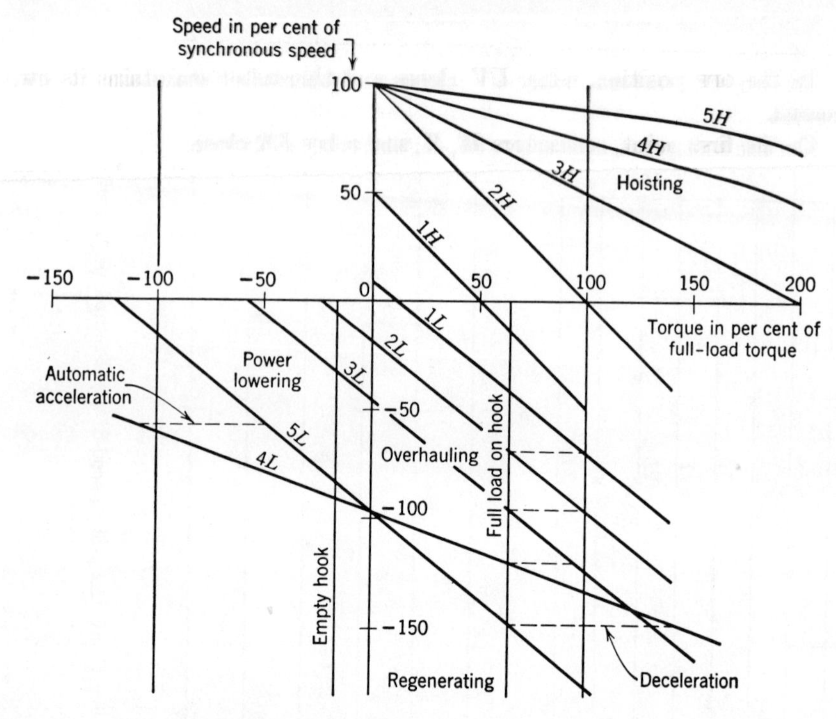

Fig. 17.12 Speed-Torque Curves for a Crane Controller Using the Unbalanced-Stator Method of Lowering

Referring to the diagram of Fig. 17.11, the following is a list of the devices used:

M	Double-pole main contactor
H	Double-pole hoist contactor
L	Three-pole lower contactor
1*L*–2*L*–3*L*	3 single-pole contactors for changing transformer taps
4*LA*	Double-pole contactor for transformer and resistor connection
2*A*–3*A*–4*A*	3 double-pole resistor contactors
10*L*–20*L*	2 overload relays
UV	Double-pole undervoltage relay
BR	Double-pole brake relay
1*CT*–2*CT*	2 condenser-type timing relays
SR	Three-phase current relay for plugging
2*CR*	Single-pole control relay
VR	Single-pole voltage or frequency relay
1*J*–2*J*	Condensers for the timing relays
11*J*	Smoothing condenser for rectifier

The hoisting operation is as follows:

In the OFF position, relay *UV* closes and thereafter maintains its own circuit.

On the first point, contactors *M*, *H*, and relay *BR* close.

The brake is released, and the motor starts with one rotor phase open.

On the second point, contactor 4*LA* closes, closing the complete rotor circuit.

On the third point, contactor 2*A* closes, increasing the torque and speed.

On the fourth point, contactor 3*A* closes, further increasing the speed.

On the fifth point, contactor 4*A* closes, short-circuiting the rotor for full motor speed.

If the master is moved rapidly, the closing of the accelerating contactors will be delayed by the timing relays 1*CT* and 2*CT*.

The lowering operation is as follows:

On the first point, contactors *L*, *M*, and relay *BR* close, followed immediately by contactors 2*A* and 1*L*. The stator of the motor is then connected to the lines through tap 2 of the transformer, and unbalanced voltages are applied to the three phases. The rotor resistor is balanced at about 50% of E/I. Relays 1*CT* and 2*CT* are energized, and relay 2*CR* is prevented from closing. The speed-torque curve for this point is curve 1*L*.

On the second point, contactor 1*L* is opened and contactor 2*L* is closed. The stator is now connected to tap 6 of the transformer, and the applied voltages are unbalanced to a different degree. The curve for this point is 2*L*.

On the third point, contactor 2*L* is opened and contactor 3*L* is closed. The stator is now connected to tap 10 of the transformer, and the unbalance of the applied voltages is again changed.

The curve for this point is 3*L*.

When the master is moved to the fourth point, contactors 3*L* and 2*A* are opened and 4*LA* is closed. The applied stator voltages are unbalanced, and the rotor circuit resistance is increased to 100% of E/I. The motor characteristic is now shown by curve 5*L*. An interlock contact of 4*LA* deenergizes the relay 2*CT*, and after a time interval 2*CT* operates, closing the circuit to 2*CR*, and through 2*CR* to 2*A*. When 2*A* closes, the resistance is again reduced to 50% of E/I and the motor characteristic is that shown by curve 4*L*. This method of transfer avoids an objectionable current inrush which would be obtained with an empty hook, if the transfer were made directly from curve 3*L* to curve 4*L*.

On the fifth point of the master, relay *2CR* and contactor *2A* are opened. This provides an increase in the lowering speed to about 150% of rated motor speed, with full load on the hook, as shown by curve *5L*.

Deceleration is accomplished by moving the master back step by step to any desired point. If the master is moved to the OFF position when the motor is running at a low speed, all relays and contactors except *UV* drop open, and the motor is disconnected and the brake set to hold the load. If the speed of the motor is not low, the relay *VR* will remain closed, maintaining the circuit to contactors *L, M, 1L,* and *2A,* and so providing a negative torque to stop the motor. At or near zero speed, *VR* opens and all contactors are deenergized.

Braking with Direct Current. The idea of applying direct current to one phase of the stator winding to obtain a braking torque when lowering a load is old in the art. The controller is of a conventional reversing design, as used to control the bridge or trolley motions of a crane, with the addition of a double-pole contactor for connecting the dc supply to one of the motor stator when lowering.

With direct current so applied, and with an overhauling load turning the motor, a voltage is generated in the rotor, and current flows in the rotor windings and through the resistor connected in the rotor circuit. The energy of the falling load is dissipated as heat in the resistor, and dynamic braking is secured.

The braking effect may be varied from the master controller in either of two ways. The rotor circuit resistance may be varied by closing the accelerating contactors, one at a time. The more usual method is to vary the voltage of the direct current which is supplied from a generator or rectifiers. Small relays controlled from the master are used for this or rectifiers. Small method permits adjustment of each speed point to the exact value desired, whereas, with the first method, the speeds are fixed by the design of the resistor, which is determined by the requirements of hoisting.

In order to develop full-load retarding torque, it is necessary to apply direct current equal to 100 to 130% of the normal stator current. The direct voltage required is determined by the current and by the resistance of the stator. A rough approximation of the dc power required would be one-eighth of the power of the hoist motor. The direct voltage might be about 32 V for a 550-V ac supply. In order to permit driving down an empty hook or a light load, one point on the lowering side of the master is arranged to disconnect the direct current and accelerate the motor to full speed on alternating current.

Ratings. For any of the above-described crane and hoist controllers, NEMA standard ratings are given in Table 17.2. These ratings are for contactors in the motor primary, or stator, winding.

TABLE 17.2

RATINGS OF CRANE CONTROLLERS

8-hour Rating (amp)	Crane Rating (amp)	Hp at 220 V	Hp at 440 and 550 V
50	67	20	40
100	133	40	75
150	200	60	125
300	400	150	300
600	800	300	600
900	1200	450	900
1350	1800	600	1200

Accelerating contactors should be equipped with blowouts and should have a crane rating of not less than the full-load secondary (rotor) current of the motor. When used for motor secondary control, the ampere rating of a three-pole ac contactor, with its poles connected in delta, is 1.5 times its standard crane rating.

The number of accelerating contactors exclusive of the plugging contactors for reversing controllers, and of the low-torque contactor for hoist controllers, is as follows:

Motor Horsepower Rating	*Minimum Number of Accelerating Contactors*
15 and less	2
16 to 75	3
76 to 200	4
Above 200	5

Electric Load Brake. The thrustor-operated brake, described in Chapter 20, may be used as an electrically operated load brake to control the speed of an overhauling load. A reversing magnetic controller, having a resistor connected in the rotor circuit and a set of accelerating contactors to commutate the resistor, is used. In hoisting, the motor of the thrustor brake is connected to the stator side of the main motor and so operates at full voltage and frequency to release the brake. The slowest lowering speed is obtained by opening the rotor circuit of the motor and connecting the thrustor motor to the rotor slip rings. Since the hydraulic pressure of the thrustor is obtained by means of a straight-blade impeller acting as a centrifugal pump, the pressure varies approximately as the square of the speed of the thrustor motor. The speed of that motor is determined by the frequency of the power supplied to it or, in other words, by the speed of the main motor. With the main motor rotor circuit open, the motor will not deliver any torque. A light overhauling load will turn the motor slowly, the frequency at the slip rings will be high, and the brake will be almost entirely released. A heavy overhauling load will turn the motor faster, the frequency at the slip rings will be lower, and the brake will be partially set. The degree of braking obtained will be determined by the load.

The second lowering point gives a higher speed by closing one phase of the rotor-circuit resistor. The motor delivers a low torque, and the net torque is the difference between the torque of the motor and that of the brake. The third point gives a still higher speed by connecting in the entire three-phase resistor and so increasing the motor torque. On the fourth and fifth points the thrustor is transferred to the stator side of the motor and the brake is completely released. The speeds will be above synchronous speed in an amount determined by the amount of resistance in the rotor circuit and by the amount of the overhauling load.

The Kraemer System. There are several methods of controlling the speed of a slip-ring motor which necessitate auxiliary machines. The general principle is the introduction of a countervoltage into the rotor circuit of the motor to increase the slip at any given load and speed. The slip energy, which is lost with resistance control, is saved with these control systems, being either returned to the line as electric energy or converted to mechanical power and applied to the shaft of the main motor. The cost of the auxiliary machines limits the use of these systems to large motors, where the energy saved will justify their cost. They are used for main mill drives in steel mills, where the motors may be from 500 to several thousand horsepower in rating, for the cutter motors on large dredges, and for similar applications.

Figure 17.13 shows the arrangement of a constant-horsepower Kraemer system. The main motor is started as a conventional slip-ring motor by the use of a resistor in the motor circuit. When the motor is up to speed it is disconnected from the starting resistor by the opening of contactor S, and the rotor is connected to the ac winding of a rotary converter by the closing of the running contactor R. An auxiliary dc

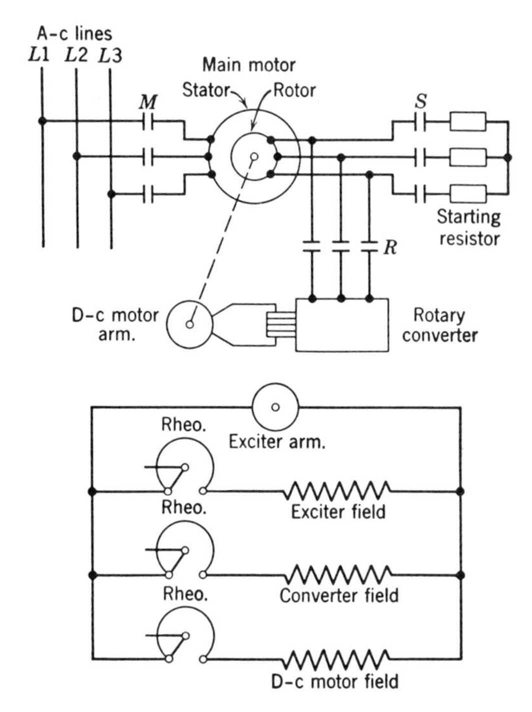

FIG. 17.13 Connections for a Constant-Horsepower Kraemer System

motor, mechanically connected to the shaft of the main motor, has its armature connected to the dc winding of the converter. The field windings of the converter and the dc motor are separately excited. The voltage applied by the converter to the main motor rotor is determined by the countervoltage of the dc motor, which may be regulated by varying the strength of the motor field. The speed of the main motor may then be controlled from the dc motor field rheostat. Some power-factor correction may be obtained by overexciting the converter field.

The dc motor, instead of being coupled to the shaft of the main motor, may be coupled to a separate generator, connected to return power to the supply lines. With this arrangement the system delivers constant torque.

With either arrangement any desired speed reduction may be obtained, the size of the auxiliary machines becoming larger as the required speed reduction increases. For example, if the system is to provide 50% reduction in speed, the dc motor must be capable of delivering half of the main motor rating at half of its speed, and it will be as large as the main motor. Most Kraemer sets are used to provide 25 to 35% speed reduction, with 50% as a practical maximum.

Since the frequency of the power supplied to the slip rings approaches zero as the speed approaches synchronism, the Kraemer system becomes unstable at speeds near synchronism. For this reason the system is usually limited to operation at speeds below synchronism.

The initial cost of the system is less when 60-hz power and high machine speeds are used. The constant-horsepower drive will cost less than the constant-torque drive, because of the additional machine required by the latter. It is general practice to bring out both ends of the rotor windings of the main motor, using six slip rings on motor and converter, as this connection results in a less costly converter. With this exception the machines for the Kraemer system are of standard design, as used for many other purposes; this is a decided advantage of the system.

Application of the Wound-Rotor Motor. The wound-rotor motor is useful where high starting torque with low starting current is desired. Heavy loads can be started slowly and smoothly, without undue line disturbance. It is also used where speed regulation is desired, as on fans, centrifugal pumps, stokers, and printing presses.

The disadvantages of the motor, as compared to the squirrel-cage motor, are the complications of machine, control, and wiring, introduced by the rotor connections, and the higher cost of the motor and control.

The motor is particularly applicable to cranes, hoists, coal-handling bridges, and similar applications, where speed control of a variety of loads is required and where overhauling loads must be safely lowered.

Design of Resistor. Slip-Ring Motor. Accelerating. Resistance is used in the secondary circuit of slip-ring motors for acceleration, plugging, and speed regulation. The ohmic value of the resistance is based on the voltage generated in the secondary windings and on the full-load secondary current. The voltage and current, being matters of motor design, vary widely in motors of the same size but of different manufacture. The data for any particular motor, obtainable from the manufacturer, will be given in volts and amperes.

E_s = voltage across the slip rings at standstill
I_s = amperes in the secondary winding at full load

The values given for the secondary volts and amperes may be checked from the equation

$$\text{Efficiency} = \frac{\text{Hp} \times 746}{E_s \times I_s \times 1.73} \qquad (17.1)$$

If the efficiency as calculated from this equation is less than 85%, or greater than 100%, the data are probably incorrect and should be checked with the manufacturer of the motor.

The controller resistance may be connected in star or delta. The resistance per phase to give 100% torque at zero speed may be determined from the equations:

$$R_{100\%} = \frac{E_s}{\sqrt{3}\, I_s} \qquad \text{(star connection)} \qquad (17.2)$$

$$R_{100\%} = \frac{\sqrt{3}\, E_s}{I_s} \qquad \text{(delta connection)} \qquad (17.3)$$

This is the total resistance per phase of the rotor plus the external resistance required to produce curve *e* of speed-torque curves in Fig. 17.14.

The resistance per phase of the rotor alone may be determined by the equation:

$$R_{\text{rotor}} = R_{100\%} \frac{S_m}{100} \qquad (17.4)$$

Where R_{rotor} is the phase resistance of the rotor and S_m is the slip in per cent at 100% torque with all external resistance cut out. Curve *a*

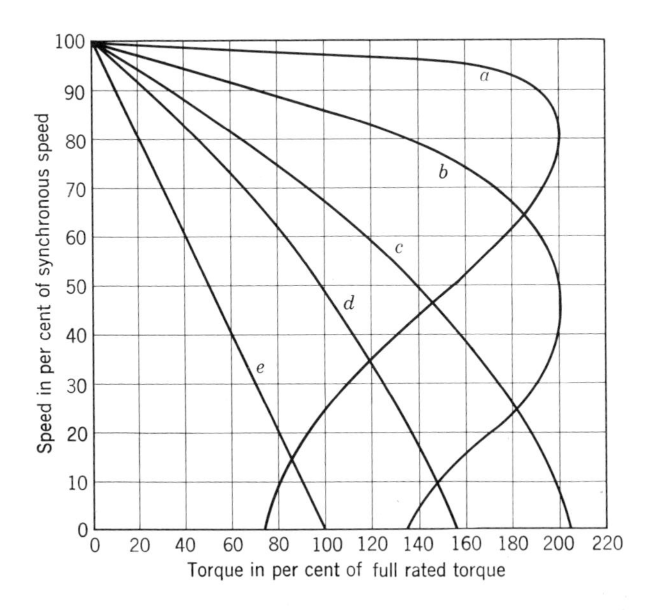

Fig. 17.14 Speed-Torque Curves of a Slip-Ring Motor

in Fig. 17.14 represents this condition and shows approximately 3% slip at 100% torque.

Equation 17.4 may be extended to determine the resistance required to provide any slip at full load by changing the terms.

$$R_t = R_{100\%} \frac{S}{100} \tag{17.5}$$

In this equation R_t is the total phase resistance required to produce slip S at full load.

Curve *b* in Fig. 17.14 crosses the 100% torque line at approximately 14% slip. By referring to equation 17.5, it can be seen that 14% $R_{100\%}$ is required to produce that slip. The external resistance of each phase is

$$R_{\text{ext}} = .14R_{100\%} - R_{\text{rotor}}$$

The same reasoning may be used to determine that curve *d* in Fig. 17.14 requires 50%$R_{100\%}$.

Figure 17.15 shows the average values of motor slip at full load for wound-rotor motors of different horsepower.

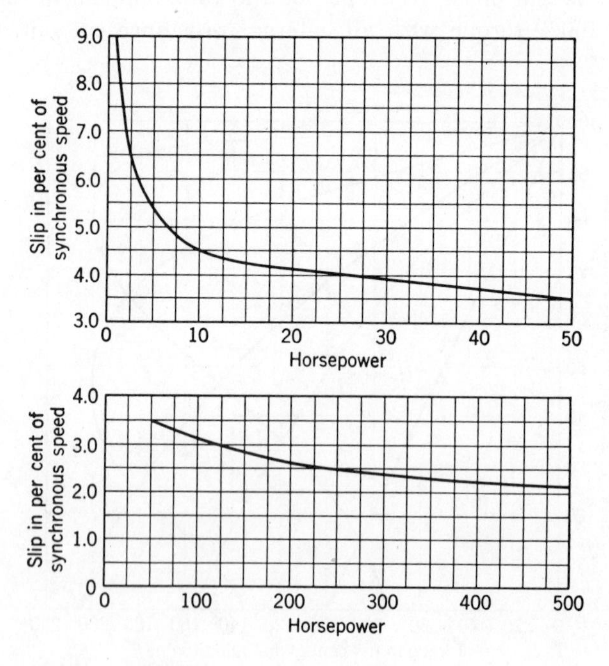

Fig. 17.15 Average Slip of Slip-Ring Motors at Full Rated Load

The resistance taper may be either calculated or determined graphically in the manner described for the shunt motor. The torque may be assumed to be proportional to the secondary current, within practical limits. On the basis of equal inrushes on each point of the controller, and of cutting out the steps when the secondary current falls to 120%, the inrushes and taper shown in Table 17.3 will apply. One hundred

TABLE 17.3

RESISTOR TAPER FOR SLIP-RING MOTORS

Number of Steps	Secondary Inrush	Primary Inrush	Taper of Resistance Steps in Per Cent of Total
1	320	300	100
2	250	230	70–30
3	210	195	55–30–15
4	180	165	40–30–20–10
5	166	155	34–26–19–13–8
6	155	150	29–23–18.5–13.5–9.5–6.5

twenty per cent is selected as a safe point to cut out the steps. If the current is allowed to drop lower, the inrushes will, or course, also be lower.

Plugging. The total resistance required for plugging is determined by equation:

$$R_p = 1.8R_{100\%}$$

This is the total resistance required to keep the primary and secondary currents within reasonable limits, because the induced voltage in the rotor at the instant of plugging is approximately double that at standstill. The plugging torque may be varied to suit the application by changing the amount of resistance. Decreasing the resistance will increase the initial plugging torque.

Speed Regulation. The resistor for speed regulation is determined in the same manner as just described.

Figure 17.16 shows typical load curves for hoist, machine, fan, and centrifugal compressor loads. A resistor which causes 50% slip at 100% torque will allow a hoist to operate at 50% speed, a machine at 58% speed, a fan at 69% speed, and a centrifugal compressor at 75% speed.

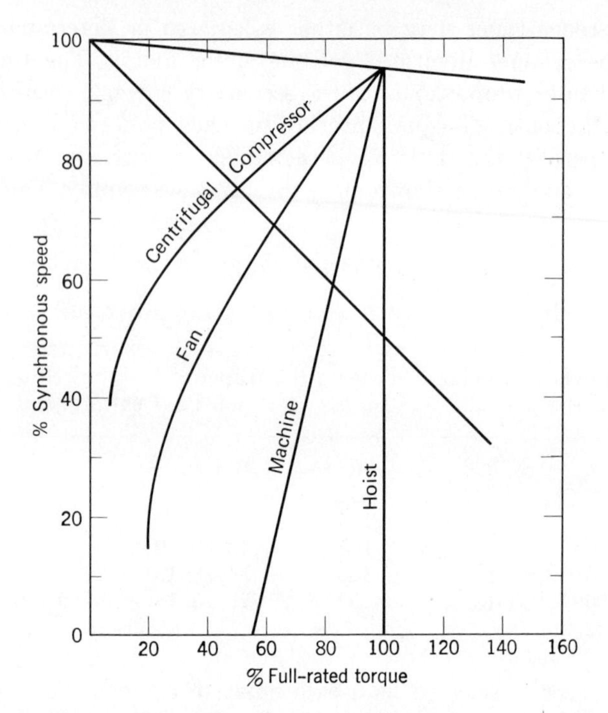

FIG. 17.16 Typical Load Curves

From this it is evident that if speed control is specified, the load curve or the type of load must be known.

Problems

1. If a wound-rotor motor has a synchronous speed of 1800 rpm and its full-load running speed is 1710 rpm, what is the per cent slip at full-load?

2. If the resistor of a wound-rotor motor controller is designed to obtain 50% speed at 50% torque, what will be the starting torque with the motor at rest?

3. The voltage measured across the slip rings of a 50-hp wound-rotor motor at standstill is 260 V. The motor is 95% efficient. What will be the rotor current in each phase when the motor is operating at full rated load?

4. A wound-rotor motor having the characteristics of Fig. 17.1 has a rotor voltage at standstill of 175 V and a rotor current of 75 amp at full-rated load. How many ohms are required in each phase of a star-connected resistor which will enable the motor to produce the maximum possible torque when starting from rest?

5. At what per cent of synchronous speed will the motor run if the resistor is left in circuit and the motor is fully loaded for a hoist load?

6. The manufacturer of a wound-rotor motor says that its rotor characteristics are 230 V and 100 amp. If this is a 50-hp motor, what is its efficiency?

7. How many ohms per phase will be required in a star-connected resistor which will limit the initial starting current to 150% of full-load current?

8. How many additional ohms will be required to limit the current to the same value when the motor is plugged?

9. How many ohms will be required in a resistor which will cause the motor to run at 80% speed when 80% loaded?

10. An 1800-rpm wound-rotor motor has secondary characteristics of 250 V and 150 amp. When it is running with a resistor in the rotor circuit, a voltmeter measures 75 V between slip rings. What is the motor speed?

11. A control manufacturer receives an order for controllers for five slip-ring motors. The rotor characteristics are given as follows:

A	25 hp, 115 V, 98 amp
B	30 hp, 182 V, 75 amp
C	50 hp, 200 V, 85 amp
D	100 hp, 385 V, 150 amp
E	200 hp, 600 V, 150 amp

Which of these may be accepted as correct, and which are in error?

12. A 150-hp slip-ring motor has rotor characteristics of 550 V and 120 amp. Calculate the ohms in each step of a two-step resistor, star-connected, which will allow a starting inrush of 250% of rated current. Use a taper of 70–30.

13. Calculate an equivalent resistor for the motor of problem 12, except using two steps each connected in delta, the two deltas being in parallel on the second step.

14. A 100-hp slip-ring motor has rotor characteristics of 550 V and 80 amp. Its resistance is 5% of E/I. Calculate the ohms required in a star-connected resistor to give 180% inrush on the first starting point.

15. For the motor of problem 14, calculate the ohms in the last two steps of a resistor which will cause the fully loaded motor to run at 85% of rated speed with the last step in circuit, and at 75% of rated speed with the last two steps in circuit for a hoist load.

16. A controller for a motor driving a printing press uses the circuit of Fig. 17.8 The motor is rated 10 hp, 440 V, three-phase, 60 Hz, and the rotor characteristics are 300 V and 15 amp. Neglecting motor resistance, calculate the ohms required in each step of a resistor which will permit a total speed reduction of 75% of rated speed for a machine load.

17. A 20-hp slip-ring motor has a controller which gives the characteristic curves of Fig. 17.1. If the rotor characteristics are 300 V and 30 amp, how many ohms will be required in each leg of a star-connected resistor which is designed to give the maximum possible starting torque on the first point?

18. A 75-hp slip-ring motor, with rotor characteristics of 400 V and 78 amp, is driving a centrifugal pump which has the load characteristics of Fig. 7.6. Calculate the ohms in each step of a star-connected resistor which will give the operating curves shown in Fig. 7.6.

19. Referring to problem 18, what current will be obtained on each step of the resistor, if the resistor is used for speed-regulating duty?

20. A 100-hp slip-ring motor has rotor characteristics of 550 V and 80 amp. The resistor which is used with it consists of three sections, each consisting of 120 cast-iron grids connected in series. The motor is rewound to have rotor characteristics of 275 V and 160 amp. How should the resistor be rearranged to suit these characteristics?

21. A factory has a 40-hp slip-ring motor which has been used to drive a machine which required speed control. The motor is to be transferred to another machine which does not require speed control, and the owner asks whether he may now use an autotransformer starter to control the motor. Can this be done, and if so what must be done to the motor?

22. A 25-Hz slip-ring motor has a slip of 5% at full load. What is the frequency of the rotor current under those conditions?

23. An 1800-rpm slip-ring motor on an elevator has a slip of 5% at full load. At what speed will it run when lowering full load, assuming 100% efficiency of the hoisting mechanism?

24. How much resistance, in per cent of E/I, would have to be used in the rotor circuit to permit the elevator to lower a full load at 150% of synchronous speed?

18

ALTERNATING-CURRENT ADJUSTABLE SPEED DRIVES

Polyphase ac motors are designed to run at or near a speed corresponding to the rotating field impressed upon the stator as described in Chapter 16. The synchronous speed of an ac motor is inversely proportional to the number of motor poles and directly proportional to the frequency applied to the stator. An induction motor will operate at less than synchronous speed when delivering rated torque, since the rotor bars or windings must cut the flux of the rotating field, causing current to flow in the rotor windings and establish the magnetomotive force necessary to rotate the rotor.

The inherent design criteria of an ac motor that causes it to follow the rotating stator field establishes the one truly practical method of controlling ac motor speed—controlling the frequency of the supply power.

AC Motor Torque. The torque produced by an ac induction motor is proportional to the magnetic flux in the air gap. For constant horsepower, torque can vary inversely with speed, but if constant torque at the output shaft of the motor is required, the flux in the air gap must be maintained constant over the speed range.

Alternating-current motors designed to run on constant voltage and constant frequency are designed with electrical and air-gap parameters consistent with torque requirements. Optimum adjustable frequency systems must be predicated on voltage as well as frequency control to provide the needed torque-producing capability at all speeds.

Rotating Alternators. An M-G set, with an adjustable speed drive establishing the speed of a rotating alternator, is a practical means of providing adjustable frequency over a limited range. A dc drive motor supplied from an adjustable voltage conversion unit is often used to obtain the desired speed range. Similarly, an eddy-current clutch or a mechanical speed changer can be used between a constant speed induction motor and the alternator to adjust the alternator speed.

In addition to the maintenance problems involved with rotating M-G sets, control of the ac output voltage amplitude and response is limited by the inherent excitation of the alternator. Most significantly, the resistance and reactance of the alternator and its motor load, limit the performance of the motor over a wide speed range.

For many years motor-generator sets have been used to supply multiple ac motors on applications such as run-out tables in the steel mills. Each roll is driven by a squirrel cage motor with large groups of these motors connected to a common alternator. The alternator is driven at a speed consistent with the mill requirements, so that each individual roll is in effect geared to all others through the adjustable frequency system. Each motor is provided with a separate disconnect switch for removal of a defective motor. A single pair of reversing contactors causes all motors to run forward or reverse as a synchronized group. The speed range, however, is limited to no more than about three or four to one.

Static Inverters. Although thyratron and mercury arc rectifiers have been used for inverters to produce adjustable frequency, the switching time of these components is too long for optimum operation. Semiconductor devices of the thyristor type (Fig. 9.1) have switching times of a few microseconds, allowing design of adjustable frequency systems with either dc to ac inversion or ac to ac conversion with a controlled frequency output. These same components are used to control the ac voltage level and thereby the torque of the ac drive.

Cycloconverter. Mercury arc rectifiers have been used as switching elements in cycloconverters on European railroad systems since shortly after World War I. However, the low efficiency, low power factor, high maintenance cost, and large physical size limited the practicability of the cycloconverter until the thyristor became available.

The cycloconverter is composed of a number of static switches interposed between the power source and the load. When properly programmed, these switches are gated in a sequence to selectively couple the power source to the load in a manner which will impress an approximate sine-wave voltage on that load. The cycloconverter illustrated by Fig. 18.1 receives power from a three-phase line and delivers single-phase power. High-speed switches gate small increments of the primary power. The output frequency wave is thus fabricated from small pieces of the input frequency, and the output is the envelope of these increments.

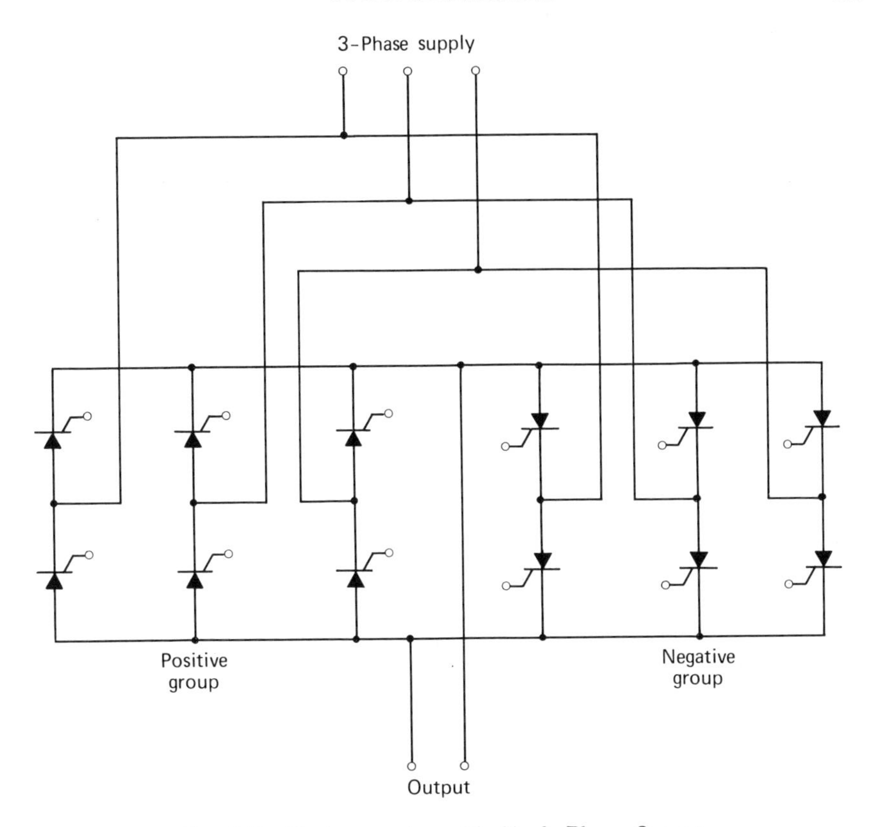

FIG. 18.1 Cycloconverter with Single-Phase Output

The cycloconverter is inherently a step-down frequency drive. From a practical standpoint the ratio of input to output frequency must be at least two. The cycloconverter can, however, operate down to zero frequency, and actually the sequencing can be reversed so that the output range on a 60-Hz, three-phase line is from −30 to +30 Hz.

Since 12 thyristors are required for single-phase output (Fig. 18.1) 36 thyristors are necessary for a full three-phase output. If a cycloconverter is applied to a motor, it is necessary to bring out all six leads for the three stator windings or to use interphase transformers to isolate inverter outputs.

Since the cycloconverter produces a voltage wave with small harmonic content, reactance in the motor will smooth the simulated sine wave and the current wave form is excellent. The advantages of a cycloconverter are, then, the minimum heating and full reversing, regenerating capabilities. As with other semiconductor inverters, efficiency is high,

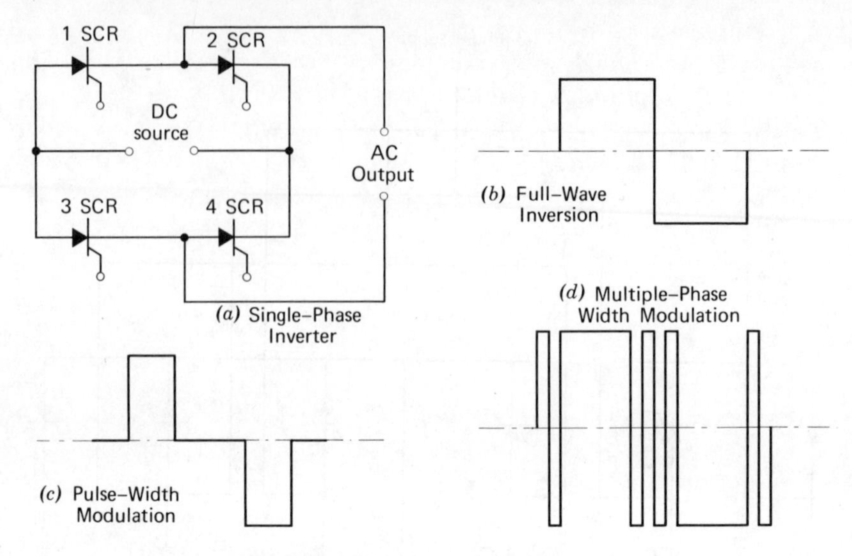

Fig. 18.2 Basic DC to AC Inversion

and power requirements are low. Major limitations are the large number of thyristors required, and the low frequency output when connected to commercial power systems. In general, these low frequencies are not adaptable to commercially available ac motors.

Inverter with DC Link. Assuming a source of dc power is available, Fig. 18.2*a* represents an inverter capable of converting the dc power to single-phase alternating current. Whenever controlled rectifiers 1SCR and 4SCR are gated on, current will flow from the direct-current source through the load in the positive direction. When 1SCR and 4SCR are turned off and 2SCR and 3SCR are turned on, current will flow through the load in the opposite direction. Since thyristors switch in microseconds, the turn-on and turn-off times are negligible. The output will in effect be a square wave.

If the pairs of thyristors are programmed to alternate on a periodic time base, the wave form will be represented by Fig. 18.2*b*. The output will be a square wave with the total area under the positive and negative half-cycles equal to the total available dc power over the same time period.

Since the control system retains full control of the interval during which each pair of thyristors is fired, it is possible not only to establish a specified frequency for the output wave form, but also to delay or advance the turn-on and turn-off point for each square wave pulse. Figure

18.2*c* represents a wave form modified in this manner. Since this technique controls voltage by modulating pulse width while also controlling frequency, it is called pulse-width modulation (PWM).

A further refinement of the pulse-width modulation concept is illustrated by Fig. 18.2*d*. By proper programming of the pairs of thyristors, positive and negative pulses of varying widths can be generated so that the resultant current flow through the load winding closely approximates a sine wave. In either form of pulse-width modulation, both the frequency and the output terminal voltage can be controlled, limited only by the minimum switching time of the thyristors.

Control of AC Output Voltage. Proper operation of an adjustable-frequency motor drive requires constant volt-seconds per half-cycle to maintain constant torque in the normal operating range. At low or high frequencies this voltage ratio must be increased to overcome losses and maintain constant flux in the ac motor air gap. To achieve this a means must be provided to control the voltage level of the dc input or the ac output. Although there are many systems for accomplishing this result, Fig. 18.3 represents, in the form of one-line diagrams, some of those most commonly used. The simplest form, Fig. 18.3*a*, provides constant-voltage direct current to the inverter from either a conventional rectifier or a dc bus. The inverter, controlled only to produce adjustable frequency, has a constant-voltage output.

An adjustable autotransformer between the inverter and the load corrects the output voltage for the necessary volt-seconds per half-cycle criteria. Since the autotransformer is operated either manually or with a pilot motor, this type of drive is not capable of high dynamic response. The most logical approach with this type of system is to utilize the voltage-adjusting autotransformer servo as the frequency reference for speed control.

Phase-controlled rectifiers or chopper-type rectifiers as used on dc adjustable-speed drives and described in Chapter 9 are used to produce adjustable-voltage direct current, which in turn is fed into an inverter for conversion to adjustable frequency (Figs. 18.3*b*, 18.3*c*). By controlling the voltage of the rectifier and the frequency of the inverter, adjustable voltage–adjustable frequency is available to the ac motor. In effect, the inverter has replaced the commutator on the dc motor to provide similar operating characteristics of ac and dc adjustable-speed drives.

Figures 18.3*d* and 18.3*e* represent two forms of pulse-width modulated drives. Direct-current power is produced by a conventional constant-voltage dc rectifier. In Fig. 18.3*d* the output of one inverter is phase-shifted with relation to the second. The resultant pulse width is estab-

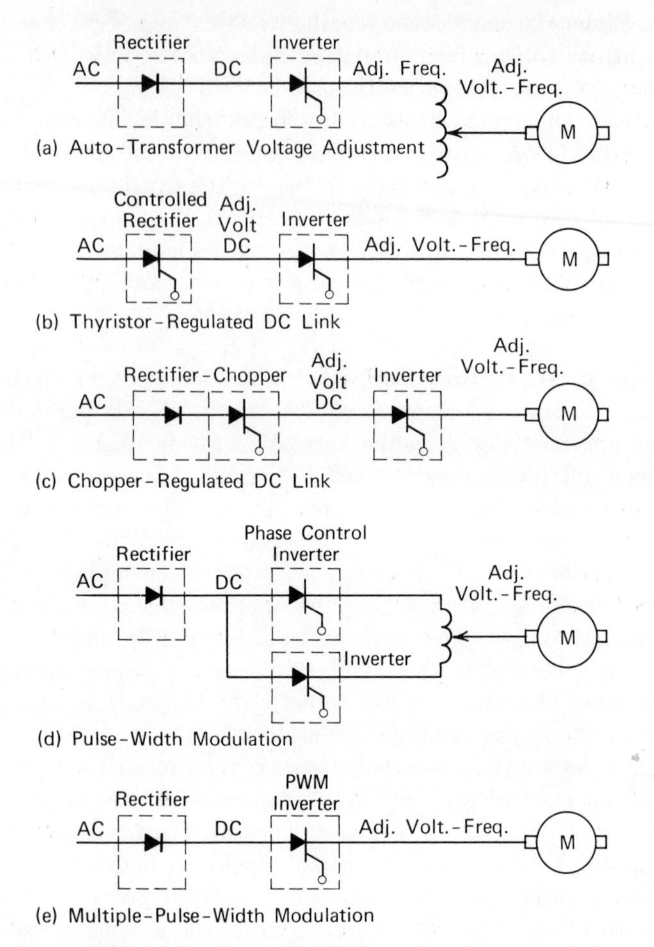

FIG. 18.3 One-Line Diagrams of Typical Adjustable-Frequency Drives

lished by the addition or cancellation of the pulses from the individual inverters depending on their phase relationships. The inverter output approximates a sine wave of controlled voltage and frequency required for ac motor drives.

The multiple-pulse-width inverter, Fig. 18-3*e*, utilizes six controlled rectifiers programmed to fire in a controlled sequence. The envelope of the multiple-pulse output approximates a sine wave. The arithmetical summation of the area encompassed by the variable length pulses in a half-cycle represents the volt-seconds per half-cycle delivered to the motor.

Three-Phase Inverter Drives. Industrial ac adjustable speed drive systems utilize three-phase induction or synchronous motors. Three single-phase inverters, as described, could be programmed for 120-deg phase displacement and could be used to supply the individual windings of an ac motor. However, from a practical standpoint, a common dc supply must be used and the motor with either wye or delta connected windings will establish common connections between phases.

In its simplest form a three-phase inverter can be considered as six switches, S4 through S9, with the operating sequences shown in Fig. 18.4. Each switch will be closed and opened sequentially, but only three switches will be closed at any instant. When a fourth switch is closed, the switch that has been closed the longest is opened. For example, when S5 is closed, S4 opens; when S8 closes, S9 opens; when S7 closes, S6 opens; when S4 closes, S5 opens.

Each output terminal is alternately connected from the positive dc line to the negative. The dwell or zero voltage produces a close approximation to a sine wave.

The wave forms illustrated in Fig. 18.5 are the line to neutral voltages impressed on the windings of a wye-connected motor. In typical circuit operation, as the frequency is decreased, the height of the wave becomes less and the width becomes greater. The wave shape does not deteriorate as the frequency is decreased.

One of the most commonly used circuits, known as the McMurray-Bedford inverter (Fig. 18.6), represents an integrated three phase drive. Thyristors 5SCR through 10SCR are high-speed semiconductor switches

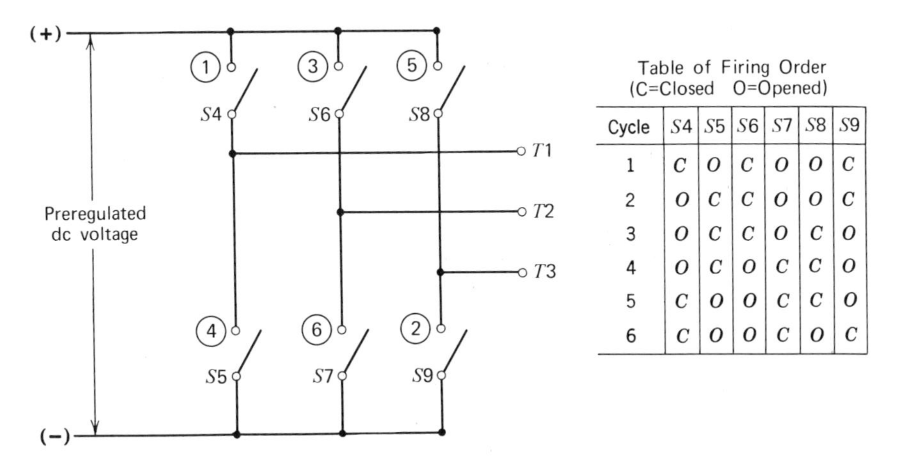

Cycle	S4	S5	S6	S7	S8	S9
1	C	O	C	O	O	C
2	O	C	C	O	O	C
3	O	C	C	O	C	O
4	O	C	O	C	C	O
5	C	O	O	C	C	O
6	C	O	O	C	O	C

FIG. 18.4 Inverter Switching Sequence

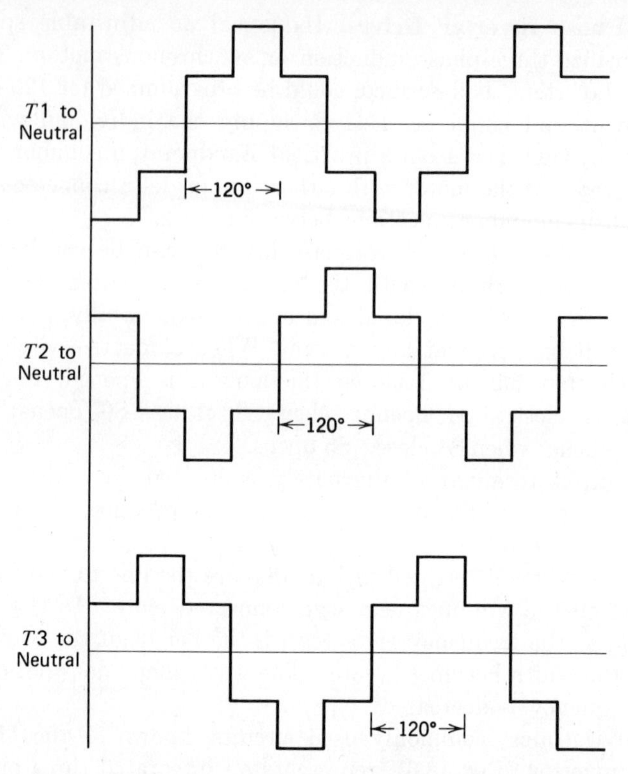

FIG. 18.5 Simulated Sine Wave Voltage Impressed Across Windings of Wye-Connected Motor or Transformer by McMurray-Bedford Inverter

fired in the proper sequence at 60-deg intervals to generate the full three-phase output. The McMurray-Bedford inverter depends upon natural commutation to turn off the thyristors. Each thyristor will continue to conduct until its complementary unit fires to eliminate the sustaining anode voltage of the first unit. If 5SCR is gated on, capacitor 1C will be essentially shorted and 4*C* will store a charge approximating the dc bus voltage. When 8SCR is gated to initiate the opposite half-cycle, the discharge of capacitor 4*C* through winding *X*1–*X*2 of transformer 1*T* will generate a transient voltage in windings *H*1–*H*2 in a direction to drive the anode voltage of thyristor 5SCR to zero. Diode 7*D* acts as a free-wheeling path to dissipate the stored energy in the motor windings during this transient condition.

Since this type of inverter depends on firing of successive thyristors to turn off the opposite-polarity, natural commutation, it cannot be used for pulse-width modulation. Voltage is controlled by regulating the am-

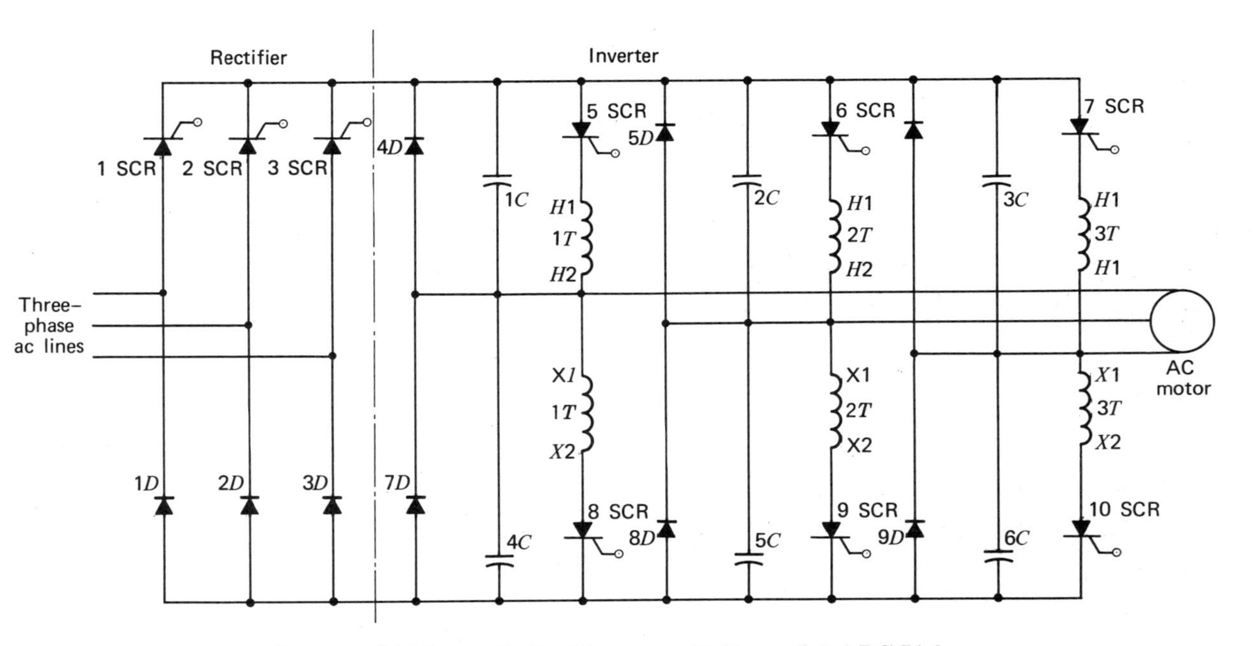

FIG. 18.6 McMurray-Bedford Inverter with Preregulated DC Link

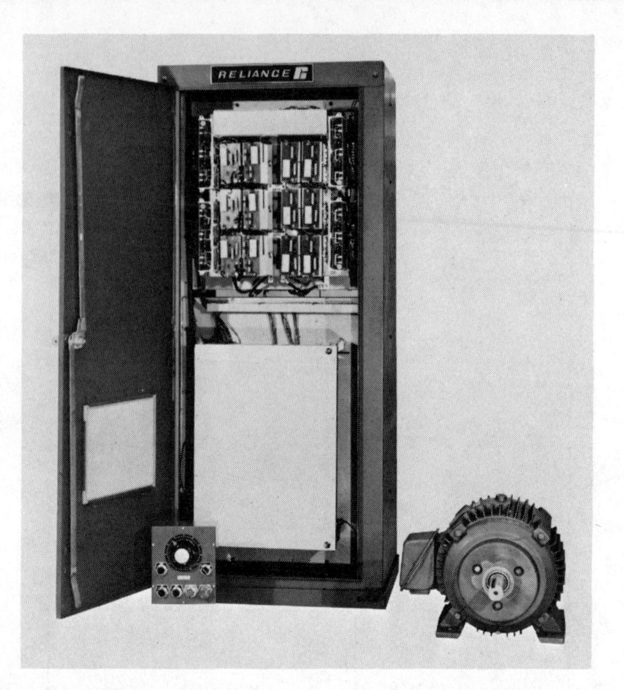

FIG. 18.7 Reliance Electric Co. Solid-State Adjustable-Speed AC Drive

plitude of the incoming dc supply. The preregulated power supply rectifier illustrated is the simplest controlled form using three diodes and three thyristors. If the rectifier bridge is bidirectional (Chap. 9), the entire rectifier-inverter combination can be made regenerative, capable of accepting pumpback current from the motor. However, a more general economical solution is to dissipate the regenerative power in resistors or combine motoring and regenerating drives on the same dc supply.

The reference for this type of inverter is an oscillator establishing a predetermined frequency and generating a train of pulses. The pulses are fed into a ring counter. The output of the ring counter is amplified with pilot thyristors and pulse transformers which fire the power thyristors in the proper order. Stability of the entire drive system must be considered. Usually the frequency is modified by the input dc voltage in a degenerative direction to maintain stability under transient load conditions.

Pulse-Width Modulation Drive. Although several designs are used to achieve pulse-width modulation, many of them use thyristors to force commutation of the basic inverter bridge. Figure 18.8 represents a circuit of this type in which commutating capacitor C is allowed to charge and discharge by the switching of controlled rectifiers 7–8–9 and 10. The charging and discharging current of the capacitor creates a voltage drop across the series reactors which dissipates the anode voltage and forces commutation of the inverter thyristors. The logic circuitry programs the switching intervals, and thereby the number and width of pulses.

Controlled rectifier 11SCR, sometimes called a crowbar element, protects the power bridge from excessive motor current. This thyristor can be programmed to turn on when there is excessive current loading causing a momentary short circuit that trips the circuit breaker and disables the power bridge. More frequently the output of a current-sensing device is used to interrupt or delay the output of the reference voltage-controlled oscillator. On this type of drive the dc supply is constant voltage from either a conventional rectifier or a dc bus.

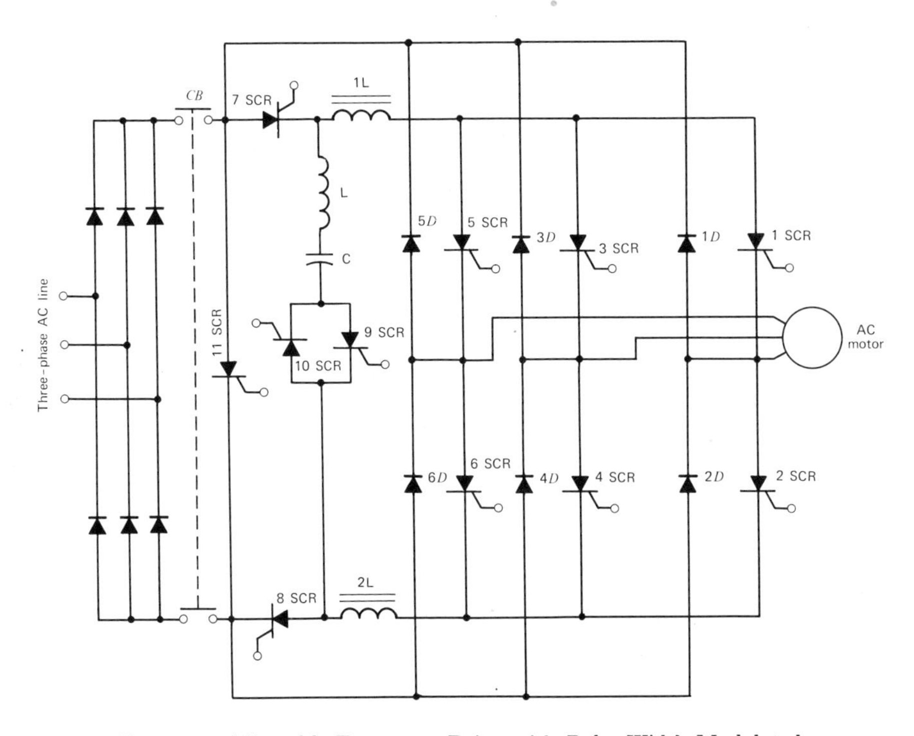

Fig. 18.8 Adjustable-Frequency Drive with Pulse-Width-Modulated Inverter and Constant Voltage Rectifier

Adjustable-Voltage Constant-Frequency Drives. Since it is necessary for a squirrel-cage induction motor to lag sufficiently far behind the rotating field to generate current in the squirrel cage, an adjustable-speed drive can be built which depends upon control of rotor slip. A drive of this type (Fig. 18.9) is relatively simple, requiring only back-to-back thyristors in the primary line supplying the motor. In each phase the thyristors are gated for a portion of each half-cycle, thus reducing the impressed voltage to that part of the cycle during which the thyristors are turned on. Since the torque of the induction motor is a function of the flux in the air gap, the torque is reduced as the square of the voltage. The adjustable voltage ac drive therefore becomes a torque-con-

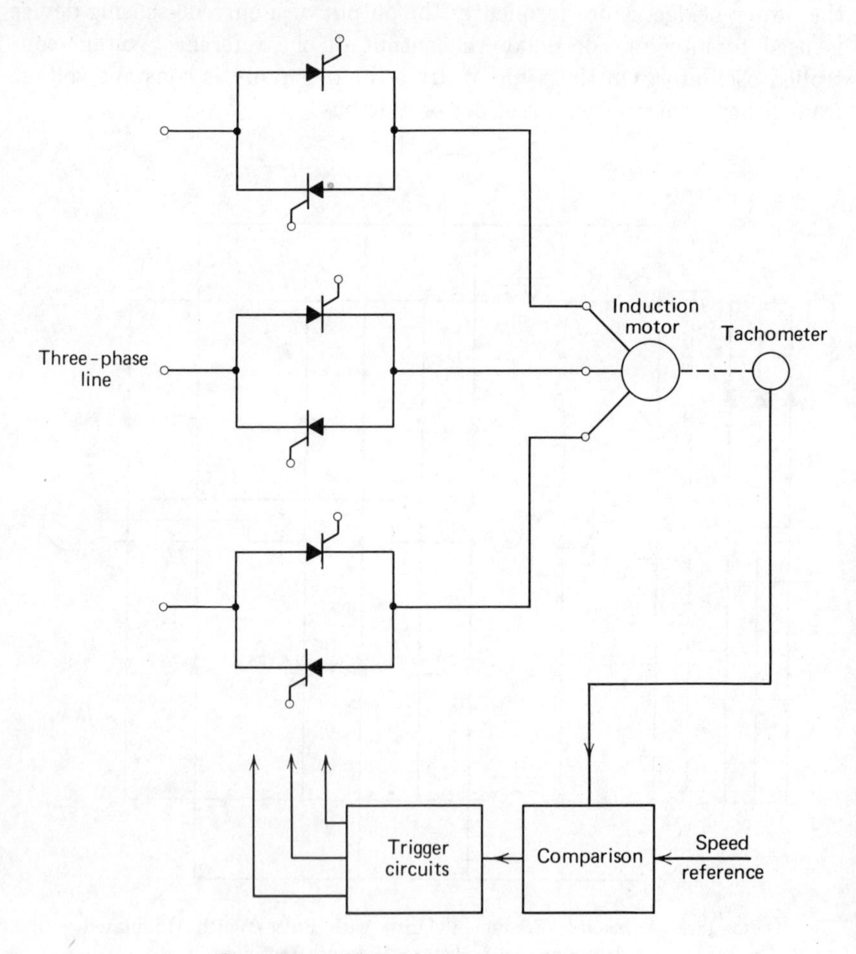

Fig. 18.9 Adjustable-Voltage AC Drive

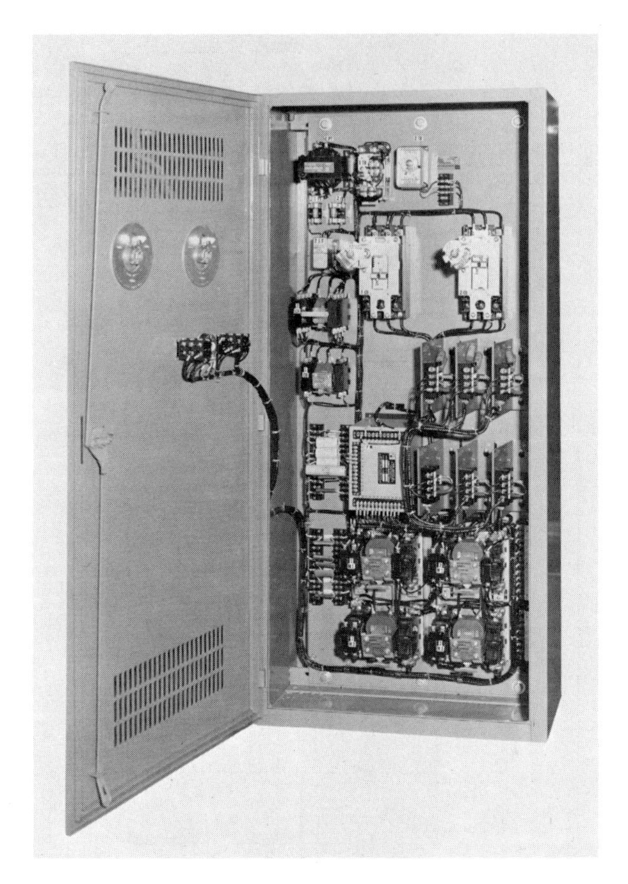

Fig. 18.10 Cutler-Hammer Adjustable-Voltage AC Drive

trolled drive. By measuring the speed of the motor with a tachometer and comparing the speed with a preset reference, an error signal can be generated which will properly gate the thyristors to create the torque necessary to maintain the desired speed. Since this type of drive has less than optimum torque-per-ampere relationships, constant torque loading will produce excessive motor heating at low speeds. The practical usage of this drive is on pump or fan loads. It is, however, a low-cost drive and quite satisfactory for this type of application.

Slip-Power Control Drive. As described in Chapter 16, the wound-rotor motor can be speed-controlled by changing the resistance values connected in the secondary circuit. Functionally, this type of control

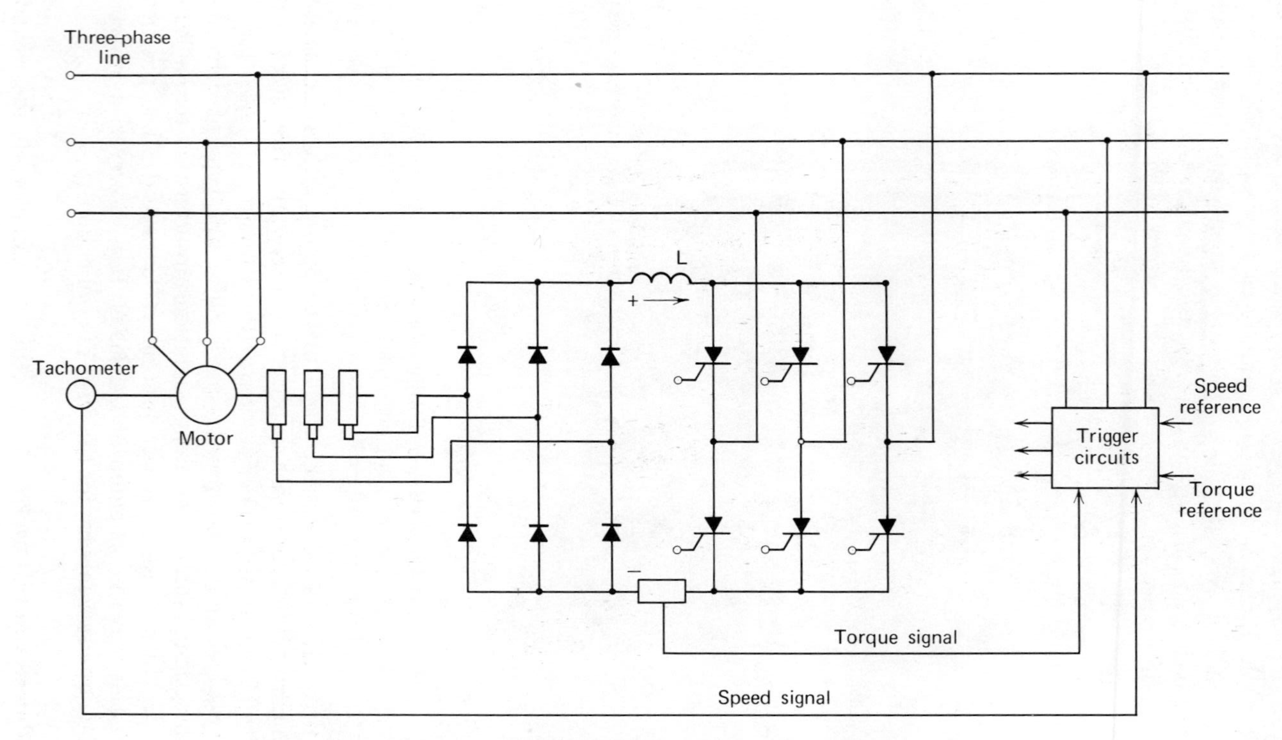

FIG. 18.11 AC Adjustable-Speed Drive by Control of Slip Power

dissipates a portion of the power transmitted across the air gap in a resistor bank. By using techniques illustrated by Fig. 18.11, the power available at the secondary terminals of a wound-rotor motor under speed-regulating conditions can be controlled and returned to the ac lines. The slip power is rectified to direct current, inverted to alternating-current at line frequency, and returned to the primary lines. The clock signal establishing the inversion rate is the ac bus frequency, and thus this type of inverter is quite straightforward. The secondary current represents torque. The entire system performs as a torque-regulated drive by utilizing the current error signal to gate the inverter thyristors and control the rate at which motor secondary power is returned to the line. Control of speed is achieved by comparing a tachometer output with a speed reference. The speed error adjusts the torque to maintain preset speed.

Problems

1. AC motor speed is equal to:

$$\frac{60 \times \text{Frequency}}{\text{Pairs of poles}}$$

Assume a 60-Hz, 1200-rpm (no load), 480-V ac motor. If this motor is supplied from an adjustable frequency inverter, what will be the motor speed when:

(a) Frequency is 120 Hz.
(b) Frequency is 30 Hz.
(c) Frequency is 60 Hz with 10% slip.

2. Briefly describe three methods of obtaining constant volts per half-cycle on an ac motor when supplied from an adjustable-frequency inverter.

3. Compare forced commutation with natural commutation described in Chapter 9.

4. Why are ac adjustable-voltage drives limited in application as compared with ac adjustable frequency?

5. Sketch a single-phase cycloconverter.

6. Name three commonly used types of controlled rectifiers.

7. Why is it advantageous to return "slip power" of a wound-rotor to the ac line, and briefly describe how this can be achieved.

19

THE SYNCHRONOUS MOTOR

Construction. Synchronous motors have an armature winding connected to an ac supply line, and a field winding connected to a dc supply line. They also have a third winding which is short-circuited. It is possible to make either the armature or the field stationary, but the common practice is to make the armature stationary and the field rotating. The armature winding is the more complex and is subject to the voltage of the ac supply, which, with large motors, may be as high as 13,200 V.

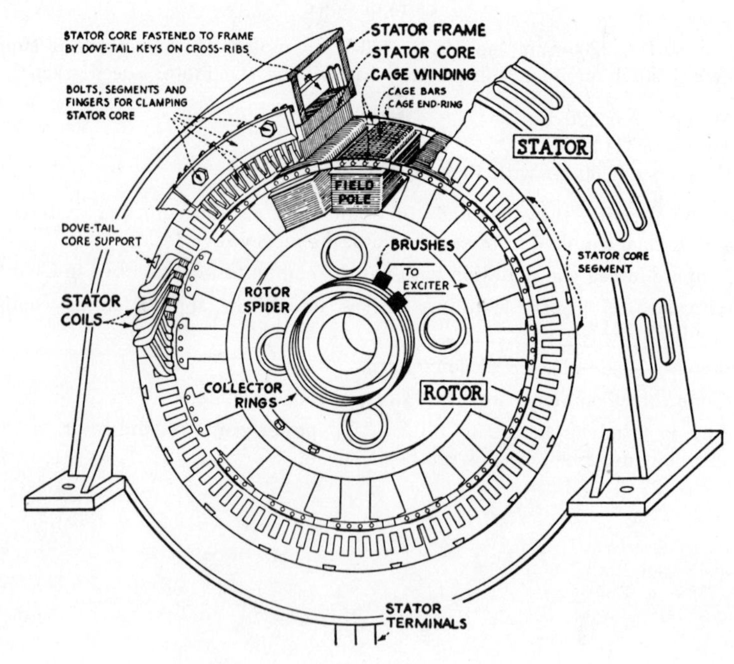

Fig. 19.1 Construction of a Synchronous Motor (Courtesy of Electric Machinery Manufacturing Company)

FIG. 19.2 Synchronous Motor (Courtesy of Electric Machinery Manufacturing Company)

1. Stator frame.
2. Core clamping plates.
3–4. Stator winding.
5. Terminal box.
6. Field winding.
7. Starting cage winding.
8. Shaft.
9. Bearing pedestal.

It is advantageous mechanically to have this winding stationary, and it is also easier to insulate it and to insulate and protect the armature leads. The field winding is energized at 115 or 230 V, and the slip rings to which the supply lines feed are subject to this relatively low voltage, and so are easy to construct and insulate.

Figure 19.1 is a vertical cross-section of a synchronous motor like that of Fig. 19.2. The outer frame is of laminated punchings, having a series of slots around the inner periphery in which the armature winding is placed. The rotating member is a spider mounted on the motor shaft, to which the cores of the field poles are keyed or bolted. The cores are usually of laminated steel, and the field coils are wound around them. Slots are provided in the ends of the field poles in a direction parallel to the motor shaft. Copper bars are placed in the slots, and the ends of the bars on each side of the poles are connected together by an end ring to form a short-circuited squirrel-cage winding. A motor constructed as described is called a salient-pole machine, and the great majority of synchronous motors follow this pattern.

It is also possible to build a motor having stationary salient-pole field windings and a rotating armature. Another possible construction is an embedded field winding wound into slots in laminated punchings like the armature windings. Either armature or field may be the rotating member. This design is seldom used for motors but is advantageous for high-speed turboalternators, as it eliminates much of the windage losses caused by salient poles.

TABLE 19.1

SYNCHRONOUS-MOTOR SPEEDS

	Speed		
Number of Poles	25 Hz	50 Hz	60 Hz
2	1500	3000	3600
4	750	1500	1800
6	500	1000	1200
8	375	750	900
10	300	600	720
12	250	500	600
14	214	428	514
16	188	375	450
18	166	333	400
20	150	300	360
22	136	273	327
24	125	250	300
26	115	231	277
28	107	214	257
30	100	200	240
32	94	188	225
36	83	167	200
40	75	150	180

Speed. The speed of a synchronous motor is determined by the number of poles for which the motor is built and by the frequency of the power supply. The equation for the speed is

$$\text{Rpm} = \frac{\text{Frequency} \times 60}{\text{Number of pairs of poles}}$$

The motor will run only at this synchronous speed. If there is a change in the load, there will be an instantaneous change in speed, lasting for a very few cycles, but the average speed will be the same. Table 19.1, showing standard speeds, is included for ready reference.

The 40-pole 25-Hz machine is the slowest standard motor of that frequency.

The 50-Hz motor is built in standard sizes to 76 poles and 79 rpm.

The 60-Hz motor is built in standard sizes to 90 poles and 80 rpm.

At any given frequency it is, of course, impossible to obtain speeds higher than those of the two-pole machine.

Torque. Table 19.2 shows the normal torque characteristics for general-purpose 60-Hz synchronous motors, as given in NEMA Motor and Generator Standards.

The locked-rotor torque is with rated voltage applied to the motor terminals. The pullout torque is with rated voltage and normal excitation supplied. The normal torques for frequencies other than 60 Hz are the same as those for 60-Hz ratings having the same number of poles.

The synchronous motor in itself has no starting torque. One purpose of the short-circuited winding in the field poles is to provide torque for starting. A motor provided with such a winding, and with the field circuit deenergized, starts by induction as a squirrel-cage motor and will reach a speed slightly below synchronism. At that point the field may be energized, and the motor will then pull into synchronism. The squirrel-cage winding also serves to damp out momentary speed changes with changes in load, and for that reason it is generally known as a damper winding. It is also called an amortisseur winding. If the resis-

TABLE 19.2

SYNCHRONOUS MOTOR CHARACTERISTICS

			Torques, Per Cent of Rated Full-load Torque		
Speed (rpm)	Hp	Power Factor	Locked-Rotor	Pull-in (Based on Normal Wk^2 of Load)	Pullout
500 to 1800	200 and below	1.0	100	100	150
	150 and below	0.8	100	100	175
	250 to 1000	1.0	60	60	150
	200 to 1000	0.8	60	60	175
	1250 and larger	1.0	40	60	150
		0.8	40	60	175
450 and below	All ratings	1.0	40	30	150
		0.8	40	30	200

tance of the damper winding is relatively low, the starting torque will be low, but the motor will approach synchronous speed rather closely. Increasing the resistance of the damper winding will increase the starting torque, but the motor will not approach synchronism so closely. The low-resistance winding is the more effective for damping.

If the torque required by the load exceeds the pullout value, the motor will drop out of synchronism. When this happens, the average torque becomes zero and the motor comes to rest. Typical speed-torque curves are shown in Fig. 19.3.

Power Factor. The power factor of a synchronous motor may be changed by varying the strength of the dc field. Normal excitation is that which produces unit power factor. Underexcitation causes the motor to take a lagging current. Overexcitation results in the motor's taking a leading current. The excitation required to produce any given power factor varies, increasing as the load increases.

Figures 19.4 and 19.5 are typical sets of curves showing the excitation required to produce desired power factors at different loads. These are called "V" curves. Lines drawn through points of equal power factor are called compounding curves.

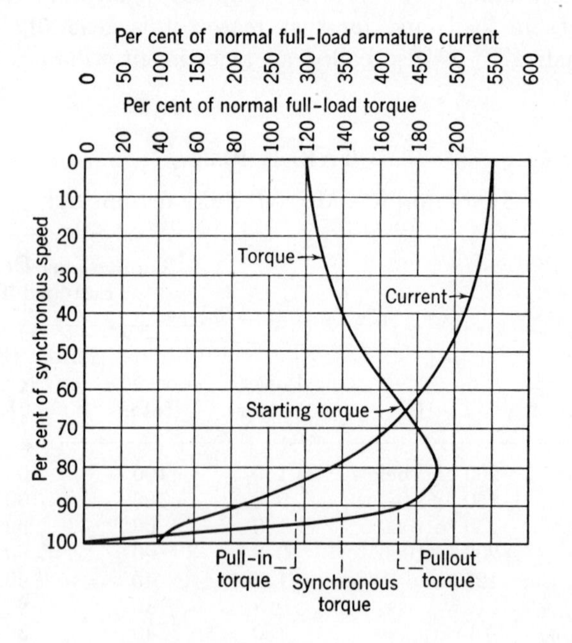

FIG. 19.3 Speed-Torque and Speed-Current Curves of a Typical General-Purpose Synchronous Motor

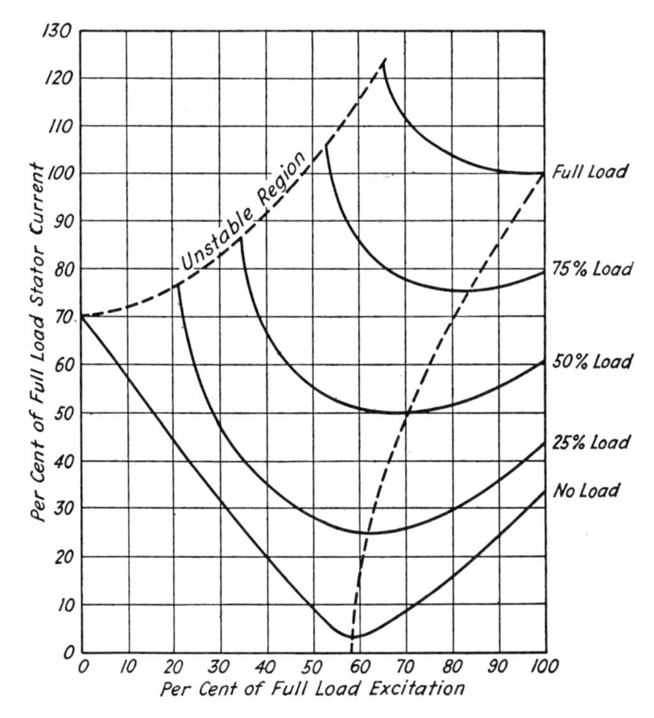

FIG. 19.4 "V" Curves of a Typical General-Purpose 100 Per Cent Power Factor Synchronous Motor

Advantages. Synchronous motors are sturdy in construction, both electrically and mechanically, and may be wound for high operating voltages. The air gap between rotor and stator may be relatively large, which decreases the chance of rotor and stator striking.

The possibility of running at a leading power factor is an important advantage. For example, a synchronous motor in a shop full of induction motors, will materially improve the power factor of the whole installation, and at the same time drive a load.

The efficiency of the synchronous motor is higher than that of the induction motor, particularly for slow-speed motors. The efficiency-load curve of a synchronous motor is relatively flat, so that the efficiency at light loads is better than that of a lightly loaded induction motor.

Constant speed may be an advantage, if that is a requirement of the application.

Low starting torque, the necessity of restarting if the motor drops out of synchronism, and the necessity of a dc supply are the principal disadvantages.

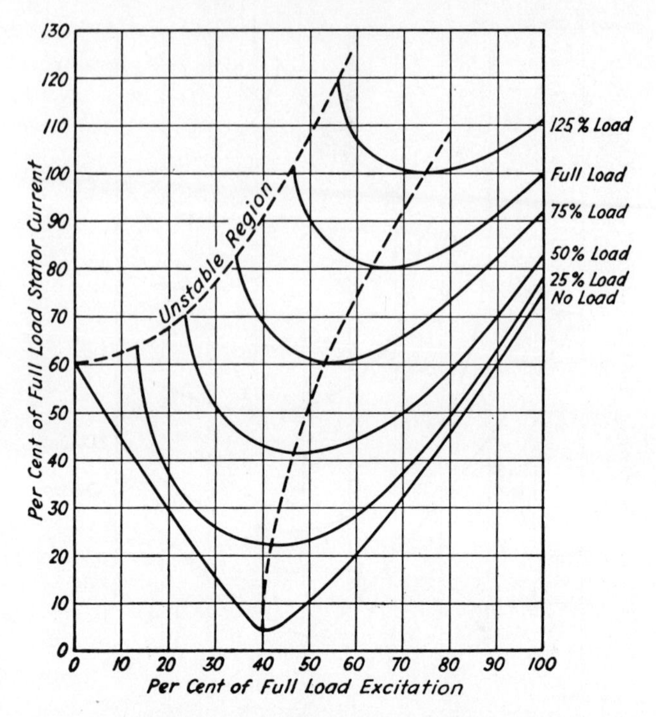

Fig. 19.5 "V" Curves of a Typical General-Purpose 80 Per Cent Power Factor Synchronous Motor

Application. Synchronous motors are ideal for constant-speed, continuous-running applications where the required starting torque is low. Typical applications are: line shafts, motor-generator sets, air and ammonia compressors, centrifugal pumps, blowers, crushers, and many types of continuous processing mills. When the required starting torque is too great for the motor, a magnetic clutch may be installed between the motor and its load. The motor is then brought up to synchronous speed while unloaded, after which the load is applied by energizing the clutch.

General Starting Method. To start a synchronous motor it must be brought up to synchronous speed, or nearly so, with the dc field deenergized, and at or near synchronism the field must be energized to pull the motor into step. A small induction motor may be mounted on the shaft of the synchronous motor for bringing it up to speed. The induction motor must have fewer poles than the synchronous motor, so that it may reach the required speed. If an exciter supplies the field and is mounted on the motor shaft, it may be used as a dc motor for starting, provided that a separate dc supply is available to energize it. However, since most synchronous motors are polyphase and are provided with a damping winding, the common practice is to start them as squirrel-cage motors, the torque being supplied by the induced current in the damper winding. Like squirrel-cage motors, they may be connected directly to the line or started on reduced voltage. When they are started on reduced voltage from an autotransformer, the usual practice is to close the starting contactor first, connecting the stator to the reduced voltage, then, at a speed near synchronism, to open the starting contactor and close the running contactor, connecting the stator to full line voltage. A short time later the field contactor is closed, connecting the field to its supply lines.

Since the shunt field winding consists of a large number of turns, special precautions must be taken to insure against a high voltage being generated in its during starting. When the field winding is stationary, as in rotary converters, switches are provided to break it up into sections and so keep down the generated voltage. This method is not practical when the field is rotating, because of the extra slip rings which would be necessary. The usual practice is to short-circuit the field through a discharge resistor, during starting. When the exciter is driven by the synchronous motor, the field may be connected to the exciter before starting and left so connected during starting, because the exciter voltage builds up slowly as the motor accelerates. Motors started in this way usually have a heavy damping winding around the field poles. When a discharge resistor is used its design must be determined by the motor manufacturer, as the ohmic value has a marked effect on the pull-in torque. A reduced-voltage synchronous-motor controller will, in general, consist of the following devices (see Fig. 19.6).

A main contactor to connect the autotransformer to the line.
A starting contactor to connect the stator to reduced voltage.
A running contactor to connect the stator to the line.
An autotransformer to supply the reduced voltage.
A contactor to connect the field to the direct-current supply.
A frequency-controlled accelerating relay.
An overload protective relay.
A relay to cut the motor off the line if it fails to pull into synchronism.
A field rheostat to adjust the excitation.
A field discharge resistor.
Relays to protect against voltage failure of either alternating or direct current.
Control circuit fuses.
Ammeters to read the alternating load current and the direct field current, to enable the operator to adjust for desired power factor.
A pushbutton station.

Additional instruments, as watt meter, power-factor meter, and voltmeter, are sometimes desired.

Instead of an autotransformer to supply the reduced voltage, any of the methods for starting squirrel-cage motors may be used. These include starting resistance in the stator circuit, starting reactance in the stator circuit, and combinations of reactance and autotransformer.

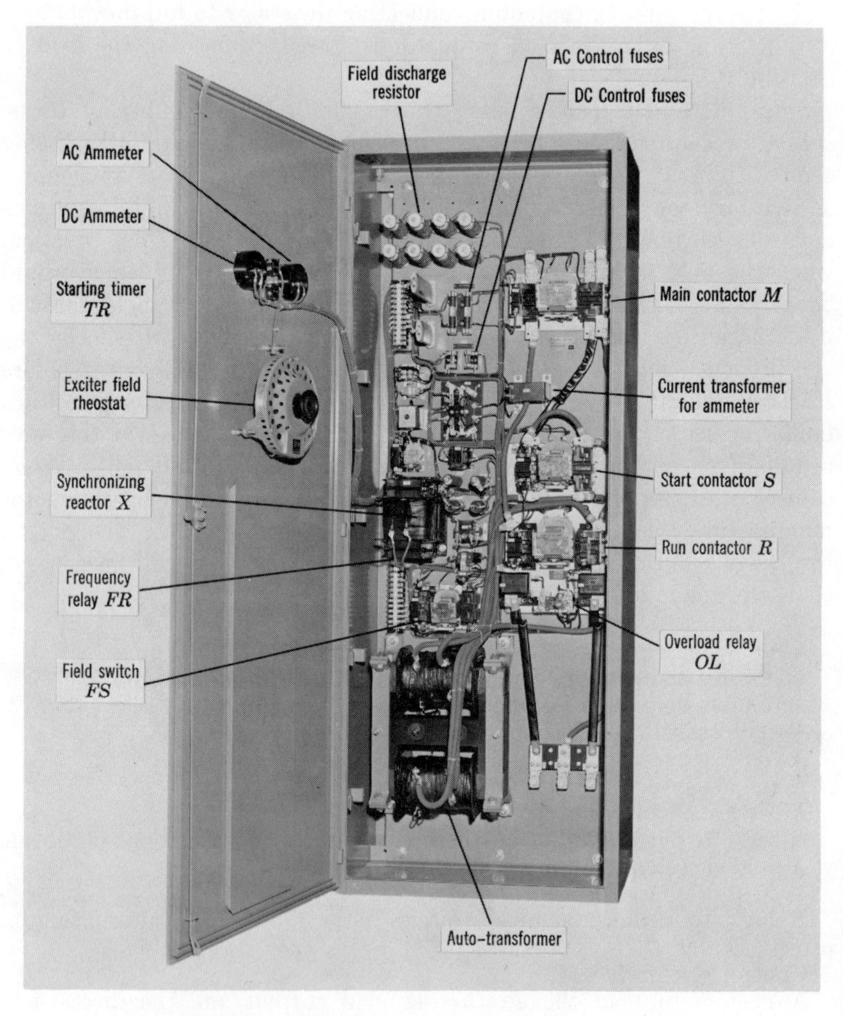

FIG. 19.6 Reduced-Voltage Synchronous-Motor Controller with Synchronism Based on Frequency

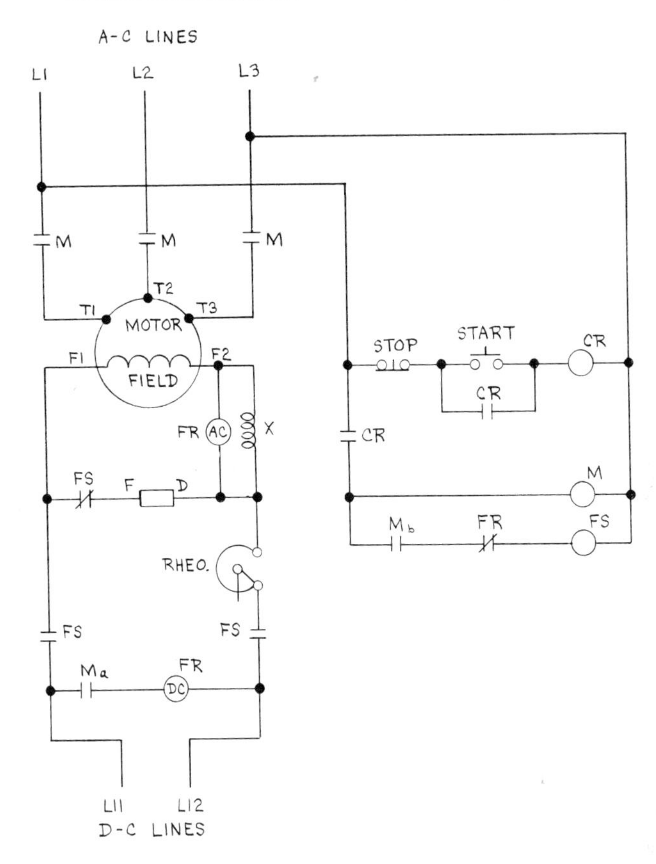

FIG. 19.7 Full-Voltage Controller with Synchronization Based on Frequency

It will be evident that the control problem specific to the synchronous motor is in the means chosen for transferring the control connections when the motor has reached a speed near synchronism.

Synchronization Based on Frequency. Figure 19.7 is a simplified diagram of a synchronous-motor controller arranged for starting the motor directly across the line and synchronizing by a relay operating at a selected frequency. A devices such as instruments and overload relays which are not pertinent to the immediate discussion have been omitted. In this diagram:

(a) M is a three-pole line contractor having two normally open interlock contacts, M_a and M_b;

(b) *FS* is a field contactor having two normally open contacts, and one normally closed contact;
(c) 1*CR* is a control relay having two normally open contacts;
(d) *FR* is a synchronizing relay having one normally closed contact;
(e) *X* is a small reactor;
(f) *F-D* is a field discharge resistor.

When the start button is pressed, the relay 1*CR* closes. One of its contacts provides a maintaining circuit for the relay, and the other contact provides a circuit to energize the line contactor *M*. When *M* closes, its main contacts close ahead of its interlock contacts and energize the stator of the motor. Current at supply frequency is induced in the field winding and flows through the discharge resistor and the coil of relay *FR*. A small proportion of this current flows through the reactor *X*, but the amount is limited because the frequency is high. Relay *FR* closes at once and is fast enough to open the circuit to relay *FS* before the interlock contact M_b closes. Let us, for the moment, neglect the interlock M_a, and the associated coil on *FR*. Since the reactance of the coil of *FR* is much lower than that of *X*, the relay will remain closed at all but very low frequencies. As the motor accelerates, the frequency of the induced current in the field winding decreases, and as it decreases an increasing amount of it flows through *X*, until at a speed close to synchronism most of the current is flowing through *X*. At this point there will no longer be enough current flowing through the coil *FR* to keep the relay armature closed, and it will oven. Field contactor *FS* then closes, energizing the field and opening the field discharge circuit, and the motor pulls into synchronism.

This is a workable scheme in itself, but the addition of another coil on *FR*, which is energized by a constant direct current when M_a closes, is a further refinement. This coil polorizes the relay, and its armature will open only when the magnetic effect of the two coils is approximately equal and in opposition. As the motor nears synchronism and the speed of the rotor approaches that of the revolving stator field, the poles of the rotor pass the stator poles relatively slowly. As each pair of poles passes, there occurs a relative position which is most favorable for synchronizing, as when a north pole of the rotor is directly aligned with a south pole of the stator. Polarizing of the synchronizing relay provides a means not only of energizing the field at a speed near synchronism but also of energizing it at a point in the ac wave which is most favorable to synchronism. If the motor should pull out of step because of an overload or for some other reason, alternating current would again be induced in the field winding, and the *FR* relay would close its armature, open contactor *FS*, and permit the motor to speed up and resynchronize.

Reduced-Voltage Starting. A reduced-voltage stator using this method of synchronizing is generally arranged so that the above-described relays control the running, or full-voltage, contactor. The field contactor is interlocked behind the running contactor and energized by the timing relay a short time after the run contactor closes. However, on resynchronizing after a pullout, the running contactor remains closed and the synchronizing device controls the field contactor. Figure 19.8 shows this arrangement.

When the start button is pressed, relay *CR* closes, followed by the starting contactor *S*. This contactor connects the autotransformer for starting, and electrical interlock on this contactor energizes the M contactor coil. This contactor closes to apply reduced voltage to the motor

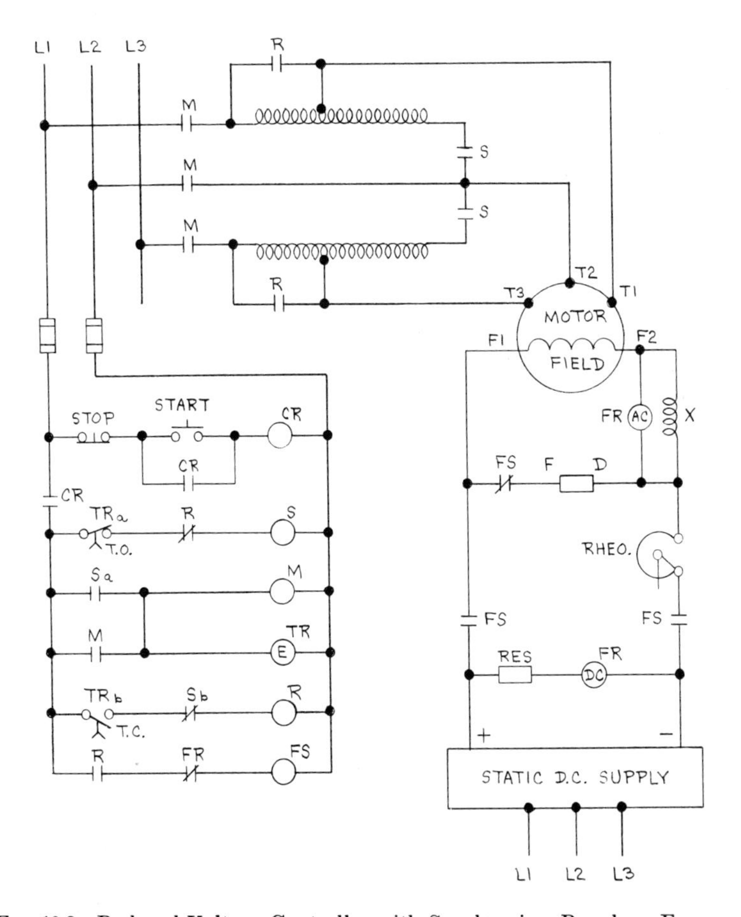

FIG. 19.8 Reduced-Voltage Controller with Synchronism Based on Frequency

stator. The *S* interlock also energizes timer *TR*, which starts timing. At the end of the timing cycle *TR*a opens and *TRb* closes. This deenergizes the *S* coil and sets up the circuit to the *R* coil. The normally closed electrical interlock *Sb* closes when *S* drops out to energize the *R* contactor. Coil *R* closes when *S* drops out to energize the *R* contactor. Coil *R* closes to connect the motor stator to full voltage.

An electrical interlock on *R* closes to set up the circuit to the field switch *FS*. The frequency relay *FR* will operate *FS* in the same manner as it does on the across-the-line starter.

Reversing and Plugging. A synchronous motor may be reversed by reversing one phase of the stator winding, in the same manner as an induction motor. Synchronous motors are not used for applications requiring rapid reversal, because such applications do not require a constant running speed and therefore they would present no advantages. Plugging is used, however, to obtain a quick stop. The field is disconnected from the power supply, and one phase of the stator is reversed. Some form of plugging relay is used to drop out the reversing contactor and disconnect the stator from the line just as the motor reaches zero speed. The method is the same as that for a squirrel-cage induction motor.

Dynamic Braking. Synchronous motors may be brought to rest quickly by dynamic braking. The dynamic-braking resistor consists of three sections, one end of each section being connected to each of the motor stator terminals. The other ends of the sections are open when the motor is running but are connected together through a double-pole normally closed contactor during braking. These spring-closed contactors are mechanically interlocked with the line contactor, so that the stator must be disconnected from the power supply before the braking circuit is closed. The field circuit remains closed, and the rotating field induces currents in the stator windings, which currents flow through the braking resistor. In this way the energy of the rotating field is dissipated as heat in the resistor, and the rotor is brought to rest. The time required to stop the motor depends on the amount of its stored kinetic energy and on the rate of dissipation, the latter being controlled by the design of the braking resistor. The resistor is designed to suit the required conditions of each application, the usual value for the stator current during braking being between two and three times normal full-load motor current. A timing relay opens the field contactor and disconnects the field after the motor has stopped.

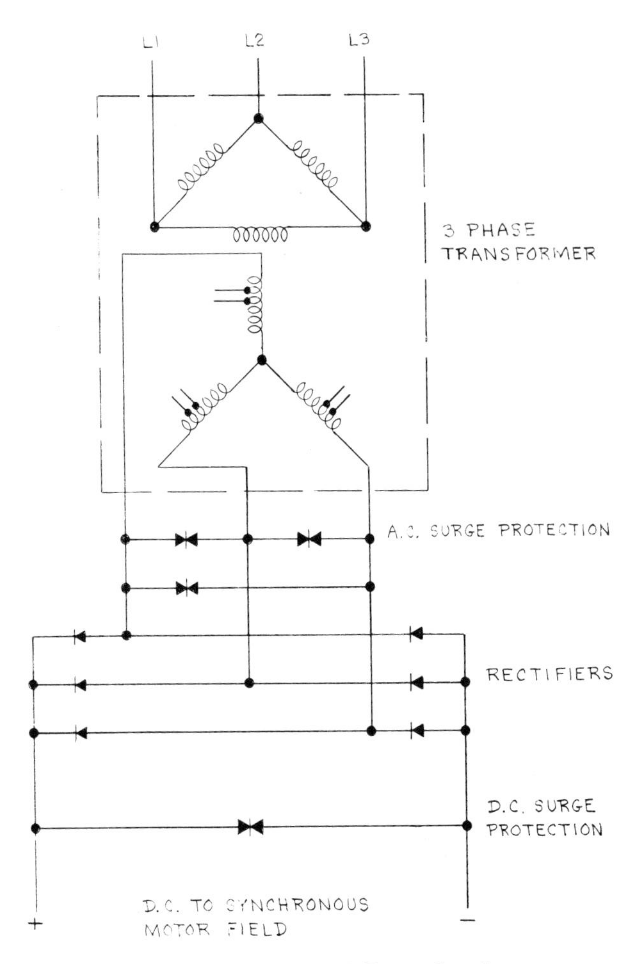

Fig. 19.9 Static DC Power Supply

Field Supplies. The dc for the synchronous motor fields has traditionally been supplied from dc shop lines or separately driven generators or exciters mounted on the motor shaft.

Silicon rectifiers have made possible small, compact static field supplies. Because of low maintenance and fairly low original cost, these have made the old methods nearly obsolete.

The static supply normally consists of three main parts. (See Fig. 19.9.)

1. A three-phase transformer with several voltage taps to supply the correct voltage.

2. A three-phase full-wave silicon rectifier bridge.
3. Voltage surge protection for the rectifiers.

Static Control. Recent advances in technology have made possible a completely static field control. These systems use the frequency of the induced field current to actuate transistorized circuitry rather than the electromechanical frequency relay described earlier. The silicon rectifiers in the static field supply are replaced by thyristors, and the transistors fire these to apply direct current to the synchronous motor field. This accomplishes the same results as the electromechanical devices described in the frequency relay synchronization except that this is completely static.

The static control has made possible the brushless synchronous motor. One big problem with the older synchronous motors is the maintenance of the slip-ring brushes of the motor and the commutator and brushes of the exciter used to supply direct current for the motor field.

Figure 19.10 shows a block diagram of one brushless synchronous motor system. The alternator and the static control are mounted on the shaft with the synchronous motor rotor. Three-phase power is connected to the stator of the synchronous motor and the field is wound on the rotor, whereas the alternator has its field on the stator and three-phase is generated in its rotor winding.

Single-phase ac, with a variac for adjustment, is connected to the alternator field through a rectifier to provide dc excitation. The output of the alternator on the rotor is rectified and connected, through the static field control, to the synchronous motor field. Since all three are rotating on a common shaft, there is no need for commutators, slip rings, or brushes.

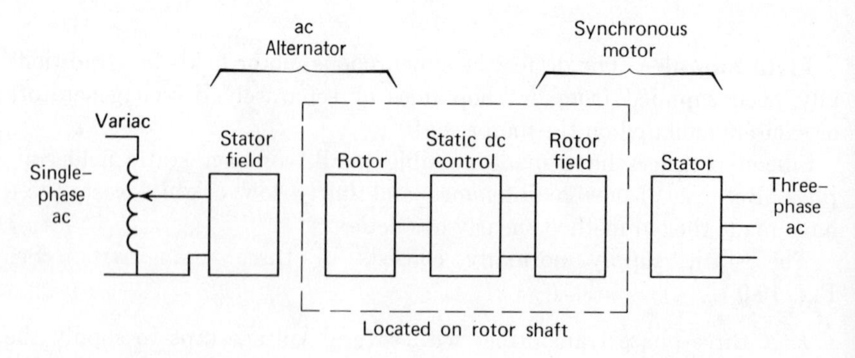

Fig. 19.10 Block Diagram of a Brushless Synchronous Motor System

Problems

1. What is the synchronous speed of a 16-pole 440-V 50-Hz synchronous motor? At what speed will it run without any load? What will be its speed when overloaded 25%? What will be its speed if the voltage drops to 400 V?

2. If the starting torque of a synchronous motor is 120% of rated torque when starting across the line, what will the starting torque be when starting with a primary-resistance starter which reduces the line voltage to 70% of normal?

3. A synchronous motor field is drawing 20 amp when connected to a 230-V supply line. If the field is disconnected, and immediately connected across a resistor of 23 ohms, what is the maximum induced voltage to which the field windings might be subjected?

4. The stored energy in the rotor of a synchronous motor is 1 million W-sec. If the motor is to be stopped in 20 sec by dynamic braking, what will be the wattage dissipated in the braking resistor at the start of the braking period?

5. A 100% power factor motor has a field winding of 30 ohms resistance, which is to be supplied from a 230-V circuit. How many ohms will be required in a field rheostat which will permit 100% power factor to be obtained with any load from zero up to full rated load? (See Fig. 19.4.)

6. How many ohms would be required in a rheostat for the motor of Fig. 19.5, the rheostat in this case permitting 80% power factor to be obtained over the range from zero load to 125% load? Field resistance 30 ohms and supply voltage 230.

7. Make an elementary diagram for a reduced-voltage autotransformer-type synchronous-motor controller, using synchronization based on frequency.

8. Make an elementary diagram for a reduced-voltage primary-resistance-type synchronous-motor controller, with dynamic braking and with synchronization based on frequency.

9. A machine is driven by a six-pole 440-V 60-Hz synchronous motor, which is supplied from a motor-generator set used for that purpose only. It is found necessary to run the synchronous motor at 660 rpm. What voltage and frequency must the generator deliver?

10. A controller like that of Fig. 19.6 has been built for a 100-hp 220-V 60-Hz synchronous motor. The owner now desires to use the controller with a 150-hp 440-V 60-Hz motor having the same field characteristics. Which of the following items will have to be changed?

M contactor coil	Fuses
S contactor coil	AC main circuit wiring
R contactor coil	Control wiring
Main contacts of *R*	*TS* relay coil
Overload relay coils	*CR* relay coil
DC ammeter	Interlock fingers on *FS*
AC ammeter	*FS* contactor coil

11. A continuously running, nonreversing mill for rolling steel slabs is subjected to a very heavy load each time a slab starts to enter the rolls. To cushion the load on the driving motor at this time, the mill is equipped wih a heavy flywheel. Two motors of identical rating are available, one a synchronous motor, and the other a wound-rotor motor. Which should be used, and why?

20

MAGNETICALLY OPERATED BRAKES

In many applications of electric motors the ability to stop quickly and accurately is important. Electrically operated brakes are widely used for that purpose, either in connection with dynamic braking or as the sole means of stopping. Brakes are also necessary on hoist, cranes, elevators, and similar machines, to hold the load after stopping.

The essential parts of a brake are the friction material, shoes or band, wheel, operating device, and mounting parts. Most brakes are electrically released and spring-set, so that the brake will be set in case of an electrical failure or a power interruption. It is occasionally advantageous to make the brake electrically set. With this type of brake it is possible to vary the applied torque by adjusting the voltage applied to the brake coil, and the rate of application can also be controlled. A typical example of this arrangement is the application of brakes to the individual rolls of a printing press.

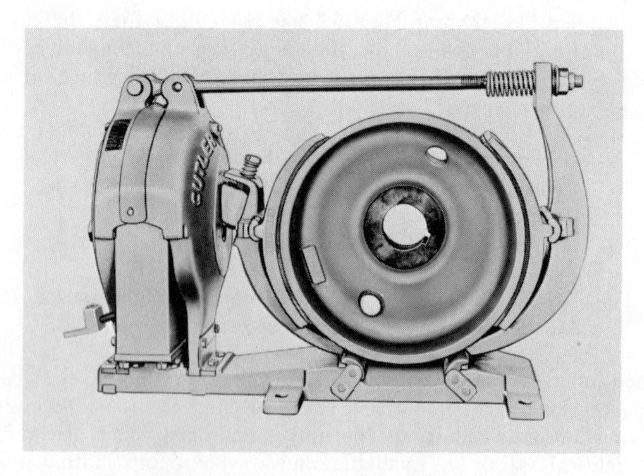

Fig. 20.1 DC Direct-Acting Shoe Brake

Fig. 20.2 Alternating-current Solenoid-type Shoe Brake

Brakes are of the shoe, the disk, or the band type. The shoe brake has the friction material mounted on two shoes, which apply on opposite sides of the brake wheel. The shoes cover approximately half of the wheel circumference. They are operated independently and are independently adjustable. The shoe brake requires only a small movement to release, the travel of the operating magnet being approximately 0.04 in. for each shoe. The short stroke gives fast operation and also reduces shock and hammer blow when releasing or setting.

The disk brake is arranged for mounting directly to the motor end bell. The brake lining is a disk which is supported by a hub keyed to the motor shaft. The disk rotates with the motor. When the brake is set, a spring pulls a stationary steel member into contact with the rotating disk. Although brakes have been built with a number of disks, the general practice is to have only one. Since the air gaps must necessarily be small, the use of more than one disk increases the chance of the brakes dragging when released.

The band brake has the friction material fastened to a band of steel which encircles the wheel and may cover as much as 90% of the wheel surface. The increased braking surface permits a lower pressure per square inch, with consequent reduction of wear of the lining. This is offset somewhat by the fact that the braking pressure is not equal over the whole band, as a wrapping action may occur when the brake sets. The band brake requires a longer stroke to release it.

Lining Materials. The basic material in all brake linings is asbestos. In all but molded types it is fashioned into a thread around a nonferrous

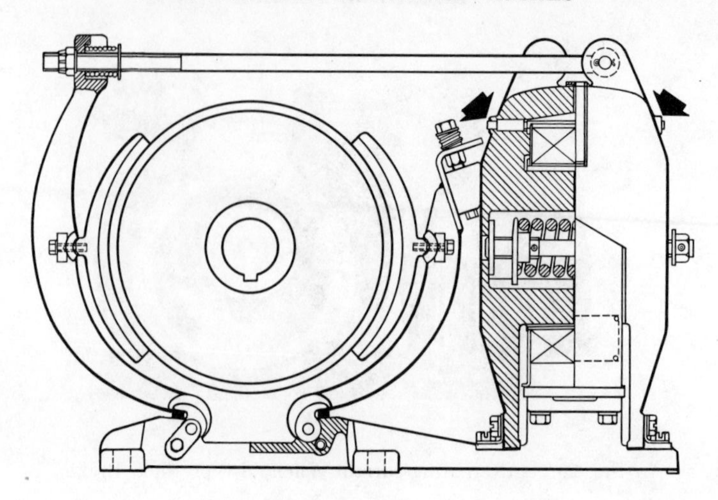

FIG. 20.3 DC Direct-Operated Shoe-Brake Construction

wire. Generally some cotton is used to make the thread more easily formed. The woven linings are made so that in effect there are several layers of woven fabric tied together by occasional strands of woof which pass through all layers. The fabric is then impregnated with a binder, which will serve as a matrix to cement the strands and fibers together. Then heat and pressure are applied to cure the binder and reduce the lining to the final shape and condition. The woven lining is reasonably flexible and can easily be formed to suit wheel surfaces. Its surface is rather rough, and the high spots must be worn down before maximum friction is delivered. In the molded type, the asbestos fibers are mixed with a binder and other constituents. Sometimes short pieces of brass wire are used to reinforce the product. The mixture is then formed in molds to the desired shapes. The molds are customarily heated to cure the binder. Molded linings are hard, dense, uniform in thickness, and smooth surfaced. They must be formed in the shape in which they will be used.

In the folded and compressed type of lining the strands are loosely woven into a rough open cloth, which is filled with the binder by a frictioning process. The friction cloth is folded together in the necessary number of sheets to build up to the desired thickness. The sheets are loosely stitched to make further handling during the subsequent operations easier. The stacks are then subjected to heat and pressure for curing and producing a uniform thickness. The folded and compressed linings are reasonably flexible and will not break while being riveted to the carrying surfaces. The surface is moderately smooth.

Fig. 20.4 Alternating-current Brake, Operated by Torque Motor

The fundamental requirement in all types is that the binder permeate the fabric and form a matrix to support the relatively weak fibers of asbestos. The binder must stand the temperature which develops when the wheel slides against the friction lining under pressure, and it must remain mechanically strong enough to support the asbestos. When linings are worked too hard, the wear rate is multiplied in all types. There is a critical temperature which, if exceeded, causes rapid failure of the

Fig. 20.5 Disk Brake

lining. It does not appear that the wear rate is especially affected at any value below the critical temperature of the particular lining. When the critical temperature is exceeded in the woven lining, the woof cuts away, and either the binder is not strong enough to hold the pieces or it vaporizes out, and strands of the warp fly out along the short loops of woof. The molded type scuffs away. The folded and compressed type separates between layers, and partial layers flake off.

The adhesion or coefficient of friction can be regulated by the manufacturer; it appears mainly to be a function of the binder in combination with the asbestos. The coefficient is not exact, and the factor depends on the condition of the metal surface which opposes the lining as well as on the lining itself. Roughly, the life is progressively shorter as the coefficient of friction increases. Temperature effects, oil effects, and water effects produce variations in results. There is a light service condition where a film of metal oxide will form on the lining face, practically eliminating the lining wear and increasing the frictional coefficient to as high as 0.8. This can be obtained with almost any type of friction material if the right conditions are provided, but the conditions represent such low energy consumption per unit of area that it is not practical to embody them in industrial apparatus. We have previously pointed out that it would require three to ten times the present accepted friction area to produce these conditions.

Brake Wheels. Materials for brake wheels must be strong and flexible but must break sharply if their strength limit is exceeded. In other words, it is desirable to have the elastic limit and the ultimate strength high and close together. The relation of diameter and length must be considered. The surface area of the wheel is fixed by the heat to be dissipated.

In operation, brake wheels heat up unequally. The source of heat is between the lining and the wheel, and the heat flows through the wheel to meet a cooling current of air at the inside of the wheel; consequently the wheel expands unequally in proportion to the inequality of temperature in the wheel. Elasticity is required to stand the inequality of expansion. The particles in the wheel should retain their strength up to fairly high temperatures, and when their strength is exceeded they should break away in infinitesimal grains and not foul the lining.

The general indicated conclusions from actual tests are that cast-iron wheels are not elastic enough for wide ranges of temperatures. Surface cracks develop, and under extreme conditions cracks will progress through the rim, caused by the inequality of stresses from the varying expansion in the rims. In ordinarily mild service the wheel surface polishes, thus producing very good conditions for long life of wheel and lining.

Another feature is that, in order to get the same rigidity, the weight must be increased approximately 50%. This adds to the flywheel effect and is undesirable. Owing to the lower breaking strength, the permissible speed of rotation is less for the cast-iron wheel than for the steel wheels.

Steel, whether of low or high carbon content, without heat treatment, under easy service conditions, with light pressure and low speeds, will ordinarily become polished and produce long wheel and lining life. There is a gradual improvement in wheel performance as the carbon content increases, but it is indicated from tests that with pressure in the range from 30 to 50 lb/in.^{2} and normal velocities, representing fair severity of service, steel without heat treatment tends to drag, and particles bed into the lining. The scoring of the wheel then progresses rapidly.

Hardened steel, even though the hardness does not exceed 40 Scleroscope, apparently produces better lining life than cast iron, in the ratio of about 4 to 3, without danger of cracking. The cost of the hardened or heat-treated wheel is certain to be somewhat higher than that for a casting, since the wheel must be machined close to finished size before it is heat-treated, and then it must finally be finished after heat treatment. Also, the final grinding operation of the hardened wheel surface requires somewhat longer than a machined finish on an untreated wheel. The heat treatment itself adds something to the cost.

Ductile iron, normalized and tempered to the required hardness, combines the excellent frictional characteristics of cast iron with the high strength of steel, while minimizing the undesirable properties of both. If a surface crack develops, ductile iron's molecular structure resists propagation of the crack in contrast to cast iron and steel in which, once formed, a crack will continue to enlarge. Another favorable characteristic of ductile iron wheels is that they are not prone to "patch weld" as are steel wheels and can withstand higher temperatures before patch welding will occur. Patch welding is a term used to describe the condition which occurs when a wheel begins to score and particles of wheel material become embedded in the friction lining. This embedded material increases scoring and heating of the wheel so that finally the instantaneous temperatures become so high that the particles weld together into the lining.

Finally, then, cast iron will be satisfactory, with any lining, for light service and limited speeds; untreated cast steel will be satisfactory, with some linings, for light service and fairly high speeds; hardened steel will be satisfactory, with any lining, for most service conditions. Ductile iron will be satisfactory with any lining for all service conditions. Lining life varies over a wide range, depending more on wheel condition than on severity of service, except that severity of service finally affects the wheel condition.

Operating Mechanisms—DC. Direct-current brakes may be operated by a solenoid or by a direct-operating magnet. A solenoid operates against a spring through suitable linkages. The torque delivered at the wheel is regulated by adjusting a spring. In the direct-operating type the armatures of the electromagnet actuate directly the brake-shoe levers. The outer armature is connected to the opposite shoe lever by a rod which passes over the brake wheel. When the brake is applied, the armatures are forced apart by a spring located in the center of the magnet field, and the brake shoes actuated by the armatures are forced against the wheel. The intensity of the braking force is regulated by varying the adjustment of the spring in the magnet field. Adjustment for wear of the lining is made by varying the length of the rod above the brake wheel and the adjustment of the position of the wedge block attached to the inner shoe lever. Provision is also made for equalizing the clearance of the two shoes.

The operating coils of either solenoid or direct-acting magnetic brake may be wound for series or shunt connection and for continuous or intermittent duty. If series-wound, the coil is connected in the motor circuit and operated by the motor current. The coil is wound to suit the horsepower and duty rating of the motor with which the brake is to be used, and for intermittent duty it is designed to release the brake on about 40% of full-load current and hold it released on about 10% of full-load current. For continuous duty the coil will require about 80% of full-load current to release the brake. Intermittent-duty series brakes are rated as either ½-hr duty or 1-hr duty, corresponding to the method of rating intermittent-duty series motors.

Shunt-wound brakes are wound for either intermittent or continuous duty. Intermittent duty is understood to mean 1 min on and 1 min off, or the equivalent, the longest continuous application of voltage not to exceed 1 hr. These brakes will release at 85% of normal voltage when adjusted for rated torque. Since shunt-wound coils have a greater inductance than series-wound coils, the shunt brakes are ordinarily not as fast in releasing. In order to increase the speed of operation, it is customary to wind the shunt coil for a voltage lower than line voltage and to use a resistance in series with it. The coil is usually wound for one-half of line voltage, but it may be wound for a voltage as low as one-tenth of line voltage in order to get extremely fast release. A relay may then be used to insert the resistance. The relay normally shorts around the resistance, and its coil may be connected in series with the brake. When voltage is applied, the inductance of the brake winding retards the rise of current. As soon as the current has built up, the relay operates and inserts the series resistance. Another frequently used method is to insert the resistance by means of a timing relay. With this arrangement the brake will release even more quickly than with a permanent series resistance. The smaller brakes, up to a 10-in.-diameter wheel, do not ordinarily require partial voltage coils, but their use is general on the larger brakes.

Series brakes have several advantages in addition to that of fast operation. The series coil, being of heavy wire, is less likely to give trouble than a fine-wire shunt coil; and since the voltage per turn is low, the insulation is not likely to break down. The coil is in series with the motor armature, so that if the armature circuit is opened the brake will set. This is an important safety feature in connection with hoists or other machines where there is an overhauling load, as a broken resistance grid or a loose wire connection might open the armature circuit and allow the load to drop. The wiring to the brake is simple, since the brake is mounted on the motor and connected to it.

Shunt brakes are used in connection with machines which have a widely varying load, when the armature current is not always great enough to keep a series brake released. It is possible to obtain some measure of protection against an open armature circuit by using a series relay with the shunt brake. The coil of the relay is connected in the armature circuit, and the contacts in series with the brake coil. When the motor is started, the inrush current will close the relay, and the relay will close the circuit to the brake.

Operating Mechanisms—AC. The three principal forms of ac brake-operating mechanisms are the solenoid type, the torque-motor type, and the thrustor type.

The ac solenoid brake is similar to the dc solenoid type, except that the solenoid frame must be laminated to reduce eddy currents. Since the ac flux passes through zero twice every cycle, the pull of the magnet is not constant. Shading coils, similar to those on ac contactors, must be used to provide a pull during the change of direction of the main flux. Even with shading coils it is difficult to design a solenoid mechanism which will be quiet in operation and free from vibration. Another disadvantage of the solenoid is that it draws a heavy current at the first application of voltage, when the magnetic gap is open. In general, solenoids are used only for the smaller brakes, although polyphase solenoids are sometimes used for the larger sizes.

The torque-motor mechanism utilizes a specially wound polyphase squirrel-cage motor, which may be stalled without injury to the windings and without drawing excessively heavy currents. The motor drives a ball jack, which translates the rotary motion of the motor to a straight-line motion and releases the brake. The motor then stalls, holding the brake in the released position until the motor circuit is opened. The brake is set by a spring, which overhauls the torque motor. The slight flywheel action of the rotor tends to eliminate shock when the brake sets. A slip clutch, constructed as a part of the ball-jack mechanism, also acts to prevent shock at the end of the brake movement. The brake is quiet in operation, since the pull of the motor is uniform, and there is no open magnetic circuit. The current taken from the line is not great and is practically uniform throughout the stroke. A small brake of this type requires 110 V-amp, compared to an inrush of 2000 V-amp for a solenoid brake of the same rating. A large size requires 1520 V-amp, compared to 21,000 for the corresponding solenoid brake. A disadvantage of this brake in comparison with the solenoid type is that the torque motor must always be operated in the same direction, which necessitates an extra relay on a reversing controller and extra trolley wires when used on a crane.

The operation of the thrustor is quite different from that of the mechanisms which have been described. The thrustor is essentially a motor-driven centrifugal pump, which operates on oil in a cylinder. The pump motor is of the polyphase-squirrel-cage type. When energized, the motor forces oil under a piston, which, in rising, moves a lever and releases the brake. To set the brake, the motor is deenergized, and a spring forces the piston to its original position. The return movement of the piston is cushioned by the inertia of the operator motor and pump, so that the brake is set without severe shock. The pump impeller has straight blades, and it pumps equally well in either direction of rotation.

Operation with Rectifiers. Since the dc operating mechanisms are much simpler than ac mechanisms, particularly in the larger sizes, there has been an increasing use of rectifiers to supply direct current for the brake on an ac installation.

Figure 20.6 shows the connections for a control of this type. The brake relay *BR* is energized by the closing of a contact in a master controller, or on the motor control panel. The closing of *BR* energizes the brake through the rectifier and also sets in motion a timing relay *TR*, which, after a short time delay to permit the brake to operate, inserts a resistor into the brake-coil circuit. Since the brake requires more current to release than it does to remain released, the current may be

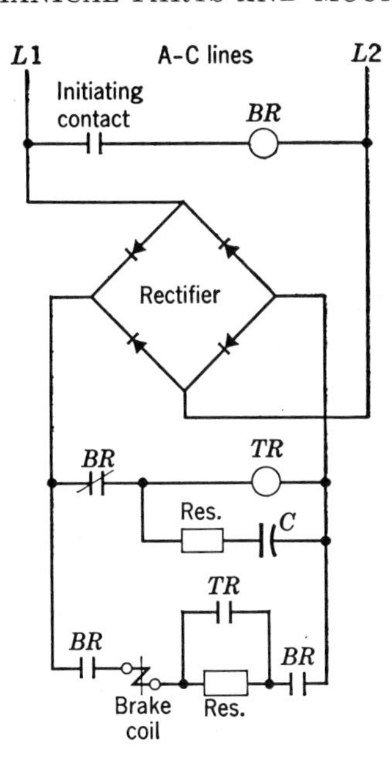

FIG. 20.6 Connections for Operation of a DC Brake on Alternating-Current

reduced in this way, permitting the use of the minimum size of rectifier. The timing relay shown is one of the condenser-discharge type. Any type giving an accurate timing of a second or so is satisfactory.

Mechanical Parts and Mounting. The frame and mechanical parts of a brake are usually made of steel or ductile iron and must be strong enough to withstand the shocks and jars of frequent operation. The structure must also be rigid so that the brake shoes and wheel will stay in proper alignment and so that side strains on the motor shaft will be avoided. Springs afford the best means of setting the brake, since they have relatively little inertia.

Brakes may be mounted either on the floor or on the frame of the motor. There is some difference of opinion regarding the preferable method, and the choice is probably best made from a consideration of the application in question. The advantages of floor mounting are:

Lower first cost
More rigid mounting
Less strain on the motor frame

The advantages of mounting on the motor are:

No special foundation is required.
The brake is automatically aligned with the motor.
The brake will not get out of alignment with the motor.
The brake and motor can be handled as a unit.
Less floor space is required.

Determination of Brake Size. There are three factors which must be considered in rating a brake, or in selecting the correct brake for a given application.

The first factor is the mechanical strength of the brake, or the retarding effort which it can deliver. This is determined by the force with which the spring holds the shoes against the wheel, the radius of the wheel to which the force is applied, and the coefficient of friction between the lining and the wheel face. The torque required of the brake is determined from the formula

$$T = \frac{5250 \text{ hp}}{N} \tag{20.1}$$

where T = torque in pound-feet
hp = rated motor horsepower
N = full-load motor speed in rpm

The brake should have a torque rating equal to, or greater than, that obtained from the formula. Where adjustable-speed motors are used, the torque should be calculated at the lowest speed.

The time required for a brake alone to stop a moving mass may be calculated from the formula

$$t = t_a + t_d = t_a + \frac{120KE}{2\pi T N_b} \tag{20.2}$$

where t = time in seconds, required to stop the moving mass after the brake is deenergized
t_a = time in seconds for the brake shoe to apply against the wheel after the brake is deenergized
t_d = time in seconds for the moving mass to stop after the brake shoes apply
KE = total kinetic energy in the moving system at the instant the brake applies
T = retarding torque in pound-feet. This torque is the sum of the brake torque and the friction torque of the moving system. In most calculations the brake torque only is considered, unless the friction torque is high
N_b = rpm of the brake wheel

For a rotating body,

$$KE = \tfrac{1}{2} I \omega^2 \tag{20.3}$$

where

$$I = \frac{WR^2}{g}$$

which is the moment of inertia of the body in pounds per square inch.

$$\omega = \frac{2\pi N}{60}$$

which is the angular velocity in radians per second. If we substitute for I and ω, the equation reduces to

$$KE = 1.7\ WR^2 \left(\frac{N}{100}\right)^2 \tag{20.4}$$

where W = weight of the rotating mass in pounds
R = radius of gyration of the mass in feet
N = rpm of the mass

For a body with linear motion,

$$KE = \tfrac{1}{2} M v^2 \tag{20.5}$$

or

$$KE = \frac{1}{2} \frac{W}{g} \left(\frac{V}{60}\right)^2 \tag{20.6}$$

where v = linear velocity in feet per second
V = linear velocity in feet per minute
g = 32.2

Equation 20.2 may be rewritten

$$t = t_a + \frac{WR^2 N_b}{308T} \tag{20.7}$$

Equation 20.7 applies to a machine having a single rotating element, or a group of elements rotating at the same speed. If the machine has elements which are rotating at speeds different from that of the brake wheel, the equivalent WR^2 of any element is

$$\text{Equivalent } WR^2 = W_1 R_1 \left(\frac{N}{N_b}\right)^2$$

where $W_1 R_1$ = the WR^2 of the element at its actual speed N.

The second factor in a brake rating is the heat-dissipating capacity of the brake wheel and the shoe lining. It is the foot-pounds of kinetic energy which can be absorbed by the wheel and lining and dissipated as heat, without causing the wheel face or the lining to become dangerously hot. The lining is a poor heat conductor, and deteriorates rapidly at temperatures above 400°F. The heat-dissipating capacity of the wheel is determined by its surface area, by the operating cycle, and by the ventilation. Where a brake is used only for holding a load, and not for stopping it, this factor need not be considered. Manufacturers do not ordinarily list figures for the heat-dissipating capacity of their brakes, since there are so many variables in the conditions encountered in service. It is customary to list limiting horsepowers with which specific brakes may be safely used, these being determined by calculation and test, and being satisfactory for most applications. Where the frequency of operation is very high, or the energy of motion is very great, the brake manufacturer should be consulted before the brake is selected.

The third factor is the duty of the operating coil or mechanism, which must be strong enough to release the brake shoes against the pressure of the brake springs, and must hold the shoes released for a period of time which depends on the duty cycle of the brake. Shunt coils are rated in voltage, for intermittent or for continuous duty. Continuous duty means that the coil will hold the brake released continuously without overheating of the coil. Series coils are rated in amperes for ½-hr or 1-hr duty, to correspond to the ratings of series motors. The ½-hr rating means that the brake coil will carry full-rated motor current for ½ hr without overheating. This is equivalent to a duty cycle of 1 min on and 2 min off (⅓ time duty), repeated continuously. Similarly, the 1-hr rating is equivalent to a duty cycle of 1 min on and 1 min off (½ time duty), repeated continuously.

Other considerations which enter into the selection of a brake are safety, ambient temperature, and maintenance. The safety consideration enters into such applications as balanced hoists, where the brake should be large enough to hold a loaded car or bucket in case a broken cable removes the counterbalance. A high ambient temperature will reduce the heat-dissipating capacity of the brake and may necessitate the selection of a larger size. Maintenance should be considered, since the severity of the service has a direct bearing on the life of the brake lining and wheel. Most brakes are designed to make adjustment simple and replacement of linings easy. Brakes which are in hard service should be inspected frequently, with particular attention to the following points.

1. Proper adjustment of the air gap between the wheel and the linings.
2. Equal clearance on the two shoes.
3. Tightening of bolts in the motor and brake bosses.
4. Grounds in the coil or leads.

TABLE 20.1

DIRECT-CURRENT BRAKE DATA

Size of Brake (in.)	Maximum Torque in Pound-Feet: 1-Hr Intermittent	Maximum Torque in Pound-Feet: 8-Hr Continuous	Maximum Torque in Pound-Feet: Series Wound ½ hr	Maximum Torque in Pound-Feet: Series Wound 1 hr	Power Required in Watts: Shunt-Wound Intermittent	Power Required in Watts: Shunt-Wound Continuous	WR^2 of Wheel	Safe Maximum Rpm	Weight of Brake (lb)	Brake Wheel Face (in.)
8	100	75	100	65	440	260	1.1	5000	150	3.25
10	200	150	200	130	500	280	3.0	4000	240	3.75
13	550	400	550	365	660	430	12	3180	465	5.75
16	1000	750	1000	650	820	490	36	2500	645	6.75
19	2000	1500	2000	1300	1240	600	70	2110	1150	8.75
23	4000	3000	4000	2600	1920	790	210	1740	1735	11.25
30	9000	6750	9000	6000	4500	2950	880	1340	3920	14.25

Typical Brake Data. Table 20.1 gives data applying to a typical line of dc direct-operated shoe brakes. Standard ratings for the smaller brakes are given below. There are no standard ratings for dc series brakes smaller than those listed in table 20.1.

DC Torque in pound-feet: *Continuous*	*DC* Torque in pound-feet: *1 hour*	*AC* Torque in pound-feet: *Continuous*	*AC* Torque in pound-feet: *Intermittent*
3	3	3	3
10	15	10	15
25	35	25	35
50	75	50	75
		125	160

Alternating-current will release at 85% of full line voltage and will operate satisfactorily at 110% of full line voltage.

Shunt brakes will release at 85% of full line voltage and will operate satisfactorily at 110% of full line voltage.

Series brakes will release at 40% of full-load motor current and will remain released down to 10% of full-load motor current.

Standardization of Brakes. A number of attempts have been made to reach a standardization of brakes, not only so far as torque is concerned, but also in wheel sizes and mounting dimensions. There has

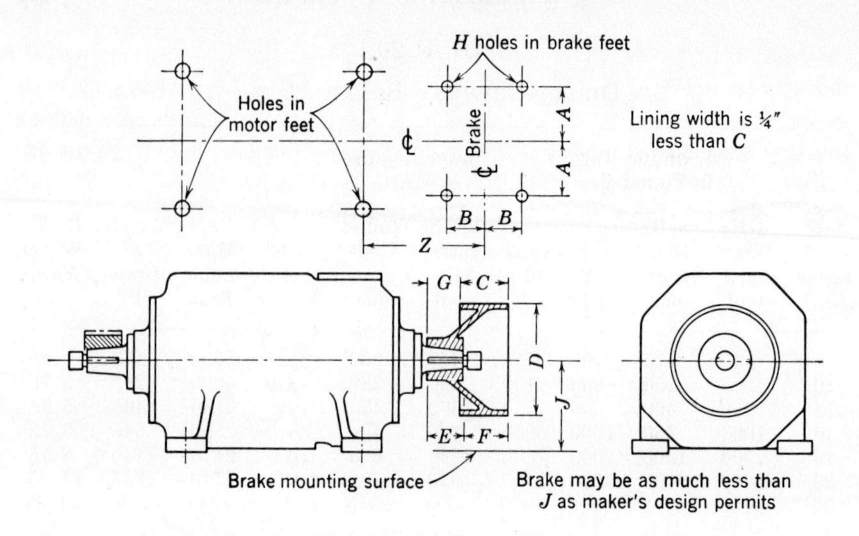

TABLE 20.2

Motor	Series Motor, lb-ft		Brake Rating, lb-ft			Mounting Dimensions					Wheel				
	0.5 hr	1 hr	0.5-hr Series, 1-hr Shunt	1-hr Series	8-hr Shunt	*A*	*B*	*H*	*Z*	Max *J*	*D*	*C*	*E*	*F*	*G*
2	46	29	100	65	75	3 1/4	2 7/8	11/16	8 1/4	7	8	3 1/4	3	2 5/8	2 3/8
602	78	49													
603	116	72	200	130	150	4	3 1/8	11/16	9 1/4	8 3/8	10	3 3/4	3 1/2	2 5/8	2 3/8
604	166	121							9 3/4						
606	337	228	550	365	400	5 3/4	4 1/2	13/16	10 1/2	9 7/8	13	5 3/4	4	3 7/8	2 1/8
608	502	350							11				4 1/2	3 3/4	2 1/2
610	765	525	1000	650	750	7 1/2	5 3/8	1 1/16	12 3/4	12 1/8	16	6 3/4	4 1/2	5 3/8	3 1/8
612	1220	830	2000	1300	1500	9 1/4	6 1/2	1 1/16	14 1/4	13 1/4	19	8 3/4	5	6 7/8	3 1/8
614	1780	1140							15 1/4						
616	2625	1750	4000	2600	3000	11 3/4	8	1 5/16	17 1/4	15 7/8	23	11 1/4	5 1/2	8 3/8	2 5/8
618	3615	2560											6		3 1/8
620	5550	3900	9000	6000	[a]	15	9 1/2	1 9/16	18 3/4	20 3/4	30	14 1/4	6 3/4	10 5/8	3 1/8
622	8460	5790	9000	6000	[a]	15	9 1/2	1 9/16	18 1/2	20 3/4	30	14 1/4	7 1/4	10 5/8	3 5/8
624	11800	8210	9000[b]	[b]	[a]	15	9 1/2	1 9/16	19 1/4	20 3/4	30	14 1/4	9 1/4	8 5/8	3 5/8

[a] No 8-hr rating established by manufacturers.
[b] Applies only to 1-hr shunt brakes.

been little progress along these lines except in the dc brakes for mill motors. In 1947 the Association of Iron & Steel Engineers, working with motor manufacturers, developed a new line of mill motors known as the 600 series of motors. Motor ratings and essential dimensions were standardized. Then, in 1948, the AISE, working with NEMA, began a program to standardize brakes for the new motors. Since all brake manufacturers were faced with the problem of redesigning their brakes to suit the new motors, whether or not there was a standardization, the value of standardization at that time was evident. These standards, which were adapted in 1952 and revised in 1957, are summarized in Table 20.2.

Problems

1. Calculate the torque of a 25-hp 600-rpm motor when it is operating at its rating.

2. If a 25-hp 600-rpm motor is used to hoist a load which requires full-rated torque when the efficiency of the hoisting mechanism is 60%, what torque must a brake supply to hold the load stationary after it is hoisted?

3. The trolley of a crane weighs 2000 lb and travels at 90 ft/min. The WR^2 of the driving motor, plus that of all rotating parts of the trolley, is 15 lb/ft^2. Calculate the torque required of a brake which will stop the motor in 10 revolutions.

4. The WR^2 of a 200-hp 450-rpm motor is 800 lb/ft^2. How long will it take to stop the motor with a brake which delivers a torque equal to 50% more than the rated torque of the motor?

5. If the motor of problem 4 is driving a direct-connected load having a WR^2 of 2000 lb/ft^2, how long will it take to stop the motor with a brake which delivers a torque equal to 50% more than the motor torque?

6. How much torque would be required of a brake which would stop the motor and load of problem 5 in the same time that the brake of problem 4 stopped the motor alone?

7. A 150-hp 460-rpm shunt motor having a WR^2 of 415 lb/ft^2 is driving a direct-connected load having a WR^2 of 500 lb/ft^2. If the motor is fitted with a brake which delivers a torque equal to the motor torque, and in addition the controller is arranged for dynamic braking which varies from three times the motor torque at the start to zero at standstill, how long will it take to stop the motor?

8. A 75-hp 1750-rpm motor having a WR^2 of 118 lb/ft^2 is driving a flywheel which is a solid cylinder having a diameter of 3 ft and a weight of 2000 lb. The gearing between the motor and the flywheel is 5 to 1. Calculate the equivalent WR^2 of the system.

9. How long will it take to stop the motor and flywheel using a brake which delivers a torque of twice the motor torque.

10. Calculate the kinetic energy in an armature having a WR^2 of 1300 lb/ft^2 and a speed of 1750 rpm.

11. If the rating of the motor of problem 10 is 700 hp, how many revolutions will be required to stop the motor after the brake sets if the brake torque is equal to the motor rated torque?

12. How many seconds will it take to stop the motor?

13. If the 700-hp 1750-rpm motor of problem 10 is driving a load having an equivalent WR^2 of 1500 including the brake wheel, calculate the torque required to reach zero speed in 15 secs, allowing 0.40 sec for the brake to set.

14. A brake which operates very infrequently can dissipate 1500 ft/lb per square inch of wheel face, per minute, without reaching a temperature which will destroy the lining material. If the width of the wheel is 0.40 times the wheel diameter, what is the wheel diameter that is required?

15. The bucket of an ore-bridge hoist is operated by two 325-hp 390-rpm motors. One called the hoist motor is geared to cables which connect to the top of the bucket. The other called the shell-line motor is geared to cables which connect to the hinged lips of the bucket, and this motor is used to open and close the bucket. In hoisting a full bucket, the motors divide the load equally. When opening the bucket, the brake on the hoist motor must hold the entire load, since the shell-line motor is paying out cable to permit the bucket to open. Calculate the torque required of the two motors in hoisting when:

The bucket weighs 45,000 lb
The ore weighs 45,000 lb
The transmission efficiency is 0.70
The hoisting speed is 172 ft/min

16. What torque is required of the brake which must hold the load while the bucket is opening?

17. When the empty bucket is lowered to pick up another load, the two motors must decelerate at the same rate. Calculate the number of revolutions of the hoist motor after the brake applies if:

The lowering speed is twice the hoisting speed
The WR^2 of the brake wheel and rotating parts is 500 lb/ft^2
The WR^2 of the motor is 1000 lb/ft^2

Note. The full weight of the bucket is on the hoisting line cables.

18. What torque is needed on the shell-line brake to stop the unloaded shell-line motor in the same number of revolutions?

19. A 280-hp 420-rpm motor having a WR^2 of 1000 lb/ft^2 is driving a direct-connected load having a WR^2 of 300 lb/ft^2. Calculate the motor torque, and select a suitable shunt-wound 1-hr intermittent-duty brake from Table 42.

20. Add the WR^2 of the wheel of the selected brake, and calculate the time to stop motor, brake, and load.

21. In order to secure faster stopping the 280-hp motor is replaced by two 140-hp motors, each having a WR^2 of 350 lb/ft^2. Calculate the torque of each motor, and select suitable shunt-wound 1-hr intermittent-duty brakes from Table 42.

22. Add the WR^2 of the wheels of the selected brakes, and calculate the time to stop motor, brake, and load. How much has the stopping time been shortened by the use of two motors?

23. A factory has a flywheel which is of irregular shape and which weighs 1000 pounds. It is necessary to determine the radius of gyration. The flywheel is clutched to a motor, accelerated to 1200 rpm, and then declutched. At the instant of declutching, a 14-inch continuous-duty brake (Table 20.1) is applied. The flywheel stops in 20 sec. What is the radius of gyration?

24. A motor-driven car in a factory weighs 10,000 lb and travels at 200 ft/min. The WR^2 of the motor brake wheel, and of all rotating parts of the car, is 150 lb/ft^2. There is a safety limit switch which cuts off power and sets the brake if the car is driven too near to the end of the track. How far from the end of the track should the limit switch be placed to insure safe stopping from full speed?

25. A machine consists of three rolls, each weighing 2000 lb. The radius of one roll is 18 in., and it is geared to the motor through a 5-to-1 gear train. The radius of the second roll is 2 ft, and it is geared through a 20-to-1 gear train. The third roll has a radius of 12 in. and is geared through a 2-to-1 gear train. The 150-hp 480-rpm motor has a WR^2 of 415 lb/ft^2. The WR^2 of the brake wheel is 313 lb/ft^2. What is the equivalent WR^2 of the whole system?

26. If the motor of problem 25 is equipped with an electronic controller which will hold the accelerating torque essentially constant at twice the full-load torque of the motor, how long will it take the motor to accelerate the system?

27. If the brake torque is 3600 lb/ft, how long will it take to stop the system?

28. If the controller is changed to a magnetic controller which permits an average accelerating torque of 50% of full-load torque, what will be the accelerating time?

29. If dynamic braking is provided to give an average decelerating torque equal to the motor full-load torque, in addition to the torque of the brake, what will be the time required to stop?

30. A motor on a crane is hoisting a load of 10 tons at a speed of 100 ft/min. The gear ratio between motor and load is 20 to 1. The efficiency of the hoist is 80%. How much torque must a brake provide to hold the load if power fails while hoisting?

21

RESISTOR DESIGN

Design Operations. The design of a resistor may be divided into three operations:

Calculation of the ohmic values
Calculation of the current-carrying capacity required
Selection of the actual materials to be used

The calculation of the ohmic values is generally not difficult. The ohms required depend upon the function of the resistor, that is, whether it is for accelerating purposes or for speed regulating or for dynamic braking. Where a number of similar resistors have to be designed, as, for instance, the accelerating resistors for a line of similar automatic starters, covering a range of horsepower, tables may be prepared which make the calculation a very simple matter. Similar tables may be prepared for the calculation of the current capacity, although the preparation of such tables is more difficult. The capacity depends upon the amount of current to be carried by the resistor and the percentage of time on in each cycle. If the current is constant, the calculation is simple, but if the current varies, the calculation becomes more involved. This type of calculation may of course be readily done on a modern digital computer. Many of the techniques shown in this chapter are valuable, however, to provide *understanding* of the resistor design process.

The selection of the actual materials to be used is a matter for which no general rules can be set down, as it depends upon the materials available. Each manufacturer of controllers has his own line of cast grids, welded steel grids, ribbon resistors, and wire-wound units, the characteristics of which are known to his engineers. After the ohms and the current capacity required have been calculated, the material is selected on the basis of minimum cost, minimum space requirement, or service requirements.

The calculation of ohmic values for various purposes is described in the chapters which cover the different types of motors and their control.

Resistor Materials. Since resistors serve so many purposes, they must be available in a wide range of ohmic values and current-carrying capacities. For low ohms and high capacity, stamped steel welded grids are generally used. For high ohms and low capacity, wire-wound units of various types are available. Resistors having a ribbon of steel or other material form an intermediate class.

Cast-Iron Grids. Cast-iron grids, as the name implies, are resistors made of cast iron, in the form of grids, and with an eye at each end for mounting. They have largely been replaced in later years with stamped or punched steel resistors, but many cast grids are still in service. The eyes are usually ground and copper-plated to insure a good connection between grids. The grids are stacked in bunches on steel mounting rods. The rods are first insulated by being covered with a mica tube or wound with an insulating material. Mica washers are inserted between the grids for insulation. Grids are made in many sizes and forms. The resistance obtainable with a single grid is about 0.125 ohm for the grid of minimum cross-section, and may go to 0.005 ohm or less for the larger cross-sections.

Cast-iron grids are cheap and strong, and, contrary to what might be expected, they are not easily damaged by corrosion. They will rust, of course, but the surface hardness produced in casting prevents rapid corrosion beyond the first layer of rust. Where protection against corrosion is important, as in marine service, the grids may be plated with zinc or cadmium or painted with aluminum paint.

If grids are to be mounted as a part of a controller, the mounting rods are made of the proper length to extend across the frame at the rear of the slate panel, or they may be fastened to brackets or cross-angles in an enclosing frame. If the grids are mounted separately from the controller, as is standard practice in steel-mill and other large installations, the rods are mounted between steel end frames. The size of the box thus formed and the number of grids in it are usually standardized by the control manufacturer, so that any box purchased will fit the mounting of any other box. Figure 21.1 shows a grid box of this type.

Punched Grids. Instead of grids being made of cast iron, they may be punched out of sheet steel, some corrosion-resistant alloy being preferable. Such an alloy may also have a negligible temperature-resistance coefficient. The grids are, of course, much stronger than cast-iron grids. Figure 21.2 shows a resistor of this type. In this design the ends of the grids are welded together to obtain a good electric contact, and the resistor becomes, in effect, a continuous strip.

FIG. 21.1 Cast-Iron Grid Box

Ribbon Resistors. Some resistor installations are subject to severe shocks, as, for example, installations on moving vehicles such as diesel-electric locomotives and road building machinery, or on Naval vessels. Grids cast of a special alloy are sometimes used on such installations, but ribbon resistor is more common. The ribbon may be a copper-nickel

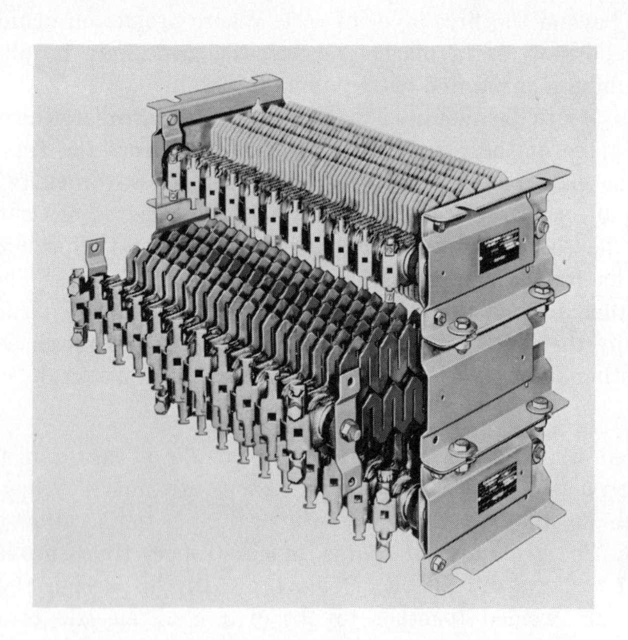

FIG. 21.2 Welded Stamped Steel Resistor Boxes Showing Draw-Out Design

FIG. 21.3 Ribbon-type Resistor

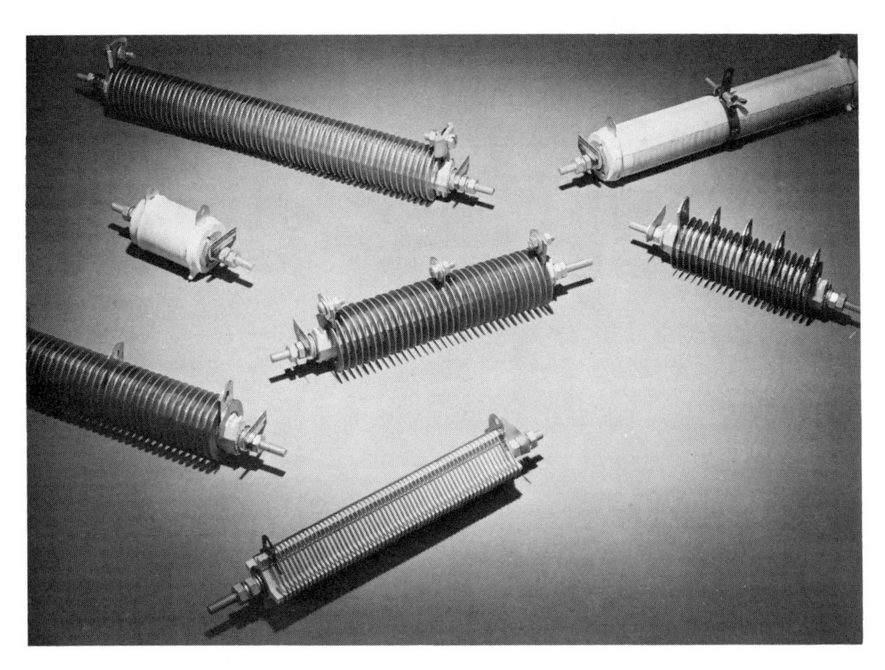

FIG. 21.4 General Electric Resistor Units: (a) Smooth-Wound with Taps; (b) Open-Wound with 3 Taps; (c) Edgewise-Wound, with Midtap; (d) Multiple Edgewise-Wound

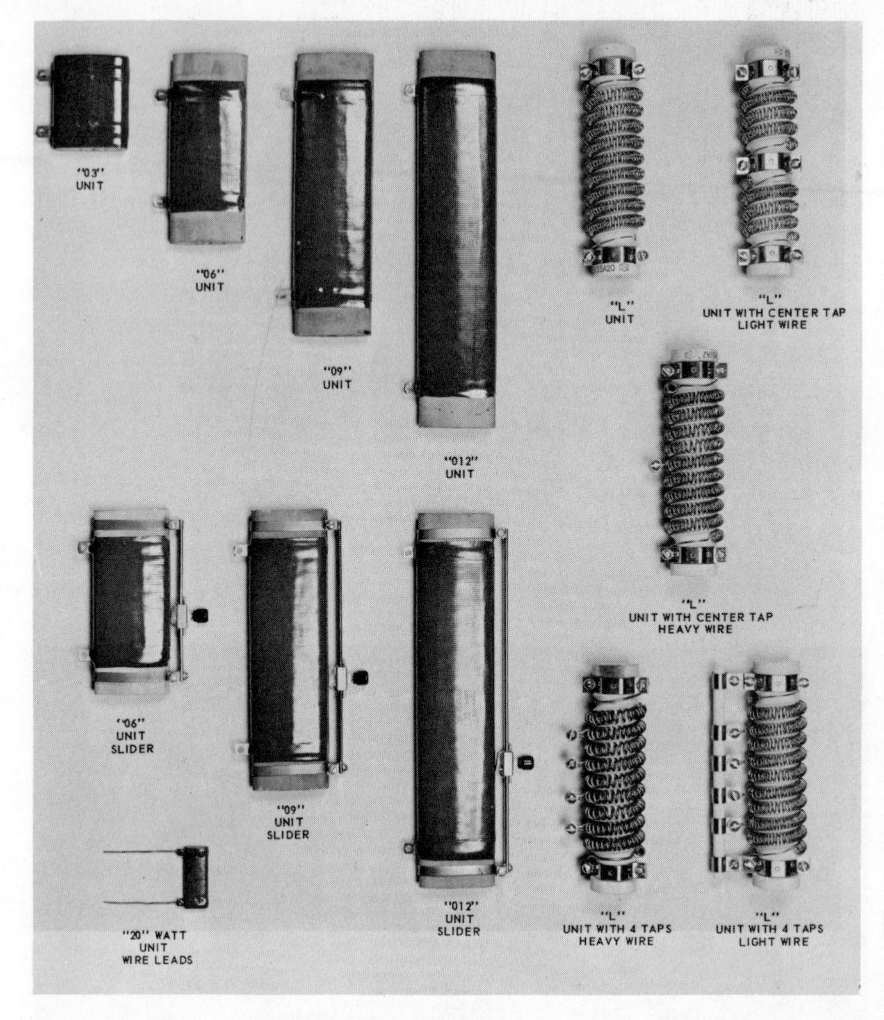

Fig. 21.5 Wire-Wound Resistor Units

alloy or stainless iron or Nichrome. The ribbon is wound edgewise, either in a spiral form on tubular porcelain insulators or between insulated mounting supports (see Figs. 21.3 and 21.4). An assembly of either type may be mounted as a part of the controller or made up in the form of the mill-type box. The principal advantage of the ribbon resistors is that they are shock-proof. In addition they are rust-proof, light in weight, and compact.

Wire-Wound Units. Wire-wound units take many forms, some of which are shown in Fig. 21.5. The wire, usually a copper-nickel alloy, is wound on the base, and the unit is covered with a cement and baked. Another method is to cover the unit with a highly heat-resistant enamel. These units are assembled on rods for mounting in a resistor compartment, or they may be mounted directly on the steel control panel. In resistance they may be wound for anything from a fraction of an ohm up to several thousand ohms per unit. The oval units are wound on ceramic bases and enameled after winding.

Figure 21.6 shows a mill-type box made up of heavy units. A base of lava has been used, cut with a spiral groove to take the wire winding. The wire itself is first wound into a spiral, which is then wound on the base. Clamps at each end of the unit serve to hold the winding in place and to provide for a connecting terminal.

A column of thin graphite disks may also serve as a resistor. When these disks are loosely stacked, the resistance of the column is high. To reduce the resistance, the disks are put under pressure. A cam provides the variation of pressure, giving equal resistance changes for equal movements of the control lever. When used for accelerating purposes, the column is finally short-circuited by a copper-to-copper contact or by a magnetic contactor.

Current Capacity. The capacity of a resistor and the amount of material required for a given application are determined by the wattage to be dissipated, the length of time on, and the cooling time or time off. On a single start, the resistor will absorb a certain amount of heat,

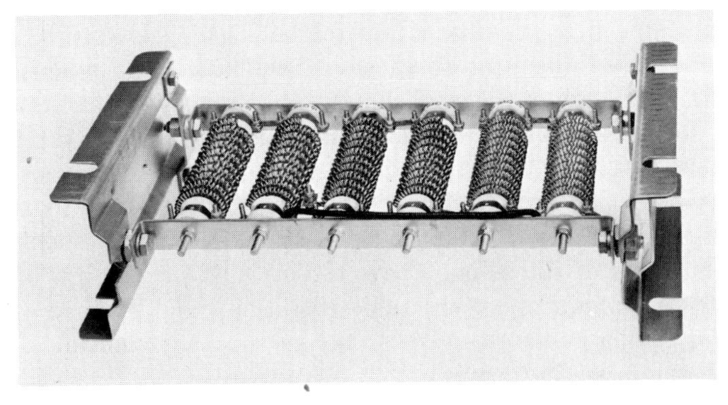

FIG. 21.6 Bank of Wire-Wound Resistor Units

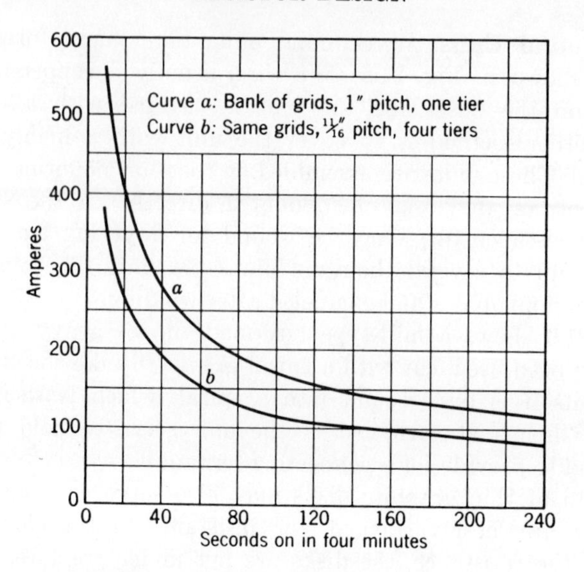

FIG. 21.7 Effect of Spacing on Grid Rating

depending upon its mass; and if the duty is infrequent, this may be the determining factor. With repeated starts, the resistor does not have time to cool fully, and the criterion is then the ability of the resistor to radiate heat to the surrounding atmosphere. In the first case the mass of material is the important factor and ventilation is not so important, but in the second case the ventilation available becomes of considerable importance. If the grids are stacked closely together on the rods, the ventilation will be reduced, and more grids will be required to dissipate the same energy. Also, if several banks of grids are stacked one above the other, the heat of each one will affect the others, and the dissipating capacity of the whole bank will be reduced (see Fig. 21.7).

In order to arrive at a rating for his resistor materials, the manufacturer must test each size of grid or unit, not only for continuous carrying capacity but also on a number of intermittent cycles. He must also determine the effects of close and wide stacking, and of mounting different numbers of stacks one above the other. To enable manufacturers to make these tests under the same conditions, and so have a common basis for rating resistors, NEMA has set up the following definitions and resistor classes.

Rating. Resistors shall be rated in ohms, amperes, and class of service.

Service Classification of Resistors

1. Resistors shall be designated by class numbers in accordance with Tables 21.1, 21.2, 21.3, and 21.4.

2. Starting and intermittent-duty resistors are primariy designed for use with motors which require an initial torque corresponding to the stated per cent of full-load current on the first point and which require an average accelerating current (rms value) of 125% of full-load current.

With a secondary resistor alternating current controller, the figures given in the tables for the per cent of full-load current on the first point, starting from rest, apply to rotor (secondary) current and to torque. The primary current will, in general, be a higher percentage of the full-load current.

3. Starting and intermittent-duty primary resistors which are designed for use with squirrel-cage motors and which meet the test described in succeeding paragraphs are included in the tables.

4. Continuous-duty resistors shall be capable of carrying continuously the current for which they are designed.

5. An adjustable speed motor having a given horsepower rating, when started with full field, generally requires a resistor having a larger ohmic value than does a constant-speed motor of the same horsepower rating.

TABLE 21.1

Class Numbers of Resistors for Nonreversing Service and Reversing Nonplugging Service Without Armature Shunt or Dynamic Braking

Approximate Per Cent of Full-Load Current on First Point Starting from Rest	Class Numbers						
	Duty Cycles						
	5 Sec On 75 Sec Off	10 Sec On 70 Sec Off	15 Sec On 75 Sec Off	15 Sec On 45 Sec Off	15 Sec On 30 Sec Off	15 Sec On 15 Sec Off	Continuous Duty
25	111	131	141	151	161	171	91
50	112	132	142	152	162	172	92
70	113	133	143	153	163	173	93
100	114	134	144	154	164	174	94
150	115	135	145	155	165	175	95
200 or over	116	136	146	156	166	176	96

TABLE 21.2

Class Numbers of Resistors For Reversing Plugging Service Without Armature Shunt or Dynamic Braking

Approximate Per Cent of Full-Load Current on First Point Starting from Rest with All Resistance in Circuit	Class Numbers			
	Duty Cycles			
	15 Sec On 45 Sec Off	15 Sec On 30 Sec Off	15 Sec On 15 Sec Off	Continuous Duty
25	151P	161P	171P	91P
50	152P	162P	172P	92P
70	153P	163P	173P	93P
100	154P	164P	174P	94P

TABLE 21.3

Class Numbers of Resistors For Dynamic Lowering Crane and Hoist Controllers

Approximate Per Cent of Full-Load Current on First Point Hoisting Starting from Rest without Armature Shunt	Class Numbers			
	Duty Cycles			
	15 Sec On 45 Sec Off	15 Sec On 30 Sec Off	15 Sec On 15 Sec Off	Continuous Duty
50	152DL	162DL	172DL	92DL
70	153DL	163DL	173DL	93DL

For this reason, the current on the first point and the capacity of the complete resistor of the same NEMA classification for an adjustable speed motor may be different from that for a constant speed motor of the same horsepower.

Note. A continuous-duty resistor shall be so designed that the controller may be operated continuously on any point when the load follows its normal speed-torque curve, except that it must not be operated continuously below the minimum speed specified.

Note I. The reader is referred to NEMA Standards for Industrial Control for further information.

TABLE 21.4

CLASS NUMBERS OF RESISTORS
FOR CONTINUOUS-DUTY SPEED-REGULATING SERVICES
WITH DIRECT-CURRENT SHUNT MOTORS
AND ALTERNATING-CURRENT WOUND ROTOR MOTORS

Per Cent Speed Reduction	Class Numbers						
	Per Cent of Rated Motor Torque at Reduced Speed						
	40	50	60	70	80	90	100
5	405	505	605	705	805	905	1005
10	410	510	610	710	810	910	1010
15	415	515	615	715	815	915	1015
20	420	520	620	720	820	920	1020
25	425	525	625	725	825	925	1025
30	430	530	630	730	830	930	1030
35	435	535	635	735	835	935	1035
40	440	540	640	740	840	940	1040
45	445	545	645	745	845	945	1045
50	450	550	650	750	850	950	1050

Note II. The stability of the motor speed obtained by simple rheostatic control is dependent upon the stability of the load on the motor. The degree of instability is directly proportional to the amount of speed reduction. Variations in load have a greater proportional effect on speed when the load is light. For these reasons Table 21.4 has not been carried beyond a speed reduction of 50% and a load torque of 40%.

Note III. With a dc shunt motor, the per cent of rated motor current which is obtained at the reduced speed is assumed to be the same as the percent of rated torque. With a dc series motor operating at less than 100% current, the per cent of torque is less than the per cent of current. With a wound-rotor motor and resistor in the rotor circuit, the per cent of rated rotor (secondary) current which is obtained at the reduced speed is assumed to be the same as the per cent of rated torque.

Note IV. A speed-regulating resistor is so designed that it may be operated continuously at any point in the speed-regulating range when the load follows its normal speed-torque curve. When additional resistance is required to obtain the starting current specified, the additional portion of the resistor shall be designed for a duty cycle selected from

Table 21.1. The resulting resistor may be completely specified by a compound number.

For example, 154/850 designates a resistor which is designed for starting and speed-regulating duty. The starting section is designed to allow 100% of full-load current on the first point, starting from rest, and a duty cycle of 15 sec on and 45 sec off. The regulating section is designed to give 50% speed reduction at 80% of rated torque and for continuous duty when the load follows its normal speed torque curve.

Note V. If the speed-regulating resistor alone is used, the approximate per cent of rated current which will be obtained on the first point of the controller, starting from rest, may be determined from the equation:

$$\text{Per cent current} = \frac{\text{Per cent torque at reduced speed}}{\text{Per cent speed reduction}} \times 100$$

Temperature Tests. When a temperature test is made on a resistor, rheostat, or dimmer at the current values, duty cycle, and elapsed time specified, the temperature rise above the ambient temperature and the methods of temperature measurement shall be in accordance with the following.

1. For bare resistive conductors, the temperature rise shall not exceed 375°C as measured by a thermocouple in contact with the resistive conductor.
2. For resistor units, rheostats, and wall-mounted rheostatic dimmers which have an embedded resistive conductor, the temperature rise shall not exceed 300°C as measured by a thermocouple in contact with the surface of the embedding material.
3. For rheostatic dimmers which have embedded resistive conductors and which are arranged for mounting on switchboards or in noncombustible frames, the temperature rise shall not exceed 350°C as measured by a thermocouple in contact with the surface of the embedding material.
4. The temperature rise of the issuing air shall not exceed 175°C as measured by a mercury thermometer at a distance of 1 in. from the enclosure.

Temperature Test—Primary Resistors for Squirrel-Cage Motors. When a temperature test is made on a general-purpose single-step primary starting resistor for a squirrel-cage motor, the resistor shall be tested with 300% of normal full-load current of the motor for which the resistor is designed, and the current shall be maintained for a duty cycle as indicated in the resistor classification. This cycle shall be repeated for 1 hr, after which period the temperature rise shall not exceed the limitations given in paragraph 4 above.

Before ribbon resistors were extensively used, the classification table was based on a 4-min cycle because practically all resistors had enough heat-absorbing capacity so that they did not reach their ultimate temperature in that time. When ribbon resistors came into general acceptance, this table no longer served, because the heat-absorbing capacity of the ribbon is relatively low. A resistor of ribbon designed for half-time on the basis of 2 min out of 4 requires practically the same amount of material as one designed for continuous duty. Since the actual operating cycle of most intermittent-duty motor-driven machines is much less than 4 min, it is practical to base the resistor on a shorter cycle. Resistors of thin ribbon, based on the short cycle, will require only about half as much material as for the same duty based on the long cycle.

Very Short Cycles. For very short cycles, in which the resistor is in circuit for only a second or two, the capacity can be calculated on the basis of heat-absorbing ability only, ventilation being neglected. As an example, suppose that it is desired to calculate the temperature rise of a ribbon resistor for dynamic braking, the peak current being 700 amp and the time 1 sec. The specific heat of the material is 0.09. The weight of the ribbon is calculated and found to be 4.8 lb, and the resistance is 0.5 ohm.

$$\text{Watts} = 700^2 \times 0.5 = 245{,}000$$

$$\text{Btu per hour} = 245{,}000 \times 3.412 = 835{,}000$$

$$\text{Btu per pound for 1 sec} = \frac{835{,}000}{4.8 \times 60 \times 60} = 48.3$$

Since 1 Btu will raise 1 lb of water 1°F, the temperature rise of the ribbon will be

$$T = \frac{48.3}{0.09} = 537°\text{F or } 298°\text{C}$$

This is below the 375°C permissible rise in accordance with NEMA and is in fact below the 300°C rise rating for coated resistors.

Intermittent-Duty Rating. The following example will illustrate a method of calculating the amount of resistor required for an intermittent-duty controller. Assume that it is desired to design a three-step class 153P resistor for a 100-hp 230-V motor whose full-load current is 375 amp. Class 153P indicates that this is a plugging controller, and that the resistor will be in circuit for 15 sec in each 60 sec.

On the basis that the maximum voltage applied (counter-emf plus line voltage) is 180% of rated voltage, and the maximum inrush for plugging is to be limited to 150% of rated current, the ohmic values are:

$$\text{Total resistance} = \frac{230 \times 1.8}{375 \times 1.5} = 0.736 \text{ ohm}$$

If the current on accelerating from standstill is permitted to go to 160% of rated,

$$\text{Accelerating resistance} = \frac{230}{375 \times 1.6} = 0.383 \text{ ohm}$$

The steps are tapered as follows.

Step	*Percentage of Total*	*Ohms*
1	48	0.353
2	34.4	0.253
3	17.6	0.130
Total	100	0.736

The time required on each point of the controller may be assumed to be in the same ratio as that of the resistance of the steps, so that the time on in each cycle will be as follows.

Step	*Percentage of Total*	*Seconds*
1	48	7.2
2	82.4	12.4
3	100	15

On the plugging step the inrush is 150% and the base current 100%. The heating of the resistor is determined by the root-mean-square current, which is

$$\text{Rms current} = \sqrt{\frac{150^2 + 150 \times 100 + 100^2}{3}}$$

$$= 126\% \text{ or } 460 \text{ amp}$$

On the other steps the inrush is 160%, and the rms current is

$$\sqrt{\frac{160^2 + 160 \times 100 + 100^2}{3}} = 131\% \text{ or } 492 \text{ amp}$$

The watts to be dissipated by each step are determined by multiplying the resistance by the square of the rms current. The wattage value is then multiplied by the time on to obtain the watt-seconds on each step. The last two steps are on part of the time at the lower current and part at the higher current.

Step	*Resis-tance*	*Seconds On*	*Rms amperes*	*Rms watts*	*Watt-seconds*
1	0.353	7.2	460	74,600	537,000
2	0.253	7.2 + 5.2	460–492	53,500 and 61,200	703,000
3	0.130	7.2 + 7.8	460–492	27,500 and 31,400	443,000

It is now necessary to refer to a table giving the wattage rating of the particular type of resistor material. Assuming that cast-iron grids are used, and that the rating of each is 600 W, on a duty of 15 sec in 60 sec, the watt-second rating is then 9000 per grid. The number of grids required for each step is determined by dividing the watt-seconds required by the step, by the grid rating.

$$\text{First step} \quad \frac{537{,}000}{9{,}000} = 60 \text{ grids}$$

$$\text{Second step} \quad \frac{703{,}000}{9{,}000} = 78 \text{ grids}$$

$$\text{Third step} \quad \frac{443{,}000}{9{,}000} = 49 \text{ grids}$$

$$\text{Total} \quad = 187 \text{ grids}$$

Since the number of grids required varies directly with the horsepower of the motor, it is possible to reduce the above calculations to a simple general form from which a similar resistor can be calculated for any size of motor.

A convenient method of expressing the ohmic value is as percentage of E/I, where E is the line voltage and I the rated current of the motor.

The first step is 0.353 ohm.

$$0.353 = \frac{KE}{I}$$

$$K = \frac{0.353 \times 375}{230}$$

$$K = 0.575, \text{ or } 57.5\% \text{ of } E/I$$

The second step is $\dfrac{0.253 \times 375}{230} = 0.41$, or 41% of E/I.

The third step is $\dfrac{0.130 \times 375}{230} = 0.21$, or 21% of E/I.

Grid tables are generally arranged to give the current-carrying capacity of the grid, for any duty cycle. The most convenient way to express the capacity required for a step is, therefore, in terms of the full rated current of the motor. It is convenient to reduce the rating to a continuous-duty basis, so that any resistor can be calculated from the continuous-duty grid table.

Assume that the type of grid to be used has a continuous rating of 125 W. The 60 grids of the first step will have a continuous rating of 60 × 125, or 7500 W. The continuous current capacity of this step will be

$$\sqrt{\frac{\text{Watts}}{\text{Resistance}}} = \sqrt{\frac{7500}{0.353}} = 143 \text{ amp}$$

Expressed in terms of rated motor current, this is

$$\frac{143}{375} = 0.39I$$

Similarly, the capacity required in the second step is

$$\sqrt{\frac{125 \times 78}{0.253}} = 196 \text{ amp} = 0.52I$$

And the capacity required in the third step is

$$\sqrt{\frac{125 \times 49}{0.130}} = 217 \text{ amp} = 0.58I$$

The complete formula for the resistance is as follows:

Step	*Ohms in Percentage of E/I*	*Current Capacity in Percentage of I*
1	57.5	39
2	41	52
3	21	58
Total	119.5	

This formula is applicable to any motor requiring a three-step class-153P resistor and using the particular type of grid on which the calculations are based.

Parallel Resistance. When the resistor steps are connected in parallel instead of in series, the distribution of the resistor material is quite different. The resistance of the first step is 119.5% of E/I. When the second step is parallel with the first step, the resultant resistance must be the same as that obtained with a series resistance when the first step is cut out, or 62% of E/I. When the third step is parallel with the first two steps, the resultant resistance must be 21% of E/I. The final contactor short-circuits all the steps. To obtain these results,

$$\text{Step 1} = 1.19E/I = 0.73 \text{ ohm}$$

$$\text{Step 2} = 1.29E/I = 0.79 \text{ ohm}$$

$$\text{Step 3} = 0.32E/I = 0.196 \text{ ohm}$$

When the first step above is in circuit, the rms current is 126%, or 460 amp. With the second step connected in, the total rms current is 131%, or 492 amp. This current divides through the two steps in inverse ratio to their resistance. With three steps in circuit the current divides through the three resistors. The first step is in circuit for the total time period, but with varying currents as the other steps are cut in. The currents and times are as shown in Table 21.5.

From these calculations it is evident that the parallel resistor requires the same number of grids as the series resistor, but that the greater part of the parallel resistor is required in the first step. This is true only in intermittent duty. On continuous regulating duty the parallel resistor will require a greater number of grids.

The current capacity required may be reduced to terms of full rated current in the same manner as with the series resistor.

The capacity of the first step is

$$\sqrt{\frac{125 \times 153}{0.73}} = 161 \text{ amp} = 0.43I$$

That of the second step is

$$\sqrt{\frac{125 \times 27}{0.79}} = 65 \text{ amp} = 0.17I$$

TABLE 21.5

	Ohms	Rms amperes	Seconds	Watt-seconds
Step 1				
On the first point	0.73	460	7.2	1,110,000
On the second point	0.73	256	5.1	244,000
On the third point	0.73	87	2.7	15,000
Total watt-seconds				1,369,000
Grids required at 9000 watt-seconds per grid				153
Step 2				
On the second point	0.79	236	5.1	224,000
On the third point	0.79	81	2.7	14,000
Total watt-seconds				238,000
Grids required				27
Step 3				
On the third point	0.196	324	2.7	55,600
Total watt-seconds				55,600
Grids required				7

And that of the third step is

$$\sqrt{\frac{125 \times 7}{0.196}} = 67 \text{ amp} = 0.18I$$

The complete formula for the parallel resistor is as follows:

Step	*Ohms in per cent of E/I*	*Current Capacity in per cent of I*
1	119	43
2	129	17
3	32	18

Regulating Duty. When the resistor is to be used for regulating duty, it must carry continuously the current required by the load. If the load is definitely known, that value should be used. Ordinarily the load will be known only to the extent of whether it is fan or machine duty. The current required may be read from the load curves of Figs. 7.8 and 7.9. The following demonstration is given to determine the relation between the material required for a series resistor and that of a parallel resistor. Assume that 80% of full-load current is required on each point of the controller and that the grids have a capacity of 125 W per grid. With the series resistor the number of grids required is as follows:

$$\text{First step} \quad \frac{0.353 \times 300^2}{125} = 254 \text{ grids}$$

$$\text{Second step} \quad \frac{0.253 \times 300^2}{125} = 180 \text{ grids}$$

$$\text{Third step} \quad \frac{0.130 \times 300^2}{125} = 94 \text{ grids}$$

$$\text{Total} \quad = 528 \text{ grids}$$

With the parallel resistance, the first step requires

$$\frac{0.73 \times 300^2}{125} = 528 \text{ grids}$$

The second step has to carry only the current through it, or 142 amp.

$$\frac{0.79 \times 142^2}{125} = 128 \text{ grids}$$

The third step has to carry 198 amp.

$$\frac{0.196 \times 198^2}{125} = 62 \text{ grids}$$

$$\text{Total} \quad = 718 \text{ grids}$$

The parallel regulating resistor, therefore, requires 36% more grids than the series regulating resistor.

Any resistor may be reduced to terms of E/I and percentage of I, but it must be borne in mind that the resultant table is applicable only to the particular type of grid upon whose characteristics it is based. Such a table, worked out for cast grids, will give entirely erroneous results if used to calculate a resistor of ribbon material or of wire-wound units.

NEMA Resistor Application Table. Table 21.7 is intended as a guide in specifying or designing resistors. The classifications are those which experience has shown to be correct for the average installation. It is recognized that there will be exceptions. The table applies to resistors composed of wire-wound units or stamped steel grids, and unbreakable ribbon resistors, provided that the time-on period does not exceed the values given in Table 21.1.

TABLE 21.6

TYPICAL RATING TABLE FOR CAST-IRON GRIDS

Grid Number	1	2	3	4	5	6	7	8	9
Grid Ohms	0.007	0.010	0.015	0.020	0.030	0.040	0.060	0.080	0.110
Per cent of Time On					Capacity in Amperes				
6.25	515	430	365	305	255	215	180	150	128
12.50	365	305	260	215	180	153	127	107	90
16.67	315	265	225	187	157	132	110	93	79
25.00	255	215	182	152	128	108	90	76	64
33.33	215	185	156	130	110	92	76	65	55
50.00	180	153	130	108	90	76	62	54	45
75.00	150	125	105	88	75	62	51	44	37
100.00	130	108	89	76	63	54	44	38	32

This table is fictitious and does not apply to the grids of any manufacturer.

TABLE 21.7

	NEMA Resistance Class		NEMA Resistance Class
Blowers		*Cement Mills (Cont.)*	
Constant pressure	135–195	Rotary dryers	145–195
Centrifugal	133–193	Elevators	135
		Grinders, pulverizers	135
Brick Plants		Kilns	135–195
Augers	135		
Conveyors	135	*Coal and Ore Bridges*	
Dry pans	135	Holding line	162
Pug mills	135	Closing line	162
		Trolley	163
By-Products Coke Plants		Bridge	153
Reversing machines	153		
Leveler ram	153	*Coal Mines*	
Pusher bar	153	Car hauls	162
Door machine	153	Conveyors	135–155
		Cutters	135
Cement Mills		Crushers	145
		Fans	134–193
Conveyors	135	Hoists, slope	172
Crushers	145	Vertical	162

TABLE 21.7 (Continued)

	NEMA Resistance Class		NEMA Resistance Class
Coal Mines (Cont.)		*Machine Tools (Cont.)*	
Jigs	135	Drills	115
Picking tables	135	Gear cutters	115
Rotary car dumpers	153	Grinders	135
Shaker screens	135	Hobbing machines	115
		Lathes	115
Compressors		Milling machines	115
Constant speed	135	Presses	135
Varying speed, plunger type	135–195	Punches	135
Centrifugal	93	Saws	115
		Shapers	115
Concrete Mixers	135		
		Metal Mining	
Cranes—General Purpose		Ball, rod, tube mills	135
Hoist	153	Car dumpers, rotary	153
Bridge, sleeve-bearing	153	Converters, copper	154
Trolley, sleeve-bearing	153	Conveyors	135
Bridge or trolley, roller-bearing	152	Crushers	145
		Tilting furnace	153
Flour Mills	135		
Line shafting	135	*Paper Mills*	
		Heaters	135
Food Plants		Calenders	154–192
Dough mixers	135		
Butter churns	135	*Pipe Working*	
		Cutting and threading	135
Hoists		Expanding and flanging	135–195
Winch	153		
Mine slope	172	*Power Plants*	
Mine vertical	162	Clinker grinders	135
Contractors' hoists	152	Coal crushers	135
		Conveyors, belt	135
Larry Cars	153	Screw	135
		Pulverized fuel feeders	135
Lift Bridges	152	Pulverizers, ball type	135
		Centrifugal	134
Machine Tools		Stokers	135–193
Bending rolls, rev.	163–164		
Non-rev.	115	*Pumps*	
Boring mills	135	Centrifugal	134–193
Bulldozers	135	Plunger	135–195

TABLE 21.7 (Continued)

	NEMA Resistance Class		NEMA Resistance Class
Rubber Mills		*Steel Mills (Cont.)*	
Calenders	155	Racks	153
Crackers	135	Reelers	135
Mixing mills	135	Saws, hot or cold	155
Washers	135	Screwdowns	153–163
		Shears	155
Steel Mills		Shuffle bars	155
Accumulators	153	Sizing rolls	155
Casting machines, pig	153	Slab buggy	155
Charging machines, bridge	153–163	Soaking pit covers	155
Peel	153–163	Straighteners	153
Trolley	153–163	Tables, main roll	153–163
Coiling machines	135	Shear approach	153–163
Conveyors	135–155	Lift	153–163
Converters, metal	154	Roll	153
Cranes, ladle, bridge, trolleys, sleeve-bearings	153–163	Transfer	153
Roller-bearing	152–162	Approach	153
Hoist	153–163	Tilting furnace	153
Crushers	145	Wire stranding machine	153
Furnace doors	155		
Gas valves	155	*Woodworking Plants*	
Gas washers	155	Boring machines	115
Hot metal mixers	163	Lathe	115
Ingot buggy	153	Mortiser	115
Kickoff	153	Molder	115
Levelers	153	Planers	115
Manipulator fingers	153–163	Power trimmer and miter	115
Side guards	153–163	Sanders	115
Pickling machine	153	Saws	115
Pilers, slab	153	Shapers	115
		Shingle machine	115

Problems

1. A 2.0-ohm grounding resistor for a transmission line must carry 2000 amp for 12 sec without exceeding a temperature rise of 900°F. The specific heat of the resistor material is 0.09. How many pounds of resistor material are required?

2. A certain cast-iron grid has a resistance of 0.01 ohm, and its heating curve shows that if it carries 290 amp for 2 min it will have a temperature rise of 500°C. At this temperature rise, how many grids will be required for a 4-ohm resistor which will carry 1600 amp for 1 min?

3. What would be the total ohmic value of a class-163P resistor for a 50-hp 230-V 180-amp series motor, neglecting motor resistance?

4. What would be the total ohmic value of a class-150 resistor for this motor, neglecting motor resistance?

5. If the resistor of problem 4 is tapered as follows, what will be the approximate rms amperes which each step must carry for the specified time cycle?

Step	*Per cent of Total ohms*
1	48
2	34.4
3	17.6

6. A dynamic-braking resistor is used to stop a certain motor once a day. The voltage is 250, the initial braking current 500 amp, duration of braking 8 sec, and the braking current decreases in direct proportion to the time. How many resistor grids will be required if the grids weigh 1.5 lb each, the specific heat of the material is 0.09, and the temperature rise is limited to 500°F?

7. A resistor which must dissipate 12,000 W is made up of units having an overall end dimension of 2 × 2 in. These units are to be mounted in an enclosure, and may be mounted in one horizontal row, or in two, three, or four rows. Determine the most economical arrangement, and calculate the cost, using the following data.

When mounted in one row each unit will dissipate 600 W, in two rows 500 W, in three rows 400 W, and in four rows 300 W. The units cost $1.50 each. The cost of the enclosing cabinet is $0.50 per inch of width, for any height.

8. Using Table 21.6, determine the size of grid and number of grids in each step of a starting resistor for a 100-hp 230-V 350-amp shunt motor, assuming the following:

Resistor class 155
4 steps of resistance
Ohmic taper 40–30–20–10% of total
Time taper same as step taper
Inrush current on each step 150% of rated current
Contactors close at 100% of rated current
Disregard motor resistance

9. Using Table 21.6, determine the size of grid and the number of grids in each step of a starting resistor for a 50-hp 230-V 175-amp shunt motor, assuming the following:

Resistor class 175
3 steps of resistance
Ohmic taper 48–35–17% of total
Time taper same as step taper
Inrush on first step 150% of rated current
Inrush on other steps 160% of rated current
Contactors close at 100% of rated current
Disregard motor resistance

10. Using the data of problem 8, calculate the ohmic values of the resistor steps, using a parallel arrangement instead of a series arrangement of the steps.

11. Calculate the size of grid and the number of grids in each step of the parallel arrangement.

12. A series speed-regulating resistor is designed to carry 200 amp continuously on any step. The ohmic value of the resistance is as follows:

Point of Regulator	*Ohms in Circuit*
First	3.0
Second	2.5
Third	2.0
Fourth	1.5
Fifth	1.0
Sixth	0.5
Seventh	0.0

Determine the size of grid, and the number required in each step.

13. Calculate the ohms, current-carrying capacity, size of grid, and number of grids in each step of this regulator, if the resistor is connected in a parallel arrangement.

14. A 250-hp 440-V three-phase 60-cycle slip-ring motor has rotor characteristics as follows:

Volts across slip rings at standstill	625
Rotor amperes per ring at full rated load	180

Calculate the ohmic values of a six-step class-165 resistor for this motor, using the data of Table 17.3. Motor resistance may be disregarded.

15. Calculate the size of grid, and the number of grids, in each step of the resistor, using the data of Table 21.6.

16. If the grids are available only in boxes of 50 grids all of one size, determine the grid sizes and the number of grids which would result in the most economical design. The grids in a box may be all in series or may be in parallel of two.

17. A 400-hp 440-V three-phase 60-cycle slip-ring motor has rotor characteristics as follows:

Volts across slip rings at standstill	800
Rotor amperes per ring at full rated load	220

Calculate the ohmic values, grid sizes, and number of grids for a five-step class-135 resistor for this motor, using Tables 17.3 and 21.6. Motor resistance may be disregarded.

18. Calculate a six-step resistor for the motor of problem 17, under the following conditions:

Motor may run continuously with the last two steps of resistance in circuit.
Motor may run half time with the last three steps of resistance in circuit.
The rest of the resistor to be starting duty only, class 145.

19. A shunt motor having an armature resistance of $0.07E/I$ has a four-step starting resistor which permits a current peak of 150% of rated current at the start, and which is tapered in the ratio of 40–30–20–10. If the accelerating contactors close when the current reaches 100% of rated current, what is the ohmic value of the resistor steps in per cent of E/I? What inrush current is obtained on each point of the controller?

20. Engineering design data for a certain crane controller resistor is as follows:

Step	*Ohms in per cent of E/I*	*Continuous Current-carrying Capacity in per cent I*
$R1$–$R2$	100	39
$R2$–$R3$	40	46
$R3$–$R4$	28	52.5
$R4$–$R5$	20	57.5
	188	

Calculate a resistor of this type for a 150-hp 230-V series motor, and determine the sizes and number of grids from Table 21.6.

TABLES OF MOTOR CURRENTS

The following tables give approximate full-load current values of various types and sizes of motors on different power systems. The National Electrical Code, Article 430, contains tables giving average current values, as do catalogs of control manufacturers. Since such data differ considerably between motors of different design and manufacture, this information must be considered as typical data only.

The actual motor name-plate current rating must always be used for the selection of running overcurrent protection. Average data is generally satisfactory for selection of all other motor circuit components and wiring when the actual nameplate data is not available.

TABLE A.1

Ampere Ratings of Two-Phase Induction Motors

For normal torque, normal starting current, squirrel-cage or wound-rotor motors.
Average values, in amperes per phase.
½–40 hp (60 Hz)

Hp	Syn. Speed rpm	Current in Amperes				
		115 V	230 V	460 V	575 V	2300 V
½	1200	3.58	1.79	.90	.72	
	900	5.02	2.51	1.25	1.00	
¾	1800	4.03	2.02	1.01	.81	
	1200	4.93	2.47	1.24	.99	
	900	5.97	2.98	1.50	1.19	
1	3600	4.76	2.38	1.19	.95	
	1800	5.28	2.64	1.32	1.06	
	1200	6.13	3.06	1.53	1.23	
	900	6.47	3.24	1.62	1.30	
1½	3600	7.22	3.61	1.81	1.45	
	1800	7.41	3.70	1.85	1.48	
	1200	8.40	4.20	2.10	1.68	
	900	10.0	5.03	2.52	2.01	
2	3600	9.6	4.81	2.41	1.92	
	1800	10.0	4.98	2.49	1.99	
	1200	11.0	5.50	2.75	2.20	
	900	12.5	6.23	3.12	2.49	
3	3600	13.6	6.81	3.41	2.72	
	1800	14.4	7.17	3.59	2.87	
	1200	15.4	7.72	3.86	3.08	
	900	17.7	8.80	4.40	3.54	
5	3600	22.0	11.0	5.50	4.40	
	1800	22.8	11.4	5.71	4.57	
	1200	24.4	12.2	6.10	4.88	
	900	27.0	13.5	6.75	5.40	
7½	3600	33.2	16.6	8.31	6.65	
	1800	33.5	16.7	8.40	6.68	
	1200	35.1	17.6	8.80	7.02	
	900	41.2	20.6	10.3	8.23	

Hp	Syn. Speed rpm	Current in Amperes			
		230 V	460 V	575 V	2300 V
10	3600	21.2	10.6	8.48	
	1800	21.8	10.9	8.75	
	1200	23.0	11.5	9.17	
	900	25.0	12.5	10.0	
	600	29.2	14.6	11.7	
15	3600	31.7	15.9	12.7	
	1800	33.0	16.5	13.2	
	1200	34.5	17.3	13.8	
	900	36.3	18.2	14.5	
	600	41.8	20.9	16.7	
20	3600	42.4	21.2	17.0	4.5
	1800	43.7	21.9	17.5	4.6
	1200	44.7	22.4	17.8	4.7
	900	47.3	23.6	18.9	5.0
	600	53.0	26.7	21.3	5.5
25	3600	51.2	25.6	20.4	5.4
	1800	54.3	27.1	21.6	5.6
	1200	56.0	28.0	22.3	5.8
	900	58.3	29.2	23.4	6.0
	600	62.4	31.2	24.9	7.0
30	1800	63.0	31.5	25.3	6.8
	1200	66.7	33.4	26.7	6.9
	900	68.7	34.3	27.5	7.1
	600	76.1	38.0	30.5	8.0
40	1800	85.0	42.5	33.9	8.7
	1200	86.0	43.0	34.3	8.9
	900	90.0	45.0	36.0	9.2
	600	98.0	49.0	39.1	10.0

Note: For high-torque squirrel-cage motors the ampere ratings will be at least 10% greater than those given above. The current in the common line of a two-phase, three-wire system is 1.4 times the phase current.

TABLE A.2

Ampere Ratings of Two-Phase Induction Motors

For normal torque, normal starting current, squirrel-cage or wound-rotor motors.
Average values, in amperes per phase.
50–500 hp (60 Hz)

		Current in Amperes						Current in Amperes			
Hp	Syn. Speed rpm	230 V	460 V	575 V	2300 V	Hp	Syn. Speed rpm	230 V	460 V	575 V	2300 V
	1800	105	52.4	41.9	10.6		1800	398	199	159	40.4
50	1200	106	52.8	42.3	10.7		1200	404	202	161	40.7
	900	110	55.0	44.0	11.3	200	900	424	212	169	42.8
	600	120	60.0	47.7	12.3		720	428	214	171	42.4
							600	431	215	172	44.1
	1800	124	62.0	49.5	12.6		450	457	228	183	46.5
60	1200	128	64.0	51.2	12.9						
	900	131	65.4	52.3	13.3		1800	495	248	198	49.8
	600	140	70.0	56.0	14.5		1200	500	250	201	50.6
							900	524	262	209	53.2
	1800	154	77.0	61.6	15.6	250	720	540	270	216	53.2
75	1200	157	78.4	62.6	15.8		600	544	272	218	52.8
	900	162	81.0	64.8	16.4		450	546	273	219	56.5
	600	172	86.0	68.8	18.2		360	585	292	234	60.6
	1800	202	100	80.6	20.4		1800	594	297	237	59.7
	1200	207	104	82.7	20.9		1200	604	302	241	60.6
100	900	212	106	84.8	21.5	300	900	625	312	250	63.6
	600	223	111	89.2	22.8		600	625	312	250	62.6
	450	251	126	101	25.8		450	660	329	263	65.8
							360	720	360	287	71.7
	1800	250	125	100	25.3						
	1200	258	129	103	25.9		1800	786	394	315	79.4
125	900	264	132	106	26.8		1200	806	403	323	81.0
	720	272	136	109	27.1	400	600	826	413	330	83.0
	600	277	139	111	28.4		450	866	433	346	86.5
	450	304	152	121	31.2		360	905	452	362	91.0
	1800	300	150	119	30.1		1800	1000	500	400	100
	1200	303	152	121	30.7		1200	970	485	388	98
150	900	314	157	126	32.0	500	600	1020	510	408	102
	720	325	163	130	32.0		450	1040	522	417	107
	600	327	164	131	33.6		360	1140	570	455	114
	450	362	181	144	36.3						

Note: For high-torque squirrel-cage motors the ampere ratings will be at least 10% greater than those given above.

The current in the common line of a two-phase, three-wire system is 1.4 times the phase current.

TABLE A.3

AMPERE RATINGS OF THREE-PHASE INDUCTION MOTORS

For normal torque, normal starting current, squirrel-cage or wound-rotor motors.
Average values, in amperes per phase.
½–40 hp (60 Hz)

Hp	Syn. Speed rpm	Current in Amperes 115 V	230 V	460 V	575 V	2300 V
½	1200	4.14	2.07	1.04	.83	
	900	5.80	2.90	1.45	1.16	
¾	1800	4.66	2.33	1.17	.93	
	1200	5.70	2.85	1.43	1.14	
	900	6.90	3.45	1.73	1.38	
1	3600	5.50	2.75	1.38	1.10	
	1800	6.10	3.05	1.53	1.22	
	1200	7.08	3.54	1.77	1.42	
	900	7.48	3.74	1.87	1.50	
1½	3600	8.34	4.17	2.09	1.67	
	1800	8.56	4.28	2.14	1.71	
	1200	9.70	4.85	2.43	1.94	
	900	11.60	5.81	2.91	2.32	
2	3600	11.1	5.56	2.78	2.22	
	1800	11.5	5.76	2.88	2.30	
	1200	12.7	6.35	3.18	2.54	
	900	14.4	7.21	3.61	2.88	
3	3600	15.7	7.87	3.94	3.14	
	1800	16.6	8.29	4.14	3.32	
	1200	17.8	8.92	4.46	3.56	
	900	20.4	10.20	5.09	4.08	
5	3600	25.4	12.7	6.34	5.08	
	1800	26.4	13.2	6.60	5.28	
	1200	28.2	14.1	7.05	5.64	
	900	31.2	15.6	7.80	6.24	
7½	3600	38.4	19.2	9.6	7.68	
	1800	38.6	19.3	9.7	7.72	
	1200	40.6	20.3	10.2	8.12	
	900	47.6	23.8	11.9	9.51	

Hp	Syn. Speed rpm	Current in Amperes 230 V	460 V	575 V	2300 V
10	3600	24.5	12.3	9.8	
	1800	25.2	12.6	10.1	
	1200	26.6	13.3	10.6	
	900	28.9	14.5	11.6	
	600	33.8	16.9	13.5	
15	3600	36.7	18.4	14.7	
	1800	38.1	19.1	15.2	
	1200	39.9	20.0	16.0	
	900	41.9	21.0	16.8	
	600	48.3	24.2	19.3	
20	3600	49.0	24.5	19.6	5.2
	1800	50.5	25.3	20.2	5.3
	1200	51.7	25.9	20.6	5.4
	900	54.6	27.3	21.8	5.8
	600	61.5	30.8	24.6	6.4
25	3600	59.2	29.6	23.6	6.3
	1800	62.7	31.3	25.0	6.5
	1200	64.7	32.3	25.8	6.7
	900	67.4	33.7	27.0	6.9
	600	71.9	35.9	28.8	8.1
30	1800	72.8	36.4	29.2	7.8
	1200	77.1	38.6	30.8	8.0
	900	79.4	39.7	31.8	8.2
	600	87.9	43.9	35.2	9.3
40	1800	98	49.0	39.2	10.0
	1200	99	49.4	39.6	10.3
	900	104	52.0	41.6	10.6
	600	113	56.5	45.2	11.5

For high-torque squirrel-cage motors the ampere ratings will be at least 10% greater than those given above.

TABLE A.4

Ampere Ratings of Three-Phase Induction Motors

For normal torque, normal starting current, squirrel-cage or wound-rotor motors.
Average values, in amperes per phase.
50–500 hp (60 Hz)

Hp	Syn. Speed rpm	Current in Amperes 230 V	460 V	575 V	2300 V
50	1800	121	60.5	48.4	12.3
	1200	122	61.0	48.8	12.4
	900	127	63.5	50.8	13.1
	600	138	69.0	55.2	14.2
60	1800	143	71.5	57.2	14.6
	1200	148	74.0	59.2	14.9
	900	151	75.5	60.4	15.4
	600	162	81.0	64.8	16.7
75	1800	178	89.0	71.2	18.0
	1200	181	90.5	72.4	18.2
	900	187	93.5	74.8	19.0
	600	199	99.5	79.6	21.0
100	1800	233	116	93.2	23.6
	1200	239	120	95.6	24.2
	900	245	123	98.0	24.8
	600	257	128	103	26.4
	450	290	145	116	29.8
125	1800	289	144	115	29.2
	1200	298	149	119	29.9
	900	305	153	122	30.9
	720	314	157	126	31.3
	600	320	160	128	32.8
	450	351	175	140	36.0
150	1800	346	173	138	34.8
	1200	350	175	140	35.5
	900	363	182	145	37.0
	720	376	188	150	37.0
	600	378	189	151	38.8
	450	418	209	166	42.0

Hp	Syn. Speed rpm	Current in Amperes 230 V	460 V	575 V	2300 V
200	1800	460	230	184	46.7
	1200	466	233	186	47.0
	900	490	245	196	49.4
	720	494	247	197	49.0
	600	498	249	199	50.9
	450	528	264	211	53.7
250	1800	572	286	229	57.5
	1200	580	290	232	58.5
	900	604	302	242	61.5
	720	625	312	250	61.5
	600	630	315	252	61.0
	450	630	315	252	65.3
	360	676	338	270	70.0
300	1800	685	342	274	69.0
	1200	696	348	278	70.0
	900	722	361	289	73.5
	600	722	361	289	72.3
	450	760	380	304	76.0
	360	830	415	332	82.8
400	1800	910	455	364	91.8
	1200	933	466	373	93.5
	600	955	477	382	96.0
	450	1000	500	400	100
	360	1050	523	418	105
500	1800	1160	578	462	116
	1200	1120	560	448	113
	600	1180	590	472	118
	450	1200	602	482	124
	360	1320	658	526	132

Note: For high-torque squirrel-cage motors the ampere ratings will be at least 10% greater than those given above.

TABLE A.5

Ampere Table For Three-Phase Synchronous Motors

Hp	Assumed Efficiency	Amperes at 100% Power Factor			
		220 V	440 V	550 V	2300 V
5	81.0	12	6	4.8	
7½	82.0	18	9	7.2	
10	83.0	23.5	11.8	9.5	
15	85.0	34.5	17.3	14.0	
20	86.0	45.5	23	18.5	
25	87.0	56	28	22.5	
30	88.0	67	33.5	27	
40	89.0	88	44	35	9
50	89.5	110	55	44	11
60	90.0	131	66	53	13.1
75	91.0	162	81	65	16.2
100	91.5	214	107	86	21.4
125	91.5	268	134	107	27
150	92.0	320	160	128	32
200	92.0	426	213	171	43
250	92.5	526	263	212	53
300	92.5	636	318	255	64
350	93.5	734	372	298	74
400	93.5	840	420	336	84
450	93.5	942	471	378	94
500	94.0	1045	523	418	105
550	94.0	1148	574	460	115
600	94.0	1250	625	500	125
650	94.5	1350	675	540	135
700	94.6	1450	725	580	145
750	94.5	1560	780	625	156
800	95.0	1660	830	665	166
900	95.0	1860	930	745	186
1000	95.0	2060	1030	825	206

Amperes given above are based on an average efficiency for given horsepower at all speeds. For instance, 25 hp amperes are based on 87% efficiency for all speeds and 1000 hp on 95% efficiency for all speeds.

For two-phase amperes multiply values in table by 0.866.

For 80% power-factor amperes multiply 100% power-factor values by 1.29.

TABLE A.6

Ampere Table For DC Motors

The following average values of full-load currents are for motors running at base speed.

Hp	120V	240V
1/4	2.9	1.5
1/3	3.6	1.8
1/2	5.2	2.6
3/4	7.4	3.7
1	9.4	4.7
1 1/2	13.2	6.6
2	17	8.5
3	25	12.2
5	40	20
7 1/2	58	29
10	76	38
15		55
20		72
25		89
30		106
40		140
50		173
60		206
75		255
100		341
125		425
150		506
200		675

References

Chestnut, Harold. *Systems Engineering Tools,* Wiley, New York, 1965.

Chestnut, Harold, and Mayer, Robert W. *Servomechanisms and Regulating System Design,* Vol. I, 2d Ed., Wiley, New York, 1959; Vol. II, Wiley, New York, 1955.

Hamilton Standard Division of United Aircraft, Windsor Locks, Conn. *Design Manual for Selected SCR Static Inverters,* Vols. I and II. Navy Department Bureau of Ships Electronics Division Contract #NObs-86900, 5 July 1962, Project Serial Number SF-013-11-05, Task 4409, April 1968.

Heumann, G. W. *Magnetic Control of Industrial Motors,* Wiley, New York, 1961.

Kintner, Paul M. *Electronic Digital Techniques,* McGraw-Hill, New York, 1968.

Siskind, Charles S. *Electrical Control Systems in Industry,* McGraw-Hill, 1963.

Index